Ecology and Ethics

This series is devoted to continuing research at the interfaces of ecology and ethics (embedded in the multiple fields of philosophy and ecology) to broaden our conceptual and practical frameworks in this transdisciplinary field. Confronted with global environmental change, the academic community still labors under a tradition of strong disciplinary dissociation that hinders the integration of ecological understanding and ethical values to comprehensively address the complexities of current socio-ecological problems. During the 1990s and 2000s, a transdisciplinary integration of ecology with social disciplines, especially economics, has been institutionalized via interdisciplinary societies, research programs, and mainstream journals. Work at this interface has produced novel techniques and protocols for assessing monetary values of biodiversity and ecosystem services, as illustrated by the Millennium Ecosystem Assessment. At the beginning of the 2010s, however, an equivalent integration between ecology and philosophy still remains elusive. This series undertakes the task to develop crucial theoretical and practical linkages between ecology and ethics through interdisciplinary, international, collaborative teamwork. It aims to establish a new forum and research platform to work on this vital, but until now insufficiently researched intersection between the descriptive and normative domains. The scope of this series is to facilitate the exploration of sustainable and just ways of co-inhabitation among diverse humans, and among humans and other-than-human co-inhabitants with whom we share our heterogeneous planet. It will address topics integrating the multiple fields of philosophy and ecology such as biocultural homogenization, Planetary or Earth Stewardship.

More information about this series at http://www.springer.com/series/11941

Ricardo Rozzi • Roy H. May Jr.
F. Stuart Chapin III • Francisca Massardo
Michael C. Gavin • Irene J. Klaver
Aníbal Pauchard • Martin A. Nuñez
Daniel Simberloff
Editors

From Biocultural Homogenization to Biocultural Conservation

Springer

Editors
Ricardo Rozzi
Department of Philosophy and Religion and
Department of Biological Sciences
University of North Texas
Denton, TX, USA

Sub-Antarctic Biocultural Conservation
Program
University of North Texas
Denton, TX, USA

Instituto de Ecología y Biodiversidad and
Universidad de Magallanes
Punta Arenas, Chile

F. Stuart Chapin III
Institute of Arctic Biology
University of Alaska Fairbanks
Fairbanks, AK, USA

Michael C. Gavin
Human Dimensions of Natural Resources
Department
Colorado State University
Fort Collins, CO, USA

Department of Linguistic and Cultural
Evolution
Max Planck Institute for the Science of Human
History
Jena, Germany

Roy H. May Jr.
Departamento Ecuménico de Investigaciones
San José, Costa Rica

Francisca Massardo
Instituto de Ecología y Biodiversidad
Puerto Williams, Chile

Centro Universitario Puerto Williams
Universidad de Magallanes
Punta Arenas, Chile

Irene J. Klaver
Department of Philosophy and Religion
University of North Texas
Denton, TX, USA

Aníbal Pauchard
Facultad de Ciencias Forestales
Universidad de Concepción
Concepción, Chile

Instituto de Ecología y Biodiversidad (IEB)
Santiago, Chile

Martin A. Nuñez
Grupo de Ecología de Invasiones, INIBIOMA
CONICET-Universidad Nacional del Comahue
Bariloche, Argentina

Daniel Simberloff
Department of Ecology and Evolutionary
Biology
University of Tennessee
Knoxville, TN, USA

ISSN 2198-9729 ISSN 2198-9737 (electronic)
Ecology and Ethics
ISBN 978-3-319-99512-0 ISBN 978-3-319-99513-7 (eBook)
https://doi.org/10.1007/978-3-319-99513-7

Library of Congress Control Number: 2018961389

This Springer imprint is published by the registered company Springer Nature Switzerland AG
The registered company address is: Gewerbestrasse 11, 6330 Cham, Switzerland

Foreword

The challenging issue of global biocultural homogenization is comprehensively presented and discussed in this important book – and its arrival is none too soon! Biocultural homogenization, as defined by Rozzi et al. in the first chapter of the book, "entails the interwoven losses of native biological and cultural diversity at local, regional, and global scales." This issue of biocultural homogenization is neither widely understood nor is its importance adequately appreciated even while it is occurring at an accelerating rate. This book will be an important aid in increasing recognition of the issue and its importance.

Homogenization is one outcome of an ever-increasing emphasis on the goal of economic efficiency, albeit it is a goal that is generally very narrowly defined. This goal of efficiency drives societies to move toward approaches focused upon production of singular outcomes, such as of food or fiber, without regard to the consequences to nature or to local cultures. Furthermore, these days the efforts toward economically efficient production of commodities are organized so as to primarily benefit global capital markets. This leads to such outcomes as the replacement (indeed, destruction) of family farms by corporate enterprises, all in the guise of economic efficiency. Of course, in turn this leads to practices which frequently have very negative effects on native biological diversity and local cultures.

This homogenization in pursuit of efficiency represents incredible threats to native biological and cultural diversity, if we truly care about such things. Homogenization is about simplification and standardization in many forms whereas nature and culture are about complexity and diversity. Approaches that incorporate complexity and diversity are not as efficient in the pursuit of many singular goals, such as production of food and fiber. However, approaches that conserve complexity and diversity are approaches that achieve multiple rather than singular objectives with their activities, reduce risks from both natural and social upheavals, and increase future societal options.

I would venture that there are powerful, fundamentally maleficent forces that specifically do not value diversity and do not wish to see it conserved. Do global capital markets really see value in diversity, other than a diversity of portfolios? Do global corporations see value in local solutions, local markets? Is there a wide appreciation that the pursuit of efficiency, of homogenization, can lead to dysfunctional outcomes for global societies? For example, is there real concern in the United States for improving the opportunities and conditions of the diversity represented by rural America? I worry a great deal about the answers to such questions as these.

Forests are the ecosystems that I am most familiar with, and they offer great examples of some of the challenges that are faced as we attempt to deal with the issue of biocultural homogenization. There has been a global movement toward the creation and management of plantations of exotic tree species in pursuit of efficient production of wood fiber, much of this in the southern hemisphere. In the last several decades, this movement has been driven by global capital markets that invest in wood production as yet another means of seeking high returns on capital. The emphasis on capital return has put an economic cap on the already highly agronomic approaches associated with plantation forestry. The collective consequences have been what I call fiber farms, which involve practices that ignore other services and goods that are provided by forest ecosystems as well as the stability of local communities and viability of other forest landownerships. The only environmental constraints on such practices are those that are imposed by legal authorities present in the regions where such plantations are grown. Usually the harvested wood goes to the global market that is willing to pay the most for it and not to a local wood processing facility, which might result in greater economic benefits for local communities. Forest landowners who wish to manage for a diversity of values are challenged because they must find markets and compete in a global wood products economy dominated by the fiber farms.

This highly simplified, homogenized approach to wood production finds support in many quarters, including a globalized economy and history, and there are many similarities here between forestry and agriculture and fisheries. I have already talked about how a capital-dominated global economy favors homogenization and the marginalization of other forest values, except where governmental authorities insist otherwise. The dominant focus of the forestry profession on wood production as the most important use of forestland has been largely congruent with the emphasis on homogenization and efficiency in pursuit of a singular outcome.[1] Local communities and governments are advised that the homogenized approach is in their best economic interest and sometimes told that this is the only real way to do sustainable forestry. (The same is presented regarding corporate agriculture and fish farms.) Forestry as a profession has failed to even conceive, let alone demonstrate to society, credible alternatives to intensive plantation management based on clearcutting and even-aged management.[2] The foresters have been abetted by the community of academic conservation biologists who argue that native biodiversity can only be conserved in preserves – areas that are set aside from human societies (as if such a

[1] Franklin, J. F., K. N. Johnson, and D. L. Johnson 2018. Ecological forest management. 646 p. Long Grove, IL, USA: Waveland Press.

[2] Bennett, Brett. 2015. Plantations and protected areas. A global history of forest management. 201 p. Cambridge, MA, USA: MIT Press.

thing was possible in the twenty-first century!). Biodiversity will not be preserved primarily by separating it from humankind but, rather, must be a part of conserved bioculture.

This book is an important contribution to the dialogue and hard work that is ultimately required to conserve as much as we can of diverse bioculture. The future of native biodiversity and local human societies are linked and face the same array of challenges. Many ideas, concepts, and examples are laid down in this volume that can move this important work forward. We are talking here about nothing less than the future of humankind – is it to be a homogenized future or one that nurtures diversity and the richness and resilience that it brings?

Let us all get on with it!

Emeritus Professor of Forest Ecosystems
University of Washington,
Seattle, WA, USA

Jerry F. Franklin

Acknowledgments

This volume conveys the work of many researchers, students, and people from different regions of the world. We are grateful to our institutions in Chile and in the USA, which have supported our editorial activities. At the Institute of Ecology and Biodiversity (IEB) and the University of Magallanes (UMAG) in Chile, we are especially thankful for the support of Andrés Mansilla, UMAG VP for Research and Graduate Studies. At the University of North Texas (UNT), the Center for Environmental Philosophy (CEP) provided support; we especially thank its director Eugene Hargrove for his constant advice. At Springer, we appreciate the encouragement and editorial guidance offered by Valeria Rinaudo, Ineke Ravesloot, and Jill Ritchie; this is the third volume of the *Ecology and Ethics* book series.

Part of the impulse for developing this book came from the field graduate course "Trends in Ecology of Plant Invasions" coordinated by Aníbal Pauchard and Ramiro Bustamante of the Laboratorio Invasiones Biológicas (LIB-IEB) located in the Chilean Cordillera of the Andes Mountains, April 2014. The course involved also the participation of three of other editors of this volume: Martin Nuñez, Daniel Simberloff, and Ricardo Rozzi. Some of the co-authors of chapters in this volume were students at that course. We are especially thankful to Brian O'Connor for his valuable editions to several manuscripts of chapters. Rozzi acknowledges the financial support of the National Commission for Scientific and Technological Research (CONICYT, Chile) grants PFB-23 and *Apoyo a Centros Científicos y Tecnológicos de Excelencia con Financiamiento* (Basal CONICYT AFB170008) awarded to IEB in Chile. This book is a contribution to the Sub-Antarctic Biocultural Conservation Program coordinated by UNT in the USA and IEB and UMAG in Chile.

Contents

Contributors

Xavier Albó Centro de Investigación y Promoción del Campesinado (CIPCA), La Paz, Bolivia

Benjamin Baiser Department of Wildlife Ecology and Conservation, University of Florida, Gainesville, FL, USA

Jeffrey A. Brown Department of Ecology, Evolution, and Natural Resources, Rutgers University, New Brunswick, NJ, USA

F. Stuart Chapin III Institute of Arctic Biology, University of Alaska Fairbanks, Fairbanks, AK, USA

Ramiro D. Crego Department of Biological Sciences, University of North Texas, Denton, TX, USA

Instituto de Ecología y Biodiversidad, Santiago, Chile

Sub-Antarctic Biocultural Conservation Program, University of North Texas, Denton, TX, USA

Silvia Regina da Lima Silva Departamento Ecuménico de Investigaciones, San José, Costa Rica

Romina D. Dimarco Grupo de Ecología de Poblaciones de Insectos, CONICET, INTA, Bariloche, Argentina

Paulo José dos Reyes (Pai Paulo José de Ogun) Babalorixa or religious leader of the Candomblé community of ILÉ AşÉ ÒGÚN ÀLÁKÒRÓ, Salvador, Brazil

Haruf Salmen Espindola Postgraduate Program in Integrated Territory Management, Universidade Vale do Rio Doce, Governador Valadares, Minas Gerais, Brazil

Franz Essl Division of Conservation Biology, Landscape & Vegetation Ecology, University of Vienna, Vienna, Austria

J. Miguel Esteban Facultad de Filosofía, Universidad Autónoma del Estado de Querétaro, Santiago de Querétaro, México

Jorgelina Franzese Laboratorio Ecotono, Instituto de Investigaciones en Biodiversidad y Medioambiente (Universidad Nacional del Comahue – CONICET), S. C. Bariloche, Argentina

Rafael A. García Facultad de Ciencias Forestales, Universidad de Concepción, Concepción, Chile

Instituto de Ecología y Biodiversidad (IEB), Santiago, Chile

Michael C. Gavin Human Dimensions of Natural Resources Department, Colorado State University, Fort Collins, CO, USA

Department of Linguistic and Cultural Evolution, Max Planck Institute for the Science of Human History, Jena, Germany

Cláudio Bueno Guerra Autonomous Environmental Consultant, Belo Horizonte, Minas Gerais, Brazil

Rafaela Guimarães Setor de Ecologia, Departamento de Biologia, Universidade Federal de Lavras, Lavras, MG, Brazil

Pandurang Hegde Appiko-Chipko Movement, Sirsi, North Kanara, Karnataka, India

George James Department of Philosophy and Religion, University of North Texas, Denton, TX, USA

Jaime E. Jiménez Department of Biological Sciences, University of North Texas, Denton, TX, USA

Instituto de Ecología y Biodiversidad, Santiago, Chile

Sub-Antarctic Biocultural Conservation Program, University of North Texas, Denton, TX, USA

Department of Philosophy and Religion, University of North Texas, Denton, TX, USA

Universidad de Magallanes, Punta Arenas, Chile

Irene J. Klaver Department of Philosophy and Religion, University of North Texas, Denton, TX, USA

Tetsuya Kono Department of Education, Rikkyo University, Tokyo, Japan

Bernd Lenzner Division of Conservation Biology, Landscape & Vegetation Ecology, University of Vienna, Vienna, Austria

Daijiang Li Department of Wildlife Ecology and Conservation, University of Florida, Gainesville, FL, USA

Julie L. Lockwood Department of Ecology, Evolution, and Natural Resources, Rutgers University, New Brunswick, NJ, USA

Fouad Makki Department of Development Sociology, Cornell University, Ithaca, NY, USA

Francisca Massardo Instituto de Ecología y Biodiversidad, Puerto Williams, Chile

Centro Universitario Puerto Williams, Universidad de Magallanes, Punta Arenas, Chile

Janet W. May Departamento Ecuménico de Investigaciones, San José, Costa Rica

Roy H. May Jr. Departamento Ecuménico de Investigaciones, San José, Costa Rica

Felipe Montoya-Greenheck Faculty of Environmental Studies, York University, Toronto, ON, Canada

Martin A. Nuñez Grupo de Ecología de Invasiones, INIBIOMA, CONICET-Universidad Nacional del Comahue, Bariloche, Argentina

Angelina Paredes-Castellanos Facultad de Filosofía, Universidad Michoacana San Nicolás de Hidalgo, Morelia, México

Aníbal Pauchard Facultad de Ciencias Forestales, Universidad de Concepción, Concepción, Chile

Instituto de Ecología y Biodiversidad (IEB), Santiago, Chile

Nahuel Policelli Grupo de Ecología de Invasiones, Instituto de Investigaciones en Biodiversidad y Medioambiente (Universidad Nacional del Comahue – CONICET), S. C. Bariloche, Argentina

Alexandria K. Poole Department of Politics, Philosophy and Legal Studies, Elizabethtown College, Elizabethtown, PA, USA

Institute of Ecology and Biodiversity, Santiago, Chile

Ricardo Rozzi Department of Philosophy and Religion and Department of Biological Sciences, University of North Texas, Denton, TX, USA

Sub-Antarctic Biocultural Conservation Program, University of North Texas, Denton, TX, USA

Instituto de Ecología y Biodiversidad and Universidad de Magallanes, Punta Arenas, Chile

Alexandrea Dagmar Safiq Department of Ecology, Evolution, and Natural Resources, Rutgers University, New Brunswick, NJ, USA

Yamila Sasal Laboratorio Ecotono, Instituto de Investigaciones en Biodiversidad y Medioambiente (Universidad Nacional del Comahue – CONICET), S. C. Bariloche, Argentina

Hanno Seebens Senckenberg Biodiversity and Climate Research Centre (SBiK-F), Frankfurt am Main, Germany

Daniel Simberloff Department of Ecology and Evolutionary Biology, University of Tennessee, Knoxville, TN, USA

Ruifei Tang CEESP (Commission on Environmental, Economic and Social Policy), IUCN, Wellington, New Zealand

Bron Taylor Department of Religious Studies, University of Florida, Gainesville, FL, USA

Kimberley Taylor Department of Ecosystem and Conservation Sciences, University of Montana, Missoula, MT, USA

Rosana Tidon Departamento de Genética e Morfologia, Instituto de Ciências Biológicas, Universidade de Brasília, Brasília, DF, Brazil

Mitsuyo Toyoda Center for Toki and Ecological Restoration, Niigata University, Sado, Niigata, Japan

Laura Zanotti Department of Anthropology, Center for the Environment, Purdue University, West Lafayette, IN, USA

Rafael Dudeque Zenni Setor de Ecologia, Departamento de Biologia, Universidade Federal de Lavras, Lavras, MG, Brazil

Chapter 1
From Biocultural Homogenization to Biocultural Conservation: A Conceptual Framework to Reorient Society Toward Sustainability of Life

Ricardo Rozzi, Roy H. May Jr., F. Stuart Chapin III, Francisca Massardo, Michael C. Gavin, Irene J. Klaver, Aníbal Pauchard, Martin A. Nuñez, and Daniel Simberloff

Abstract Biocultural homogenization entails interwoven losses of native biological and cultural diversity at local, regional, and global scales. It is a driver and a product of complex and pervasive losses of biological and cultural diversity; however, it is not yet widely recognized to its full extent. In this book we show how the processes of biological and cultural homogenization are intricately interrelated. A guiding theme is the conceptual framework of the biocultural ethic and its "3Hs" model, which facilitates understanding how some life habits that are being global-

R. Rozzi (✉)
Department of Philosophy and Religion and Department of Biological Sciences, University of North Texas, Denton, TX, USA

Sub-Antarctic Biocultural Conservation Program, University of North Texas, Denton, TX, USA

Instituto de Ecología y Biodiversidad and Universidad de Magallanes, Punta Arenas, Chile
e-mail: Ricardo.Rozzi@unt.edu

R. H. May Jr.
Departamento Ecuménico de Investigaciones, San José, Costa Rica

F. S. Chapin III
Institute of Arctic Biology, University of Alaska Fairbanks, Fairbanks, AK, USA
e-mail: terry.chapin@alaska.edu

F. Massardo
Instituto de Ecología y Biodiversidad, Puerto Williams, Chile

Centro Universitario Puerto Williams, Universidad de Magallanes, Punta Arenas, Chile
e-mail: Francisca.Massardo@umag.cl

M. C. Gavin
Human Dimensions of Natural Resources Department, Colorado State University, Fort Collins, CO, USA

Department of Linguistic and Cultural Evolution, Max Planck Institute for the Science of Human History, Jena, Germany
e-mail: michael.gavin@colostate.edu

R. Rozzi et al. (eds.), *From Biocultural Homogenization to Biocultural Conservation*, Ecology and Ethics 3, https://doi.org/10.1007/978-3-319-99513-7_1

ized can lead to homogeneous habitats with detrimental consequences for many human and other-than-human co-inhabitants. The 3Hs conceptual framework enables a visualization of the interrelations between the homogenization of habits and habitats and the consequences it has for the well-being or the displacement of human and other-than-human co-inhabitants. In this way, it can inform and provide insights for decision-making in environmental policies, development, and educational programs, in order to foster processes of biocultural conservation and avoid pressing social and environmental injustices conveyed by current processes of biocultural homogenization.

Keywords Biocultural ethics · Biotic homogenization · Environmental justice · Land grabbing · Sustainability

1.1 Introduction

Biocultural homogenization is a wicked problem of the Anthropocene. It involves complex interdependencies, and, despite being pervasive, it is not yet widely recognized as such. *Homogenization* means the generation of sameness. The word combines the Greek terms *homos* (ὁμός) and *génesis* (γένεσις), which mean "same" and "creation," respectively. *Biocultural* indicates a combination of biological and cultural factors. During the past three decades, biotic homogenization has been investigated by ecologists (McKinney and Lockwood 1999; Simberloff and von Holle 1999; Olden and Rooney 2006), while cultural homogenization has been researched by social scientists or humanists (Schaedel 1979; Petitat 1987; Quijano 2000; Rizvi and Lingard 2000). Biocultural homogenization interrelates these two processes (Rozzi 2012).

The process of biocultural homogenization entails simultaneous and interlinked losses of native biological and cultural diversity at local, regional, and global scales (Rozzi 2013). The massive replacement of native biota and cultures by cosmopoli-

I. J. Klaver
Department of Philosophy and Religion, University of North Texas, Denton, TX, USA
e-mail: Irene.Klaver@unt.edu

A. Pauchard
Facultad de Ciencias Forestales, Universidad de Concepción, Concepción, Chile

Instituto de Ecología y Biodiversidad (IEB), Santiago, Chile
e-mail: pauchard@udec.cl

M. A. Nuñez
Grupo de Ecología de Invasiones, INIBIOMA, CONICET-Universidad Nacional del Comahue, Bariloche, Argentina

D. Simberloff
Department of Ecology and Evolutionary Biology, University of Tennessee, Knoxville, TN, USA
e-mail: dsimberloff@utk.edu

tan species, languages, and cultures disrupts coevolutionary interrelationships between local cultures and their habitats. We argue that a person who is mainly exposed to globally homogeneous habits is more likely to build globally homogeneous habitats; at the same time, globally homogeneous urban and rural habitats reinforce globally homogeneous life habits and mindsets. To recognize these positive and wicked feedbacks between cosmopolitan habits and habitats, and their consequences for human and nonhuman co-inhabitants, Ricardo Rozzi (2001, 2012) coined the term *biocultural homogenization*.

This book was initially conceived as an elaboration of the chapter "Biocultural Ethics: From Biocultural Homogenization to Biocultural Conservation" published by Rozzi in the first volume of the *Ecology & Ethics* series, *Linking Ecology and Ethics for a Changing World: Values, Philosophy, and Action* (Rozzi et al. 2013). Now, in this third volume, we interrelate the processes of biological and cultural homogenization by using the conceptual framework of the biocultural ethic and its "3Hs" model (Rozzi 2013). On the one hand, the 3Hs model facilitates the visualization of, and understanding about, how a *habit* (e.g., focusing on monospecific plantations or damming waterways) can lead to homogeneous *habitats* (e.g., monocultures or infrastructure such as dams or aqueducts) with detrimental consequences for many human and other-than-human *co-inhabitants*. On the other hand, as Irene Klaver (2018) emphasizes in her chapter, the 3Hs focus enables critical analyses of such homogenizing habits and can help reorient them toward habits that could instead promote processes of biocultural conservation.

This book introduces novel concepts, methods, and case studies to tackle complex processes of biocultural homogenization that permeate the Anthropocene but that have not yet been analyzed in this perspective. Our main goal is to explore and suggest conceptual and practical avenues for transitions from biocultural homogenization toward biocultural conservation. In 2014, Aníbal Pauchard, Martin Nuñez, Daniel Simberloff, and Rozzi met at the international course "Trends in Ecology of Plant Invasions" in the Andes Cordillera of Chile. We realized that biocultural homogenization incorporates three key additions to the more familiar and better documented process of biotic homogenization.

First, biocultural homogenization broadens the expression "a few winners replacing many losers" coined by Michael McKinney and Julie Lockwood (1999) beyond the scope of biological diversity toward cultural and socioeconomic diversity (Rozzi 2012). The latter includes a plethora of local communities with their idiosyncratic languages and worldviews. These cultural traditions are embedded in unique life habits coupled with specific habitats. These local communities influence, and are influenced by, biocultural landscapes inhabited by a diversity of living beings with whom human communities share their material lives as well as their rituals and symbolic lives. Hence, the replacement of many local cultures and their languages by a few widespread languages and global life habits promotes not only large-scale cultural homogenization but also biocultural homogenization.

Second, humans are both the direct and indirect drivers of biocultural homogenization. This contrasts with the case of invasive exotic species, which have become the direct drivers of habitat transformations and reduction of communities of native co-inhabitants. For example, in Part II of our *Biocultural Homogenization* book, Crego et al. (2018) show how today in South American Patagonia North American beavers are transforming habitats by reducing the native forest cover and North American minks are transforming the communities of co-inhabitants by devastating populations of endemic bird species. The original cause for this process was the introduction of beavers and minks for a particular practice or habit: fur trade. However, today fur trade has disappeared, and these alien mammals have become invasive in most Patagonian habitats. In the case of biocultural homogenization, most practices involve active, although not always intentional, participation by human agents; hence, the ultimate solution would depend on the disposition to change life habits.

Third, life habits are shaped by a wide variety of factors, some of which are internal to the individual and many of which are influenced by the complex and dynamic social-economic-ecological systems in which all individuals live. Reorienting life habits requires negotiation among different stakeholders in challenging scenarios of economic and political power at local, national, and global scales, including large-scale land acquisitions. Actors external to local communities include, among others, both large preservationist organizations and multinational corporations that displace local communities. As one example, in his chapter analyzing processes of biocultural homogenization in Ethiopia, Fouad Makki (2018) criticizes the privatization of nomadic pastoralists and peasant land and village commons, which is presented today by global discourses as the inescapable prerequisite for bringing the benefits of the new green revolution to Africa.

To broaden the spectrum of perspectives on biocultural homogenization and to address pressing global and local socio-environmental problems from standpoints of different disciplines and regions of the world, Rozzi invited environmental philosopher Irene Klaver, conservation biologist Francisca Massardo, liberation theologian Roy May, ecosystem ecologist Terry Chapin, and biocultural conservation scientist Michael Gavin to join the editorial team. Together we summoned 46 contributors from Africa, Asia, Europe, and North and South America, who provide a balanced set of viewpoints from the Global South and North. To facilitate the incorporation of these biocultural viewpoints into research, policy, and educational domains, we have organized the book into three complementary parts: (I) Biocultural Homogenization, (II) Biotic Homogenization, and (III) Biocultural Conservation.

1.2 Part I: Biocultural Homogenization

In Part I, we aim to show how tackling biocultural homogenization requires nonlinear thinking. The systemic, dynamic, and contextual model of the 3Hs of the biocultural ethic (sensu Rozzi 2013) enables an understanding of the interdependencies among habitats, life habits, and communities of co-inhabitants, which helps to identify opportunities for transitioning from biocultural homogenization to conservation. Tipping points exist in decision-making processes of socio-environmental, economic, and educational policies and actions. These processes tend either to favor conservation or restoration of native habitats and associated life habits of local human communities and biota or, on the contrary, to cause their disappearance and replacement by habitats that are built in similar fashion around the world. Proponents of large-scale projects usually argue that they can replace local practices and infrastructures in order to provide services more effectively. This may be true in some cases. However, usually biocultural homogenization, driven by cloned habitats and linked to life habits embedded in global development policies, conveys benefits for relatively few co-inhabitants.

Co-inhabitant is a central notion to support Rozzi's (2018a) critique of biocultural homogenization. When the majority of human beings and other living beings are seen as co-inhabitants, active subjects with an intrinsic value, biocultural homogenization becomes a question of socio-environmental justice. When local communities are forced to migrate to other places with promises of a prosperous future made by development projects and policies, most often human lives and a myriad of other living beings are sacrificed. Also lost are the relationships of co-inhabitation that humans have established with other-than-human co-inhabitants in specific socio-ecosystems. As alternatives to homogenizing practices, many resilient communities are currently promoting initiatives to restore (or conserve) co-habitation relationships in rural and urban areas.

The concept of *riversphere* and the metaphor of *meandering*, proposed by Irene Klaver (2018), serve to illustrate the case of re-creating a resilient community even in the context of highly homogenized contemporary cities. The Los Angeles River in California was channelized in the 1950s, and co-inhabitation relationships embedded in material culture and everyday symbolic expressions disappeared. Today, however, the river and some of these relationships of co-inhabitation are partially restored. The Los Angeles River meanders throughout history, and meandering is a metaphor that helps us to understand the wicked problems of biocultural homogenization. Klaver associates the sinuous movement of rivers carving landscapes with forms of non-linear and non-deterministic thinking. With a meander model of the riversphere similar to the model of the "3Hs" of biocultural ethics, she first visualizes how a sociocultural economic practice or *habit* (damming, channeling, selling, and diverting waterways) leads to homogeneous *habitats* (infrastructure, paved-over or concrete "riverbeds," and aqueducts) with a consequent reduction of communities of *co-inhabitants*. Conversely, she proposes that the 3Hs model could also "enable a reorientation towards re-connecting to rivers and re-valuing, re-vitalizing, and re-imagining riverine relations within processes of biocultural conservation and cultural diversification. Such a new cultural habit, including a biocultural mentality, would diversify habitats and broaden the spectrum of co-inhabitants' survival and well-being" (p. 50).

In the chapter "Biostitutes and Biocultural Conservation: Empire and Irony in the Motion Picture *Avatar*," Bron Taylor (2018) portrays the polarization between mentalities of biocultural homogenization and conservation. James Cameron's science fiction film *Avatar* has been praised as a substantive defense of biocultural conservation and, at the same time, criticized as another expression of biocultural homogenization because it simplifies the polarization between local communities and corporative capitalism. To tackle this polarization, Taylor problematizes the role of scientists whose naive objectivity helps to create a mentality where local communities would have no choice but to cede their territories, leading to homogenization of habitats through land-grabbing processes. Taylor uses the notion of *biostitutes* (a term that blends the words biologists and prostitutes) to critique how some scientists "serve, willingly if sometimes ambivalently, a capitalistic economic system that seeks to maximize profit and places little if any value on the habitats and cultures it must exploit to continue with its expansionist raison d'être" (pp. 76–77). Taylor points out that the film exposes how these scientists can become catalyzers for homogenization of life habits by "befriending" indigenous people, "learning from and about them" in order to be able to mitigate the changes being forced upon them. Much as Klaver points out forms of nonlinear and nondeterministic thinking, Taylor highlights ambiguities involved in biocultural homogenization. The series of case studies and concepts presented in our book elucidates how the multivariate roles and intentions of actors involved in development and/or conservation projects cannot be reduced to simple, dichotomous polarizations. The film *Avatar* has been criticized for its standard neocolonialist plot structure, in which local communities need help from "a hero" to survive. However, this film has the merit of triggering discussions about the ambiguous and complex nature of biocultural homogenization, including multifaceted roles played by scientists and other actors in this process (see Taylor 2013).

A current case study of collision of viewpoints and interests is presented by Fouad Makki (2018) in "The Political Ecology of Land Grabs in Ethiopia." The confluence of the 2008 world economic crisis with global food and energy crises has generated a frenzy of large-scale land acquisitions across the Global South. In this chapter, Makki criticizes how "the privatization of peasant holdings and the village commons is viewed as the inescapable prerequisite for bringing the benefits of the new green revolution to Africa" (p. 84). The World Bank and other international financial institutions are promoting a narrative of *terra nullius* to designate so-called "underutilized" spaces as ideal for large-scale commercial development. This ongoing policy seriously threatens native habitats and customary life habits of peasants and nomadic agropastoral communities for whom transhumance is linked to ownership of livestock and communal appropriation of pastures. Today a distorted representation of the Ethiopian habitats and habits as "unproductive" hides two facts: (i) these habitats include the world's second largest forest reservoir of overland mammal migrations and (ii) the life habits in the *res communes* support equitable and sustainable agroecological practices.

A recent case where concerns of local communities and technical recommendations were not seriously considered by a megaproject is presented by Haruf Espindola and Claudio Guerra (2018). In 2015, the largest socio-environmental disaster in Brazil's history occurred in the Rio Doce basin, when a dam belonging to the world's second largest mining company ruptured. Espindola and Guerra point out that the company framed the disaster as an accident, while those whose daily lives and habitats were destroyed called it a crime.

Roy May (2018) presents the case of a hydroelectric dam project in Latin America that exposes the difficulties for local communities to be heard. In 2016,

Berta Cáceres, an indigenous Lenca environmentalist from Honduras who was awarded the prestigious Goldman Environmental Prize in 2015, was murdered as retaliation for her opposition to the dam. In her prize acceptance speech, Berta affirmed that "the Lenca people are ancestral guardians of the rivers." However, guardians of the environment are severely repressed worldwide. In 2017, another Goldman Prize winner, a Tarahumara indigenous environmentalist, Isidro Baldenegro López, was shot to repress his leadership in the struggle to protect the pine-oak forests of Mexico's Sierra Madre mountain range. Since 2002, across the globe, at least 1176 environmental leaders have been murdered. May concludes that "the constant intimidation and murder of defenders of land rights of indigenous and peasant people and the imposition of economic enterprises inappropriate to ecological conditions, demonstrates the persistent coloniality that affects nature and people" (p. 109).

Coloniality and biocultural homogenization are interwoven. Coloniality involves displacements of both local cultures and local biota. Bernd Lenzner and collaborators (2018) undertake a historical perspective to analyze Europe's central role in introducing species into regions outside their native range worldwide. They highlight that from the fifteenth to the nineteenth centuries, Europe's processes of global trade implied "alien species introductions [that] lead to the breakdown of biogeographic barriers, thereby promoting a homogenization of the world's biota" (p. 125). In the twentieth century, expansions of world markets and global connectivity intensified these homogenization processes with profound effects on the displacement of species around a continuously more globally interconnected world.

Miguel Esteban (2018) undertakes a historical perspective to show how European Renaissance and Modern art have contributed to cultural homogenization by uniformizing the construction of images of the fauna discovered in Africa, Asia, and the Americas. Isolated from their habitats, animals were painted against European symbolic backgrounds. To illustrate this point, Esteban focuses on Albrecht Dürer's famous portrayal of the Rhinoceros, a "pictorial construction of the otherness of exotic animals [that] reaffirmed the beneficial exceptionalism of Europe and, consequently, reinforced the legitimacy of Western colonization of a wild and alien nature, waiting to be reduced and converted into merchandise" (p. 137). He proposes that if we aim to reorient these trends of biocultural homogenization toward biocultural conservation, then it is critical "to denounce the construction of homogeneous biocultural habitats based on habits such as visual production and the consumption of images" (p. 137). Angelina Paredes and Rozzi (2018) undertake this task by deconstructing the "imperial eye" under which Latin American landscapes and inhabitants are perceived and portrayed. A critique of this exoticism enables a better appreciation of Latin American artistic, literary, and political concepts and movements that have revalued local geographies and co-inhabitants, including the political subjects, their identities, and their cultures. A major contribution offered by Latin American environmental arts, schools of thought, and movements is an alternative eco-epistemology and aesthetic hermeneutic to both recover and build a new understanding of nature as a subject, where we co-inhabit.

Part I concludes with a critical analysis of biocultural homogenization associated with the taxonomic bias found in the work of one of the principal European modern philosophers. Rozzi (2018b) focuses on the eighteenth-century Scottish empiricist philosopher, David Hume. In his complete works are 510 mentions of animals; among them 97% are vertebrates, and only two species, horses and dogs, account for more than 50% of the occurrences. This narrow spectrum leaves out the most diverse groups of animals worldwide: invertebrates, which encompass 95% of the known species. It is noteworthy, however, that although in his complete works, Hume included only three references to invertebrates, one of them portrayed an oyster as a sentient organism. In this way Hume stated a relevant attribute that is common to vertebrates and invertebrates and provides philosophical foundations for the ethical consideration of all living beings. Moreover, this finding in Hume's work shows how within dominant Western modern thinkers we can also find "fractures" or alternative perspectives for overcoming taxonomic biases and other drivers of Eurocentric biocultural homogenization.

1.3 Part II: Biotic Homogenization

Part II focuses on biotic homogenization, analyzing and proposing novel links with cultural and social dimensions. To undertake this task, it presents perspectives and case studies from South and North America, which approach biotic homogenization at different scales: from species assemblages to ecoregions. Daniel Simberloff (2018) introduces the concept of bioculture defined as "human populations, other species, habitats, and 'the totality of interaction biological and cultural relationships'" (p. 207). With his ecological perspective, Simberloff's definition is complementary to the one given to this term by humanists who have proposed that "culture and history must be rethought with an understanding of their inextricable, if highly variable, relation to biology" and that "biology, as a science, cannot exist outside culture; culture, as a practice, cannot exist outside biology" (Davis and Morris 2007, p. 411 and p. 418, respectively). These ecological and humanist perspectives broaden the biocultural approach to overcome the split between nature and culture and the onto-epistemological divide between biology and culture (cf. Rozzi 2013). Simberloff then problematizes the discussion by arguing that nonnative species can also be part of biocultures and may even replace native species to form new biocultures. He invites us to investigate the extent to which biotic homogenization can produce characteristically distinct biocultures and to identify traits or characteristics that could be assessed for comparing ancient and newer biocultures. From an ethical point of view, he cautions that there is nothing necessarily xenophobic about the goal to preserve traditional biocultures. This goal could be motivated by general principles equivalent to those that support the right of human societies to maintain their cultural distinctness (UNESCO 2001) or by understanding the intrinsic value of native co-inhabitants, their life habits, and habitats (Rozzi 2012).

Martin Nuñez and collaborators (2018) tackle the question of why some nonnative, even invasive, species become deeply integrated into local and regional cultures, while others do not. A striking example is provided by *Eucalyptus* trees from Australia that were planted in California at the beginning of the nineteenth century and later became invasive. In the early twentieth century, a major school of art focused on *Eucalyptus*. Despite the demonstrated negative impact on other native species and other ecological problems caused by these invasive trees, there is a strong opposition to removing them (Nuñez and Simberloff 2005). Nuñez et al. identify five key factors that make people support or reject management of invasive species: arrival time, economic impact, aesthetic value, effect on human health, and origin of nonnative species and of human immigrants. They conclude that local support is critical for achieving effective management plans and propose that educational programs take into account these factors when explaining the problems that invasive species produce.

Ramiro Crego and collaborators (2018) offer a complementary approach by focusing on multispecies invasive assemblages. They illustrate how a North American temperate-subarctic assemblage of three mammal species (beavers, muskrats, and minks) was introduced into the southern end of South America in sub-Antarctic Tierra del Fuego in the mid-twentieth century, because of a Northern Hemisphere habit: fur trade. They show how biocultural interactions dynamically respond to changes in cultural and economic habits and to subsequent reassembling of introduced species associated with changes in the ecological interactions among introduced and native species. In a parallel case study but with plants, Rafael García and collaborators (2018) describe the introduction of Northern Hemisphere tree species of the genus *Pinus* or pines into temperate South America, which is characterized by the dominance of broadleaf tree species and has no native species of *Pinus*. Despite the demonstrated ecological problems caused by these species and the invasive character they have acquired, commercial *Pinus* plantations continue to expand. García et al. conclude that to mitigate the impacts caused by pines, it is important to implement comprehensive landscape planning and to understand better how pine plantations could coexist in diverse landscapes without affecting and damaging other land uses.

In tropical latitudes, Rafael Zenni and collaborators (2018) examine biotic homogenization in one of the world's biodiversity hotspots: the Cerrado, a vast savannah of grassland, woods, and dry forest located mostly in central Brazil and parts of Bolivia and Paraguay. It is characterized by rich biodiversity and high endemism but since the 1970s has been subjected to increasing development and economic pressures that are converting this unique ecoregion into agriculture and pasture lands, as well as urban areas and infrastructure. Zenni et al. review the evidence of a myriad of biological invasions that are underway in the Cerrado, from grasses to insects, which threaten biodiversity. They plead for greater social awareness about the rapid biotic homogenization in the Cerrado and urge collaboration with management plans to mitigate its impact.

Unfortunately, biotic homogenization not only affects heavily anthropogenicized ecosystems, it also affects relatively pristine protected areas. National parks face many of the same threats to biodiversity as non-protected areas, including nonnative species invasion. Li and collaborators (2018) present a comprehensive analysis of taxonomic and phylogenetic homogenization across US national parks, based on species composition for bird and plant assemblages (considering and excluding nonnative species). They find significant patterns of homogenization and show that these patterns can be complex because taxonomic and phylogenetic homogenization can be decoupled (based on their research on birds). The authors appeal to the important role that national parks have played for cultural identity, biodiversity conservation, education, and recreation in the USA, in order to call for the protection of these areas from threats such as biotic homogenization. In the last chapter of Part II, Alexandrea Safiq and collaborators (2018) measure biotic homogenization at the scale of species assemblages by tracking taxonomic changes between 1994 and 2015 on the Atlantic coast of Florida, USA. They track taxonomic changes of fish assemblages and find that sites closer to populated coastlines or that have been subjected to substantial disturbance events are more likely to show homogenization. Interestingly, protected coastal sites show little evidence of biotic homogenization, and the authors propose that societal values as well as divers' practices and experience can help avoid homogenization. Therefore, protected sites or areas play a role in controlling biotic homogenization, and with biocultural education, these areas could be sustainably used for ecotourism and other regulated activities.

1.4 Part III: Biocultural Conservation

The chapters in Part III offer conceptual frameworks, cultural traditions, and practical applications to biocultural conservation. Conceptual frameworks include the foundations of the biocultural ethic, the United Nations agenda, and the indigenous worldviews seldom included in public policies. The cultural traditions include the Candomblé of Brazil, the aesthetics of European gardens and their relationship with Taoism and illustration, and the thought and action of the theology of liberation. The practical applications include local community movements in Brazil, India, and Inner Mongolia, China, and conservation and participatory restoration projects in Costa Rica and Japan, respectively. The objective of Part III is to provide a diversity of approaches to biocultural conservation, combining work from local to global scales.

Rozzi (2018c) begins by proposing that the conservation of habitats and life habits is so critical today that it constitutes an ethical imperative that should be incorporated into government policies as a matter of socio-environmental justice. To implement this ethical imperative, it is essential to reorient global society toward fostering a bioculture that effectively integrates ontological, ecosocial, and ethical foundations into education, policies, and governance. The transformation of the prevailing educational and policy-making systems will require a great degree of participation by intellectuals, communities, and social movements of the Global North and of the Global South, the West, and the East. This greater participation will help to remove the mantle of a universal discourse (with its forms of governance and education) that has denigrated and made invisible the multiplicity of vernacular worldviews, local knowledge, language, practices, and ecological values that still exist and are defended by diverse communities across the planet. Rozzi's criticism is that the Modern Era has been erected on supposed universal knowledge, thus conveying a univocal sense of reality that is presented as epistemologically and technologically superior and that has been institutionalized through its iconic higher education institution: the *uni-versity*. Epistemological homogenization is the basis of biocultural homogenization.

Alexandria Poole (2018) criticizes the 17 Sustainable Development Goals (SDGs) recently launched by the United Nations (UN) General Assembly in its Resolution of September 25, 2015, entitled "Transforming Our World, the 2030 Agenda for Sustainable Development." This document is proposed to be "a plan of action for people, planet, and prosperity." According to Poole, although the 2030 Agenda intends to account for the shortfalls found in the original UN Millennium Development Goals, the "SDGs still neglect fundamental qualities of cultural sovereignty that are key for maintaining sustainable practices, values, and lifestyle habits" (p. 315). To solve this omission, Poole proactively argues that the UN should consider including an eighteenth SDG that acknowledges biocultural heritage. With a complementary approach, based on his long-term work with base communities in Bolivia, Xavier Albó (2018) interprets the concept of "living well," *suma qamaña* in Aymara or *buen vivir* in Spanish as the fundamental moral logic that guides the life habits of Aymara and other Andean cultures and that even has been incorporated into the national constitutions of Bolivia and Ecuador.

Laura Zanotti (2018) focuses on the biocultural heritage and conservation movement of the indigenous Mẽbêngôkre-Kayapó people in the Brazilian Amazon. Zanotti points out the tensions between indigenous and conservation views and between Kayapó biocultural perspectives and national and international sociopolitical and institutional contexts. Regarding the first tension, she states that "while river and riparian habitats are central to Kayapó livelihoods and political goals, Kayapó understandings of aquatic landscapes are not central to national and international visions of sustainable Amazonian futures" (p. 343). Regarding the second tension, she focuses on the construction of the third largest hydroelectric dam in the world, Belo Monte. Despite 30 years of conflicts, the demonstrated negative impacts it would have on the Kayapó indigenous people and other local communities and their habitats, and over the opposition of the Inter-American Commission on Human Rights that challenged the constitutionality of the project, Belo Monte Dam was inaugurated in 2016. This case alerts us to the serious constraints facing biocultural conservation projects today.

A different outcome and approach are discussed by Felipe Montoya-Greenheck (2018) based on a participatory action research (PAR) project in the Ngöbe Indigenous Territory in southern Costa Rica. He concludes that by the end of the project, they "had more questions than answers" (p. 372). However, they found obvious differences between the Ngöbe-controlled lands, which remain mostly covered by old-growth forests with only small cultivated patches for family orchards, and the neighboring lands, which are mostly in the hands of corporate agro-industrial producers and nonindigenous rural communities that clear-cut the forest to "clean" it for planting maize or introduced grasses for livestock grazing. Montoya-Greenheck concludes that it would not be so difficult "if the process of biocultural homogenization would be simply the advance of one aesthetic favoring standardization over another aesthetic with a penchant for diversity" but explains that "the issue is much more complex and broader in scope, … it is a matter of ethics, of right and wrong. Questions like who gets to live and who must die are a matter of socio-environmental justice (Rozzi 2013) and are intimately bound within the homogenization-diversity clash" (p. 376).

Egba, egba, enigba lati bereFeran aye, anytime is the time to begin to love nature, is the Yoruba proverb that opens the chapter on Candomblé, an African-origin religion widely practiced in Brazil. Silvia Regina da Lima Silva interviews *Babalorixá* Paulo José dos Reyes (2018) to uncover the intimate and unique relation of Candomblé with nature. They concisely introduce a historical and cultural account of the Yoruba people who arrived in Brazil during the final period of slavery (eighteenth and nineteenth centuries). To overcome biocultural homogenization, it is indispensable to listen to and to dialogue with worldviews and cultural practices such as those associated with Candomblé. This intercultural hermeneutic exercise implies moving from the "only one truth" paradigm to an understanding that there are "other paradigms," a "diversality" as opposed to the "universality" (sensu Mignolo 2011). Focusing on a Candomblé community that was recently legally recognized as "*quilombo* land," they conclude that it is possible to learn new and different forms of living; moreover, "the relationship to the earth, the natural environment, needs to be understood from other paradigms" (p. 389).

Based on their long-term experience with *liberation theology* in Latin America, Roy and Janet May (2018) assert that "this theology is a force for cultural diversity and local respect in an ever-increasingly homogenized world order" (p. 393). Solidarity, cultural identity, and diversity are core values of liberation theology that today are expressed through a cultural aesthetics embodied in liturgies, hymnody, poetry, and visual arts. These art forms are ways of protesting injustice and reinforcing resistance and hope for marginalized and oppressed communities.

National-level policy changes can also contribute to, or hinder, biocultural conservation efforts. Ruifei Tang and Michael Gavin (2018) analyze how policies of the Chinese government have driven profound changes in land and resource ownership, which in turn impacted social-ecological conditions in Inner Mongolia from the 1960s to the early 2000s. Government actions forced the privatization of land and resources, removed the rights of communal organizations to government committees, and centralized social services, thereby undermining traditional practices and knowledge systems in the grasslands of Inner Mongolia. This shift also had negative ecological consequences leading to desertification processes and the loss of saxaul (*Haloxylon ammodendron*), a keystone shrub species in the Gobi Desert. However, recent conservation initiatives have triggered positive changes in grassland management. These are based on local knowledge, revitalization of the traditional practices of Mongolian herders, as well as allowing support from conservation and development organizations that have added other forms of knowledge and sources of funding at regional, national, and international scales. Based on these conservation case studies, Tang and Gavin conclude that greater adherence to principles of biocultural conservation (sensu Gavin et al. 2015, 2018) may lead to interrelated positive social and ecological outcomes.

A different case of biocultural conservation has taken place in the Western Himalayas. Pandurang Hegde and George James (2018) portray the Chipko Movement, an iconic grassroots environmental movement of India initiated in the 1970s. This movement has inspired other conservation initiatives, such as the Appiko Movement in the forested region of southwestern India. Both *Chipko* and *Appiko* mean "to hug," referring in this case to literally hugging trees and inspiring a strategy of non-violent resistance to protect biodiversity, traditional life habits, and local access to forested habitats.

The Japanese word *saisei* resonates with the biocultural ethic. *Saisei* means more than ecological restoration; it means a revitalization of life habits and habitats. Mitsuyo Toyoda (2018) demonstrates how re-vitalizing the commons in estuarine ecosystems has stimulated local participation, from children to elders. She introduces an innovative case study of restoration of "secondary nature" in rural and suburban coastal ecosystems rather than untouched nature in remote areas. This case highlights how local ecological knowledge is dynamic because it is based on the new tradition of oyster fishing that began only in the twentieth century. At the same time, these revitalizing practices maintain the traditional Japanese spirit of "commons" as expressed by the word *sato*, which signifies the close interrelationship of nature and culture and the importance of shared management of natural resources. This notion is known globally through *satoyama*, meaning commonly used forest, but it reaches more broadly to *satoumi* referring to commonly used coastal resources and *satogawa* to commonly used rivers. The notion of *sato* has encouraged people to reinterpret the values of commonly used natural resources while taking into consideration current pressing socio-environmental issues.

The Garden as a Representation of Nature: A Space to Overcome Biocultural Homogenization? This is the question posed by Tetsuya Kono (2018) in the closing chapter. By comparing French formal gardens and Japanese *Daimyo gardens*, he explores how gardens can play an important role for achieving biocultural conservation or, instead, how gardens can be an expression of and contribute to biocultural homogenization. In European and Asian garden traditions, there are both remarkable similarities and contrasts about the design, motivations, and values associated with the gardens. Kono proposes that "a good garden of today must neither be a garden which represents a universe rationally organized from a privileged viewpoint such as Versailles, nor a garden which represents an ever-fertile farm and orchard useful only for human beings, nor a garden which represents Taoist paradise such as Daimyo Tei-en, but a garden which represents sustainability and coexistence of nature and human society" (p. 474). He concludes that "a better understanding of the aesthetic, political, and cultural values of gardens in these traditions can serve as a basis to better foster biocultural conservation and prevent biocultural homogenization" (p. 460).

1.5 Concluding Remark

This book's ultimate objective is to stimulate a research agenda in biocultural homogenization and ways to reorient this process toward biocultural conservation. Through its concepts and case studies, this volume invites readers to consider further the wealth of biocultural worldviews and practices existing across the globe. These biocultural worldviews and practices markedly contrast with the monoculture of consumption established by a global hegemonic economic discourse. As stated by Rozzi (2018c) in chapter 19, "The current trend of disconnection of global society from biocultural diversity represents an anomalous life habit. This trend needs to be reoriented in light of the values and practices that are still alive in a plethora of cultures and people who have vital awareness of ways of co-inhabiting with diverse co-inhabitants" (p. 311). Toward this end, this volume is a first attempt. We hope, however, that it will stimulate further research about the interrelations between the homogenization (and or conservation) of native habitats and the life habits of the diverse co-inhabitants that co-inhabit them. Such a research agenda could serve as a bridge for considering the global, as well as the local, interests of the communities that inhabit the heterogeneous regions of the planet. The visualization of the interrelations between the homogenization of habits and habitats and the consequences it has for the well-being or the displacement of human and other-than-human co-inhabitants can inform and provide insights for decision-making in environmental policies, development, and educational programs. If the essays in this volume contribute to this visualization and associated actions, this book will have fulfilled its purpose.

References

Albó X (2018) *Suma qamaña* or living well together: a contribution to biocultural conservation. In: Rozzi R, May RH Jr, Chapin FS III, Massardo F, Gavin M, Klaver I, Pauchard A, Nuñez MA, Simberloff D (eds) From biocultural homogenization to biocultural conservation. Ecology and ethics, vol 3. Springer, Dordrecht, pp 333–342

Crego RD, Rozzi R, Jimenez JE (2018) Fur Trade and the biotic homogenization of sub-polar ecosystems. In: Rozzi R, May RH Jr, Chapin FS III, Massardo F, Gavin M, Klaver I, Pauchard A, Nuñez MA, Simberloff D (eds) From biocultural homogenization to biocultural conservation. Ecology and ethics, vol 3. Springer, Dordrecht, pp 233–243

Davis LJ, Morris DB (2007) Biocultures manifesto. New Lit Hist 38(3):411–418

dos Reyes PJ, da Lima-Silva SR (2018) Candomblé in Brazil: the contribution of African-origin religions to biocultural diversity in the Americas. In: Rozzi R, May RH Jr, Chapin FS III, Massardo F, Gavin M, Klaver I, Pauchard A, Nuñez MA, Simberloff D (eds) From biocultural homogenization to biocultural conservation. Ecology and ethics, vol 3. Springer, Dordrecht, pp 379–391

Espindola HS, Guerra CB (2018) The ongoing danger of largescale mining on the Rio Doce: an account of Brazil's largest biocultural disaster. In: Rozzi R, May RH Jr, Chapin FS III, Massardo F, Gavin M, Klaver I, Pauchard A, Nuñez MA, Simberloff D (eds) From biocultural

homogenization to biocultural conservation. Ecology and ethics, vol 3. Springer, Dordrecht, pp 97–108
Esteban M (2018) Dürer's Rhinoceros: biocultural homogenization of the visual construction of nature. In: Rozzi R, May RH Jr, Chapin FS III, Massardo F, Gavin M, Klaver I, Pauchard A, Nuñez MA, Simberloff D (eds) From biocultural homogenization to biocultural conservation. Ecology and ethics, vol 3. Springer, Dordrecht, 137–165
García A, Franzese J, Policelli N, Sasal Y, Zenni R, Nuñez MA, Taylor K, Pauchard A (2018) Non-native pines are homogenizing the ecosystems of South America. In: Rozzi R, May RH Jr, Chapin FS III, Massardo F, Gavin M, Klaver I, Pauchard A, Nuñez MA, Simberloff D (eds) From biocultural homogenization to biocultural conservation. Ecology and ethics, vol 3. Springer, Dordrecht, pp 245–263
Gavin MC, McCarter J, Mead ATP, Berkes F, Stepp JR, Peterson D, Tang R (2015) Defining biocultural approaches to conservation. Trends Ecol Evol 30(3):140–145
Gavin MC, McCarter J, Berkes F, Mead ATP, Sterling EJ, Tang R, Turner NJ (2018) Effective biodiversity conservation requires dynamic, pluralistic, partnership-based approaches. Sustainability 10(6). https://doi.org/10.3390/su10061846
Hedge P, James G (2018) Challenging biocultural homogenization: experiences of the Chipko and Appiko movements in India. In: Rozzi R, May RH Jr, Chapin FS III, Massardo F, Gavin M, Klaver I, Pauchard A, Nuñez MA, Simberloff D (eds) From biocultural homogenization to biocultural conservation. Ecology and ethics, vol 3. Springer, Dordrecht, pp 427–442
Klaver I (2018) Re-claiming Rivers from homogenization: meandering and riverspheres. In: Rozzi R, May RH Jr, Chapin FS III, Massardo F, Gavin M, Klaver I, Pauchard A, Nuñez MA, Simberloff D (eds) From biocultural homogenization to biocultural conservation. Ecology and ethics, vol 3. Springer, Dordrecht, pp 49–69
Kono T (2018) The garden as a representation of nature: a space to overcome biocultural homogenization? In: Rozzi R, May RH Jr, Chapin FS III, Massardo F, Gavin M, Klaver I, Pauchard A, Nuñez MA, Simberloff D (eds) From biocultural homogenization to biocultural conservation. Ecology and ethics, vol 3. Springer, Dordrecht, pp 459–474
Lenzner B, Essl F, Seebens H (2018) The changing role of Europe in past and future alien species displacement. In: Rozzi R, May RH Jr, Chapin FS III, Massardo F, Gavin M, Klaver I, Pauchard A, Nuñez MA, Simberloff D (eds) From biocultural homogenization to biocultural conservation. Ecology and ethics, vol 3. Springer, Dordrecht, pp 125–135
Li D, Lockwood JL, Baiser B (2018) Taxonomic and phylogenetic homogenization across US National Parks – The role of non-native species. In: Rozzi R, May RH Jr, Chapin FS III, Massardo F, Gavin M, Klaver I, Pauchard A, Nuñez MA, Simberloff D (eds) From biocultural homogenization to biocultural conservation. Ecology and ethics, vol 3. Springer, Dordrecht, pp 275–288
Makki F (2018) The political ecology of land grabs in Ethiopia. In: Rozzi R, May RH Jr, Chapin FS III, Massardo F, Gavin M, Klaver I, Pauchard A, Nuñez MA, Simberloff D (eds) From biocultural homogenization to biocultural conservation. Ecology and ethics, vol 3. Springer, Dordrecht, pp 84–95
May RH Jr (2018) Land grabbing and violence against environmentalists. In: Rozzi R, May RH Jr, Chapin FS III, Massardo F, Gavin M, Klaver I, Pauchard A, Nuñez MA, Simberloff D (eds) From biocultural homogenization to biocultural conservation. Ecology and ethics, vol 3. Springer, Dordrecht, pp 109–123
May RH Jr, May JW (2018) Latin American theology of liberation and biocultural conservation. In: Rozzi R, May RH Jr, Chapin FS III, Massardo F, Gavin M, Klaver I, Pauchard A, Nuñez MA, Simberloff D (eds) From biocultural homogenization to biocultural conservation. Ecology and ethics, vol 3. Springer, Dordrecht, pp 393–403
McKinney ML, Lockwood JL (1999) Biotic homogenization: a few winners replacing many losers in the next mass extinction. Trends Ecol Evol 14:450–453
Mignolo W (2011) The darker side of western modernity: global futures, decolonial options. Duke University Press, Durham

Montoya-Greenheck F (2018) Biocultural diversity and *Ngöbe* people in the South Pacific of Costa Rica. In: Rozzi R, May RH Jr, Chapin FS III, Massardo F, Gavin M, Klaver I, Pauchard A, Nuñez MA, Simberloff D (eds) From biocultural homogenization to biocultural conservation. Ecology and ethics, vol 3. Springer, Dordrecht, pp 361–378
Nuñez MA, Simberloff D (2005) Invasive species and the cultural keystone species concept. Ecol Soc 10(1). https://www.ecologyandsociety.org/vol10/iss1/resp4/
Nuñez M, Dimaraco RD, Simberloff D (2018) Why some exotic species are deeply integrated into local cultures while others are reviled. In: Rozzi R, May RH Jr, Chapin FS III, Massardo F, Gavin M, Klaver I, Pauchard A, Nuñez MA, Simberloff D (eds) From biocultural homogenization to biocultural conservation. Ecology and ethics, vol 3. Springer, Dordrecht, pp 219–231
Olden J, Rooney T (2006) On defining and quantifying biotic homogenization. Glob Ecol Biogeogr 15:113–120
Paredes-Castellanos A, Rozzi R (2018) Biocultural exoticism in the feminine landscape of Latin America. In: Rozzi R, May RH Jr, Chapin FS III, Massardo F, Gavin M, Klaver I, Pauchard A, Nuñez MA, Simberloff D (eds) From biocultural homogenization to biocultural conservation. Ecology and ethics, vol 3. Springer, Dordrecht, pp 167–183
Petitat A (1987) School and the production of society. Br J Sociol Educ 8:379–390
Poole AK (2018) The UN sustainable development goals and the biocultural heritage lacuna: where is goal number 18? In: Rozzi R, May RH Jr, Chapin FS III, Massardo F, Gavin M, Klaver I, Pauchard A, Nuñez MA, Simberloff D (eds) From biocultural homogenization to biocultural conservation. Ecology and ethics, vol 3. Springer, Dordrecht, pp 315–331
Quijano A (2000) Coloniality of power and eurocentrism in Latin America. Int Sociol 15:215–232
Rizvi F, Lingard B (2000) Globalization and education: complexities and contingencies. Educ Theory 50:419–426
Rozzi R (2001) Ética ambiental: raíces y ramas latinoamericanas. In: Primack R, Rozzi R, Feinsinger P, Dirzo R, Massardo F (eds) Fundamentos de Conservación Biológica: Perspectivas Latinoamericanas. Fondo de Cultura Económica, Mexico, pp 311–359
Rozzi R (2012) Biocultural ethics: the vital links between the inhabitants, their habits and regional habitats. Environ Ethics 34:27–50
Rozzi R (2013) Biocultural ethics: from biocultural homogenization toward biocultural conservation. In: Rozzi R, Pickett STA, Palmer C, Armesto JJ, Callicott JB (eds) Linking ecology and ethics for a changing world. Ecology and ethics, vol 1. Springer, Dordrecht, pp 9–32
Rozzi R (2018a) Biocultural homogenization: a wicked problem in the Anthropocene. In: Rozzi R, May RH Jr, Chapin FS III, Massardo F, Gavin M, Klaver I, Pauchard A, Nuñez MA, Simberloff D (eds) From biocultural homogenization to biocultural conservation. Ecology and ethics, vol 3. Springer, Dordrecht, pp 21–47
Rozzi R (2018b) Biocultural homogenization in modern philosophy: David Hume's noble Oyster. In: Rozzi R, May RH Jr, Chapin FS III, Massardo F, Gavin M, Klaver I, Pauchard A, Nuñez MA, Simberloff D (eds) From biocultural homogenization to biocultural conservation. Ecology and ethics, vol 3. Springer, Dordrecht, pp 185–204
Rozzi R (2018c) Biocultural conservation and biocultural ethics. In: Rozzi R, May RH Jr, Chapin FS III, Massardo F, Gavin M, Klaver I, Pauchard A, Nuñez MA, Simberloff D (eds) From biocultural homogenization to biocultural conservation. Ecology and ethics, vol 3. Springer, Dordrecht, pp 303–314
Safiq AD, Lockwood JL, Brown J (2018) Homogenization of fish assemblages off the coast of Florida. In: Rozzi R, May RH Jr, Chapin FS III, Massardo F, Gavin M, Klaver I, Pauchard A, Nuñez MA, Simberloff D (eds) From biocultural homogenization to biocultural conservation. Ecology and ethics, vol 3. Springer, Dordrecht, pp 289–300
Schaedel RP (1979) From homogenization to heterogenization in Lima, Peru. Urb Anthropol 8:399–420
Simberloff D (2018) Nature, culture, and natureculture: the role of nonnative species in biocultures. In: Rozzi R, May RH Jr, Chapin FS III, Massardo F, Gavin M, Klaver I, Pauchard A, Nuñez MA, Simberloff D (eds) From biocultural homogenization to biocultural conservation. Ecology and ethics, vol 3. Springer, Dordrecht, pp 207–218

Simberloff D, Von Holle MB (1999) Positive interactions of nonindigenous species: invasional meltdown? Biol Invasions 1:21–32

Tang R, Gavin MC (2018) The dynamics of biocultural approaches to conservation in Inner Mongolia, China. In: Rozzi R, May RH Jr, Chapin FS III, Massardo F, Gavin M, Klaver I, Pauchard A, Nuñez MA, Simberloff D (eds) From biocultural homogenization to biocultural conservation. Ecology and ethics, vol 3. Springer, Dordrecht, pp 405–425

Taylor B (ed) (2013) Avatar and nature spirituality. Wilfrid Laurier University Press, Waterloo

Taylor B (2018) Biostitutes and biocultural conservation: empire and irony in the motion picture avatar. In: Rozzi R, May RH Jr, Chapin FS III, Massardo F, Gavin M, Klaver I, Pauchard A, Nuñez MA, Simberloff D (eds) From biocultural homogenization to biocultural conservation. Ecology and ethics, vol 3. Springer, Dordrecht, pp 71–82

Toyoda M (2018) Revitalizing local commons: a democratic approach to collective management. In: Rozzi R, May RH Jr, Chapin FS III, Massardo F, Gavin M, Klaver I, Pauchard A, Nuñez MA, Simberloff D (eds) From biocultural homogenization to biocultural conservation. Ecology and ethics, vol 3. Springer, Dordrecht, pp 443–457

UNESCO (2001) UNESCO Universal Declaration on Cultural Diversity, UNESCO. http://portal.unesco.org/en/ev.php-URL_ID=13179&URL_DO=DO_TOPIC&URL_SECTION=201.html

Zanotti L (2018) Biocultural approaches to conservation: water sovereignty in the Kayapó Lands. In: Rozzi R, May RH Jr, Chapin FS III, Massardo F, Gavin M, Klaver I, Pauchard A, Nuñez MA, Simberloff D (eds) From biocultural homogenization to biocultural conservation. Ecology and ethics, vol 3. Springer, Dordrecht, pp 343–359

Zenni RD, Guimarães R, Tidon R (2018) Biotic homogenization of the South American Cerrado. In: Rozzi R, May RH Jr, Chapin FS III, Massardo F, Gavin M, Klaver I, Pauchard A, Nuñez MA, Simberloff D (eds) From biocultural homogenization to biocultural conservation. Ecology and ethics, vol 3. Springer, Dordrecht, pp 265–274

Part I
Biocultural Homogenization

Chapter 2
Biocultural Homogenization: A Wicked Problem in the Anthropocene

Ricardo Rozzi

Abstract Wicked problems are difficult or impossible to solve. To overcome this dilemma, I propose an approach to better understand and contribute to solve a wicked problem that is particularly pervasive in the Anthropocene: biocultural homogenization. This approach can also help to guide biocultural conservation processes. Toward this aim, I propose to use the "3Hs" conceptual framework of the biocultural ethic that values the vital links among unique life *habits* of *co-inhabitants* who share specific *habitats*. A first outcome is the identification of feedback processes entailing interwoven losses of biological and cultural diversity. Then, I organize these feedback processes in a hierarchical sequence of increasing structural complexity. Analysis of these processes offers a theoretical framework for understanding the interrelations between the homogenization of habits and habitats and the consequences of biocultural homogenization for the lives of diverse human and other-than-human co-inhabitants. Co-inhabitants are subjects (not objects). They co-constitute their identities and share habitats that they co-structure through co-inhabitation relationships of complementarity and reciprocity. Habitats are the condition of possibility for the existence and well-being of the co-inhabitants. I have introduced the notion of co-inhabitants as an ethical justification to oppose biocultural homogenization and to demand biocultural conservation in terms of socio-environmental justice.

Keywords Co-inhabitants · Ethics · Habits · Traditional ecological knowledge · Environmental justice

---

R. Rozzi (✉)
Department of Philosophy and Religion and Department of Biological Sciences, University of North Texas, Denton, TX, USA

Sub-Antarctic Biocultural Conservation Program, University of North Texas, Denton, TX, USA

Instituto de Ecología y Biodiversidad and Universidad de Magallanes, Punta Arenas, Chile
e-mail: Ricardo.Rozzi@unt.edu

R. Rozzi et al. (eds.), *From Biocultural Homogenization to Biocultural Conservation*, Ecology and Ethics 3, https://doi.org/10.1007/978-3-319-99513-7_2

2.1 The Anthropocene and Biocultural Homogenization

The Anthropocene represents a new geological era in which the degree of human influence over the biosphere exceeds that of other natural forces. The term makes explicit that we are experiencing a shift from an *anthropocentric* modernity toward an *anthropogenic* global civilization. Today, the connection between human beings and the biosphere has a decisive material impact on the destiny of humans and other-than-human beings even in remote places of the planet. The connection between citizens and ecosystems, between humans and other-than-human beings (other animals, plants, oceans, mountains, spirits), is no longer a bizarre question confined to mythological or symbolic stories. To realize that modern society is anthropocentric is not a great surprise, but to realize that global society is an anthropogenic agent with a power of change that surpasses the power of the geological forces that shape the planet Earth should shake us.

In the Anthropocene the old dichotomies between "biophysical" and "cultural" dimensions of reality vanish. Consequently, it becomes necessary to adopt a biocultural lens to understand processes and patterns of co-inhabitation immersed in the biological and cultural diversity of the biosphere. This biocultural lens yields a new understanding about our position amidst the Anthropocene that transcends the purely descriptive plane: affirming that human agency has become the main force shaping the face of the Earth immediately raises questions of ethics.

In this chapter, I focus on the "3Hs" conceptual framework of the biocultural ethic that values the links between the unique life *habits* of the *co-inhabitants* who share specific *habitats*. The conservation of these habitats is a necessary condition for the continuity of life habits that contribute to the well-being and identity of the co-inhabitants. These vital interrelationships among the "3Hs" of the biocultural ethic are omitted in the culture and modes of production that prevail in global society, because they are based on a reduced group of biological species and life habits. This *bio-cultural reduction* leads to processes of *biocultural homogenization*.

Biocultural homogenization is both a driver and a product of complex and interwoven losses of biological and cultural diversity at local, regional, and global scales, which are taking place at accelerating rates in the Anthropocene. Despite its pervasiveness, the problem of biocultural homogenization is not fully understood, and its impact not widely recognized. The biocultural ethic addresses the severe socio-environmental injustices associated with biocultural homogenization and demands valuing and respecting the vital links among the specific life habits of co-inhabitants (human and other-than-human) that share habitats.

2.2 Biocultural Homogenization and the "3Hs" Model of the Biocultural Ethic

Wicked problems are difficult or impossible to solve (Rittel and Webber 1973). I undertake the challenge to characterize better the wicked problem of biocultural homogenization, and to propose solutions to it, by using the 3Hs conceptual

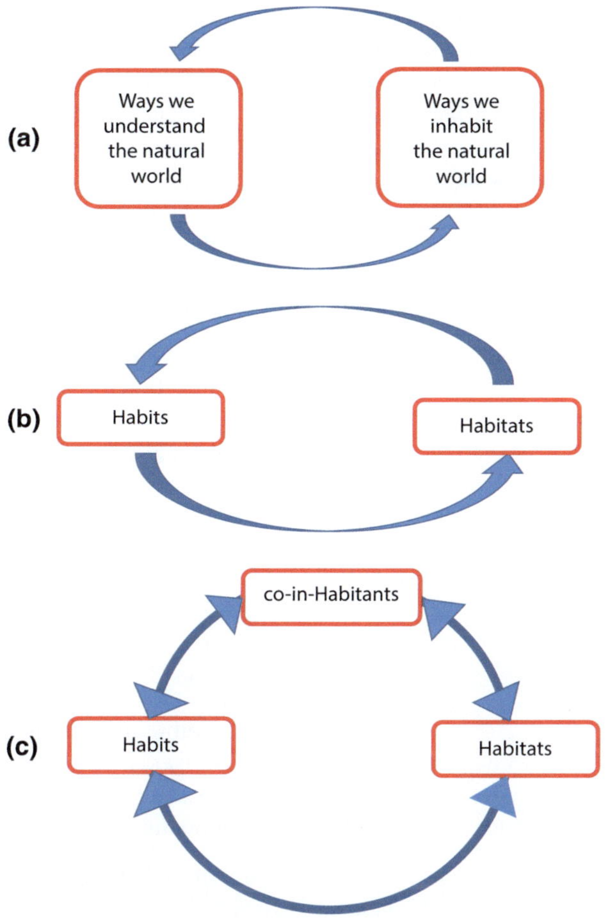

Fig. 2.1 Biocultural homogenization includes feedback processes entailing interwoven losses of biological and cultural diversity at three levels of increasing structural complexity. (**a**) At the level of life habits, there are reciprocal influences between reductionist ways of understanding and inhabiting the biosphere. (**b**) At the level of life habits and habitats, the homogenization of habits drives the homogenization of habitats; in turn, the homogenization of habitats reinforces the establishment of globally homogenized life habits. (**c**) At the level of life habits, habitats, and co-inhabitants, the reciprocal influences between the homogenization of habits and habitats have consequences for the lives of human and other-than-human co-inhabitants

framework. I begin by identifying feedback processes entailing interwoven losses of biological and cultural diversity at three levels. These levels are organized in a hierarchical sequence of increasing structural complexity (Fig. 2.1).

(A) At the level of life *habits*, there are feedbacks between reductionist ways of understanding and inhabiting the biosphere. These feedbacks between mind-sets and life practices underlie globalizing processes of technological change and market economy and their consequent negative socio-ecological externalities.

(B) At the level of *habits* and *habitats*, there are feedbacks between the homogenization of life habits associated with the exponential growth of the markets of the global north that are disseminated throughout the global south and homogenizing unique local habitats throughout the planet. The homogenization of these habitats, in turn, reinforces the establishment of global life practices and mind-sets (i.e., globally homogenized life habits).

(C) At the level of life *habits*, *habitats*, and *co-inhabitants*, feedbacks provoke displacements, even extermination of local populations of human and other-than-human beings. Biodiversity, biocultural memory, ecological knowledge, values, practices, and living conditions are all eroded.

Understanding the interrelations between the homogenization of habits and habitats, and the consequences that biocultural homogenization has for the well-being of human and other-than-human co-inhabitants, could orient decision-making in environmental policies, development, and educational programs in a way that contributes to a biocultural conservation that is socially and environmentally just. To understand the 3Hs conceptual model of the biocultural ethic, however, first it is important to, concisely, define the meaning I assign to each of the "Hs": *habitats* (Box 2.1), *habits* (Box 2.2) and *co-inhabitants*.

Box 2.1 Habitat

In the conceptual framework of the biocultural ethic the notion of *habitat is understood* as a three-dimensional concept. Changes in one dimension imply changes in the other ones (Rozzi 2015). The three complementary dimensions of the biocultural concept of habitat are the following:

1. *Biophysical* dimension, which scales up from local ecosystems to the global *biosphere* (*sensu* Vladimir Vernadsky, see Huggett 1999). In dictionaries, the definition of "habitat" is centered on the biophysical dimension: "the natural place of growth or occurrence of a species; the locality in which a plant or animal naturally grows or lives" (OED 1980). In ecology, the concept of habitat is often characterized with an ecosystem perspective as the place where an organism or a community of organisms lives, including all living and nonliving factors or conditions of the surrounding environment. A host organism inhabited by parasites is a habitat as much as a grove of trees is a terrestrial habitat or a small pond is an aquatic habitat (see Allen and Hoekstra 2015).
2. *Cultural* and *symbolic-linguistic* dimension, which scales up from vernacular languages to the global *logosphere* (*sensu* Michael Krauss 2007; i.e., the symbolic planetary web of *logos* or words, languages, and narratives). The sphere of the languages, or language habitats, houses a sphere of thought or intellect (in Greek *noos*), which was characterized by Vernadsky as the *noosphere*. In 1926, he proposed that the *noosphere* is generated by

(continued)

Box 2.1 (continued)

the set of living beings endowed with intelligence and their interactions with the environment in which they live (see Vernadsky 2007). In the 1970s, research about the interactions between language and the environment became active giving origin to the field of ecolinguistics (Haugen 1972; Fill and Mühlhauser 2001; Mühlhauser 2003; Bang et al. 2007). As Humberto Maturana has emphasized, humans inhabit languages, and we become who we are through "languaging" (Maturana et al. 1995). *Language- habitat* is a core concept of ecolinguistics (Bang and Trampe 2014). The language-habitat concept is useful for understanding the sorts of biocultural homogenization that have as a cause and/or a consequence the drastic losses of the diversity of languages that are taking place today (see Krauss 1992; Poole 2018; Rozzi 2018a).

3. *Sociopolitical*, *institutional*, and *technical* dimension, which scales up from local institutions to the global *technosphere* (*sensu* Zev Naveh and Arthur Lieberman 1990). In the twentieth century, the technosphere was defined by Naveh and Lieberman as that part of the physical environment affected through building or modification by humans. In the twenty-first century, ecologists Charles Redman and Thaddeus Miller (2015) have extended the meaning of the term *technosphere* to integrate infrastructure (physical and organizational) and technological systems. The constructed habitat or technosphere includes both the physical structures (e.g., roads, bridges, water supply, sewers, or electrical networks) and the organizational structures of services provided and the institutions. Redman and Miller (2015) contend that "those concerned with sustainability … must more robustly account for the centrality of technology in human-environment interactions, adjusting our conceptual frameworks to explore socio-eco-technological systems (SETS)" (p. 270).

All three dimensions of habitats –biophysical, linguistic, and technological including both physical and organizational infrastructure (e.g., institutions and policies)– have decisive influences on (and, in turn, are influenced by) life habits (Rozzi 2015).

Box 2.2 Life Habits

The notion of *habit* has become a lively topic of debate in various contemporary fields of theoretical and applied research, due to concern for the need to generate new habits in relation to sociopolitical and socio-environmental issues, such as neo-racism or climate change (Bennett 2016). However, the meaning of habit is often limited to the behavioral or psychological spheres.

(continued)

Box 2.2 (continued)

Indeed, the definition offered by the Oxford English Dictionary (OED 1980) is constrained to "an established disposition or tendency to act in a certain way, especially one acquired by frequent repetition of the same act until it reaches almost or quite involuntarily, an established practice, custom, use." This definition reduces the concept of habit to psychology and automatic behavior (behavior that is not consciously controlled). It is a reductionist definition that represents to a great extent a legacy of the mind-body dualisms that prevailed in philosophy from Descartes to Kant. In contrast, in the biocultural ethic I assign the term *habit* a broader connotation that considers its multiplicity of meanings through history and across disciplines.

The concept of habit is not limited to the behavioral or psychological spheres but includes biological, sociological, neurological, epistemological, phenomenological, ontological, and moral dimensions (see Sparrow and Hutchinson 2015). Regarding the latter, the term habit has an ethical connotation linked to the Aristotelian notions of virture and living well. In his *Nicomachean Ethics*, Aristotle stated that "neither by nature… nor contrary to nature do virtues arise in us; rather we are adapted by nature to receive them, and are made perfect by habit" (2009, II, 1, 1103a 23–26). The process of character formation is based on the practice of habits, which are the foundation for Aristotle's virtue ethics. To be fully human, we rely on a "first nature" (biological) and a "second nature" (social, cultural) that is learned and practiced. Today, in order to confront processes of biocultural homogenization, this ethical dimension of the habit helps us to understand the sociocultural roots of unsustainable lifestyles, as well as to reorient them toward sustainable habits.

In the twentieth century, French sociologist Pierre Bourdieu examined how life habits are learned through socialization processes that are influenced by family, social class, socioeconomic status, language, and culture. Bourdieu (1990) defined *habitus* as "embodied history, internalized as a second nature and so forgotten as history – is the active presence of the whole past of which it is the product" (p. 56). He critically observed that the habitus seemed innate, but that it is actually formed from schemes of perception and valuation of a social structure.

The meaning I attribute to the concept of *habit* is close to that of the Latin word *habitus*. Historically, habitus has captured the intentional and intelligent dispositions that are part of practical reason (Crossley 2013), and with Bourdieu's work has clearly addressed its social conditions. However, I have chosen the term *habit*, instead of *habitus*, for two reasons. First, habitus has acquired a meaning that is deeply marked by Bourdieu's sociology work. Second, I am interested in recovering the multiple meanings of the term habit, including their extension and reinterpretation with an ecological and evolutionary sense (Odling-Smee et al. 2003, Laland et al. 2016). This is, a

(continued)

Box 2.2 (continued)
modus vivendi involving ways of life that are more complex than mere instincts and that take place in multiple species. A broader biocultural meaning of the concept of habit emerges from an understanding that integrates Bourdieu's sociological approach of habitus, neo-Aristotelian schools of virtue ethics, and new findings of ecological and ethological sciences. This biocultural meaning of habit enables contextual and systemic analyses, based on the 3Hs model, to examine human globalized life habits that are driving the rapid socio-environmental changes of the Anthropocene, and to assess the consequences they have for the majority of human and other-than-human co-inhabitants. *Co-inhabitant* is a core concept of the biocultural ethic, and it is a word that is not included in English dictionaries. Consequently, I will explain the meaning I give to the word, and why I have decided to use it, in a separate section of this chapter (see Sect. 2.4). However, before I will address the main aim of this chapter: to show how the 3Hs model helps to identify the causes of biocultural homogenization and the results that derive from this process. In short, how it helps to visualize the process, the agents, and the victims. This understanding could help alleviate the problem of biocultural homogenization through reorientation of public policies and educational programs that contribute, instead, to generating processes of biocultural conservation (this is the general goal of our *From Biocultural Homogenization to Biocultural Conservation* book).

2.3 The 3Hs Model and Three Levels of Feedbacks in Biocultural Homogenization

2.3.1 Feedbacks of Biocultural Homogenization at the Level of Habits

A first level of biocultural homogenization is that of life habits. It derives from the feedbacks that are established between two realms of life habits: the ways of understanding and of inhabiting the world (Fig. 2.1a). My thesis is that *the ways of understanding the natural world influence, and are influenced by, the ways of inhabiting the natural world* (Rozzi 1999). Therefore, if the spectrum of ways of understanding the natural world is reduced, probably the spectrum of practices and ways of inhabiting the natural world will also be reduced and vice versa. Such reduction has been driven by a conception of the natural world dominated by a narrow economic vision that has become globalized. Both the spectrum of values and the spectrum of ways of inhabiting the natural world have been reduced under the governance of a narrow set of goals defined by global markets. This reduction has fostered one-dimensional life habits centered on consumerism and economic growth that have alienated society from knowledge, appreciation, and respect for the biological and cultural diversity with which we co-inhabit.

To reverse this trend, it is urgent to redirect global society toward habits of life (*modus vivendi*) that aim for equity and welfare of the diverse humans and other living beings. For such a reorientation a double task is necessary. First, re-vitalize the understanding of the broad spectrum of values hosted by the ecological worldviews and practices of numerous local communities that inhabit heterogeneous regions of the planet (Callicott 1994). Second, strengthen policies that enforce socio-environmental justice by setting limits to the globally established modes of production that function at the expense of the lives of local communities of human and other-than-human beings (Rozzi 2015, Makki 2018).

2.3.2 Feedbacks of Biocultural Homogenization at the Level of Habits and Habitats

A second level of biocultural homogenization derives from feedbacks that are established between life habits and the transformation of habitats (Fig. 2.1b). My thesis is that *life habits influence, and are influenced by, the habitats where they take place*. Therefore, if the spectrum of life habits is reduced, the spectrum of habitats will probably also be reduced and vice versa. One-dimensional life habits will be in line with the replacement of a diverse mosaic of biocultural landscapes by a homogeneity of landscapes that favor monocultures. Not only agricultural practices will be homogenized but also the architecture of cities as well as the infrastructures of coastal, fluvial, and terrestrial rural areas.

To reverse the trend toward biocultural homogenization, it is again necessary to assume a double task at this second level of complexity linking life habits and habitats. First, it is urgent to protect biocultural habitats and to revitalize awareness and understanding about the multidimensionality of habitats. Habitats entail biophysical, linguistic-cultural, and institutional realms that have historical and socio-ecological dynamics at different spatial and temporal scales. The protection of these unique biocultural habitats favors the expression of unhomogenized biocultural life habits. The continuity of these habits depends on the conservation of these habitats; at the same time, the continuity of traditional life habits favors the conservation of native habitats (Box 2.3). On the contrary, if native habitats are replaced by monocultures, both native habitats and local practices are suppressed. This process often involves displacements, usually involuntary, of local communities (Box 2.4). Therefore, the second task is to reinforce a sense of socio-environmental justice that has the capacity to condemn the forced displacement of local communities and the oppression, even the extermination, of communities of co-inhabitants.

Box 2.3 Biocultural Conservation: Habitats and Life Habits in Germany's First Protected Area

The reciprocal links between habitats and life habits can be appreciated in the historical dynamics of the first protected area in Germany: the Drachenfels (Dragon's Rock) (Fig. 2.2). It was established during the 1830s as a natural monument (*Naturdenkmal*) to protect the ruined Drachenfels castle that was built in the twelfth century on the summit of a hill on the banks of the Rhine south of Bonn. Later, the area was extended to include the surrounding hills in the nature protection area (*Naturschutzgebiet*) in Siebengebirge. Both the hills of the Siebengebirge and the ruin of Drachenfels had, however, a high symbolic value in the context of romanticism and the search for national identity in Germany, which at that time was divided into many small, more or less independent states (Jax and Rozzi 2004). The Drachenfels *Naturdenkmal* shows how in Germany the conservation movement began not as a movement to protect "wild" landscapes but as *Heimatschutz* (Dominick 1992, Knaut 1993), involving the protection (*Schutz*) of biocultural landscapes that formed traditional habitats or homeland (*Heimat*).

Fig. 2.2 Biocultural conservation in Germany's first protected area. **Above:** Biocultural habitats including the Drachenfels castle (left) and a mosaic of native forests, rocks, and vineyards (right). **Below:** The protection of habitats has enabled the continuity of traditional life habits such as viticulture, as well as more recently developed practices such as tourism involving navigation, hiking, wine tasting, and visits to the restored castle. Photographs by Kurt Jax and Wolfgang Schumacher

(continued)

Box 2.3 (continued)

The rock and the ruins gained popularity in the romantic era, and a neo-Gothic castle (Schloss Drachenburg) was built lower on the hill in 1882. A rack railway was also built at the end of the nineteenth century to satisfy growing demands from tourism. Today, tourism activities are complementary to viticulture practices on the slopes that are maintained and harvested by hand, since they are too steep for machinery. The case of this German protected area could apply to other cases that illustrate the intimate bonds between life habits and habitat conservation, and how these links are dynamic, and can satisfy multiple needs.

Box 2.4 Biocultural Homogenization: Replacement of Mangrove Habitats and Life Habits by Shrimp Farms

The disruption of vital links between native habitats and life habits of local communities takes place in a variety of global development projects that overlook socio-ecological problems associated with biocultural homogenization. A notorious example is the industrial farming of Ecuadorian shrimp, famous in today's international cuisine. Commercial cultivation of two species of shrimp, *Penaeus stylirostris* and *P. vannamei*, began in Ecuador in 1968 and in 1983, respectively. Since then Ecuador has become the world's principal producer of shrimp. This monoculture industry yielded such a large environmental impact that in 1995 the extension of shrimp ponds surpassed that of mangroves along the Ecuadorian coast (Suárez and Ortiz 2006) (Fig. 2.3).

In tropical regions of the world, mangroves are key habitats that act as "ecosystem membranes" between terrestrial and marine ecosystems, recycling nutrients and regulating hydrological flows (Anthony and Gratiot 2012). Their massive conversion to shrimp ponds dramatically increases the levels of sedimentation in coastal waters and the loss of nutrients that are limited in tropical soils. The destruction of mangroves, diversion of streams and rivers, and pollution of estuarine ecosystems drastically affect the diversity and population levels of species of algae, fish, crustaceans, and mollusks, all of which depend on mangroves at some phase of their life cycles (Mera 1999) as well as the health of humans who traditionally gather and consume shrimp, crabs, oysters, and other coastal organisms (Hagler 1997).

(continued)

Box 2.4 (continued)

Fig. 2.3 Biocultural homogenization (BH) driven by the replacement of mangrove habitats by shrimp farms in Ecuador, Northwestern South America. **Left:** Pictures illustrate the mangrove habitats where life habits of fishermen (above) and *concheras* (below) take place. *Concheras* are women who collect *conchas* or shellfish for subsistence in mangroves. If the mangrove habitats are destroyed, their *conchera* life-habit cannot continue. **Right:** Pictures illustrate a nearby area after replacement of mangroves by shrimp farms. The graph shows the area of shrimp ponds exceeded the area of mangroves in 1995. Photographs by Luis Suárez and Ricardo Rozzi

The shrimp industry causes serious social problems by limiting the access of local communities to coastal natural resources and by increasing income differences between a few rich people and a growing number of poor people (Veuthey and Gerber 2012). Local communities, particularly *concheras* of Ecuadorian and Central American coastal communities, have resisted the shrimp industry since the 1970s. *Concheras* are mostly African-American women who collect *conchas* or shellfish for subsistence in mangroves. These women have attempted to stop deforestation of mangroves, risking their lives by lying down in front of bulldozers that are used to create the shrimp ponds. On March 11, 1999, a *conchera* wrote that:

> we are struggling for something which is ours, our ecosystem, but not because we are professional ecologists but because we must remain alive, because if the mangroves disappear, a whole people disappears, … I do not know what will happen to us if the mangroves disappear, we shall eat garbage in the out-skirts of the city of Esmeraldas or in Guayaquil, we shall become prostitutes… (in Martínez-Alier 2001, pp. 715–6).

(continued)

Box 2.4 (continued)
The *conchera* expresses a clear understanding about the vital links between the conservation of mangrove habitats and the life habits and well-being of her community. Her criticism also makes obvious that large-scale natural resource exploitation models generally satisfy the needs of consumerist societies in distant places and not those of local people. More than 90% of the shrimp produced and exported by companies based in Ecuador are consumed only by people of three regions: the USA (51%), Japan (27%), and the European Union (17%) (Suárez and Ortiz 2006).

The export boom of Ecuadorian shrimp has not only provoked drastic habitat losses and degradation, but it also had a less well known "side effect": it disrupted life habits and brought a reduction in the quality of life of local people inhabiting the coastal regions of all Northwestern South America. The case of Ecuadorian shrimp could apply to numerous analogous cases throughout the planet that forcibly displace local communities that have sustainable life habits in their native habitats.

2.3.3 Feedbacks of Biocultural Homogenization at the Level of Habits, Habitats, and Co-inhabitants

A third level of biocultural homogenization derives from the feedbacks that are established among the transformation of life habits and habitats and the impacts of these transformations on communities of human and other-than-human co-inhabitants (Fig. 2.1c). My thesis is that *life habits and habitats influence and are influenced by the co-inhabitants with whom we share habitats*. The conservation of habitats fosters the well-being of human and other-than-human co-inhabitants whose life habits are interrelated (Box 2.5). On the contrary, if the spectrum of habitats is reduced, the spectrum of co-habitants and their life habits will probably also be reduced. Often, the destruction and replacement of habitats imply that the diversity of original co-inhabitants is eliminated and replaced by a few co-inhabitants associated with widespread monocultures (Box 2.6).

The three proposed levels of feedbacks that generate biocultural homogenization do not exhaust the complexity of this process. However, they contribute to visualizing the interrelations between the homogenization of life habits and habitats and the consequences this homogenization has for co-inhabitants. The notion of co-inhabitants is a key ethical construct for opposing biocultural homogenization. Hence, I need to explain what I mean by this notion and why I have decided to use the term co-inhabitants.

Box 2.5 Biocultural Conservation: Native Habitats, Co-Inhabitants, and Life Habits in the Subantarctic Archipelagoes of Chile

The vital links among habitats, habits, and co-inhabitants are embedded in a coevolutionary history in the archipelagoes of Cape Horn at the southern end of South America. The evergreen beech (*Nothofagus betuloides*) provides critical foraging and nesting habitat for the Magellanic Woodpecker (*Campephilus magellanicus*) (Fig. 2.4). The woodpecker controls the populations of insect larvae that infect the beech. Healthy beeches also enable an essential life habit of the Yahgan indigenous people (Rozzi et al. 2010). The tree's bark is used to build canoes to navigate the waters of Cape Horn. Hence, the evergreen beech is a biocultural keystone species, which enables a bird species and a human culture to inhabit the landscapes and seascapes of Cape Horn by being part of a community of co-inhabitants. Trees, woodpeckers, and humans participate in webs of ecological interactions that make possible co-inhabitation and co-constitute biological and cultural diversity.

Fig. 2.4 Biocultural conservation in the UNESCO Cape Horn Biosphere Reserve in subantarctic archipelagoes of Chile. **Left:** *Habitats* include the subantarctic evergreen rainforests dominated by the evergreen beech tree *Nothofagus betuloides*, on mountainous archipelagoes crossed by intricate fjords and channels. **Center:** *Co-inhabitants* include the Magellanic Woodpecker (*Campephilus magellanicus*), the largest woodpecker of South America, and an endemic endangered species. This bird depends on *Nothofagus* for foraging and nesting. In the past, Yahgan indigenous people gather barked of *N. betuloides* trees to build canoes. **Right:** Life *habits* include picking tree bark from only one side of the trunk so that the tree stays alive, and building canoes with this bark to navigate fjords and channels. Today, bark canoes are built as handcrafts for educational and ecotourism purposes. Photographs by Paola Vezzani, Jordi Plana, and Ricardo Rozzi

(continued)

Box 2.5 (continued)

The Magellanic Woodpecker is called *lana* by the indigenous Yahgan people. This name derives from the Yahgan word *lan*, which means tongue. The name alludes to the habit of this woodpecker of extending its long tongue to extract larvae from the holes it pecks in the trunk of old growth *Nothofagus* trees. The scientific name of the bird also alludes to its feeding *habit* and *habitat*: *Campephilus* means "larvae (*Campe* in Greek) lover" (*Philo* in Greek), and *magellanicus* (Latinized) indicates that it inhabits the Magellanic forests. Its English common name, Magellanic Woodpecker, also defines the identity of this bird by its *habit* of pecking wood in the Magellanic subantarctic forests (Rozzi et al. 2008). Hence, the intimate links between the habitats and the habits of this bird inhabitant are recognized both by the ornithological knowledge and by names in indigenous, scientific, and English languages.

The Magellanic Woodpecker belongs to the same genus of the two largest woodpecker species known worldwide, the Imperial Woodpecker (*Campephilus imperialis*) and the Ivory-billed Woodpecker (*C. principalis*). These two species inhabited the forests of North America and today are presumed to be extinct due to the destruction of their habitats and hunting pressures. Like its congeneric species from the temperate forests of the northern hemisphere, the Magellanic Woodpecker could become extinct if the relationships between its habits and habitats are not understood, are not valued, and are not incorporated into current conservation policies.

The understanding of this type of complex relationships among trees, humans, and birds is now of critical importance because globalization policies frequently overlook and override local specific biocultural interrelationships, which make it possible for human and other-than-human life to be unique and sustainable.

Box 2.6 Biocultural Homogenization: Replacement of Native Temperate Forests by *Eucalyptus* Monospecific Plantations

The temperate forests of Chile in Southwestern South America are classified as a world "biodiversity hotspot" because of their high species diversity and high endemism (Armesto et al. 1998, Arroyo et al. 2004). However, since the 1970s these native forests have been subjected to an accelerated process of replacement by monocultures of Monterrey pine (*Pinus radiata*) and *Eucalyptus* species (Echeverria et al. 2006) (Fig. 2.5). Since 1974, these monotypic plantations with exotic fast-growing tree species have been subsidized by Government Decree Law 701 that provides 75% to 90% of the cost of planting and that gives tax exemptions for the products obtained from these plantations (Pastor-Barrué 2004).

(continued)

Box 2.6 (continued)

Fig. 2.5 Biocultural homogenization (BH) driven by the replacement of native temperate forests by *Eucalyptus* monospecific plantations on Chiloé Island, Chile, Southwestern South America. **Left:** Pictures illustrate the diversity of forest *habitats* associated with small forestland ownership; *habits* that include gathering plant fibers, berries, and medicinal plants. The communities of co-inhabitants include indigenous Mapuche families; the *foye* (*Drimys winteri*), an endemic tree species sacred for the Mapuche people; and the *chukao* (*Scelorchilus rubecula*), an endemic terrestrial bird species. **Right:** Pictures illustrate a nearby area after replacement of the native forests by *Eucalyptus* plantations. The sign reads: "Entrance Forbidden" (*Entrada Prohibida* in Spanish); access to local people is not allowed inside this plantation. Photographs by Iván Díaz, Paola Vezzani, and Ricardo Rozzi

The massive substitution of native forests by exotic monocultures causes severe losses of biodiversity, soil erosion, and compactness and is detrimental to the hydrological cycle, provoking floods during winter and droughts during summer (Salas et al. 2016). A strong migration of rural population to urban centers is promoted because of the following: (1) Small owners have sold their lands to the tree companies; the consequent concentration of the land ownership has entailed displacement of local communities. (2) Forestry plantations require less labor than other agriculture, and labor is required only during intermittent years for planting, thinning, and cutting. (3) Most of the labor force comes with companies from other regions of Chile. (4) Other multiple uses and values of native forests are eliminated.

The living conditions of Mapuche indigenous communities have declined with the transformation of native forests into homogenous plantations of exotic trees. Today, Mapuche resistance movements demand recovering access to the land, restoring native forest and its natural water cycles that have been disrupted significantly by the forestry plantations, and autonomy and opportunities for framing their own development strategies (Torres-Salinas et al. 2016). The Mapuche environmental justice movement offers an example to the world of how to defend biocultural conservation as an alternative to biocultural homogenization.

2.4 Co-inhabitants

I began using the term *co-inhabitant* motivated by my fieldwork experiences, where I was impressed by the birds, the intentionality of their actions, the surprises of their behaviors, and the care they showed in their family groups (Rozzi et al. 2005, 2006, 2010). These life habits took place in the same habitats that I inhabited. From this experience, I understood the meaning of being co-inhabitants, sharing the same habitat. Just as the concept of *companion* alludes to sharing bread (from Latin, *cum* = with; *panis* = bread), the term co-inhabitant refers to sharing a habitat. Sharing the habitat implies an ecological-evolutionary process, and taking care of the habitat implies an ethical duty. Therefore, the concept of co-inhabitant has a double meaning; it is both descriptive and normative.

The concept of co-inhabitant is also consistent with ecological worldviews of many native peoples, where the birds are companions with whom a habitat must be shared. In addition, in Native American stories, there is often a sense of genealogical kinship. For some native cultures, such as the Koyukon in North America (Nelson 1983) or the Fuegian Yahgans in South America (Rozzi 2004), birds were once humans in ancestral times. For other Native American cultures, such as the Ojibwa in North America (Callicott and Nelson 2004) and the Mapuche of South America (Rozzi 2004), some humans are considered to be descendants of birds. In these indigenous worldviews, birds and humans share common ancestors; hence, birds and humans can be considered as "cousins," an expression used by Charles Darwin in his *On the Origin of Species*. Indeed, Native American ecological knowledge and scientific Western knowledge converge in the notion of evolutionary kinship. The ethical implications of the notion of kinship implicit in Darwin's evolutionary theory were elaborated in the mid-twentieth century by Aldo Leopold (1949), who wrote in his essay on the passenger pigeon (*Ectopistes migratorius*):

> It is a century now since Darwin gave us the first glimpse of the origin of species. We know now what was unknown to all the preceding caravan of generations: that men are only fellow voyagers with other creatures in the odyssey of evolution. This new knowledge should have given us, by this time, a sense of kinship with fellow-creatures; a wish to live and let live; a sense of wonder over the magnitude and duration of the biotic enterprise. (p. 117)

From this evolutionary and ecological foundation, Leopold extends the domain of the community of moral subjects to include the totality of beings with which we co-inhabit. In his work, Leopold portrays waters, soils, plants, and animals as "companions of a biotic community" or as "traveling companions in the odyssey of evolution." The notion of co-inhabitant is complementary to the ecological-evolutionary notion of companion used by Leopold. Both notions imply an extension of our ethical considerations beyond the human species to other-than-human co-inhabitants (Box 2.7). However, the notion of co-inhabitant makes explicit the ethical duty of taking care of the shared habitats.

The notion of co-inhabitant has interrelated ethical and ontological implications, which are relevant to critiquing forms of thought that contribute to biocultural

Box 2.7 *Other-Than-Human* Beings

I use the expression "other-than-human" beings instead of the more common expression "non-human" beings for three reasons. First, to avoid a simplistic dichotomous thinking between humans and non-humans. This dichotomy generates a fracture between humans and other beings; it is anthropocentric and hierarchical because it distinguishes between humans and the "rest."

Second, the concept of "other-than-human" beings refers to a set of beings that exist at different scales and levels of organization in the natural world, some of which are not commensurable with humans. "Other-than-human" beings can include biotic beings (humans, other animals, plants) and abiotic beings (rivers, rocks, glaciers). The multiple scales and levels of organization of "other-than-human" beings are understood not only from an ecological scientific point of view but also from other ecological worldviews. For example, the Aymara and Quechua Native American people in the high Andes of South America, the sun, moon, lightning, frost, hail, mountains, lakes, and springs, in addition to plants, animals, and humans, are subjects (Rozzi 2001). In addition, their interconnectedness with the land, air, and oceans is acknowledged through symbolic interactions and caring practices (May 2015, 2017).

Third, the expression "other-than-human" enables us to understand that these beings inhabit not only biophysical reality but also the images, symbols, and values of each culture. Therefore, they are co-inhabitants in the minds as well as in the biophysical habitats. Habitats include biophysical as well as linguistic domains of reality. Both domains influence (and are influenced by) cultural expressions, which take place in dreams (during the oneiric phases of our lives) and in the perceptions, association of ideas, and social practices (during the Waking phase of our lives) (Rozzi 2012).

This understanding of the concept of other-than-human beings is consistent with recent policy documents that aim to overcome dichotomies between humans and nature which have been associated with colonial practices and are responsible for some of the negative anthropogenic impacts that drive the Anthropocene. For example, the preamble of the Political Constitution of the Plurinational State of Bolivia approved in 2009 states that:

> In ancient times mountains arose, rivers moved, and lakes were formed. Our Amazonia, our swamps, our highlands, and our plains and valleys were covered with greenery and flowers. We populated this sacred Mother Earth with different faces, and since that time we have understood the plurality that exists in all things and in our diversity as human beings and cultures. Thus, our peoples were formed, and we never knew racism until we were subjected to it during the terrible times of colonialism.

homogenization. The currently prevailing conceptual framework of ecosystem services holds that human subjects administer goods and services provided by ecosystem objects and processes. Consequently, the only subjects (active agents with their own interest) are humans (see Naeem 2013). Biodiversity and ecosystems are viewed as passive objects without intentionality or interests. Under the perspectives of ecosystem services, these objects are managed by a utilitarian ethics that supposes an ontological split between human subjects and nature objects. This ontological split has a long history in Western philosophy and underlies the anthropocentric concept of sustainable development envisioned by the Brundtland Commission report, *Our Common Future* (WCED 1987). Environmental philosopher Irene Klaver (2013, p. 93) has pointed out that "the dualism between subject and object has been pervasive, deeply imbedded in Western thought, and at the root of a variety of interlocking dualisms, such as activity (or agency) versus passivity, resonating in culture versus nature. A dualistic mind-set comes with a value attribution, with an implied sense of superiority (culture, agency) versus inferiority (nature, passivity) and hence an implied legitimation for use, domination and exploitation."

In contrast to utilitarian ethics, the concept of co-inhabitant considers all living beings as active subjects with their own interests (Rozzi 2013). This statement is supported by concepts associated with the notion of co-inhabitant, which have ancient roots in Western philosophy. Pre-Socratic philosophers and Aristotle considered that all living beings had souls (Lat. *anima*), which meant spirit and, in turn, spirit (Lat. *spiritus*) meant breathing (see Rozzi 2015). In modernity, the utilitarian school of ethics has based moral considerability on the capacity to feel pain. However, the founder of utilitarianism, Jeremy Bentham, wrote in 1780 that "the day will come when humanity will spread its mantle over everything it breathes" (Bentham 1988, p. 98). Bentham's remark opens the possibility of grounding moral considerability on the capacity of breathing, which would imply a moral consideration for all multicellular living beings.

Contemporary sciences have demonstrated that eukaryotes, including all multicellular organisms and unicellular organisms of the Kingdom Protista, use the same cellular respiration processes. The same set of biochemical reactions take place in the cells of organisms that require oxygen to convert the energy of nutrients into molecules of adenosine triphosphate (ATP) (Mazzarello 1999). This vision of the contemporary sciences resonates with the assertions of Aristotle and of Native American people for whom all living creatures, domestic and wild, are perceived as having "spirit" that must be respected (May 2017). As an older woman from a Quechua community, who grows potatoes in the high Andes of Peru said: "You must respect (*respetar*) the potato really really well. If you treat her badly, she will give, … a smaller harvest… You must care for potatoes with affection" (in Angé et al. 2018, p. 34). According to Quechua people, potatoes are able to perceive whether humans respect them and can react to lack of respect by stopping their own growth process. In this sense, the subjectivity of the potato is linked to a particular type of intentionality (Angé et al. 2018). Reproduction of the potato requires a reciprocal circulation of respect, which is not just a normative stipulation. Respect is also an affection that frames human interactions with plants and animals. Plants like potatoes and animals like llamas are not "mere natural resources" but rather co-inhabitants that participate in rituals, farming and husbandry practices, and everyday life of Quechua communities (Mamani-Bernabé 2015, Rozzi 2015). For other Native American cultures, such as the Kayapó in Amazonia, plants and animals are considered friends with whom they engage as co-inhabitants in interrelated processes of production, exchange, and consumption (see Zanotti 2018).

According to the *Oxford English Dictionary* (OED 1980), words that we can associate with the concept of co-inhabitants were used for the first time in English in the sixteenth century. The term *co-inhabitors* was used by English poet and soldier Sir Philip Sydney in *The Countess of Pembroke's Arcadia* (1580), where he wrote: "all mankind being as it were *co-inhabitors* or world-citizens togeither [sic]" (in OED 1980, p. 475). In the seventeenth century, Henry More, English philosopher and theologian, used the word in *The Immortality of the Soul*, where he wrote "Our being co-inhabitants of the same element, the Earth" (in OED 1980, p. 475). However, both Sydney and More used these words only in reference to humans. Based on conceptual foundations of Western civilization, contemporary sciences, and Native American ecological worldviews, I use the word co-inhabitants in a broader sense to emphasize three essential attributes of the diverse beings that share a habitat:

(i) Co-inhabitants are subjects not objects.
(ii) Co-inhabitants co-constitute their identities and their well-being by dwelling with other human and other-than-human beings.
(iii) Co-inhabitants share habitats that they co-structure through co-inhabitation relationships. They establish ecological relationships of complementarity and reciprocity that occur through exchanges of matter and energy. Therefore, the care and conservation of habitats is the condition of possibility for the existence and well-being of the diverse co-inhabitants.

2.5 Concluding Remarks: A Biocultural Cascade Effect

Co-inhabitant is a central notion for ethically questioning the process of biocultural homogenization. When diverse human beings and other living beings are seen as active subjects with an intrinsic value, as co-inhabitants, then biocultural homogenization becomes a question of socio-environmental justice. Biocultural homogenization does not only imply monocultures or globalized modes of living, but also implies an ecocide that is rapidly expanding around the planet. The majority of diverse human and other-than-human co-inhabitants are absent from the mind-sets, discourses, and moral consideration in globalized society, economy, and governance. When they are present, they are habitually perceived and, valued as mere "human resources" or "natural resources" to be exploited.

To address this ethical dimension of biocultural homogenization, I propose a change of language and use the notion of co-inhabitant that questions reducing living (and nonliving) beings to mere resources. With the theoretical framework of the biocultural ethic, I aim to contribute to *decolonize* social mind-sets dazzled by the paradigm of one-dimensional economic growth. Noncapitalist dimensions of human existence have been eliminated under the prevalence of this paradigm (Brand and Wissen 2013). The one-dimensional man portrayed by Herbert Marcuse in the 1960s has come to dominate. As Marcuse (1991) argued, consumerism is a form of social control. The one-dimensional economic growth paradigm is presented as governed by democratic processes, but, in reality these processes are controlled by plutocratic powers (Rozzi 2015).

In order to reverse the sequence of feedback processes underling biocultural homogenization, as a foundational step, we need to change the language. It is indispensable to overcome the mind-set dominated by a one-dimensional instrumental way of thinking. Complementarily, it is indispensable to (re)incorporate a plethora of marginalized concepts, values, and ways of thinking and living. Reintegrating this multi-dimensionality into educational, political, and decision-making spheres de-homogenizes the mind-sets and life habits of globalized society and could reorients society toward biocultural conservation. In summary, I propose a *biocultural cascade* effect triggered by changing the conceptual languages, which, in turn, could modify habits of thinking and acting, thereby modifying the impacts on the habitats and the co-inhabitants with whom we share these habitats.

2.5.1 Bringing Back Pluri-Cultural Concepts of Human and Other-Than-Human Co-inhabitants

As stated above, the concept of co-inhabitant is deeply rooted in Western civilization, in scientific ecological thinking, and in Native American worldviews and practices. These three cultural sources offer languages and world views based on which we can criticize the exploitation of biocultural diversity as a mere "resource."

Humans were not viewed as separate from nature in early Western Civilization. The fact that our species *Homo sapiens* is part of ecosystems is implicit in the etymological origin of the word human, which comes from the Latin word *humus* that means soil. This understanding is also found in the origins of the Judeo-Christian tradition. In Genesis, the name of the first human being is Adam, which derived from the Hebrew *adamah*, which means *humus* or cultivable soil or "earth" (May 2015, Callicott 1994). In the biblical text, both the name and the material origin of the first human being are associated with Earth: "Then the Lord God formed man (*adam*) from the dust of the earth (*adamah*), and breathed into his nostrils the breath of life; and was the man became a living being" (Genesis 2:7 NRSV).

Humans are not separated from nature according to recent scientific discoveries. Contemporary biogeochemical sciences corroborate the early Western philosophical and religious concepts. Ecologists William Schlesinger and Emily Bernhardt (2013) point out that six elements (H, C, N, O, P, and S) are the major constituents of all living beings, humans included, and account for 95% of the mass of the biosphere. Moreover, they highlight how living organisms transform the Earth and its habitats by creating *humus*. Schlesinger and Bernhardt invite the reader to consider the increase in the total mass of organic matter that must have occurred during the Carboniferous period, about 300 million years ago (p. 6). This biogeochemical perspective resonates with the biocultural ethic's affirmation that habitats are formed by communities of co-inhabitants (Rozzi 2012). In the mid-twentieth century, ecologist Aldo Leopold (1949) had already proposed to substitute the purely economic image of the land with an ecological one: "Land, then, is not merely soil (p. 216); it is a fountain of energy flowing through a circuit of soils, plants, and animals. Food chains are the living channels which conduct energy upward; death and decay return it to the soil" (p. 204). Leopold affirms that the land ethic enlarges the boundaries of the community of beings that deserve moral considerability "to include soils, waters, plants, and animals, or collectively: the land."

Humans are not separated from nature in the ecological worldviews and practices of Native American people. Many Native American cultures call themselves people of the land (Rozzi 2013). In northern Brazil, the concept of co-inhabitant finds an exemplary convergence with the Kayapó concept of *ômbiqwa-ô-toro*: plants that are "good friends" or "good neighbors" with respect for each other. The Kayapó know that, when planted together, some combinations of plant species are more prosperous. Synergistic cultivation of plant species requires complex cultivation patterns and is characterized in terms of "plant energy." In this way, a "Kayapó garden" is created through careful consideration of complex combinations of "plant energies." Planting practices based on plant energies can be compared to the ecological principles that today guide the "new agroforestry" of Western science (Posey 1985). Rooted in their ecological worldview, the Kayapó people practice itinerant horticulture in Amazonia, achieving a high productivity of fruits, seeds, and tubers. Additionally, monkeys and other animals are attracted to these habitats. The temporal sequence in the sowing of a variety of crops closely mimics the sequence of the ecological succession of the plants and other organisms that co-inhabit with the Kayapó (Posey 1983). The resulting biocultural landscape with its mosaic of habitats brings well-being to both human and other-than-human co-inhabitants. The Kayapó now inhabit a reserve that includes a mosaic of tropical habitats, ranging from dense forests to vast grasslands, intertwined with varied waterscapes. As Laura Zanotti (2018) discusses, the Kayapó are currently facing growing challenges to protect their habitats against illegal logging, ranching, gold mining, and the construction of large dams.

The previous examples of philosophical, religious, scientific, and Native American ecological worldviews provide conceptual foundations to reconceptualize the relationship between global society and the biosphere in terms of co-inhabitation, and to demand changes in the language used in policies to implement effective conservation programs and to address complex socio-environmental problems. Scientists such as Clive Spash and Iulie Aslaksen (2015) have warned about the shortcomings of an economic discourse that is being increasingly used by ecologists and conservation biologists who conceptualize biodiversity and ecosystems as goods and services that can be represented by monetary values in policy processes. Spash and Aslaksen caution that this narrow instrumentalist approach denies value pluralism and incommensurability and call for re-establishing an ecological discourse in biodiversity and broader environmental policies. It is relevant to note that in 1982 the UN World Charter for Nature stated that:

> every form of life is unique, warranting respect regardless of its worth to man [sic], and, to accord other organisms such recognition, man must be guided by a moral code of action. (UN 1982, p. 2)

In this chapter I have criticized the reduction of language and "naturalization" of economic growth mind-sets with the aim to bring back into global society (its culture, science, and policy) concepts that have been marginalized, and to introduce the concept of co-inhabitant. By considering co-inhabitants as subjects (not mere objects), global society is invited to establish relationships of reciprocity. The reconceptualization of the relationship between global society and the biosphere in terms of co-inhabitation represents a paradigm shift that transforms the utilitarian ethics of ecosystem services and broadens associated conceptual frameworks, such as *Nature Contributions to People* (NCP) currently used by the Intergovernmental Platform on Biodiversity and Ecosystem Services (IPBES) (Pascual et al. 2017). The co-inhabitation paradigm fosters an ethics of reciprocity that requires adding the concept of *People's Contribution to Nature*. In this way, a biocultural cascade effect triggered by a change of language can reorient habits of thought and action, and their impacts on habitats. The protection of habitats is assumed as an ecological and ethically just action for the wellbeing of life in its biological and cultural diversity. This is not just an academic exercise, but it is a crucial matter of life or death for myriads of co-inhabitants in all corners of the planet. As Berta Caceres stated: "Let us wake up, humankind. We are out of time. We must shake our conscience free of the rapacious capitalism, racism… and patriarchy that will only assure our own self-destruction… Let us build societies that are able to coexist in a dignified way…in a way that protects life" (in May 2018, p. 110).

2.5.2 Reorienting Habit-Habitat Biocultural Homogenizing Feedback Processes

The conceptual model of the "3Hs" reveals that in human cultures there are tendencies that lead towards biocultural homogenization and others that lead towards biocultural conservation. The multi-potentiality of the mentalities and life practices of human individuals and societies, and their dynamic and fluctuating nature, are illustrated in the boxes (2.1, 2.2, 2.3, 2.4) of this chapter. Part I of this book further examines the dialectical tensions between both biocultural tendencies (homogenization and conservation) and suggests how they contribute to critical analysis and forms of action. To favor the transformation of a one-dimensional mentality underlying the processes of biocultural homogenization, Part I begins with the chapter by Irene Klaver (2018) who coined the metaphor of *meandering* to establish an analogy between the biophysical behaviors of rivers with a biocultural mentality of nonlinear thinking. This way of thinking would facilitate reverting linear thinking that leads to building one-dimensional habitat, to favoring instead multidimensional thinking for restoring diverse and dynamic socio-ecosystems. Haruf Espindola and Claudio Guerra (2018) criticize selective ignorance by some scientists associated with large development projects. Through this attitude, these scientists have accomplices of environmental disasters that have negatively affected the lives of myriads of human and nonhuman co-inhabitants, leading to oppressive biocultural homogenization. Other scientists, in contrast, have criticized large development projects with drastic socio-environmental consequences and have proposed measures of ecological restoration and protection of indigenous territories favoring biocultural conservation. Roy May (2018) documents how ignoring traditional people's rights and resorting to intimidation and even murder are strategies used to grab land and impose economic models that generate biocultural homogenization. On the other hand, murdered environmental leaders such as Berta Caceres (May 2018) and Chico Mendes (da Rocha and Valenti 2015) have been recognized for their work defending rivers and forests. Berta received the Goldman Award, and the Ministry of Environment in Brazil is named after Chico Mendes. Based on the film Avatar, Bron Taylor (2018) discusses the multifaceted motivations of scientists and the transformation of thinking processes and ethical attitudes that move away from biocultural homogenization trends to biocultural conservation actions.

In processes of biocultural homogenization, co-inhabitants are harmed directly if they are eliminated from their biophysical habitats and indirectly if they are eliminated or distorted in the representations and social mindsets. Ricardo Rozzi (2018b) presents a paradox in the work of the Scottish philosopher David Hume, one of the pillars of modern philosophy. On the one hand, Hume's philosophical imagination of animals is concentrated on only two species, horses and dogs. The most diverse group of animals, the invertebrates, is almost completely omitted. On the other hand, when Hume mentions an oyster, he refers to this animal as having the ability to feel and reason. Through analogical thought, Hume provided the theoretical basis for Darwin who conceived a common evolutionary origin and a common biological nature for all living beings, thus overcoming the Cartesian dualism between humans and other living beings. A paradigmatic case of Eurocentric distortion in the representation of animals is presented by Miguel Esteban (2018). Esteban analyzes the influential painting by Albrecht Durer of a Rhinoceros that was distortedly portrayed with a military breastplate and has been reproduced hundreds of times through the history of European art and biology teaching. Referring to Marshall McLuhan's concept of "The Gutenberg Galaxy," Esteban discusses the effects of mass media, especially the printing press, on European culture and how this has shaped global society's consciousness. Today, homogenized representations are effectively communicated to mass culture in order to keep it distracted, so as not to challenge the *status quo* and demand change. Counteracting this trend, Angelina Paredes-Castellanos and Ricardo Rozzi (2018) show how some Latin American painters and poets are denouncing the exoticization of Latin America. For example, in his painting *Biodiversity*, the Costa Rican artist Joaquín Rodríguez del Paso criticizes the exotic woman served for pleasure and economic exploitation by portraying a "wild woman" of the New World seated on a crocodile and served on a platter.

Homogenized mind-sets with distorted representations of the global South induce consumerism of biocultural diversity in the upper and middle classes of the global North and a growing neocolonialist process of land grabbing. Bernd Lenzner and collaborators (2018) observe that transnational land purchase has emerged as an important new feature in globalized land use and that large-scale land acquisitions have been shown to be mainly directed toward the global South. Land grabbing and other forms of concentration of land ownership are major drivers of the rapid rates of rural-urban migration in Africa, Asia, and Latin America that have been expanding toward the end of the twentieth century (Rozzi 2015). Based on his work in Ethiopia, Fouad Makki (2018) considers that land grabs associated with new forms of industrial agriculture constitute an archetype of biocultural homogenization. The extensive introduction of genetically uniform mono-crops has become the major threat to local crop varieties and traditional practices. Concerns of local communities have been overridden by global development policies coupled with the federal state in Ethiopia.

When local communities are forced to migrate to other places with promises of a prosperous future made by development projects and policies, most often human lives and a myriad of other living beings are sacrificed. Also lost are the relationships of co-inhabitation that humans have established with other-than-human co-inhabitants in specific socio-ecosystems. As alternatives to homogenizing practices, many resilient communities are currently promoting initiatives to restore (or conserve) co-inhabitation relationships in rural and urban areas. The 3Hs model of the biocultural ethic offers a conceptual framework to understand the complexities of biocultural homogenization. It also aims to identify avenues for environmental justice policies, institutional platforms, and actions that can promote biocultural conservation as an alternative to biocultural homogenization.

As argued above, to reorient the feedback relationships between life habits and habitats from biocultural homogenization toward biocultural conservation, it is necessary to rectify the one-dimensional lifestyle of consumerism and accumulating capital and to halt processes of land grabbing that override the rights and needs of local communities. In summary, in this chapter, I have offered the 3Hs conceptual model to deconstruct the wicked problem of biocultural homogenization. To read this book, and more importantly to disentangle and rectify cases of biocultural homogenization, I have suggested ways to:

(i) Broaden, with the theoretical framework of the biocultural ethic, the concepts of habitats, life habits, and co-inhabitants.
(ii) Broaden the spectrum of co-inhabitants, as well as of life habits and habitats, that are included in decision-making processes.
(iii) Advance the understanding about the interrelationships among life habits, habitats, and co-inhabitants.
(iv) Understand better the complexities of multidimensional and multi-scale processes involved in the wicked problem of biocultural homogenization.

(v) Structure a heuristic framework to assist decision-making processes in assessing the consequences that current or future development projects, under rapidly changing socio-environmental scenarios, might have for the conservation or destruction of habitats, life habits, and welfare of co-inhabitants.

References

Allen TF, Hoekstra TW (2015) Toward a unified ecology. Columbia University Press, New York

Angé O, Chipa A, Condori P, Ccoyo AC, Mamani L, Pacco R, Quispe N, Quispe W, Sutta M (2018) Interspecies respect and potato conservation in the Peruvian Cradle of domestication. Conserv Soc 16:30–40

Anthony EJ, Gratiot N (2012) Coastal engineering and large-scale mangrove destruction in Guyana, South America: averting an environmental catastrophe in the making. Ecol Eng 47:268–273

Aristotle (2009) The Nicomachean Ethics. Brown L (ed), (trans: Ross D). Oxford University Press: Oxford, UK Ulrich Brand

Armesto JJ, Rozzi R, Smith-Ramírez C, Arroyo MTK (1998) Effective conservation targets in South American temperate forests. Science 282:1271–1272

Arroyo MTK, Marquet P, Simonetti J, Marticorena C, Cavieres L, Rozzi R (2004) Central Chile/Valdivian Patagonian forests. In: Mittermeier R et al (eds) Hotspots revisited. CEMEX, Mexico City, pp 99–103

Bang JC, Trampe W (2014) Aspects of an ecological theory of language. Lang Sci 41:81–92

Bang JC, Door J, Steffensen S, Nash J (2007) Language, ecology, and society: a dialectical approach. Continuum, London

Bennett T (2016) Mind the gap: toward a political history of habit. Comparatist 40:28–55

Bentham J (1988) The principles of morals and legislation. Prometheus Books, New York

Bourdieu P (1990) The logic of Practice (Nice R, trans.). Polity, Cambridge

Brand U, Wissen M (2013) Crisis and continuity of capitalist society-nature relationships: The imperial mode of living and the limits to environmental governance. Rev Int Polit Econ 20(4):687–711

Callicott JB (1994) Earth's insights: a multicultural survey of ecological ethics from the Mediterranean basin to the Australian outback. University of California Press, Berkley

Callicott JB, Nelson MP (2004) American Indian environmental ethics: an Ojibwa case study. Pearson, New York

Crossley N (2013) Habit and habitus. Body Soc 19(2–3):136–161

da Rocha FJR, Valenti-Possamai F (2015) Chico Mendes and José Lutzenberger: ecosystem management at multiple scales of government. In: Rozzi R, Chapin FS III, Callicott JB, Pickett STA, Power ME, Armesto JJ, May RH Jr (eds) Earth stewardship: linking ecology and ethics in theory and practice, Ecology and ethics, vol 2. Springer, Dordrecht, pp 419–430

Dominick RHI (1992) The environmental movement in Germany. Prophets and pioneers 1871–1971. Indiana University Press, Bloomington

Echeverría C, Coomes D, Salas J, Rey-Benayas JM, Lara A, Newton A (2006) Rapid deforestation and fragmentation of Chilean temperate forests. Biol Conserv 130:481–494

Espindola HS, Guerra CB (2018) The ongoing danger of large-scale mining on the Rio Doce: an account of Brazil's largest biocultural disaster. In: Rozzi R, May RH Jr, Chapin FS III, Massardo F, Gavin M, Klaver I, Pauchard A, Núñez MA, Simberloff D (eds) From biocultural homogenization to biocultural conservation, Ecology and ethics, vol 3. Springer, Dordrecht, pp 97–108

Esteban M (2018) Dürer's Rhinoceros: biocultural homogenization of the visual construction of nature. In: Rozzi R, May RH Jr, Chapin FS III, Massardo F, Gavin M, Klaver I, Pauchard A, Núñez MA, Simberloff D (eds) From biocultural homogenization to biocultural conservation, Ecology and ethics, vol 3. Springer, Dordrecht, pp 137–165

Fill A, Mühlhäusler P (eds) (2001) The Ecolinguistics reader. Continuum, London
Hagler M (1997) Shrimp – the devastating delicacy. Greenpeace Reports, Washington DC
Haugen E (1972) The ecology of language. Stanford University Press, Stanford
Huggett RJ (1999) Ecosphere, biosphere, or Gaia? What to call the global ecosystem. Glob Ecol Biogeogr 8:425–431
Jax K, Rozzi R (2004) Ecological theory and values in the determination of conservation goals: examples from temperate regions of Germany, USA and Chile. Rev Chil Hist Nat 77:349–366
Klaver IJ (2013) Environment imagination situation. In: Rozzi R, Pickett STA, Palmer C et al (eds) Linking ecology and ethics for a changing world: values, philosophy, and action, Ecology and ethics, vol 1. Springer, Dordrecht, pp 85–105
Klaver IJ (2018) Reclaiming rivers from homogenization: meandering and Riverspheres. In: Rozzi R, May RH Jr, Chapin FS III, Massardo F, Gavin M, Klaver I, Pauchard A, Núñez MA, Simberloff D (eds) From biocultural homogenization to biocultural conservation, Ecology and ethics, vol 3. Springer, Dordrecht, pp 49–69
Knaut A (1993) Zurück zur Natur! Die Wurzeln der Ökologiebewegung. Jahrbuch für Naturschutz und Landschaftspflege (Supplement) 1:1–480
Krauss M (1992) The World's language in. Crisis Language 68:4–10
Krauss M (2007) Mass language extinction and documentation: the race against time. In: Miyaoka O, Sakiyama O, Krauss ME (eds) The vanishing languages of the pacific rim. Oxford University Press, New York, pp 3–24
Laland K, Matthews B, Feldman MW (2016) An introduction to niche construction theory. Evol Ecol 30(2):191–202
Lenzner B, Essl F, Seebens H (2018) The changing role of Europe in past and future alien species displacement. In: Rozzi R, May RH Jr, Chapin FS III, Massardo F, Gavin M, Klaver I, Pauchard A, Núñez MA, Simberloff D (eds) From biocultural homogenization to biocultural conservation, Ecology and ethics, vol 3. Springer, Dordrecht, pp 125–135
Leopold A (1949) A sand county almanac and sketches here and there. Oxford University Press, New York
Makki F (2018) The political ecology of land grabs in Ethiopia. In: Rozzi R, May RH Jr, Chapin FS III, Massardo F, Gavin M, Klaver I, Pauchard A, Núñez MA, Simberloff D (eds) From biocultural homogenization to biocultural conservation, Ecology and ethics, vol 3. Springer, Dordrecht, pp 83–95
Mamani-Bernabé V (2015) Spirituality and the Pachamama in the Andean Aymara worldview. In: Rozzi R, Chapin FS, Callicott JB, Pickett STA, Power ME, Armesto JJ, May RH Jr (eds) Earth stewardship: linking ecology and ethics in theory and practice, Ecology & Ethics Book Series. Springer, Dordrecht, pp 65–76
Marcuse H (1991) One-dimensional man: studies in ideology of advanced industrial society. Routledge, London
Martínez-Alier J (2001) Ecological conflicts and valuation: mangroves versus shrimps in the late 1990s. Environ Plan C 19:713–728
Maturana H, Mpodozis J, Letelier JC (1995) Brain, language, and the origin of human mental functions. Biol Res 28:15–26
May R (2015) Andean Llamas and Earth Stewardship. In: Rozzi R, Chapin FS III, Callicott JB, Pickett STA, Power ME, Armesto JJ, May RH Jr (eds) Earth stewardship: linking ecology and ethics in theory and practice, Ecology and ethics, vol 2. Springer, Dordrecht, pp. 77–86
May RH Jr (2017) Pachasophy: landscape ethics in the Central Andes Mountains of South America. Environ Ethics 39:301–319
May RH Jr (2018) Land grabbing and violence against environmentalists. In: Rozzi R, May RH Jr, Chapin FS III, Massardo F, Gavin M, Klaver I, Pauchard A, Núñez MA, Simberloff D (eds) From biocultural homogenization to biocultural conservation, Ecology and ethics, vol 3. Springer, Dordrecht, pp 109–123
Mazzarello P (1999) A unifying concept: the history of cell theory. Nat Cell Biol 1(1):E13–E15
Mera V (1999) Género, manglar y subsistencia, Ediciones Abya-Yala, Quito

Mühlhäusler P (2003) Language environment. In: Environment of language: a course in Ecolinguistics. Battlebridge, London
Naeem S (2013) Ecosystem services: is a planet servicing one species likely to function? In: Rozzi R, Pickett STA, Palmer C et al (eds) Linking ecology and ethics for a changing world: values, philosophy, and action, ecology and ethics, Ecology and ethics, vol 1. Springer, Dordrecht, pp 303–321
Naveh Z, Lieberman AS (1990) Landscape ecology. Springer, New York
Nelson RK (1983) Make prayers to the raven. A Koyukon view of the northern forest. University of Chicago Press, Chicago
NRSV (1989) New revised standard version of the bible. Division of christian education of the national council of the churches of christ in the USA.
Odling-Smee FJ, Laland KN, Feldman MW (2003) Niche construction: the neglected process in evolution. Princeton University Press, Princeton
OED (1980) Oxford English dictionary, 1st edn. Oxford University Press, New York
Paredes-Castellanos A, Rozzi R (2018) Biocultural exoticism in the feminine landscape of Latin America. In: Rozzi R, May RH Jr, Chapin FS III, Massardo F, Gavin M, Klaver I, Pauchard A, Núñez MA, Simberloff D (eds) From biocultural homogenization to biocultural conservation, Ecology and ethics, vol 3. Springer, Dordrecht, pp 167–183
Pascual U, Balvanera P, Díaz S, Yagi N (2017) Valuing nature's contributions to people: the IPBES approach. Curr Opin Environ Sustain 26:7–16
Pastor-Barrué M (ed) (2004) Forêts et développement durable au Chili. Indianité Mapuche et mondialisation. Presses Universitaire du Mirail, Toulouse
Poole AK (2018) The UN sustainable development goals and the biocultural heritage lacuna: where is goal number 18? In: Rozzi R, May RH Jr, Chapin FS III, Massardo F, Gavin M, Klaver I, Pauchard A, Núñez MA, Simberloff D (eds) From biocultural homogenization to biocultural conservation, Ecology and ethics, vol 3. Springer, Dordrecht, pp 315–331
Posey D (1983) Indigenous knowledge and development: an ideological bridge to the future. Ciência e Cultura 35(7):877–894
Posey D (1985) Indigenous management of tropical forest ecosystems: the case of the Kayapó Indians of the Brazilian Amazon. Agrofor Syst 3:139–158
Redman CL, Miller TR (2015) The Technosphere and earth stewardship. In: Rozzi R, Chapin FS III, Callicott JB, Pickett STA, Power ME, Armesto JJ, May RH Jr (eds) Earth stewardship: linking ecology and ethics in theory and practice, Ecology and ethics, vol 2. Springer, Dordrecht, pp 269–279
Rittel HW, Webber MM (1973) Dilemmas in a general theory of planning. Policy Sci 4(2):155–169
Rozzi R (1999) The reciprocal links between evolutionary-ecological sciences and environmental ethics. Bioscience 49(11):911–921
Rozzi R (2001) Éticas ambientales latinoamericanas: raíces y ramas. In: Primack, R., R. Rozzi, P. Feinsinger, R. Dirzo, F. Massardo y colaboradores, Fundamentos de Conservación Biológica: Perspectivas Latinoamericanas. Fondo de Cultura Económica, México, pp 311–362
Rozzi R (2004) Implicaciones éticas de narrativas yaganes y mapuches sobre las aves de los bosques templados de Sudamérica austral. Ornitol Neotrop 15:435–444
Rozzi R (2012) Biocultural ethics: the vital links between the inhabitants, their habits and regional habitats. Environ Ethics 34:27–50
Rozzi R (2013) Biocultural ethics: from biocultural homogenization toward biocultural conservation. In: Rozzi R, Pickett STA, Palmer C, Armesto JJ, Callicott JB (eds) Linking ecology and ethics for a changing world: values, philosophy, and action, Ecology and ethics, vol 1. Springer, Dordrecht, pp 9–32
Rozzi R (2015) Implications of the biocultural ethic for earth stewardship. In: Rozzi R, Chapin FS III, Callicott JB, Pickett STA, Power ME, Armesto JJ, May RH Jr (eds) Earth stewardship: linking ecology and ethics in theory and practice, Ecology and ethics, vol 2. Springer, Dordrecht, pp 113–136

Rozzi R (2018a) Biocultural conservation and biocultural ethics. In: Rozzi R, May RH Jr, Chapin FS III, Massardo F, Gavin M, Klaver I, Pauchard A, Núñez MA, Simberloff D (eds) From biocultural homogenization to biocultural conservation, Ecology and ethics, vol 3. Springer, Dordrecht, pp 303–314

Rozzi R (2018b) Overcoming biocultural homogenization in modern philosophy: Hume's Noble Oyster. In: Rozzi R, May RH Jr, Chapin FS III, Massardo F, Gavin M, Klaver I, Pauchard A, Núñez MA, Simberloff D (eds) From biocultural homogenization to biocultural conservation, Ecology and ethics, vol 3. Springer, Dordrecht, pp 185–204

Rozzi R, Draguicevic JM, Arango X, Sherriffs M, Ippi S, Anderson C, Acevedo M, McGehee S, Plana J, Cortés E, Massardo F (2005) Desde la ciencia hacia la conservación: el programa de educación y ética ambiental del Parque Etnobotánico Omora. Ambiente y Desarrollo XXI (2):20–29

Rozzi R, Massardo F, Anderson C, Heidinger K, Silander J Jr (2006) Ten principles for biocultural conservation at the southern tip of the Americas: the approach of the Omora Ethnobotanical Park. Ecol Soc 11(1):43 [online] URL: http://www.ecologyandsociety.org/vol11/iss1/art43/

Rozzi R, Arango X, Massardo F, Anderson C, Heidinger K, Moses K (2008) Field environmental philosophy and biocultural conservation: the Omora Ethnobotanical Park educational program. Environ Ethics 30(3):325–336

Rozzi R, Massardo F, Anderson C, McGehee S, Clark G, Egli G, Ramilo E, Calderón U, Calderón C, Aillapan L, Zárraga C (2010) Multi-ethnic bird guide of the Sub-Antarctic forests of South America. UNT Press – Ediciones Universidad de Magallanes, Denton/Punta Arenas

Salas C, Donoso PJ, Vargas R, Arriagada CA, Pedraza R, Soto D (2016) The forest sector in Chile: an overview and current challenges. J For 114:562–571

Schlesinger WH, Bernhardt ES (2013) Biogeochemistry: an analysis of global change. Academic Press, New York

Sparrow T, Hutchinson A (2015) A history of habit: from Aristotle to Bourdieu. Lexington Books, Lanham

Spash CL, Aslaksen I (2015) Re-establishing an ecological discourse in the policy debate over how to value ecosystems and biodiversity. J Environ Manag 159:245–253

Suárez L, Ortiz D (2006) Producción de camarones y destrucción de manglares en Ecuador. In: Primack R, Rozzi R, Feinsinger P, Dirzo R, Massardo F (eds) Fundamentos de Conservación Biológica: Perspectivas Latinoamericanas. Fondo de Cultura Económica, Mexico City, pp 195–197

Taylor B (2018) Biostitutes and biocultural conservation: empire and irony in the motion picture avatar. In: Rozzi R, May RH Jr, Chapin FSIII, Massardo F, Gavin M, Klaver I, Pauchard A, Núñez MA, Simberloff D (eds) From biocultural homogenization to biocultural conservation, Ecology and ethics, vol 3. Springer, Dordrecht, pp 71–82

Torres-Salinas R, García GA, Henríquez NC, Zambrano-Bigiarini M, Costa T, Bolin B (2016) Forestry development, water scarcity, and the Mapuche protest for environmental justice in Chile. Ambiente & Sociedade 19(1):121–144

UN (1982) United Nations world charter for nature. United Nations, New York

Vernadsky VI (2007) La Biosfera y la Noosfera. Ediciones IVIC, Caracas

Veuthey S, Gerber JF (2012) Accumulation by dispossession in coastal Ecuador: shrimp farming, local resistance and the gender structure of mobilizations. Glob Environ Chang 22(3):611–622

World Commission on Environment and Development [WCED] (1987) Our common future. Oxford University Press, Oxford

Zanotti L (2018) Biocultural approaches to conservation: water sovereignty in the Kayapó lands. In: Rozzi R, May RH Jr, Chapin FSIII, Massardo F, Gavin M, Klaver I, Pauchard A, Núñez MA, Simberloff D (eds) From biocultural homogenization to biocultural conservation, Ecology and ethics, vol 3. Springer, Dordrecht, pp 343–359

Chapter 3
Reclaiming Rivers from Homogenization: Meandering and Riverspheres

Irene J. Klaver

Abstract Here I develop a model around two key riverine components: meandering and riverspheres. I show how an analysis of their conceptual and material workings, and their interactive dynamics, facilitates a *re*valuing, *re*imagining, and *re*vitalizing of rivers and thus contributes to biocultural conservation and cultural diversification. Meandering and riversphere are presented as a functional, dynamic, nondeterministic model for moving beyond the confines of positivist constructs and assumptions about rivers and how we might live well with them as urban citizens, equitable and sustainable. The Meander River and the Los Angeles River afford a space for exploring rivers in human affairs. The Meander, once a geographic space critical to great historical movements and now nearly erased from the cultural imagination, serves as a profound metaphor upon which to build new old ways of thinking. The little Los Angeles River, once nearly forgotten by the very city that derived its existence from it, flows as an example of how rethinking and reimagining can lead to re-rivering and the redefining of a riversphere.

Keywords Meandering · Riverspheres · Urban rivers · *Mētis* · Los Angeles River

"There it is – take it!" With this legendary concise dedication speech, William Mulholland inaugurated the brand-new Los Angeles Aqueduct in 1913 (Mulholland 2000, p. 246). With these words, Mulholland had initiated not only an aqueduct but also an era, a new mentality, a modern lifestyle, a homogenization of water into a resource, and a stratification of rivers into aqueducts and reservoirs. Large-scale water infrastructure projects became paradigmatic of the twentieth century. Wildly diverse water bodies – rivers, streams, lakes, wetlands, estuaries, and aquifers – were pumped, piped, stratified, dammed, diverted, and converted, yielding a staggering loss of ecological and cultural diversity. Just in the realm of dam building, the World Commission on Dams (WCD) estimates that between 1945 and 2000,

---

I. J. Klaver (✉)
Department of Philosophy and Religion, University of North Texas, Denton, TX, USA
e-mail: Irene.Klaver@unt.edu

R. Rozzi et al. (eds.), *From Biocultural Homogenization to Biocultural Conservation*, Ecology and Ethics 3, https://doi.org/10.1007/978-3-319-99513-7_3

40–80 million people worldwide were evicted to accommodate large dam developments – the majority being indigenous peoples and ethnic minorities (Johnston 2012, p. 304; WCD 2000). Add to this, at least 427 million river-dependent people whose downstream way of life has basically been obliterated by the effect of hydro-development (Richter et al. 2010), driving them to the homogenization of poverty. One can also see water development as one of the major factors in the global decline of both ecological diversity and cultural diversity (Johnston 2012, p. 305). Irrigated agriculture and thirsty cities have dammed rivers or rerouted them into complex river-linking schemes and changed natural lakes and aquifers into shrinking remnants of themselves, heavily impacting biocultural and hydro-ecological diversities. Water flowed into the twenty-first century homogenized as a marketable and transferable economic commodity.

This homogenization and utilitarian approach to water stands as a powerful and useful exemplar of Ricardo Rozzi's (2013) 3Hs model of biocultural ethics: it shows a *habit* (damming, canalizing, selling, and diverting waterways) that leads to homogenous *habitats* (infrastructure, paved-over or concrete "riverbeds," and aqueducts) with a consequent reduction of communities of *coinhabitants*. A 3Hs focus enables a reorientation toward *re*connecting to rivers and *re*valuing, *re*vitalizing, and *re*imagining riverine relations within processes of biocultural conservation and cultural diversification. Such a new cultural habit, including a biocultural mentality, would diversify habitats and broaden the spectrum of coinhabitants' survival and well-being.

In the following I develop conceptual tools to facilitate the emergence of such a new cultural relation in the context of urban rivers in the wealthier or so-called developed world. The twenty-first century is the century of the city. In 2007, the global urban population, for the first time in history, surpassed the rural population. According to the 2014 United Nations report *World Urbanization Prospects*, 66 percent of the world's population is projected to be urban by 2050. The report emphasizes that sustainable development challenges will concentrate in cities and will require integrated policies (UNDESA 2014, pp. 1–7). These trends compel us to rethink urbanism in terms of cities as agents of change rather than mere engines of growth – change for greater social justice and environmental sustainability. How we imagine cities and envision "urban citizenship" (Amin et al. 2000) around water bodies is of critical importance: living at the waterfront has become thoroughly commodified through high-priced property values resulting in gentrification, with cultural homogenization in its wake.

Toward facilitating rethinking rivers, I develop a model around two key riverine components: meandering and riverspheres. I show how an analysis of their conceptual and material workings, and their interactive dynamics, facilitates a *re*valuing, *re*imagining, and *re*vitalizing of rivers and thus contributes to biocultural conservation and cultural diversification. This corresponds with Rozzi's 3Hs model of biocultural ethics but is explicitly rooted in the workings of rivers and the diverse cultural relations to rivers. Thus, urban *re*connecting to the city's river is co-constituted by a joint agency of the river and urban citizens.

3.1 Meandering and Riversphere

Meandering refers to the sinuous movement of rivers carving through, and hence creating, a landscape. Because of the complexity of this sinuosity, meandering also stands as a symbol for prototypically nondeterministic systems. Important to my argument for the revaluing of meandering is the etymological root of the word in an actual river, the Meander – now Büyük Menderes – River in Anatolia, Turkey (Fig. 3.1).

In a meandering of history, the Meander River played major roles in antiquity and then all but disappeared from the cultural imagination (Klaver 2014b, 2016). From early modernity onward, rivers were homogenized; meanders were engineered away to facilitate major modern projects, such as commercial river transportation, property boundary determinations, and city planning. By then, meandering had acquired a negative connotation, synonymous with aimless wandering, ambling along a winding path, and rambling through a long-winded argument.

In the course of the latter part of the twentieth century, however, one can see an implicit re-evaluation of meandering (Klaver 2014b, 2016). New understandings of chaos and complexity became widely accepted and led to a revaluing of complexity in the sciences and in the cultural imagination. Meandering as a metaphor for a different sort of thinking is founded in and summarizes the nondeterministic models used in many fields of science that were once the hallmark of linear, positivist thinking.

Meandering makes room for thinking in terms of atmosphere, the less quantifiable three-dimensional aspects of rivers. Rivers are more than blue lines on a map,

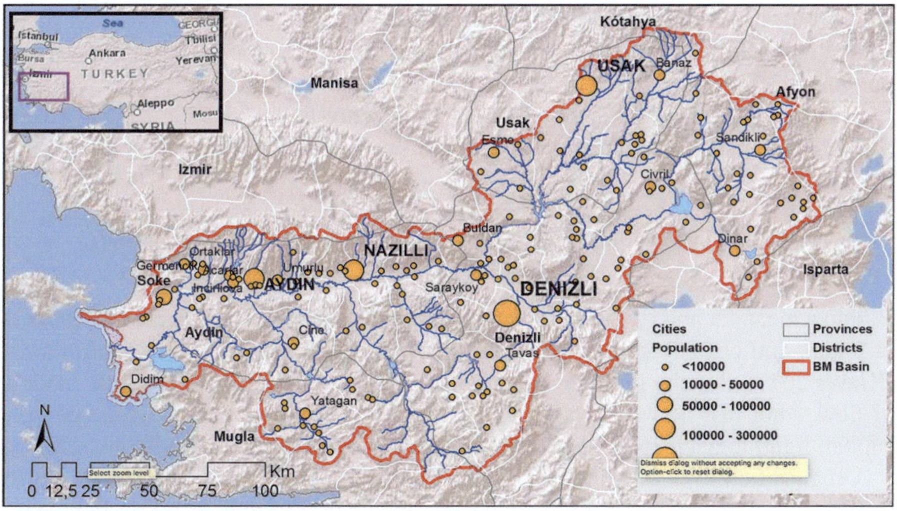

Fig. 3.1 Map of Büyük Menderes River Basin and its main cities. (Source: Handbook of monitoring in the Büyük Menderes River Basin 2015, p. 9)

more than their basins, their watersheds, or drainage areas. They influence the geology, the air, and soil around them, life around them, and cultures around them (Klaver 2012). They create their own hydrosphere, biosphere, and atmosphere. They form intricate networks of relations, conditions of possibilities. I specify the concept of riverine atmosphere as "riversphere" to examine rivers as places of multi-scalar and multi-vector connectivity and complexity.

My sense of riversphere resonates with Gernot Böhme's (1993) concept of atmospheres:

> Atmospheres are indeterminate above all as regards their ontological status. We are not sure whether we should attribute them to the objects or environments from which they proceed or to the subjects who experience them. We are also unsure where they are. They seem to fill the space with a certain tone of feeling like a haze. (Klaver 2012, p. 114)

Riversphere also resonates with the notion of ambiance, the cosmopolitan and open ambiance of a city (Amin et al. 2000), and with Nikhil Anand's concept hydraulic citizenship (2017). The notion of riversphere as atmosphere adds social, political, cultural, aesthetic, and emotional dimensions to our thinking about rivers and cities. Riversphere is a thick concept. It enriches the conceptualization of rivers in the cultural imagination, intertwining hydrological, biological, ecological knowledge and experience with lived experience, social cultural and political activities, storytelling, etc. In *Hydraulic City: Water and the Infrastructures of Citizenship in Mumbai*, Nikhil Anand emphasizes the power of stories:

> stories have multiple vocalities and multiple sites of production. Unlike discourses, stories are particularly attendant to the diverse locations at which human agency is thwarted or dreams are partially realized. Stories are unstable. … The telling of stories is always a political act. (Anand 2017, pp. vii–viii)

Anand develops a notion of hydraulic citizenship predicated upon the deep intertwinement, the entanglement, of the dynamic infrastructural water flows in pipes and pumps, with citizens, technicians, politicians, and plumbers: a complex vibrant mix of stories, theories, facts, and experiences.

Theories of complexity are well suited to a twenty-first-century era of high technology, globalization, and urbanization. As John Law and John Urry (2005) state:

> With its many convergent, overlapping and irreversible interdependencies "globalization" is remaking "societies" but not in a linear, closed and finalized form. We might see the growth and spreading of theories of complexity as part of, and simultaneously helping to enact, the very processes of global change. (p. 404)

Within a meander and riversphere approach, geometrical and homogenizing models of nature and city planning give way to models of complexity and indeterminacy (Klaver 2017), thereby giving room to biocultural conservation, to multiple models of flow – not just flows of water but of sediments, animals, plants, soils, people, capital, light, luggage, tourists, money, exchanges, and experiences.

Urbanization in a globalizing world comes with many forms of injustices, also the urban reconnecting to rivers in the developed world (Kibel 2007). In the meander model for reclaiming rivers from homogenization, I emphasize the importance of the river as a space for the commons in the wake David Harvey's (2008) concept of right to the city.

The question of what kind of city we want cannot be divorced from that of what kind of social ties, relationship to nature, lifestyles, technologies and aesthetic values we desire. The right to the city is far more than the individual liberty to access urban resources: it is a right to change ourselves by changing the city. It is, moreover, a common rather than an individual right since this transformation inevitably depends upon the exercise of a collective power to reshape the processes of urbanization (p. 23). The crucial word here is "common." The center of Harvey's right to the city is the shift from an individual right to a common right. Water affords the materiality, medium, and framework or model to think this common, to think the "with."

Where rivers had been foregrounded as anchors of civilization since ancient times, they became backgrounded in the era of modernity, specifically in the industrial era (Klaver 2012, pp. 15–19). Polluted, diverted, and dammed rivers "disappeared" not only in the cultural imagination, but they were literally paved over or dried up because of impoundments or relegated to the unsavory side of town. Once the backbones of towns they became backsides: the unhealthy and unsafe zones, poor people's areas, harbors, dumping grounds, and sites of water on fire. In the second half of the twentieth century, a shift in mentality emerged with the rise of an environmental movement, culminating in growing environmental legislation, including the 1972 Clean Water Act. These developments facilitated a *re*newed foregrounding of rivers in city planning. Rivers meandered back into the cultural imagination (Klaver 2013, 2014b).

Climate change-driven floods and droughts in the twenty-first century have put water back on the map in bold. The cultural realm has been inundated with water: a flood of water-related books, advertisements, brand names, real estate ventures, art projects, and movies – including a James Bond film *Quantum of Solace*. Water has even streamed into the stock market: water stocks, as in infrastructure projects, desalination plants, and bottled water companies, have soared. Where in the "developed world," water had become rather backgrounded in the course of the twentieth century, often invisible in its infrastructural existence, it has come back with a vengeance and has become a solid part of our social-political, economic world and our cultural imagination. Many cities in the industrial world are designing projects to *re*connect to "their" river. This bespeaks a shift in what I call environmental imagination: a socially, culturally, and environmentally *re*valuing of rivers (Klaver 2013, 2014a). River reconnection projects are often driven by real estate and chamber of commerce interests; it is especially in this context that a right to the river is called for to counter the injustices of gentrification, commodification, homogenization, and surveillance (Kibel 2007; Klaver 2018a, b).

In the following I explore the reconnecting of cities to their rivers as a meandering movement, a movement of the again and the *re-*. I first trace the value shifts in the notion of meandering and *re*connect meandering to the history of the Meander River in Anatolia, the Asian part of Turkey. I then turn to an exemplary case of urban *re*connecting to one's river: the City of Los Angeles' *re*vitalizing of the Los Angeles

River. I show how in the *re*conceptualization of the Los Angeles River as well as in its realization, meandering abound, as does the prefix *re-*, the again and again.

How do we *re*consider our situation, *re*imagine our future, facilitate a change in mentality, and foster an environmental imagination? I argue that the prefatory syllable *re-* functions as catalyzer; it prompts the movement of the again: *re*think, *re*connect, *re*build, and *re*configure. The prefix *re-* conveys the creative capacity of meandering and reflects the trajectory from biocultural homogenization to biocultural conservation.

3.2 Meandering

Linearity has been the privileged paradigm of progress and its leading model of efficiency; its concomitant mind-set has been goal-oriented or teleological.[1] Meandering, convoluted and seemingly undirected, is seen as not just the opposite of efficiency but as being in its way.

In the course of the second half of the twentieth century, nonlinear systems had become widely accepted in the sciences – physics, mathematics, and engineering. Einstein's theory of relativity and Heisenberg's uncertainty principle redefined a linear Newtonian world. Complexity, chaos theory, and nondeterministic as well as stochastic nonlinear modeling become the state of the art in many fields, including in analyzing the behavior of large-scale natural or social systems in ecology, economy, and politics. Analyses of both practices and systems highlight the importance of field-dependency, of a larger context. These dynamics and an increasingly complex society in terms of media, globalization, and education led to an acceptance of complexity in the cultural imagination and a revaluation of meandering.

Meandering in its material movement conveys the nature of the nonlinear, symbolically and metaphorically. It allows for ambiguity, uncertainty, and hybridity, for that which cannot easily be measured or replicated in laboratory experiments. Its activity of sedimentation and reactivation is based in the unpredictable workings of the material realm not ruled by structures of scheduled time. Meandering is messier than the straight line. It entails a rethinking of progress through complexity.

Revaluing meandering has a train of effects on a variety of concepts and practices. Meandering facilitates a different way of thinking about efficiency, acknowledging that it might be more efficient in the long term to take more time and explore possibilities, just as a river does when it meanders through a basin. It is a slower process than water running through a concrete channel, because it takes more factors into consideration.

Meandering as a method, as a mental strategy, privileges exploration: a messy process, learning from mistakes, and following contingent relations. Many human

---

[1]The Meander and Metis sections are based upon previous writings. See Klaver 2014b, 2016, 2017, 2018a, b

practices develop in sinuous ways: learning through failing, honing a skill, building experience, facing unexpected challenges, and starting anew. Meandering foregrounds the searching in the notion of research. Meandering invokes a model of engineering in terms of ingenuity, a bricolage, and tinkering that acknowledges and interacts with various kinds of knowledge and expertise that is capable of adjusting itself to local situations and demands.

Meandering holds much in common with *metis* – a term describing practical, even cunning, intelligence in ancient Greek culture. *Mêtis* stands for resourcefulness, practical effectiveness, and experiential wisdom. Homer's Odysseus is described as *polymêtis* – experienced, crafty, wily, and cunning.

Meandering invokes, elucidates, and hints at a different imagination, another mind-set, a new epistemological and ontological model, and a cultural and political framework that diversifies what counts as expertise, knowledge, politics, progress, and efficiency. Meandering does not allow for simple analyses or reductionist geopolitical frameworks; it demands ongoing political deliberation. It bespeaks the social-political necessity of taking time to explore terrains, to elucidate attributes, relations, problems, and solutions, as a gateway to new constructs of imagination, to a capacity to aspire (Appadurai 2004).

The movement of meandering echoes an ongoing beginning and reveals how beginning works. Beginning does not take place in a vacuum, is not *a creatio ex nihilo*, but is always building on past experience, and, at the same time, can entail a break with this experience. The emergence and fading of the Meander River in the cultural imagination can itself be seen as a meandering: an appearance and disappearance of the very river that left its indelible mark on human culture by giving its name to the process in which it disappeared again. The self-referential character intensifies the complexity of the process.

3.3 The Meander River

The Meander River (Anatolia, Turkey) once formed a mercantile and military conduit between Europe, North Africa, and Asia. Herodotus mentions the Meander's winding ways, and Strabo has given us the meaning of meandering as wandering. The earliest mentions of the Meander are found in Homer and Hesiod between approximately 750 and 650 BCE, in which the Meander region is portrayed as rather backward.

Not long thereafter the Greeks settled the Ionian Coast including the Meander Delta. The Meander Valley became the vital trading route between the Mediterranean and Asia and emerged as a region of high cultural significance.

We might call this the first meandering of the Meander River in the cultural imagination. From a "rural backwater" to the most precious gateway to the east: "vast caravans of wood, wheat and spices, marble and ivory" followed its course (Seal 2012, pp. 11–12). Trade and armies traversed the basin. The city at the high headwaters of the Meander River, Dinar (Celaenae in the fifth century BCE), was of strategic importance: its pass connected east and west. Xerxes' Persians headed west in 481 BCE to conquer the Greeks; 150 years later Alexander the Great headed east from Macedonia to conquer the Persians. These classic power shifts between the east and the west kept meandering along the river that gave the process its name (Fig. 3.2).

Near the Meander's mouth on the Aegean Sea was the prosperous port of Miletus. In the sixth and fifth centuries BCE, it was a cultural center, booming and bustling with celebrated musicians, poets, engineers, mapmakers, and philosophers such as Thales, Anaximander, and Anaximenes. Aristotle called Thales of Miletus the first Greek philosopher. Thales considered water to be the beginning, an originating, and guiding principle or *archê* (Fig. 3.3).

The Meander River had created a fertile valley. However, in another meandering twist of history, the very same agricultural development that made the region prosper and provided food for military and mercantile caravans enhanced erosion and silt formation in the basin, and the once so powerful harbor city Miletus became a landlocked town. Over the centuries, the Miletus Bay silted up with alluvial deposits from the very river that nurtured its importance. The economy of the once-prominent harbor city collapsed. Nowadays, the ruins of the city lie some 10 kilometers from the Aegean Sea.

Fig. 3.2 M. Pouqueville, "milet et cours du Meandre," Paris: Firmin Didot frères, 1835. Personal copy of lithograph

Fig. 3.3 Miletus Bay silting evolution map. (Eric Gaba, Wikimedia, 2009)

3.4 Meander and Metis

The very twisting and wandering character for which meandering became so well-known bespeaks a way of thinking that has been long ignored, belittled, and even considered counterproductive, precisely because it connotes complexity and multiplicity instead of linearity and unity. In its polymorph character, adjusting itself to the circumstances, meandering is structurally comparable to the ancient Greek notion of applied or real-world and practice-based intelligence, *mêtis*.

In Greek mythology Meander and Metis (*M τις*) were brother and sister. According to Hesiod, Thetis and Oceanus had 3000 sons, river-gods or *Potamoi*, and 3000 daughters, the *Oceanid*, each of them patroness for a specific spring, river, or lake. Only the foremost were mentioned by name – among the sons, Meander, and among the daughters, Metis (Fig. 3.4).

Metis was, initially, an important deity, the first spouse of Zeus and represented wisdom, skill, craft, and cunning – a highly praised combination. However, Zeus, fearing her powers and her offspring, swallowed her, but she had already conceived Athena, who was born fully armed from Zeus' forehead. Metis faded from Greek mythology, eclipsed by her daughter, Athena, goddess of wisdom. Metis symbolized cunning intelligence in politics, practice-based knowledge in military art and

Fig. 3.4 Winged goddess thought to be Metis, in a scene depicting the birth of Athena. Detail on black-figure amphora from 550 to 525 BCE in the collection of the Louvre. (Photograph by Marie-Lan Nguyen)

medicine, and the skills of the artisan crafts; all these forms of experiential wisdom, rooted profoundly in the intimacy of specificity, were called *mêtis*.

In *Cunning Intelligence in Greek Culture and Society*, Detienne and Vernant (1978) argue that *mêtis* escapes simple definition – it "always appears more or less below the surface, immersed as it were in practical operations" (p. 3). Its way of knowing, its kind of intelligence, and

> its field of application [are] the world of movement, of multiplicity, and of ambiguity. It bears on fluid situations which are constantly changing and which at every moment combine contrary features and forces that are opposed to each other (p. 20).

According to Detienne and Vernant, *mêtis* is "at the heart of the Greek mental world in the interplay of social and intellectual customs where its influence is sometimes all-pervasive" (p. 3). However, despite its pervasiveness, *mêtis* is never explicitly thematized or analyzed in ancient Greek philosophical texts. While there are many treatises about logic, there are none about *mêtis*. The intellectual world of classic Greek philosophy, in contrast to its everyday mental world, was a dualistic world with a dichotomy between being and becoming, the intelligible and sensible, and the unchanging one and changing multiple. In this framework of thought, there was no place for *mêtis*, which "is characterized precisely by the way it operates by continuously oscillating between two opposite poles" (p. 5). The mode of thinking of *mêtis* does resonate profoundly with pre-Socratic philosophers, specifically with Heraclitus.

The Meander confounded early lawyers concerned with boundaries and scientists concerned with the mechanisms of meandering streams. Meander symbolized irregularity, complexity, ambiguity, and instability. In the latter part of the twentieth century, precisely these "meandering" qualities brought out the value of multiple perspectives in arts and sciences; the weak ontology of becoming became as valuable as the traditionally more privileged strong ontology of being; the inductive, analogical, and emergent as valuable as control and generalizability (O'Connor et al. 2003, p. 99). The understanding of probability and complexity provided new forms of explanation and new ways to operate even within fields long founded on "ideal" characteristics and laws. The meander came to be seen as an irregular waveform, at once subject to and generating random processes and forms.

Various characterizations of ingenuity and of emergent and analogical thinking bear deep resemblance to the *mêtis* of antiquity. Dreyfus and Dreyfus (1986) speak of *expertise* in terms of "intuition [that] is the product of deep situational involvement and recognition of similarity" and note: "how experience-based holistic recognition of similarity produces deep situational understanding" (pp. 29, 32). Similar concepts characterize the notion of ingenuity and engineering design: explicitly pragmatic, contingent, visual in character, satisficing, messy, holistic, whimsical, and learning from failure (O'Connor et al. 2003, p. 104).

Hapgood describes the first phase of engineering design as a "metaphorical traversal through solution space," in which "failure, imagination, and stuckness" are at play. The traversal and design process is "idiographic and unpredictable" and often beset with "painful trials or iterations." For Hapgood the engineer is a "tinkerer who engages in activities within an artistic and subjective context" (Hapgood 1993, p. 96). O'Connor and Wyatt use the term "thinkering" to blend Hapgood's tinkering together with Dreyfus's deep situational involvement into "engineering discovery by doing" (O'Connor and Wyatt 2004, p. 12).

James Scott's seminal work, *Seeing Like a State* (1998), demonstrated the significance of *mêtis* for the social sciences and fields, such as geography and architecture. He invokes the term *mētis* "to conceptualize the nature of practical knowledge and to contrast it with more formal, deductive, epistemic knowledge" (p. 6).

> There may be some rules of thumb, but there can be no blueprints or battle plans drawn up in advance; the numerous unknowns in the equation make a one-step solution unimaginable. In more technical language, such goals can only be approached by a stochastic process of successive approximations, trial and error, experiment, and learning through experience. The kind of knowledge required in such endeavors is not deductive knowledge from first principles but rather what Greeks of the classical period called *mētis* (…). Usually translated, inadequately, as "cunning," *mētis* is better understood as the kind of knowledge that can be acquired only by long practice at similar but rarely identical tasks, which requires constant adaptation to changing circumstances. (pp. 177–78)

In cogent prose Scott describes how this kind of knowledge had become backgrounded in modernity with devastating consequences. "The utilitarian commercial and fiscal logic that led to geometric, mono-cropped, same-age forests also led to severe ecological damage" (p. 309). It is this trajectory from homogenization to biocultural conservation into which I insert the importance of *mētis* again, this time in consort with meander.

3.5 Re-meandering

Re-meandering has become a popular practice in ecological restoration, even in places where there never were meanders. Rivers are resurfacing in the public imagination as places to congregate, and as cultural and ecological corridors, creating a cultural rejuvenation around urban renewal projects. Also in rural areas, river restoration is underway: the re-meandering of watercourses and restoring of floodplains are being carried out – sometimes even by the same engineering firms that straightened the waterways in the early or mid-twentieth century. New adaptive management regimes are seeking to work *with*, not against, rivers.

Meandering is dependent on the complex interaction of many material vectors. It is a symbol for how power operates in the everyday, lateral traversing, picking up material and depositing, reactivating in the process. Meandering stands for an ethics of adjustment, a politics of engagement, enabling and ongoing deliberation, a sense of experiment: tinkering, "thinkering," emergent, and transient. Meander brings the social, political, technological, and natural together in an ongoing dynamic. The Law of the Meander is not the straight line but the sinuous back and forth, symbolized linguistically by prefix *re-*, the notion of the again and again, the experience one gets in *mêtis*, the exploration through wandering, and the essay in Montaigne's original sense of trial and attempt.

Meandering seems to be a slower process than the straight line of progress; yet this is only the case for the simply defined objective. Meandering proceeds by covering more ground, percolating into deeper depths, listening to the murmurs of more voices, being what it is when and where it is observed. Meandering makes room for what cannot easily be measured, what does not want to be measured, for the slow, and for the workings of the material realm not ruled by the structures of scheduled time.

The Meander River no longer functions as a Mediterranean thoroughfare, but the notion of meandering has re-emerged as valuable. Meandering is not a symbol for closure or homogenization but one of ongoing change and exchange, of biocultural conservation.

Perhaps no story better entwines re-rivering, meandering, and environmental imagination of riversphere than that of the Los Angeles River.

3.6 Returning to (Reconsidering) Mullholland: Reimagining and Revitalizing Los Angeles Riversphere

"There it is – take it!" Proclaimed Mulholland to 40000 citizens of Los Angeles as water rushed into the San Fernando Valley on that November day in 1913.[2] This was not the water of the Los Angeles River. This water came from afar, from the Owens River Valley, carried across 220 miles of desert through the pipes of the brand-new Los Angeles Aqueduct. Mulholland had supervised the Aqueduct's construction (Mulholland 2000, p. 246.) (Fig. 3.5).

"There it is – take it!" was the inauguration of an aqueduct and of an era, a new mentality, a modern lifestyle. The new supply of water precipitated an era of explosive growth in Los Angeles, transforming it from a stagnating industrial town into a megalopolis, the City of Dreams. The trajectory of water and growth in Los Angeles is an exemplary case of the trajectory of modernity. It exemplifies Ricardo Rozzi's 3Hs model of the intricate relations between habits, habituation, and inhabitants (Rozzi 2013). Los Angeles is paradigmatic for a new modern mentality: between

Fig. 3.5 The Opening of the Los Angeles Aqueduct November 1913. (Source: USC Digital Library http://digitallibrary.usc.edu/cdm/singleitem/collection/p15799coll65/id/8248/rec/21)

[2]This section is based upon previous writing with J. Aaron Frith on the history of water supply in Los Angeles. See Klaver and Frith 2014

Fig. 3.6 *Channelizing the Los Angeles River*. After devastating floods in 1934 and 1938, the US federal government embarked on a massive flood control program in Los Angeles in the 1950s, straightening and channelizing the Los Angeles River with 3.5 million barrels of cement and 147 million pounds of steel. With concrete lining its banks for 94% of its course, the river, once the lifeblood of the city, became a "fifty-one-mile storm drain." (Source: Image courtesy of Brian C. O'Connor, 2014)

1850 and 1970, it took a strictly utilitarian approach to water management, viewing water as a resource to be used as fuel for the urban "growth machine" (Fulton 2001). The city embraced a policy of "urban water imperialism," importing new water supplies from well beyond its city limits (Hundley 1992, p. 120). According to Reisner (1987), "The Owens River created Los Angeles, letting a great city grow where common sense dictated that one should never be" (p. 106) (Fig. 3.6).

Water made Los Angeles – water mainly imported from rivers elsewhere. Waters of the Owens River, the Feather River, and the Colorado River were redirected in such great quantities that the Colorado River no longer reached the ocean and Owens Lake became a dust bowl. The Los Angeles River – the reason why the city was where it was – had been straightened into a flood control channel, a 51-mile-long concrete scar (Price 2008, p. 547) (Fig. 3.7).

The Los Angeles River meanders throughout the story – a small unimposing river, outgrown, overgrown, and straightened and ditched by the city to which it gave birth – then reentering the cultural imagination as a character in its own right (Fig. 3.8).

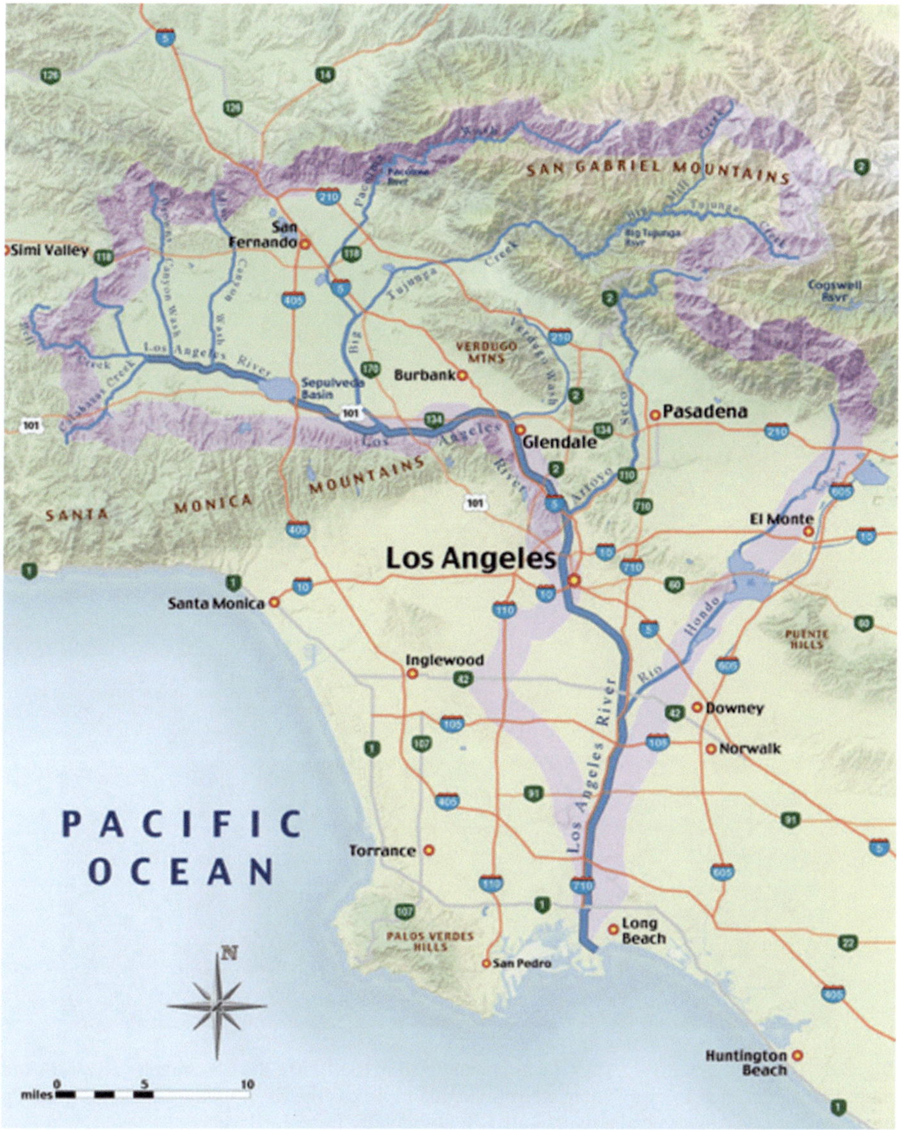

Fig. 3.7 *The Los Angeles River and Greater Metropolitan Los Angeles*. Fifty-one miles long, the Los Angeles River drains the Santa Monica, San Gabriel, and Santa Susana mountain ranges, passing through Glendale, Downtown Los Angeles, and East LA before emptying into San Pedro Bay just west of Long Beach. (Source: EnviroReporter.com LLC 2006–2013)

The Los Angeles River presents us with a green-gray hybrid infrastructure that questions strict separations between human built/technology and nature, between various social-economic cultures, and between different practices. This very hybridity gives promise of twenty-first-century urban rivers made to re-meander,

Fig. 3.8 *The Forsaken River*. After it was channelized in the 1950s, the Los Angeles River looked more like a "deserted freeway" than a river. Unable to access the river, many Angelenos grew up unaware that Los Angeles had a river at all. (Photograph by Irene J. Klaver)

re-creating a sense of water as common and public space, a riversphere, which expresses a sense of culture as a capacity to aspire (Appadurai 2004). It is this re-rivering, the meandering back of rivers into a riversphere, that put rivers back on the map of our imagination.

In the 1980s Angelenos began to reimagine Los Angeles's original water supply, the Los Angeles River. Many did not even know they had a river in their midst, even though they had, as most Americans, seen striking images of it – in the movies. In

Them! (1954), James Arness encountered giant ants on its concrete banks; in *Grease* (1978), John Travolta drag raced across its paved bed; and in *Terminator 2*, future California governor Arnold Schwarzenegger escaped a killing machine from the future by fleeing down its concrete channel on a motorcycle (Gumprecht 1999, p. 244). People had no clue that this flood control channel was in fact a river. They did not see it as a river; they did not experience it as a river.

In 1985, a group of citizens began to re-envision the Los Angeles River. Poet Lewis MacAdams became fascinated with the river. His work attracted like-minded people, and together they founded the Friends of the Los Angeles River (FoLAR) "to bring the River back to life" (Gottlieb and Azuma 2007, p. 27).

This resonated with an emerging cultural imagination infused with environmental consciousness (Klaver 2014a) and laid the groundwork for a grassroots effort to revitalize the Los Angeles River. Working through FoLAR, community activists began building bike paths, planting trees, and lobbying for riverside parks, all with the intent of reclaiming the river, to make the Los Angeles River "a place of community and ecological revitalization" (Gottlieb and Azuma 2007, p. 24).

By the turn of the twenty-first century, the movement to revitalize the Los Angeles River had drawn the interest of academics and city planners, who crafted plans for an urban revitalization program centered on the river. FoLAR partnered with the Urban and Environmental Policy Institute at Occidental College to produce a series of events called "Re-Envisioning the Los Angeles River," which considered the possibilities of renewal from a multidisciplinary perspective (Gottlieb and Azuma 2007, pp. 32–33). Voters approved bond measures for park and recreational development along the river, and the city established the Ad Hoc Committee on the Los Angeles River to coordinate the efforts of community activists, business leaders, and city leaders to remake the river (City of Los Angeles 2007). Again, it was the inauguration of an era, a new mentality, a re-envisioned lifestyle, this time the era of the river (Fig. 3.9).

Fig. 3.9 *Reimagining the Los Angeles River*. In the 1980s, community activists began to reimagine the Los Angeles River, urging city officials and fellow Angelenos to reconnect with the river and to reclaim it as a place of community and revitalization. Their efforts culminated in 2007 when the City of Los Angeles issued the *Los Angeles River Revitalization Master Plan*. Above, a promotional brochure explains the Plan's "bold vision for transforming the river.". (Source: Los Angeles River Project, City of Los Angeles, Department of Public Works, Bureau of Engineering, 2008)

In 2007, the Committee issued the "Los Angeles Revitalization Master Plan," that re-envisioned the river as the "green spine" of the City. Renewing the river would renew the fragmented city: "the revitalized River would foster community identity and civic pride, thereby bringing communities together" (City of Los Angeles 2007, pp. 3–4). Its promotional brochure explains the Plan's "bold vision for transforming the river" and in large font with "re-" italicized: *RE*CLAIM, *RE*VITALIZATION, *RE*IMAGINE OUR RIVER, OUR FUTURE.

Once forgotten, the Los Angeles River re-emerged as a model of urban reclamation and sustainability. Revitalization promised "flood control, cleaner water and cleaner air and desperately needed neighborhood parks, wetlands, and wildlife habitat," not to mention increased "local water supplies" (Price 2008, pp. 551–52).

Even the US federal government recognized the river's potential to re-create public space and reforge communal ties. In 2010, the Environmental Protection Agency (EPA) officially designated the Los Angeles River a "traditional navigable waterway," affording its additional protection under the 1972 Clean Water Act. The Los Angeles River and its community had removed the crust of concrete and invisibility and presented an exemplar of revitalization and renewal.

As urban planner and writer John Arroyo (2010) emphasizes in his thesis *Culture in Concrete: Art and the Re-imagination of the Los Angeles River as Civic Space*, "artists have taken to the River as a creative venue. Their actions have redefined the River and have allowed us (and impel us) to re-imagine the River as the civic space" (p. 3). They have flourished in "the un-designed, un-planned, and the spontaneous nature of the River space" independent of any formal urban planning or intervention (p. 3).

River revitalization plans often come with gentrification and a complex redrawing of the public and private: when the old abandoned, neglected, polluted, dangerous riverside becomes "beautified," poor neighborhoods all too often are elbowed out to make place for a new upper middle-class population. As Kibel (2007) states succinctly, with any of these projects, one needs to "consider the questions of who makes decisions about our urban rivers (…) and who ultimately benefits from or is burdened by these decisions" (p. 15). The danger of commodification is a sanitized and controlled space – lacking the conceptual and social "messiness" of abandoned and waste places, which function as meandering space. Foucault calls these places "heterotopias": "unique, nontraditional, and differentiated 'other places' where the constraints of typical regulations and rules were suspended" and which therefore entice the imagination, spontaneous reactions, and transformative powers (Arroyo 2010, p. 66).

It is too soon, yet, to say which ways the reimagined Los Angeles River will flow. Price (2008) sums it up nicely: "the revitalized river will be a product of continuous compromise and negotiation" (p. 552). In fact, this is the definition of the politics of an engaged community that fosters the ongoing deliberative process of civic life, engaging in ongoing debate, meandering like a river. In all its hybridity, the Los Angeles River crosses boundaries of race, class, and human and physical geography, concrete, and earth. As a re-meandering but still mainly concrete river, it is the symbol of a new twenty-first-century paradigm of hybridity in water management and environmentalism. As such, the river provides a platform for the city to foster a sense of culture as a capacity to aspire (Appadurai 2004).

In conclusion, I return to the pre-Socratic philosopher Heraclitus to summarize our meandering through human relations to rivers. In two fragments Heraclitus observes (Kahn 1979, pp. 52–3): "as they step into the same rivers, other and still other waters flow upon them," and – one of his most famous sayings – "one cannot step twice into the same river." The philosopher reminds us that while we can step into the same location, the same coordinates in a three-dimensional space, we would at different times encounter different water, different sedimentary material being carried and being deposited, and even different materials in different formations on the river bottom. Likewise, other components of the riversphere will have changed, some nearly imperceptibly and some perhaps beyond recognition. While we may not be able to return any river or system of rivers to some former configuration, the words of Heraclitus can help to remind us that we might take steps that reduce homogenization that enable distinctly "other and still other waters" flow through our riverspheres.

The story arc of the Meander River gives us a vivid long view of a river in relation to humans, giving passage to great armies, bearing witness to the beginnings of philosophy, confounding lawyers and geographers, and all but disappearing from memory, yet showing us the value of sinuosity and "messiness." The story arc of the Los Angeles River enables us to reframe our relationships with our rivers and to revitalize the rivers and ourselves. Such stories revalue the practice of *mētis* and meander alike, and with the vigor of renewed imagination, re-rivered rivers meander back as experiential places.

References

Amin A, Massey D, Thrift N (2000) Cities for the many not the few. Policy Press, Bristol

Anand N (2017) Hydraulic city: water and the infrastructures of citizenship in Mumbai. Duke University Press, Durham

Appadurai A (2004) The capacity to aspire: culture and the terms of recognition. In: Rao V, Walton M (eds) Culture and public action. Stanford University Press, Stanford, pp 59–84

Arroyo J (2010) Culture in concrete: art and the re-imagination of the Los Angeles River as civic space. M.C.P. thesis, Massachusetts Institute of Technology. www.arroyoseco.org/arroyojcthesis.pdf. Accessed 16 Nov 2015.

Böhme G (1993) Atmosphere as the fundamental concept of a new aesthetics. Thesis Eleven 36:113–126

City of Los Angeles (2007) The Los Angeles River revitalization master plan. http://boe.lacity.org/lariverrmp/CommunityOutreach/masterplan_download.htm. Accessed 16 Nov 2015

City of Los Angeles (2015) Oversight and access. http://www.lariver.org/About/OversightandAccess/index.htm. Accessed 16 Nov 2015

Damasio A (2012) Self comes to mind: constructing the conscious brain. Vintage Books, New York

Detienne M, Vernant J-P (1978) Cunning intelligence in Greek culture and society (trans: Janet Lloyd) University of Chicago Press, Chicago

Dreyfus HL, Dreyfus SE (1986) Mind over machine, the power of human intuition and expertise in the era of the computer. Basil Blackwell, Oxford

Fulton WB (2001) The reluctant metropolis: the politics of urban growth in Los Angeles. Johns Hopkins University Press, Baltimore
Gottlieb R, Azuma AM (2007) Bankside Los Angeles. In: Kibel PS (ed) Rivertown: rethinking urban rivers. MIT Press, Cambridge, pp 23–46
Gumprecht B (1999) The Los Angeles River: its life, death and possible rebirth. Johns Hopkins University Press, Baltimore
Hapgood F (1993) Up the infinite corridor: MIT and the technical imagination. Addison-Wesley Publishing Company, Reading
Handbook of monitoring in the Büyük Menderes River Basin (2015) TR2009/0327.02-02/001-Technical Assistance for Capacity Building on Water Quality Monitoring. National Programme for Turkey 2009. http://www.suyonetimi.gov.tr/Libraries/su/TAWQM_20150414_Handbook_EN_3.sflb.ashx. Accessed 28 June 2018
Harvey D (2008) The right to the city. New Left Review II (53):23–40
Hundley N (1992) The great thirst: Californians and water, 1770s–1990s. University of California Press, Berkeley
Johnston BR (2012) Water, culture, power: hydrodevelopment dynamics. In: Johnston BR, Hiwasaki L, Klaver IJ, Ramos Castillo A, Strang V (eds) Water, cultural diversity & global environmental change: emerging trends, sustainable futures? UNESCO International Hydrological Program, The Hague, pp 295–318
Kahn CH (1979) The art and thought of Heraclitus. Cambridge University Press, Cambridge
Kibel PS (2007) Bankside urban: an introduction. In: Kibel PS (ed) Rivertown: rethinking urban rivers. The MIT Press, Cambridge, pp 1–22
Klaver IJ (2012) Placing water and culture. In: Johnston BR, Hiwasaki L, Klaver IJ, Ramos Castillo A, Strang V (eds) Water, cultural diversity and global environmental change: emerging trends, sustainable futures? UNESCO International Hydrological Program, The Hague, pp 9–29
Klaver IJ (2013) Environment imagination situation. In: Rozzi R, Pickett ST, Palmer C, Armesto JJ, Callicott JB (eds) Linking ecology and ethics for a changing world: values, philosophy, and action. Ecology and ethics, vol 1. Springer, Dordrecht, pp 85–105
Klaver IJ (2014a) Landscapes of environmental imagination: ranging from NASA and Cuyahoga images to Kiefer and O'Keeffe paintings. In: Drenthen M, Keulartz J (eds) *Environmental aesthetics: crossing divides and breaking ground*. Fordham University Press, New York, pp 135–153
Klaver IJ (2014b) Meander(ing) multiplicity. In: de Châtel F, Holst-Warhaft G, Steenhuis T (eds) *Water scarcity, security and democracy: a Mediterranean mosaic*. Cornell University Press, the Atkinson Center for a Sustainable Future, and Global Water Partnership-Mediterranean (GWP-Med), Ithaca, pp 36–46
Klaver IJ (2016) Re-Rivering Environmental Imagination: Meander Movement and Merleau-Ponty. In: Bannon BE (ed) *Nature and experience: phenomenological approaches to the environment*. Rowman and Littlefield International Limited, London, pp 113–127
Klaver IJ (2017) Indeterminacy in place: rivers as bridge and meandering as metaphor. In: Donohoe J (ed) Phenomenology and place. Rowman and Littlefield International Limited, London, pp 209–225
Klaver IJ (2018a) Engaging the Water Monster of Amsterdam: Meandering toward a Fair Urban Riversphere. In: Stefanovic IL (ed) Forthcoming in The Wonder of Water: Lived Experience, Policy and Practice. University of Toronto Press, Toronto
Klaver IJ (2018b) River's paradox, river's promise: meandering and river spheres. Open rivers: paradoxes of water, no. 4. http://editions.lib.umn.edu/openrivers/article/
Klaver IJ, Aaron Frith J (2014) A history of Los Angeles's water supply: towards reimagining the Los Angeles River. In: Tvedt T, Oestigaard T (eds) A history of water, series III, vol. 1: water and urbanization. I.B. Tauris, London, pp 520–549
Law J, Urry J (2005) Enacting the social. Econ Soc 33(3):390–410
Mulholland C (2000) William Mulholland and the rise of Los Angeles. University of California Press, Berkeley

O'Connor B, Copeland JH, Kearns JL (2003) Hunting and gathering on the information savanna. Scarecrow Press, Lanham

O'Connor B, Wyatt RB (2004) Photo provocations, thinking in, with, and about photographs. Scarecrow Press, Lanham

Price J (2008) Remaking American environmentalism: on the banks of the L.A. River. Environ Hist 13:536–555

Reisner M (1987) Cadillac desert: the American West and its disappearing water. Penguin Books, New York

Reissmann E (2012) The origin of the meander blogmymaze. http://blogmymaze.wordpress.com/2012/05/20/the-origin-of-the-meander/. Accessed 16 Nov 2015

Richter BD, Postel S, Revenga C, Scudder T, Lehner B, Churchill A, Chow M (2010) Lost in development's shadow: the downstream human consequences of dams. Water Alternatives 3(2):14–42

Rozzi R (2013) Biocultural ethics: from biocultural homogenization toward biocultural conservation. In: Rozzi R, Pickett ST, Palmer C, Armesto JJ, Callicott JB (eds) Linking ecology and ethics for a changing world: values, philosophy, and action. Ecology and ethics, vol 1. Springer, Dordrecht, pp 9–32

Seal J (2012) Meander: east to west, indirectly, along a Turkish River. Bloomsbury, New York

Scott JC (1998) Seeing like a state: how certain schemes to improve the human condition have failed. Yale University Press, New Haven

Strabo S (1924) The geography of Strabo (trans: Jones HL). Harvard University Press, Cambridge

UNDESA (United Nations, Department of Economic and Social Affairs), Population Division (2014) World urbanization prospects: the 2014 revision, highlights (ST/ESA/SER.A/352). United Nations, New York. https://esa.un.org/unpd/wup/Publications/Files/WUP2014-Highlights.pdf/ Accessed 10 Aug 2016

World Commission on Dams (2000) Dams and development: a new framework for decision-making. Earthscan Publications, London. Also available from http://www.dams.org/report/contents.htm

Chapter 4
Biostitutes and Biocultural Conservation: Empire and Irony in the Motion Picture *Avatar*

Bron Taylor

Abstract The motion picture *Avatar*, directed by James Cameron, metaphorically spotlighted the long processes on Earth wherein powerful societies invade territories inhabited by indigenous peoples, damaging or destroying their societies and environments. The film, which set box office records around the world, was beloved by theatergoers and passionately debated. Some averred that Hollywood productions including *Avatar*, despite expressing sympathy for indigenous peoples, actually denigrate them, sometimes in racist ways. Others argued that *Avatar* and some other films express an unwarranted romanticism toward aboriginal societies and erode an understanding of their diversity and, thus, promote an insidious homogenization and cultural erosion. Yet Cameron self-consciously chose the genre of the blockbuster Hollywood film to spotlight and condemn, and urge resistance to, colonial processes of cultural and biological simplification, which have, of course and tragically, often included genocides and extinctions. Cameron did so while implicitly raising the possibility of blending in salutary ways the knowledge systems of aboriginal peoples with those of the invaders. Thus and ironically, he deployed one of the world's art forms that has been cited as an important instrument of cultural homogenization in an effort to promote respect for and political solidarity with aboriginal cultures and their putatively superior ecological practices.

Keywords Biocultural conservation · Biocultural simplification · Biodiversity · Cultural diversity

> *Avatar* asks us to see that everything is connected, all human beings to each other, and us to the Earth. And if you have to go four and a half light years to another, made-up planet to appreciate this miracle of the world that we have right here, well, you know what, that's the wonder of cinema right there, that's the magic. (Associated Press 2010)

B. Taylor (✉)
Department of Religious Studies, University of Florida, Gainesville, FL, USA
e-mail: bron@ufl.edu; http://www.brontaylor.com

R. Rozzi et al. (eds.), *From Biocultural Homogenization to Biocultural Conservation*, Ecology and Ethics 3, https://doi.org/10.1007/978-3-319-99513-7_4

With these words, director James Cameron described his artistic intention when creating *Avatar*. The statement's evocative operative terms – interconnection, miracle, wonder, and magic – suggest that the filmmaker's objectives were aesthetic, ethical, environmental, and even spiritual. He sought to move people. In many ways, he did.

One way he moved them was bodily, into theaters, in droves. A master of the blockbuster film genre, having directed *Terminator*, *Aliens*, and *Titanic*, within 2 years of its December 2009 release, *Avatar* earned $2.8 billion. Nearly three quarters of this gargantuan amount came from outside of the United States. Had not the Chinese government cut its run short because it feared the film might precipitate resistance to its development projects and resettlement schemes, this figure would have been far greater (Stanton 2010). Depending on the time frame and analytics used, the film would be somewhere between the first and fifth most profitable films of all time. Cameron obviously met the profit-making intention that also animated his effort in the film.

The film's commercial success has been all the more remarkable when considering its subversive, even radical environmental themes. *Avatar* presented an anti-imperialist view at variance with now global narratives of economic and social progress fueled by economic growth and technological innovation. It also expressed forms of eco-spirituality that were not only heterodox, but heretical, to Western religious mainstreams. Indeed, *Avatar* promoted the idea that entangled in the worldviews, ethics, and spiritualities of indigenous peoples are ecological understandings and environmentally sustainable lifeways. Consequently, the film argued indigenous societies deserve respect and protection. Moreover, the film suggested that if modern industrial societies are to flourish, they must learn from the very indigenous societies they have threatened, suppressed, stolen from, and sometimes annihilated.

That these were among the film's teachings might not be obvious. As I noted in *Avatar and Nature Spirituality* (2013a), the film is like a Rorschach test, and reactions to it tend to be informed by the backgrounds of the observer. Some conservative Christians, for example, thought the film was peddling spiritually dangerous paganism (Douthat 2010; Rizzo 2010).[1] Some on the left side of the political spectrum could not get beyond their distaste for the enormous profits Cameron had made from the film to even notice the themes that cohere with their own political views or even thought it was deeply racist (Žižek 2010). Some on the right viewed the film as anti-American, anti-military, and subversive anti-capitalist propaganda (Boaz 2010). Some feminists argued that film's characters reinforced sexism or even misogyny (Klassen 2013). Those with indigenous backgrounds were variously

[1] Evangelical Christians affiliated with the Cornwall Alliance, for further example, released a 12-part DVD series titled *The False World View of the Green Movement* and a subsequent segment, *From Captain Planet to Avatar: The Seduction of Our Youth*, which attacked the film and other programs they feel threaten Christian faith. See http://www.resistingthegreendragon.com/. Cf. http://www.CornwallAlliance.org, and especially Beisner (n.d.); Wanliss 2011.

critical, appreciative, or enthusiastic, or they were ambivalent, having more than one such reaction to the film (Good Fox 2010; Justice 2010, 2013).

Although the film precipitated diverse reactions, one of its dimensions drew little attention, namely, the aspects of the film that made connections between biological and cultural diversity. As explicated in detail in this volume and earlier by Rozzi (2013), the erosion of biological and cultural diversity go hand in hand. Such biocultural homogenization reduces our planet's genetic, species, and cultural diversity, and with such processes, much ecological knowledge is lost, including understandings about how to live in a way that does not degrade ecosystems and biodiversity. Conversely, biological and cultural diversity tend to be mutually reinforcing, helping to ensure the resilience and flourishing of both environmental and social systems.

No doubt, many missed these themes in *Avatar*, which, following the blockbuster formula, had many distractions – science fiction intrigue, beautiful imagery, novel special effects, a love story, and dramatic battles with explosions and wrenching violence. A brief synopsis of the film, however, can spotlight the dimensions that express and promote understandings of and respect for biocultural diversity.[2]

4.1 Synopsis

Avatar is set on Pandora, a moon circling a gaseous planet in the Alpha Centauri star system, but the planet and the entire story are a metaphor for the human history of Earth, which has involved the imperial expansion of large and powerful societies at the expense of small-scale indigenous ones.

The film explains that humans had seriously degraded their home planet and were now colonizing the galaxy, which depended on a rare mineral found on Pandora that was to be procured by the Resource Development Administration (RDA), a mining company, and its mercenary army.[3] These imperial humans had been waging a campaign to subjugate tall, blue, human-resembling (and tail-wagging) hunters, the Na'vi, who were the moon's aboriginal inhabitants. The staging of the film, of course, is reminiscent of a host of colonial histories in which invading forces want or need "resources" in places they do not inhabit, consider themselves and their lifeways superior to the inhabitants and ways of life currently practiced there and therefore have no qualms about making these places, both biologically and culturally, like their home places. Whether on Earth or Pandora, the

[2] I have provided a more detailed synopsis in the introduction to *Avatar and Nature Spirituality* (2013c). This book of essays wrestles with the intense criticism, praise, and ambivalence that the film evoked.

[3] The terminology in the company name is worth unpacking: the word "resource" reflects an instrumental approach that values nature (in this case Pandoran nature) only to the extent that it is valuable to the invading humans; moreover, the word "development" is ironic because the mineral sought will not be used for any Pandoran development but only for the purposes of the imperial humans.

original inhabitants of a place rarely welcome invaders who seek to radically change their lives and habitats. Recognizing this, the RDA deployed two entwined strategies in its campaign, one social and the other military.

The social strategy was scientific and led by Dr. Grace Augustine, who had founded a school on the moon, learned the language of the Na'vi as well as much from them, including the ways their biological and cultural understandings were interconnected. Augustine's subordinates were in awe of her; one gushed that she had "written the book" on Pandoran ecology. Although her discipline was never clearly specified she was, apparently, an anthropologist with a specialty in ethnobiology.

Augustine's primary passion was to learn about the moon's environment and the Navi's environmental and social systems, but she was also expected to provide information that could be used to convince the Na'vi to relocate away from their forested landscapes, which they considered sacred. The ironically named mineral, unobtanium, was under the forest floor. If peaceful means were to fail, the RDA and its military planned to subjugate the Na'vi by force and take the mineral without their consent.

The military strategy involved bringing soldiers and sophisticated weaponry to Pandora to help secure the unobtanium. One of the soldiers was a former marine named Jake Sully. He was assigned to work with Augustine but specially charged to learn of Na'vi vulnerabilities to ensure military success if (as expected) military action would become necessary. Augustine, Sully, and some others utilized Na'vi bodies called "avatars." These were created by innovative technologies that allowed the minds of humans to inhabit avatars, which had been made with their own DNA. This was done in such a way that their Na'vi bodies resembled their human forms. In their Na'vi bodies, which could survive in the Pandoran atmosphere, they could better relate to the Na'vi and attempt to build trust with them.

The imperial forces did not anticipate, however, that Augustine and Sully would come to love the Na'vi, their culture, and the forest. In Augustine's case, she was impressed with Na'vi botanical knowledge and intrigued by their perception of the moon as a Gaia-like, organic, bio-neurological network, whom they personified as the goddess Eywa. Yet Augustine sought to understand this network scientifically, not religiously. Eventually, Sully more viscerally came to a Na'vi-like perception, in no small part due to his relationship with Neytiri, a Na'vi female who was, in a unique twist to the white man meets native princess story, a badass huntress and apprentice shaman to her own mother, Mo'at. Neytiri mentored Sully (at the insistence of Mo'at) and taught him how to perceive the vital energies and value of the forest; indeed, she helped him to find his way toward interspecies empathy (Sideris 2013).

But the RDA and its military, based on information obtained from Augustine, learned that the Na'vi considered the gigantic, sequoia-resembling "Hometree" to be sacred and concluded that its destruction would break their will to fight. From Sully, the RDA learned the structural vulnerabilities of the tree. With such strategic knowledge gained, the RDA decided that the time had come to force the Na'vi to leave the Hometree, which grew atop the supposedly unattainable mineral. Learning the attack was imminent, however, Sully warned the Na'vi, pleading for them to leave before what he said would be a devastating attack. When Neytiri asked how he knew this, he confessed that he had initially arrived in support of the invader's mission but explained that he had fallen in love with the forest, the Na'vi, and Neytiri, and now wanted to help them avoid disaster. Despite his confession and apology, Neytiri felt betrayed, and, in anguish, she denounced Sully, who was then seized, together with Augustine, by Na'vi warriors, moments before the attack on Hometree commenced. Nevertheless, as the battle began, Mo'at intervened and freed Sully, urging him to help, if he really had become one of them. Despite the brave resistance of the Na'vi warriors and Sully's efforts, Hometree was destroyed, and the Na'vi were scattered in fear and despair.

Realizing that Sully and Augustine had become traitors, RDA operatives pulled them out of their Avatar bodies, put them back into their human ones, and jailed them. This detainment was, like the previous one, also short-lived. Sully and Augustine were freed by another badass female, this one human, the helicopter pilot Trudy Chacón, who had earlier disobeyed orders to attack Hometree. Chacón and a few others, who had worked closely with Augustine, had been moved by the Na'vi and became species traitors, defending the Na'vi and their biocultural systems. During their escape, however, Augustine was mortally wounded.

After their escape, Sully and Augustine managed to get to a remote station where they were transported back into their Avatar bodies. Sully then brought the wounded Augustine to the Well of Souls, the moon's most sacred place and Pandora's neural hub. There, Mo'at orchestrated a ceremony to tap into Eywa's healing energies, but Augustine was too badly injured to be saved. Before she died, Augustine told Sully that she regretted that she "always held back" when it came to the Na'vi and said she was proud of him for fully giving them his heart. Moments later, with her final breath, she exclaimed about Eywa "she's real," suggesting that she not only had come to understand this nature goddess as a metaphor for biocomplexity and interdependence but had finally grasped a deeper, ontological reality.

Sully had indeed embraced Na'vi spirituality, if awkwardly: knowing the next battle would be in defense of the Well of Souls itself, he went to the willow-like Mother Tree in the Well of Souls, where the spirits of the ancestors were still to be found, and prayed to Eywa for help, explaining that the invading earthlings had already destroyed their world and, given the chance, they would do the same to Pandora. Soon after, the RDA's mercenaries attacked, and despite great heroism by the Na'vi, now led by Sully, it appeared that they would soon destroy the Well of Souls and its Mother Tree and, thus, any hope for the survival of the wondrous biocultural diversity of Pandora. All things seemed lost. Then, suddenly, the fiercest Pandoran creatures appeared as if out of nowhere, routing the invaders, leading Neytiri to exclaim, incredulously, that Eywa had heard and responded to Sully's prayer.

Soon after their defeat, the imperial humans were marched off and forced to leave Pandora, forced to return to their decimated home planet. The final scene was of another ceremony at the Well of Souls. This one was successful, transforming Sully from human/avatar hybrid into a fully Na'vi being. The film concluded by symbolically reinforcing the "wake up" theme, which had been woven into important moments throughout the film, as Sully's eyes suddenly opened to his new life with the Na'vi in the Pandoran forest.

4.2 Analysis

Those who create motion pictures, even when they have an underlying moral message, dare not be too didactic, lest they lose their audiences. It is unsurprising, therefore, that some of what Cameron was trying to convey was so subtly expressed that many would not notice. He did, however, underscore his intentions in post-release interviews, explaining that he sought to change the consciousness of his audience by awakening their emotional connections to nature and by kindling respect for indigenous peoples and their rights (Cameron and Dunham 2012). Moreover, he asserted, humanity has much to learn from indigenous societies, including their value systems and spiritualities, which have "allowed them to live in harmony with nature for a long time" (Suozzi 2010).

Cameron expressed such views in the film through the character Augustine who effused about the intelligence and knowledge of the Na'vi, contrasting this with the learning disabilities of the invaders and by emphasizing how important it was for Sully to learn the intimate details of Pandora's biocultural systems. Cameron appears to have been informed by contemporary scientific understandings to the effect that the worldviews, languages, ceremonies, and values of indigenous peoples are often entangled with ecological knowledge and sustainable lifeways. This is part of the reason why cultural and biological diversity are properly understood to be mutually dependent, and so both need to be protected if either is to flourish. In scholarly terms, Cameron appears to have been promoting, if in a simplified manner, an understanding of the value of what scholars have come to call "traditional ecological knowledge" or "indigenous knowledge systems" (Berkes 2005, 2008 [1999]; Ellen et al. 2000; Kimmerer 2013; Laudine 2009; Nelson 1999; Nelson and Shilling 2018; Posey 1999; Rozzi et al. 2010; Stepp et al. 2002; Whyte 2013, 2018; Williams and Baines 1993).

Although Augustine herself appeared to have gained such an understanding, she was also complicit with the invading forces and seemed resigned that the Na'vi had no choice but to cede their territory. She seemed to think that by befriending the Na'vi and learning from and about them, she might be able to mitigate the changes that were being forced upon them. Yet her scientific training, including its taboo against being political or allowing one's emotions to skew one's research, led her to refrain from fully empathizing with them, entering their world, and incorporating their understandings. Put harshly, Augustine was a *biostitute* – a little-known term environmental activists deploy against corporate and governmental scientists whom they charge with taking money in exchange for misrepresenting biological facts. Such scientists serve, willingly if sometimes ambivalently, a capitalistic economic system that seeks to maximize profit and places little if any value on the habitats and cultures it must exploit to continue with its expansionist raison d'être.

For Cameron, Augustine was clearly "on the wrong side, she's one of the invaders" (Cameron and Dunham 2012, p. 191). Yet he also depicted her as being conflicted and as having good intentions. She did seek to understand and translate the spiritual ecology of the Na'vi into Western cognitive frames. Her approach appeared to be akin to ethnobiology, which fuses anthropology and botany and strives to glean from indigenous interlocutors a deeper understanding of biota than would be possible with Western lenses alone. There was a moment in the film where she stressed that their knowledge was not some kind of "pagan Voodoo," suggesting they had valuable knowledge. Perhaps she thought if she could convince the RDA's decision-makers, they might go easier on the Na'vi.

Nevertheless, like most Western scientists, Augustine appeared to have been reluctant, and indeed unable, to integrate Na'vi understandings as her own. This is unsurprising given the anthropological taboo against "going native," being affectively and intellectually subsumed by a culture being studied, thus losing one's supposed objectivity. But at the moment of her death, Augustine regretted that she had not fully immersed herself in the Na'vi way. And with this apparent, last-moment openness, she had an experience that suddenly convinced her that the Na'vi way of knowing and their understandings were as real, if not more so, than her scientific training would allow. For his part, Sully stated early in the film that his mind was, essentially, a "blank slate." Of course, Sully was a creature of his culture and previous life experiences, and as a former marine, he initially had no qualms about obeying orders and participating in the RDA's imperialist enterprise. But in that moment in the film, Sully's claim about being a blank slate suggested that because he had no such *scientific* indoctrination, he was better able to learn to incorporate his affective and sensual experience in the forest and with the Na'vi into his own understandings, which enabled him to be drawn into the Na'vi way of knowing and being.

The film is not, of course, an academic treatise on adaptive management and resilience theory. Nor does it address whether and how to blend, or find a rapprochement, between indigenous knowledge systems and those of Western science. Nor does it directly address the processes of cultural homogenization that so often accompanies the invasion by powerful societies of less powerful ones (perhaps because the air is poisonous to the invading humans and, thus, they want to raid Pandora for its mineral wealth but do not, apparently, plan to stay long term). But metaphorically, the film exposes the often fraught and tragic relationships between colonizers, scientists, and other agents of expansionist empires and those peoples who are in the way of their presumptions and ambitions. It also shows how sometimes, in some places, some from expansionist empires are moved and even converted, in part or whole, to the understandings and lifeways of those they came, directly or indirectly, to exploit. Sometimes, some of them become allies. And sometimes, native and non-native alliances secure concessions and, more rarely, even outright victories, from their adversaries (Davis 2010; Taylor 1995).

The film also raises, if implicitly and with little development, important epistemological and ethical conundrums faced by open-hearted people who become entangled in colonial processes. Most basic is the question as to whether and if so how it is ever morally permissible to be involved in such processes at all. What if any epistemological, political, economic, or ethical frameworks can justify the invasion of lands inhabited by others? Short of this, is it ever morally permissible to participate in such processes, to have good intentions in so doing, for example, by ameliorating suffering and disruption or by bringing things one thinks are valuable (such as understandings, technologies, arts, philosophies, and religions)? Whatever one's role or place during colonial processes, it is imminently human to be curious about the others one encounters in these situations. It is only human to want to learn interesting and valuable things that these others know. How and in what ways should those with different understandings relate to one another? What are the possibilities of respectful, mutual learning? Is this even possible given the differences in power held by colonizers and colonized, those with and without economic and political power? If it is, toward what ends?[4]

Unfortunately, usually, these encounters take place in situations of vastly unequal economic and military power, and often if not usually, those with the greater power (or the most powerful within these groups) are not interested in mutual learning, respect, and reciprocity, but in gaining access to the less powerful groups' knowledge and resources for their own commercial or martial purposes. On Earth these encounters have rarely been characterized by the powerful individuals and group meeting at cultural boundaries exercising epistemological humility and a desire to test one's own understandings through an open-minded, respectful, long-term, and dialogical process. This is also why many indigenous groups have grown suspicious of anthropologists and others who seek them out, knowing that their knowledge might be used against them or profits made without honest negotiations and fair trade for what they know and can produce.

Viewed charitably, *Avatar* suggests the possibility of respectful, reciprocal relations while also stating that those from dominant societies would first have to wake up, and open themselves up, to dramatic perceptual and affective change, as well as efforts to dramatically change all socioeconomic systems that fetishize economic growth and expansion. Viewed historically, there are now examples of scientists, conservationists, and indigenous rights advocates from powerful groups who are working *together* with indigenous peoples in the cause of biocultural conservation (and survival) (Gadgil et al. 1993; Berkes et al. 2000, 2003; Berkes and Folke 2002; Gunderson and Holling 2002; Rozzi 2013, 2015a, b). Some of the contributions in this volume spotlight such efforts.

Viewed uncharitably, the film is another example of the ways the increasingly global entertainment industry promotes a global monoculture and distracts most people from seeing and understanding the cataclysmic effects of the global decline of biocultural diversity. Such criticisms must be taken seriously. Yet despite criticisms and ambivalence about *Avatar* from those long engaged in the struggle to

[4] For an analysis of the fraught and fragile relationships between indigenous and environmental activists, as they try to work out their relationships, including whether and how to share cultural understandings and ceremonies, see Taylor (1997).

defend Earth's diverse cultures and habitats, many indigenous people and their allies have appreciated the filmmaker's intentions and, even if critical or ambivalent about some aspects of the film, have recognized that rarely have such messages and understandings been presented by artists and never to the kind of global audience that Cameron's film attracted (Good Fox 2010; Justice 2010, 2013; cf. Taylor 2013b). And there is evidence that some audience members were moved by the film and inspired to become more active environmental activists and indigenous allies (Barnhill 2013; Greenberg 2013; Haluza-Delay et al. 2013; Holtmeier 2013; Istoft 2013; Taylor 2013b).

Cameron himself has stated that, although he long considered himself an environmental activist, these commitments were dramatically strengthened through the process of making the film and from its aftermath, when he became more personally connected to indigenous thinkers, activists, and allies, coming to know in more depth the challenges they face. He joined and used his celebrity to bring attention to a number of campaigns, including the (eventually unsuccessful) resistance to the gigantic Belo Monte Dam in the Brazilian Amazon, which destroyed the habitat of a number of indigenous tribes (Barrionuevo 11 April, 2010) (Fig. 4.1). Although victories to protect indigenous cultures and their habitats continue to be few, often times artistic works – including documentaries, animated and theatrical films,

Fig. 4.1 Shortly after the premiere of the movie *Avatar*, its director James Cameron traveled to the Kayapo territory to express his support for the conservation of its culture and its habitats. Before bidding farewell to the Kayapo elders, Cameron expressed that "Rivers and forests have the moral right to continue to exist as they have for thousands of years, [...] and I believe that you have the moral right to exist as you have done for thousands of years" (The Guardian, April 17, 2010). (Photo Kim Youngsang)

novels, music, and photography – are cited by activists as inspirations, sometimes decisive ones, for their lifework. Examples include *Harry Potter* and J.R.R. Tolkien's Middle-Earth novels, children-targeted animations including *Captain Planet*, *Avatar: The Last Airbender*, and some produced by the Disney corporation (Taylor 1993, 2005, 2010). There is, moreover, at least some evidence that "blockbuster films and documentaries are playing an increasingly important role in global environmental justice 'struggles' or 'cosmopolitics'" (Adamson 2012, p. 146; Ivakhiv 2013). As I argued in an earlier analysis of the film and its impact, although subjected to much criticism, some of which was compelling, *Avatar* nevertheless fostered discussion of the links between environmental and cultural destruction, evoked sympathy "for the many victims of these processes," and provided "social and environmental activists educational openings and recruitment opportunities for their causes" (Taylor 2013b, p. 322).

I thought then, and still think, that *Avatar* reflects and promotes "global awareness of the value of both biological and cultural diversity and of the ways all of today's dominant civilizations continue to erode them. At the same time, it also reflects diverse new ways that people today express and promote reverence for life" (Taylor 2013b, p. 325). It is no small irony that one of the greatest forces of global cultural homogenization, a Hollywood-funded blockbuster motion picture, would become the world's most-viewed critique of the ways in which the world's dominant empires have precipitated the ongoing, global process of biocultural simplification. What remains to be seen is whether the diverse actors who have come to such understandings – scientists, philosophers, activists, and perhaps some who have been moved by *Avatar* – will succeed in effectively blending the best of the world's diverse knowledge systems and thus contribute significantly to the conservation of Earth's remaining biocultural diversity.

References

Adamson J (2012) Indigenous literatures, multinaturalism, and Avatar: the emergence of indigenous cosmopolitics. Am Lit Hist 24(1):143–167

Associated Press (2010) James Cameron's 'Avatar' wins big at golden globes. Access Hollywood, 17 January. http://www.accesshollywood.com/james-camerons-avatar-wins-big-at-golden-globes_article_27831

Barnhill DL (2013) Spirituality and resistance: Avatar and Ursula Le Guin's the word for world is forest. In: Taylor B (ed) Avatar and nature spirituality. Wilfrid Laurier University Press, Waterloo, pp 221–240

Barrionuevo A (2010) Tribes of Amazon dind an ally out of "Avatar". New York Times (online) at http://www.nytimes.com/2010/04/11/world/americas/11brazil.html, A1

Beisner E (n.d.) The competing world views of environmentalism and Christianity, Burke: Virginia, Cornwall Alliance (online) at http://www.Cornwallalliance.Org/Articles/Read/the-Competing-World-Views-of-Environmentalism-and-Christianity/, Accessed Feb 2012

Berkes F (2005) Traditional ecological knowledge. In: Taylor B (ed) Encyclopedia of religion and nature. Continuum International, London/New York, pp 1646–1649

Berkes F (2008 [1999]) Sacred ecology: traditional ecological knowledge and resource management. 2nd ed (New York: Routledge). Original edition, Taylor & Francis, Philadelphia

Berkes F, Folke C (2002) Linking social and ecological systems. Cambridge University Press, Cambridge

Berkes F, Colding J, Folke C (2000) Rediscovery of traditional ecological knowledge as adaptive management. Ecol Appl 10(5):1251–1262

Berkes F, Colding J, Folke C (2003) Navigating social-ecological systems: building resilience for complexity and change. Cambridge University Press, Cambridge

Boaz D (2010) The right has 'Avatar' wrong. Los Angeles Times (online, opinion) at http://articles.latimes.com/2010/jan/26/opinion/la-oe-boaz26-2010jan26

Cameron J, Dunham B (2012) James Cameron: interviews. Jackson. University Press of Mississippi

Davis L (2010) Alliances: re/envisioning indigenous-non-indigenous relationships. University of Toronto Press, Toronto/Buffalo

Douthat R (2010) Heaven and nature. New York Times (online) at http://www.nytimes.com/2009/12/21/opinion/21douthat1.html

Ellen R, Parkes P, Bicker A (eds) (2000) Indigenous environmental knowledge and its transformations. Harwood Academic Publishers, Amsterdam

Gadgil M, Berkes F, Folke C (1993) Indigenous knowledge for biodiversity conservation. Ambio 22:151–156

Good FJ (2010) Avatars to the left of me, Pandora to the right: an indigenous woman considers James Cameron's *Avatar*. Goodfox: culture, politics, Indian country (online blog) at http://juliagoodfox.wordpress.com/2010/01/21/48_avatar, n.p. Accessed 22 Apr 2012

Greenberg JH (2013) *Avatar* and Artemis: indigenous narratives as neo-romantic environmental ethics. In: Taylor B (ed) Avatar and nature spirituality. Wilfrid Laurier University Press, Waterloo, pp 201–219

Gunderson LH, Holling CS (2002) Panarchy: understanding transformations in systems of humans and nature. Island Press, Covelo

Haluza-Delay R, Ferber MP, Wiebe-Neufeld T (2013) Watching *Avatar* from 'Avatar Sands' land. In: Taylor B (ed) Avatar and nature spirituality. Wilfrid Laurier University Press, Waterloo, pp 123–140

Holtmeier M (2013) Post-Pandoran depression or Na'vi sympathy: *Avatar*, affect, and audience reception. In: Taylor B (ed) Avatar and nature spirituality. Wilfrid Laurier University Press, Waterloo, pp 83–93

Istoft B (2013) Avatar fandom, environmentalism, and nature religion. In: Taylor B (ed) Avatar and nature spirituality. Wilfrid Laurier University Press, Waterloo, pp 61–82

Ivakhiv AJ (2013) Ecologies of the moving image: cinema, affect, nature. Wilfrid Laurier University Press, Waterloo

Justice DH (2010) James Cameron's Avatar: missed opportunities. First peoples: new directions in indigenous studies (online) at http://www.firstpeoplesnewdirections.org/blog/?p=169

Justice DH (2013) Afterword: considering the legacies of avatar. In: Taylor B (ed) Avatar and nature spirituality. Wilfrid Laurier University Press, Waterloo, pp 337–352

Kimmerer RW (2013) Braiding sweetgrass: indigenous wisdom, scientific knowledge and the teachings of plants, 1st edn. Milkweed Editions, Minneapolis

Klassen C (2013) Becoming the 'nobel savage': nature religion and the other in avatar. In: Taylor B (ed) Avatar and nature spirituality. Wilfrid Laurier University Press, Waterloo, pp 143–159

Laudine C (2009) Aboriginal environmental knowledge: rational reverence, vitality of indigenous religions series. Ashgate, Farnham/Burlington

Nelson MK (1999) Original instructions: indigenous teachings for a sustainable future. Inner Traditions/Bear & Co, Santa Fe

Nelson MK, Shilling D (eds) (2018) Traditional ecological knowledge: learning from indigenous practices for environmental sustainability. Cambridge University Press, Cambridge

Posey DA (ed) (1999) Cultural and spiritual values of biodiversity. United Nations Environmental Programme, Nairobi

Rizzo A (2010) Vatican Slams 'Avatar'. Huffington Post (online) at http://www.huffingtonpost.com/2010/01/12/vatican-slams-avatar-prom_n_419949.html
Rozzi R (2013) Biocultural ethics: from biocultural homogenization toward biocultural conservation. In: Rozzi R, Pickett STA, Palmer C, Armesto JJ, Callicott JB (eds) Linking ecology and ethics for a changing world. Ecology and ethics, vol 1. Springer, Dordrecht, pp 9–32
Rozzi R (2015a) Earth stewardship and the biocultural ethic: Latin American perspectives. In: Rozzi R, Chapin FS III, Callicott JB, Pickett STA, Power ME, Armesto JJ, May RH Jr (eds) Earth stewardship: linking ecology and ethics in theory and practice. Ecology and ethics, vol 2. Springer, Dordrecht, pp 87–112
Rozzi R (2015b) Implications of the biocultural ethic for earth stewardship. In: Rozzi R, Chapin FS III, Callicott JB, Pickett STA, Power ME, Armesto JJ, May RH Jr (eds) Earth stewardship: linking ecology and ethics in theory and practice. Ecology and ethics, vol 2. Springer, Dordrecht, pp 113–136
Rozzi R, Massardo F, Anderson C, McGehee S, Clark G, Egli G, Ramilo E, Calderón U, Calderón C, Aillapan L, Zárraga C (2010) Multi-ethnic bird guide of the Sub-Antarctic forests of South America, 2nd edn. UNT Press, Denton, Texas
Sideris L (2013) I see you: interspecies empathy and avatar. In: Taylor B (ed) Avatar and nature spirituality. Wilfrid Laurier University Press, Waterloo, pp 241–260
Stanton P (2010) China pulls Avatar from their cinemas fearing civil unrest. Moviefone, 19 January 2010 (online) at http://blog.moviefone.com/2010/01/19/china-bans-avatar-from-their-cinemas-fearing-civil-unrest/
Stepp JR, Wyndham FS, Zarger RK (2002) Ethnobiology and biocultural diversity: proceedings of the seventh international congress of ethnobiology. International Society of Ethnobiology, Athens (Distributed by the University of Georgia Press)
Suozzi MA (2010) Avatar: it's a complete reboot of how we see things. Interpress Service (online) at http://ipsnews.net/news.asp?idnews=51222, 28 April 2010
Taylor B (1993) Evoking the ecological self: art as resistance to the war on nature. Peace Rev 5(2):225–230
Taylor B (ed) (1995) Ecological resistance movements: the global emergence of radical and popular environmentalism. State University of New York Press, Albany
Taylor B (1997) Earthen spirituality or cultural genocide?: radical environmentalism's appropriation of native American spirituality. Religion 17(2):183–215
Taylor B (2005) Disney worlds at war. In: Taylor B (ed) Encyclopedia of religion and nature. Continuum International, London/New York
Taylor B (2010) Dark green religion: nature spirituality and the planetary future. University of California Press, Berkeley/Los Angeles
Taylor B (ed) (2013a) Avatar and nature spirituality. Wilfrid Laurier University Press, Waterloo
Taylor B (2013b) Epilogue: truth and fiction in Avatar's cosmogony and nature religion. In: Taylor B (ed) Avatar and nature spirituality. Wilfrid Laurier University Press, Waterloo, pp 301–336
Taylor B (2013c) Introduction: the religion and politics of avatar. In: Taylor B (ed) Avatar and nature spirituality. Wilfrid Laurier University Press, Waterloo, pp 13–22
Wanliss J (2011) Resisting the green dragon: dominion or death. Cornwall Alliance, Burke
Whyte KP (2013) On the role of traditional ecological knowledge as a collaborative concept: a philosophical study. Ecol Process 2(7):1–12
Williams NM, Baines G (1993) Traditional ecological knowledge: wisdom for sustainable development. Centre for Resource and Environmental Studies, Australian National University, Canberra
Whyte KP (2018) What do indigenous knowledges do for indigenous peoples? In: Nelson MK, Cambridge SD (eds) Traditional ecological knowledge: learning from indigenous practices for environmental sustainability. Cambridge University Press
Žižek S (2010) Return of the natives. New Statesman (online), at http://www.newstatesman.com/film/2010/03/avatar-reality-love-couple-sex, 4 March 2010. Online: http://www.newstatesman.com/film/2010/03/avatar-reality-love-couple-sex

Chapter 5
The Political Ecology of Land Grabs in Ethiopia

Fouad Makki

Abstract The intersection of the 2008 financial crisis with ongoing food and energy crises unleashed a tidal wave of large-scale land acquisitions. By enclosing the village commons and evicting farmers from their land, the land grabs are accelerating global trends toward the consolidation of chemical-intensive industrial farming that facilitates biocultural homogenization. This article examines this dynamic in relation to contemporary Ethiopia and a specific context of state formation. By critically analyzing the designation of common lands as empty or underutilized spaces awaiting redemptive development, it highlights the ways in which the deployment of *terra nullius* narratives serves to efface the distinctive ecologies and socio-spatial dynamics of the zones designated for enclosures.

Keywords Land grabs · Agricultural commercialization · Enclosures · Displacement and dispossession

The confluence of the 2008 world economic crisis with global food and energy crises has generated a frenzy of large-scale land acquisitions across the Global South. In part an expression of the perceived need to alleviate food and energy crises through increased production of staples and agrofuels, the land grabs were also a manifestation of the search for new zones of valorization by a footloose speculative capital following the meltdown of the banking and securities markets. The global land rush is in this sense a symptom of a synchronized global socio-ecological crisis and the desire by international development agencies and national states to market the village commons and farmlands as lucrative sites of investment for the production of grains and agrofuels. The upshot has been a protracted and contested rearrangement of land tenures and a consequent acceleration of global trends toward the consolidation of a corporate food regime (Akram-Lodi et al. 2007; Araghi 2000).

This is a revised version of an essay that originally appeared in *Rural Sociology* 79, Winter 2014.

F. Makki (✉)
Department of Development Sociology, Cornell University, Ithaca, NY, USA
e-mail: fmm2@cornell.edu

R. Rozzi et al. (eds.), *From Biocultural Homogenization to Biocultural Conservation*, Ecology and Ethics 3, https://doi.org/10.1007/978-3-319-99513-7_5

In this new global conjuncture, the social universe of African peasants is once again being portrayed and castigated as a residual zone of "backwardness" that prevents the rationalizing potential of the competitive market from adapting African agriculture to new productivity-enhancing technologies and property regimes. The privatization of peasant holdings and the village commons is viewed as the inescapable prerequisite for bringing the benefits of the new green revolution to Africa. As Kofi Annan, former UN Secretary General and Chair of the Alliance for a Green Revolution in Africa, expressed it, "we have arrived at the tipping point, and are now taking Africa's Green Revolution to scale" (AGRA 2012). This formulation echoes the revival of an ideologically charged discourse about the "tragedy of the commons." Implicit in this neo-Malthusian idea is an assumption that agro-industry represents a benign extension of expert knowledge, improved seeds, and agrochemical inputs to African farming. In this essay, I explore the implications of this claim through an analysis of agricultural commercialization and enclosures in contemporary Ethiopia where a spatially differentiated dynamic of tenure rearrangements is currently unfolding. While in the densely populated highland regions, an extensive process of smallholder commercialization is fostering what Sara Berry (1993) calls "accumulation without dispossession"; in the sparsely populated lowland peripheries, villagers and indigenous communities are being evicted from their ancestral land to make way for large-scale enclosures.

This latter dynamic of "development by dispossession" forms part of a longer historical and global process of colonial and capitalist enclosures, and that have been crucial to the spread and implantation of market imperatives as the regulative mechanisms of social life. The uprooting of peasants from the soil and the emergence of a regime of private property constituted the sociological content of primitive or original accumulation. But capitalist enclosures were not just about dispossessing peasants and reducing them to wage laborers. The separation of the peasants from the land simultaneously alienated the land from the peasants and made possible their mutual reproduction in the form of "fictitious commodities" (Polanyi 1944, pp. 75–76). By dismantling the village commons and deracinating the commoner, enclosures disenchanted and reordered the culturally varied modes of the social interchange with nature, accelerating the metabolic rift that is arguably one of the sources of the contemporary global ecological crisis (Foster 2000).

The advance of agrarian and industrial capitalism in England, where between 1793 and 1815 "more than 6.5 million acres of common fields were enclosed by acts of Parliament," provides a paradigmatic illustration of the twined colonial-capitalist dimensions of this process of primitive accumulation (McNally 1988, p. 11). Robin Blackburn has suggested that the super profits generated by New World slavery served to augment enclosures within England so that the number of enclosure bills before Parliament closely followed trends in plantation profitability (1997, p. 577). The settler colonies, where a more violent and extensive process of enclosures unfolded, also served as escape valves for European commoners evicted by the privatization of land and unabsorbed by industrial expansion. Commoners in both the metropole and the colonies were henceforth vilified as "a mischievous race of people" and the commons maligned as a breeding ground for "barbarians" (Thompson 1966, p. 219).

The mid-twentieth-century collapse of the European colonial empires and the extension of an interstate system based on the norm of the nation-state did not erect an insuperable barrier to the structural extension of development by dispossession. Inequalities of wealth and power could now be expressed in a seemingly more neutral language of development and progress that harkened back to Lockean ideas of improvement. Social spaces previously considered static and primitive were recast as undeveloped zones awaiting redemption and valorization. By the 1980s, in the context of the debt crisis, IMF-imposed structural adjustment programs rolled back what little autonomous leverage postcolonial states exercised in determining their development strategy, essentially relegating their role to maintaining macroeconomic stability through budgetary discipline and the privatization of public assets. Subsidies to domestic agriculture were drastically reduced or dismantled as states were forced to open their protectionist gates to what Farshad Araghi (2000) has called the great global enclosures of our times.

5.1 From Old to New Enclosures

The Bretton Woods institutions, the World Bank in particular, are today in the forefront of promoting a narrative of *terra nullius* to designate "underutilized" or "unproductive" spaces as ideal for large-scale commercial development. Against an earlier consensus on the inverse relationship between farm size and productivity that served as the "economic rationale for redistributive land reforms" (Dyer 1996, p. 103), the bank has increasingly placed its institutional weight behind the establishment of large-scale mechanized agriculture, albeit one in which foreign investors are regulated by a modest "code of conduct" (De Schutter 2011).

A 2009 World Bank publication entitled *Awakening Africa's Sleeping Giants* posited the existence of a vast underused land reserve that could be "tapped to produce food, agricultural raw materials, and biofuels feedstocks, not only for Africa but also for other regions" (World Bank 2009, p. 175). A year later, the bank released a companion report classifying countries in terms of yield gaps, defined as the difference between the attained and possible productivity of land. It found these gaps to be especially large for sub-Saharan Africa where no country appeared to be realizing even 50% of its potential yield (Deininger and Byerlee 2011, p. 182). Both reports concluded that "in this region, low population densities and low mobility prevail, which suggests that agricultural intensification will require larger farm sizes" (Hall 2011, p. 7).

One of the states identified by the World Bank for this form of redemptive development is Ethiopia. In 2004, the bank prepared a confidential report on Ethiopia identifying "areas of growth potential, where increased public investment in specific geographic and development areas might make an optimal contribution to economic growth". The study classified the country into 4 zones using 51 indicators. While the first and second zones covered the densely populated highland regions, the third and fourth zones were located in the more sparsely populated western and eastern lowlands, described as having a high potential for irrigated farming and commercial livestock production. In its essentials, the report represented a virtual blueprint for the Ethiopian state's spatially differentiated program of agricultural commercialization (in zones 1 and 2) and enclosures (zones 3 and 4).

The development strategy that eventually crystallized was labeled "Agricultural Development-Led Industrialization." Its formative premise was that the expanded surplus derived from the commercialization of smallholder farming would fund an industrialization drive and provide the material basis for a "democratic developmentalist state" (Zenawi 2006). This new orientation was initially understood to be consistent with the 1975 land reform that was reaffirmed in Article 40 of the 1994 Constitution. Gradually, however, under pressure from newly emergent middle classes, diaspora capital, and international donors, the government embraced the idea of greater reliance on market mechanisms as the key to the development of agrarian capitalism (Jemma 2001; Kibret 1998). The radical land reform and the nationalization of land in 1975 proved instrumental to this reorientation in two distinct ways. By dispossessing the landed nobility and breaking the political grip of the *ancien régime*, the revolution destroyed the social and political power of the dominant agrarian classes that had constrained the spread of market-mediated production and exchange relations. At the same time, however, by nationalizing all the land, the reform had effectively dispossessed peasants of their customary access to land and endowed the state with extraordinary plenipotentiary powers to determine its allocation and use. This crucial legacy of the first republic enabled the post-1991 second republic to pursue an accelerated program of commercialization and enclosures with the ostensible aim of increasing agricultural production, supplying the raw material needs of an emergent commercial-industrial sector, and securing the cheap wage food critical to regulating the price of labor. In this sense, the new development strategy sought to internalize within the framework of the national state the international division of labor that had been critical to industrialization in the West (Friedmann 2005; McMichael 2005).

The form market dependence takes varies across the political space of Ethiopia. In the densely populated highlands, where small farm sizes and an intensive form of plough-cultivation have long prevailed, the strategy of smallholder commercialization has registered real if uneven gains in productivity while keeping foreign capital investments to small enclaves specializing in floriculture or horticulture production (Lavers 2011; MOFED 2006, p. 3). While this form of commercialization is consistent with the provisions of the 1975 land reform, some of its key provisions are being diluted in the name of devolving the administration of land to the constituent regional governments. Titling and certification of use rights has moreover been widely implemented, and peasants can now lease up to 50% of their plot for limited periods of time. This has fostered the emergence of an illicit land market that is compounding social inequalities in the context of heightened demographic pressures and diminishing plot sizes. As a result, in some districts of the Tigray and

Amhara regions, a fifth to a quarter of rural households are today engaged in contract farming (Gebreselassie 2006).

There is no doubt that given the high population-to-land ratio, traditional patterns of smallholder cultivation have their own intractable limits. But the strategy of market-oriented specialization generates its own problems, placing at risk long-established survival strategies and production for use. The trend toward the commodification of land is reconstituting relationships of reciprocity with profound implications for agrarian socio-ecological reproduction in the context of climate change and the steady depletion of critical renewable resources. The promotion of hybrid seeds, chemical fertilizers, and microcredits are likewise drawing peasants into new forms of indebtedness that can serve as levers of dispossession in times of natural or social crisis (Berhanu 2009:751). The imperative to increase productivity need not entail a headlong lurch into the commodification of land and labor. As an extensive study by scientists from 60 countries recently concluded, smallholder agroecological farming has the potential to make agriculture more resilient and capable of improving livelihoods (IAASTD 2009).

5.2 Enclosures and the Dar Ager

In contrast to the strategy of smallholder commercialization in the central highlands, state policy in the surrounding lowlands relies on a more naked assertion of state power to reallocate land and determine its use. This strategy is reproducing in new capitalist form older core-periphery dynamics within the Ethiopian social formation. Historically, the western and eastern lowlands were zones of plunder and pillage, a virtual no-man's land, from the perspective of state builders. This was nowhere more so than the southwest peripheries that supplied much of the gold, ivory, and enslaved captives that sustained highland state formation (Donham 1986; Makki 2011). The delimitation of state boundaries in the early part of the twentieth century did little to change this socio-spatial hierarchy, and as late as 1955, the Imperial Constitution simply designated the communal and pastoral commons as state domains (Helland 2006, p. 14). The 1975 land reform partially redressed this historic neglect by stipulating that "nomadic people shall have possessory rights over the lands they customarily use for grazing," a provision that was however little honored by state actors (Rahmato 2007).

The contemporary emphasis has been on turning the lowlands into sites of large-scale, export-oriented agriculture. This orientation is informed by a faith in economies of scale, for which large inputs of machinery, capital, and chemicals are a necessary prerequisite. Consequently, instead of a strategic alliance between smallholders and the state, as envisioned for the highlands, the strategic alignment in the lowlands involves a pact between the state and large-scale foreign and domestic investors. The emerging social relationship is therefore one of vertical polarization rather than horizontal smallholder differentiation. Toward this end, the government has provided special incentives to investors and has undertaken extensive infrastructural projects to facilitate large-scale agricultural commercialization. A 45,000-kilometer network of roads currently connects all the regions to the capital, investment codes have been liberalized to reduce the capital requirement for foreign investors, and provisions have been put in place to allow unrestricted repatriation of profits and asset sales (FDRE 2003; EIU Unit 2008). As a result, foreign direct investment has steadily been growing since 2004, when it reached US$545 million, with the share of the agricultural sector accounting for 32% of the total.

To manage this flow of foreign investments, in January 2009 the government established an Agricultural Investment Support Directorate as the principal agency for negotiating lease agreements on land above 5000 hectares. A Federal Land Bank was set up to serve as a repository of land designated for leasing, and by 2011, around 3.5 million hectares—a landmass equivalent to 50% of the total land currently under smallholder cultivation—had been transferred to the Land Bank (MOARD 2009, p. 11). Expectations are that another 1.5 million hectares will be transferred to the land bank in the coming years. Almost all the land so far allocated to investors has come from the four administrative regions of the western and eastern lowlands: Benishangul-Gumuz; Southern Nations, Nationalities, and Peoples; Gambella; and Afar. Reliable figures on how much of this land has been leased or cultivated are hard to come by, but a 2011 World Bank report maintains that between 2004 and 2008, prior to the big global land rush, 406 investors had been granted a total of 1.2 million hectares (Deininger and Byerlee 2011, p. xxxii). The investors represent a mix of private entrepreneurs, national states, and international agribusinesses producing cut flowers, rice, cotton, soybeans, and various biofuels such as jatropha and oilseeds. While the largest international investors are Indian firms, there are also Dutch, German, Israeli, Italian, and Chinese firms operating in different parts of the country. The bulk of the capital formation so far, however, is accounted for by Ethiopian nationals, suggesting an emergent pattern of class formation that is being fostered by the transition to a market economy (Githinji and Mersha 2007).

To realize this program of large-scale agricultural investments, the government has embarked on a vast program of land clearance and the resettlement of indigenous communities in designated villages. Federal proclamations 455/2005 and Council of Ministers Regulations 135/2007 have empowered regional and district administrations to resettle peasants on the basis of a vaguely defined "public purpose." Available studies indicate that almost a third of these reallocations have so far "benefited private investments rather than the public" (Deininger and Byerlee 2011, p. 105). Given the disastrous experience of the forced resettlement programs of the 1980s, the current program emphasizes its ostensibly voluntary and intraregional features. But while the displaced have not been completely deprived of access to land, by relocating them from the riverbanks that are being made available to foreign investor, the program deprives indigenous peoples of their extensive grazing and common land. According to Umod Ubong Olom, the president of the Gambella Regional Administration, "the region has prepared 3.2 million hectares of land" for agricultural investors and "resettled more than 30 thousand scattered households of the region into 43 resident villages." In the 2013 budget year, he continued, it was planning to settle another 10,688 households, which amounts to a staggering two thirds of the total number of households in the region (quoted in Tariku 2013). Similarly, in the Somali region of the eastern lowlands, some "150,000 households were resettled in 21 woredas [districts] in the villagization program carried out in the past four years," and plans are afoot to "villagize 43,000 households this Ethiopian budget year" (Walta 2013). In all likelihood, these resettlement programs will accentuate the general effects of dispossession and enclosures, and even if the promised schools, clinics, and the provision of clean water were to materialize, they will do little to mitigate the structural vulnerability engendered by the concentration of land in the hands of a few large-scale producers. The widespread replacement of local food crops by export crops is only one of the symptoms of this structural precarity.

The *res communes* of the lowlands, where little if any certification and titling has thus far taken place, are not afforded even the minimal protections stipulated by federal law. A recent review of existing laws concluded that "communal holding rights may be subject to privatization at any time without the consent of the communities concerned, and that there are no clear legal provisions as to whether or not communally held land is to be compensated" (Tamrat 2010, p. 14). Those being displaced find themselves caught between different forms of marginalization and exclusion as they are compelled to seek seasonal employment on new commercial farms where their environmentally attuned knowledge is invariably discounted. A new mobile labor force is consequently emerging that will likely expand into a floating rural labor force as only a small share of them can hope to secure full-time employment in the large mechanized farms (Deininger and Byerlee 2011, p. 64; Oakland Institute 2013, p. 35).

Besides cheap land and labor, another resource that has attracted corporate investors to Ethiopia is undoubtedly the availability of water for large-scale irrigation schemes. With a national territory that covers 12 river basins, Ethiopia is relatively well endowed with water. But high dependence on rain-fed agriculture and wide spatial and temporal variations in rainfall patterns have traditionally restricted farmers to one harvest per year, exposing them to a precarious existence in the face of frequent dry spells. In the second half of the 1990s, rainfall variability was estimated to have pushed 12 million people below the "absolute poverty" line. The key decision-makers—federal planners, regional administrations, and international funding agencies—all concur on the need to expand irrigation in order to reduce the risks associated with rainfall variability. Estimates are that up to 2.7 million hectares of land have irrigation potential but that fewer than 300,000 hectares are as yet developed. Yet the irrigation programs so far conceived are large-scale projects and require an extensive enclosure of water resources that are likely to marginalize smallholders and induce water-related social conflicts. In the Awash River Basin where 150,000 hectares have been brought under commercial cultivation, for example, the Afar pastoralists have been experiencing a complex set of transformations restricting their mobility and use of dry and wet grazing areas (Helland 2006; Markakis 1998). The Qoqa Dam constructed in the Upper Awash constructed to regulate the water has also served as a mechanism to monitor the Afar pastoralists, who are increasingly excluded from the flood plains reserved for commercial farming. This has exacerbated conflicts across the valley as enclosure-induced resettlement modifies the customary norms that once regulated inter-clan socio-spatial relations (Kassa 2001; Rahmato 2007).

5.3 *Terra Nullius* and Social Ecology

In order to justify these large-scale enclosures of land and water, the government, the World Bank, and agribusiness investors have increasingly deployed virtual notions of *terra nullius*. Meles Zenawi, the former prime minister and architect of its development strategy, repeatedly insisted that only empty and underutilized land in the lowlands was being made available to investors. The rationale he provided was the relatively sparse population of these regions, averaging about 30 persons per square kilometer, and the fact that agropastoralists occupy large expanses of land even though they account for only 11% of the total population. But this one-dimensional and quantitative view of the social ecology of the lowlands is deeply misleading as it obscures the specific, environmentally adaptive ways in which space is socially constituted, just as much as social relations are spatially constructed (Massey 2005).

Across the span of human history, communities have organized and mobilized space in distinctive ways, and the modes of socially appropriating space have varied in time and place (Lefebvre 1991; Smith 1984). The spatial organization of pastoralist societies, which occupy different parts of land at different times of the year, is profoundly different from the ways in which space is conceptualized and organized in agrarian empires, where power tends to coagulate into centers and thin out at its peripheries. Agrarian empires have in turn a different conception of space from that characteristic of modern nation-states defined by fixed boundaries (Anderson 1991). Discourses of *terra nullius* invariably flatten and distort this variable production of space, willfully obfuscating the particular ways in which the social relations of village or pastoral communities are spatialized.

In the nomadic and agropastoral zones, for instance, transhumance is generally characterized by household ownership of livestock and communal appropriation of pastures. The migratory cycles are typically wide-ranging so that wherever water fails to come to the land, people and animals move to water so land is never a fixed possession. Social organization in these societies is consequently spatially extensive rather than intensive. Even among the more settled communities along the forested western lowlands, where shifting cultivation and agropastoralism are combined with gaming and fishing, spatial relations can be understood as relatively extensive. The livelihood of the indigenous communities in these lowlands is determined by the availability of pasture and water, and access to the village or pastoral commons is central to their subsistence and survival strategies. Unlike sedentary village communities, whose spatial relations are oriented toward the control and use of land and labor within restricted physical localities, the social and cultural orientations of agropastoral communities are altogether much wider. Their kinship patterns emphasize the breadth rather than the depth of their connections, and marriages are made to safeguard or pass resources in a territorial system of horizontal ties. The social and cultural markers of their social world are altogether set much wider (James 1986, pp. 119–147).

The negation of these ecologically adapted forms of producing and inscribing space, and the representation of the lowland expanses as unproductive or underutilized, is but a prelude to a project of reconstituting them as emptiable spaces. That is space emptied of social and cultural meaning and fetishized into a commodity amenable to utilitarian calculation and rationalization. Modernization efforts of this sort have historically been accompanied by discourses that represent "traditional" social spaces as stagnant, a virtual *terra nullius* outside the invigorating dynamics of industry and modernity. Their redemption entails a process of disenchantment and cultural erasure, a reconfiguring of the different social forms through which the metabolism with nature is regulated. The critical precondition for this remaking of heterogeneous social spaces in order to construct the abstract space-time homogeneity of commodity production and circulation is the sweeping away of customary relations and practices, a process that is being advanced in Ethiopia today through enclosures and the fixing of mobile communities in place.

If the discourse of *terra nullius* typically ignores the specific ways in which space is socially constructed, the natural habitat is likewise reductively conceived as an inert stage for the unfolding of human history. And the same mechanisms that seek to instrumentally rationalize land by disembedding it from its social integuments, also abstract it from the wider fauna and flora constitutive of the ecosystem, turning nature itself into a fetishized and commodifiable resource. The biodiversity of the regions targeted for enclosures is consequently represented as external to the development process and assigned an essentially passive role or conceived as merely "natural resources" for exploitation. In these circumstances, the shift from smallholder-based polyculture to a capital- and chemical-intensive monoculture is likely to accelerate biodiversity loss while shifting the burden of environmental costs onto the marginalized communities.

This dynamic of expropriation is perhaps most evident in the southwestern areas of Gambella Region, a prime zone of large-scale land alienations. The region borders the dense carbon-rich rain forest of the southwest, which is a source of livelihood for indigenous communities and wildlife. This forest zone constitutes the second largest area of overland mammal migration in the world and is believed to be one of the most important biodiversity regions in the world, in large part because it lies at the intersection of the Sudan-Guinea savannah biome and the Somali-Masai biome. An important ecological feature of the region is its vast wetland areas and the 4 river systems that drain the wider basin and sustain over 110 fish species, 6 of which are endemic to the region. With little regard to this ecological heritage and the vital ecosystem services that wetlands and forests provide, the regional government of Gambella has ceded parts of the "formally designated national park, protected area and wildlife sanctuary" to private investors (Rahmato 2011, p. 17). This systematic disregard builds on a history of environmental degradation that has, since 1990, resulted in the destruction of an estimated 100,000 hectares of forest in the region, with incalculable loss of biodiversity and species extinction (Behailu et al. 2011, p. 18). Similar ecological threats exist in the other regions designated for enclosures, including the Benishangul-Gumuz administrative region, where the Millennium Dam—projected to be the largest in Africa—is currently under construction near the headwaters of the Blue Nile and the fertile area of the Lower Omo Valley, which was designated a UNESCO World Heritage Site because of its unique cultural and ecological landscape (Oakland Institute 2013).

The extension of industrial farming has placed smallholder agroecology and the genetically heterogeneous seed and crop varieties complex ecology of Ethiopian farming, making it more prone to the vagaries of climate change. This is perhaps most evident in the growing influence of both DuPont's Hi-Bred Pioneer Seeds that supply hybrid maize, sunflower, sorghum, alfalfa, and a variety of vegetable seeds to local farmers and Heineken Brewery's key role in the formation of an Ethiopian malt barley supply chain located primarily in the Arsi region. These seeds require high maintenance and are input intensive, and while the health and environmental risks associated with them have not received adequate study in the Ethiopian context, existing investigations have already documented the health effects on the predominantly low-waged female labor force of chemical-intensive agriculture (Sulaiman and Wageyehu 2008). Enabled by the land grabs, these new forms of industrial agriculture constitute an archetype of biocultural homogenization. According to Ethiopia's Institute for Biodiversity Conservation, "the extensive introduction of genetically uniform and improved varieties has resulted in monocropping and has been the major threat to local crop landraces/farmers' varieties. Market oriented and cash crop cultivation has ignored farmers' varieties and the traditional as well as changing needs of communities" (IBC 2009, p. 57).

These concerns have received little if any serious consideration from the federal state beyond the formal affirmation that all land leases above 5,000 hectares are required to undergo environmental impact assessments, a stipulation that is "often waived as sunset clauses" for lack of institutional capacity to enforce them (Deininger and Byerlee 2011, p. 121). There is also little sign of a willingness to reconsider the large-scale leasing of land for the production of biofuels despite recent studies suggesting that more carbon dioxide is being released into the atmosphere from the production of some biofuels (Searchinger et al. 2008, pp. 1238–1240). According to the MELCA Mahiber, a local nongovernmental organization that has worked with indigenous communities to promote sustainable and mixed land-use planning, by 2008 some 50 investors had been registered to cultivate biofuels on 1.65 million hectares, 300,000 hectares of which had already been leased (MELCA 2008).

The pursuit of this form of development rests on the belief that the establishment of large-scale mechanized farms, for all their social or ecological costs, will ultimately generate a dynamic of self-sustaining growth essential for securing the general welfare of Ethiopians. But there are serious reasons to doubt this claim. The strategy of agriculture-led industrialization is premised on the belief that the surplus generated from commercialized farming will enable a concerted industrialization drive without the political risks associated with the eviction of the mass of highland peasants and the resulting proliferation of urban slums. But this conviction presupposes mutually reinforcing agricultural and industrial sectors within a purely *national* framework, one in which changes in rural productivity will directly stimulate industrial growth and the expansion of the domestic market. So long as the strategy depends on transnational agribusinesses, however, there is no reason to expect that a synergistic relationship between agriculture and industry will emerge. Corporate agribusinesses operate on an international scale, and newly incorporated production sites in Ethiopia will constitute the lower rungs of global value and commodity chains.

In the name of expanding food production and generating the surplus necessary for industrialization, extensive tracts of farmland are being annexed, rivers are being redirected, and local ecosystems are being reconfigured. This high modernist project of agricultural modernization is premised on the establishment of large-scale farms and the widespread application of modern methods of farming. It requires as a precondition the prior conquest of nature and mastery over "traditional communities." From this perspective, the *res communes* constitute no more than a vast reservoir of backwardness and an impediment to modernization. A recent report by the High-Level Panel of the UN Committee on World Food Security has questioned the likelihood that large-scale intensive agrochemical farming will in fact improve rural livelihoods: "Evidence from this land rush to date shows very few such cases. Rather, large scale investment is damaging the food security, incomes, livelihoods and environment for local people" (FAO 2011, p. 8). Viewed from the vantage point of the disinherited, this technoscience of development—and the combination of technocratic hubris and cultural contempt it entails—appears as a gigantic project to foreclose the capacity of peasant communities to fashion alternative futures. The persistent polarity between state and market that continues to frame development policies continues to leave out alternatives based on the self-defined needs of peasant communities, who are conspicuously absent from the cacophony of voices in the public sphere. Their inclusion can arguably shift the axis of the debate in the direction of a more equitable and sustainable social re-embedding of land and the promotion of agroecological farming.

Acknowledgments The author thanks Irene Klaver.

References

AGRA (Alliance for a Green Revolution in Africa). (2012). African Green Revolution Forum. Tanzania, Arusha. http://www.agrforum.com/images/stories/downloads/AGRF_InterimBrochure.pdf

Akram-Lodhi AH, Borras SM Jr, Kay C (eds) (2007) Land, poverty and livelihoods in an era of globalization: perspectives from developing and transition countries. Routledge, New York

Anderson B (1991) Imagined communities. Verso, London

Araghi F (2000) The great global enclosure of our times: peasants and the Agrarian question at the end of the twentieth century. In: Magdoff F, Foster JB, Buttel FH (eds) Hungry for profit: the agribusiness treat to farmers, food and environment. Monthly Review Press, New York, pp 145–160

Behailu H, Tsedal M, Jemal A, Katckos G, Weshik R, Asere G, Haile T, Banjo A, Wasihu Z, Uomode U (2011) Action plan of adaptation to climate change. Gambela People's Regional State Land Use Allotment Strategy, Ethiopia

Berhanu A (2009) The Ethiopian extension and the farmer: a view from the farm. In Ege S, Aspen H, Teferra B, Bekele S (eds) Proceedings of the 16th International conference of Ethiopian studies. Trondheim, Norway

Berry S (1993) No condition is permanent: the social dynamics of Agrarian change in sub-Saharan Africa. University of Wisconsin Press, Madison

Blackburn R (1997) The making of new world slavery. Verso, London

De Schutter O (2011) How not to think of land-grabbing: three critiques of large scale investments in farmland. J Peasant Stud 38(2):249–279

Deininger K, Byerlee D (2011) Rising global interest in farmland: can it yield sustainable and equitable benefits? The World Bank, Washington, DC

Donham D (1986) Old Abyssinia and the new Ethiopian empire: themes in social history. In: Donham D, James W (eds) The southern marches of imperial Ethiopia: essay in history and social anthropology. Cambridge University Press, Cambridge

Dyer G (1996) Output per acre and size of holding: the logic of peasant agriculture under Semi-Feudalism. In: Bernstein H, Brass T (eds) Agrarian questions: essays in appreciation of T. J. Byres. Frank Cass, London

EIU (Economist Intelligence Unit) (2008) *Country profile: Ethiopia*, 3 March

FAO (UN Food and Agricultural Organization) (2011) Land tenure and International investment in agriculture: a report by the high-level panel of experts on food security and nutrition of the committee on World Food Security. FAO, Rome

FDRE (2003) Council of Ministers Regulation on Investment incentives and investment areas reserved for domestic investors. Regulation No. 84, Federal Negarit Gazeta. Addis Ababa, Ethiopia

Foster JB (2000) Marx's ecology: materialism and nature. Monthly Review Press, New York

Friedmann H (2005) Feeding the empire: pathologies of globalized agriculture. In: The empire reloaded: socialist eegister, pp 124–143

Gebreselassie S (2006) Land, land policy and smallholder agriculture in Ethiopia. Future Agricultures Briefing, Institute of Development Studies, Brighton

Githinji MWA, Mersha G (2007) Untying the Gordian Knot: The Question of Land Reform. In: Akram-Lodhi AH, Borras SM Jr, Kay C (eds) Land, poverty and livelihoods in an era of globalization: perspectives from developing and transition countries. Routledge, New York

Hall R (2011) The many faces of the investor rush in Southern Africa: towards a typology of commercial land deals. ICAS Review Paper Series No. 2. Initiatives in Critical Agrarian Studies (ICAS), Land Deal Politics Initiative (LDPI) and Transnational Institute (TNI)

Helland J (2006) Pastoral land tenure in Ethiopia. Paper presented at the symposium "At the frontier of land issues: social embeddedness of rights and public policy." Campus ENSAM-INRA, Montpellier, France. Retrieved October 21, 2013. https:/www.mpl.ird.fr/colloque_foncier/Communications/PDF/Helland.pdf

IAASTD (2009) Global report: agriculture at a crossroads. International Assessment of Agricultural Knowledge, Science and Technology for Development, Washington, DC

IBC (Institute of Biodiversity Conservation) (2009) Convention on biological diversity Ethiopia's 4th country report. IBC, Addis Ababa

James W (1986) Lifelines: exchange marriage among the Gumuz. In: Donham D, James W (eds) The southern marches of imperial Ethiopia. Cambridge University Press, Cambridge

Jemma H (2001) The debate over Rural land tenure policy options in Ethiopia: review of the post-1991 contending views. Ethiop J Dev Res 23(2):35–84

Kassa G (2001) Resource conflict among the afar of Northeast Ethiopia. In: Salih M, Dietz T, Abdel Ghaffer M (eds) African pastoralism: conflict, institutions and government. Pluto Press, London, pp 145–171

Kibret H (1998) Land reform: revisiting the public versus private ownership controversy. Ethiop J Econ 7(2):46–64

Lavers T (2011) 'Land grab' as development strategy? The political economy of agricultural investment in Ethiopia. J Peasant Stud 39(1):105–132

Lefebvre H (1991) The production of space. Blackwell, Cambridge

Makki F (2011) Empire and modernity: dynastic centralization and official nationalism in late imperial Ethiopia. *Camb Rev Int Aff* 24(2)

Markakis J (1998) Resource conflict in the horn of Africa. Sage, London

Massey D (2005) For space. Sage, London

McMichael P (2005) Global development and the corporate food regime. In: Buttel F, McMichael P (eds) New directions in the sociology of global development. Elsevier Press, Oxford

McNally D (1988) Political economy and the rise of capitalism: a reinterpretation. University of California Press, Berkeley

MELCA (2008) Rapid assessments of biofuels development status in Ethiopia and proceedings of the national workshop on environmental impact assessment and biofuels. Addis Ababa, Ethiopia

MOARD (Ministry of Agriculture and Rural Development) (2009) Agricultural investment potential of Ethiopia. Addis Ababa, Ethiopia

MOFED (Ministry of Finance and Economic Development). (2006) Ethiopia: building on progress: a plan for accelerated and sustained development to end poverty. Addis Ababa, Ethiopia

Oakland Institute (2013) Omo: local tribes under threat: a field report from the Omo valley. The Oakland Institute, Oakland

Polanyi K (1944) The great transformation: the political and economic origins of our time. Beacon, Boston

Rahmato D (2007) Customs in conflict: land tenure issues among pastoralists in Ethiopia. Forum for Social Studies, Addis Ababa, Ethiopia.

Rahmato D (2011) Land to investors: large-scale land transfers in Ethiopia. In: Land Governance for Equitable and Sustainable Development. http://www.landgovernance.org/system/files/Ethiopia_Rahmato_FSS_0.pdf. Accessed 22 August 2011

Searchinger T, Heimlich R et al (2008) Use of U.S. Croplands for Biofuels Increases Greenhouse Gases Through Emissions from Land-Use Change. *Science* 319:1238–1240

Smith N (1984) Uneven development: nature, capital and the production of space. Blackwell, Oxford

Sulaiman H, Wageyehu A (2008) Social and environmental costs of economic growth. Paper presented at the conference on "Environmental Sustainability and/or Economic Growth?" October 30–31.

Tamrat I (2010) Large-scale agricultural investment in Africa: the case of Ethiopia. Paper presented at the World Bank conference on Land Policy and Administration, April 26–27

Tariku M (2013) Federalism and equitable development: the case of Gambella Region. http://www.waltainfo.com/index.php?option=com_content&view=article&id=6970:federalism-and-equitable-development-the-case-of-gambela-region&catid=82:articles-and-features-&Itemid=402. Accessed on 14 January 2013.

Thompson EP (1966) The making of the English working class. Victor Gollancz, London

Walta (2013) Chief of Somali Regional State praises six-month performance. Jijiga March 7. Retrieved March 14, 2013. http://www.waltainfo.com/index.php/explore/7531-chief-of-somali-regional-state-praises-six-month-perforance

World Bank (2009) Awakening Africa's sleeping giant: prospects for commercial agriculture in the Guinea Savannah Zone and beyond. World Bank, Washington, DC

Zenawi M (2006) Speech to the Africa Task Force, Brooks World Poverty Institute, Manchester University, England, August 3

Chapter 6
The Ongoing Danger of Large-Scale Mining on the Rio Doce: An Account of Brazil's Largest Biocultural Disaster

Haruf Salmen Espindola and Cláudio Bueno Guerra

Abstract Bento Rodrigues was a community with a deep sense of belonging. On November 5, 2015, its people had to watch their houses, stories, and ways of life destroyed in a matter of minutes after the fatal rupture of the Samarco/Vale/BHP tailings dam. Almost a year later, September 24, 2016, people said a last good-bye to the place during the feast of Our Lady of Mercy among its ruins. Bento Rodrigues would be covered by water after the construction of the S4 dike, to block the tailings still coming down from the Fundão dam.

Keywords Biocultural ethics · Biocultural homogenization · Governance · Socio-environmental justice · River

6.1 Introduction

On November 5, 2015, the largest socio-environmental disaster in the history of Brazil struck the Rio Doce basin. A major tailings dam of one of Brazil's mining-metallurgical giants ruptured, devastating the river basin in its wake. The companies involved hastened to call, treat, and frame the disaster as accident, while many of those whose daily lives and habitats were affected – if not destroyed – fought to call the disaster a crime. We present an account of the background and consequences of this event, we sketch a larger biocultural context, and we zoom in on the radical divergence in experiencing and framing of such a large-scale disaster.

H. S. Espindola (✉)
Postgraduate Program in Integrated Territory Management, Universidade Vale do Rio Doce, Governador Valadares, Minas Gerais, Brazil
e-mail: Haruf@univale.br

C. B. Guerra
Autonomous Environmental Consultant, Belo Horizonte, Minas Gerais, Brazil

R. Rozzi et al. (eds.), *From Biocultural Homogenization to Biocultural Conservation*, Ecology and Ethics 3, https://doi.org/10.1007/978-3-319-99513-7_6

We present the crucial role of three giants of the mineral sector: Samarco Mineração S.A. (Samarco), Vale S.A. (Vale), and BHP Billiton Brasil Ltda. (BHP) in the disaster. Our research points to the responsibility of these three companies and specifically of their decision-makers, who, out of concern for shareholder income, prioritized economic-financial results without much regard for the risks to local populations and the environment.

We examine the two opposing ethics implied by seeing the disaster as an accident or as a crime. There is the explorationist view of the companies, mainly considering market value, ignoring the local value of historical, cultural, and ecological heritage and their co-inhabitants, and the view of territorial belonging, the sense of value of communities, and their immersive interrelationships between habits and habitats that shape the identity and well-being of the co-inhabitant (Rozzi 2013).

We present our work as engaged academics and as concerned members of the community harmed by the disaster. We hope our account will contribute substantively to the discussion on the impacts of large mining projects on river basin habitats and cultures.

6.2 The Rio Doce Basin and Its Mining-Metallurgical Territorial System

The Rio Doce basin is the size of Portugal, and has a population of 3.5 million inhabitants, most of them living in relatively small communities of fewer than 20,000 people. The river is 875 km long and plays an important role in providing water for domestic, agricultural, and industrial uses as well as electric power generation. The Rio Doce and its tributaries also convey domestic wastes and industrial effluents.

The Rio Doce is the primary connection between a rich cultural tapestry of regions, histories, peoples, and economies within this territory. Its waters link and influence the economy, public health, fishing, leisure, and most other cultural manifestations (Guerra and Torres 2012). Its basin is characterized by great intra-regional diversity of its geomorphology, hydrology, ecosystems, socioeconomic structures, and cultural traditions. Approximately 90% of the Rio Doce basin belongs to the Atlantic Forest biome.

The basin has seen two significant historical eras of colonization. At the end of the seventeenth century and the beginning of the eighteenth century, the Western Highlands were colonized when gold was found in the area. Most of the basin was colonized in the first three decades of the twentieth century through many different activities, from small farms, cattle ranches, and gold mines to coffee plantations and waves of Italian and German immigration in the mountain region of the state of Espírito Santo. These diversified occupation processes resulted in remarkable intra-regional differences; different origins and spatial distribution yielded diverse habits and habitats.

International exploration in the beginning of the twentieth century revealed large reserves of iron ore in the Rio Doce basin. The basin became the target of the interests of foreign capital and the State. After three decades of controversy between these two economic forces over iron ore control, a compromise was agreed upon that state capital would dominate the mining business and private capital the steel industry. In 1942, the Brazilian government, with the support of the United States and Great Britain, created the Companhia Vale do Rio Doce to extract and export iron ore from the mountains of Minas Gerais State. In 1997, this company was privatized and its name changed to Vale S.A. – it became the second largest mining company in the world.

The mining-metallurgical complex (export of iron ore and steel plates) was built with direct investment of the Federal Government and the government of Minas Gerais State. The alliance between the Brazilian Government and large private capital was fundamental in this scenario. Thus this powerful alliance was actively linked to local, state, and federal political actors, which gave legitimacy to government priorities and smothered any criticism of negative impacts caused by huge companies operating at the local level. As a consequence, the national developmental project of the Brazilian elites overlapped with the interests of the local establishment.

In the Rio Doce basin, between 1930s and 1970s, the exogenous interests associated with mining and steelmaking were decisive for the Brazilian Government to build the infrastructure of communication and energy and to eradicate or control major diseases, such as malaria and yellow fever. Again, local urban elites became frontline advocates of the presence of large corporations as they saw them as the source of progress.

The mining-metallurgical complex in the Rio Doce basin comprises the world's large-scale steel and pulp industries, as ArcelorMittal, Aperam South America, Usiminas, Fibria, and CENIBRA (Celulose Nipo-Brasileira S.A.) are located on the banks of the Rio Doce and its main tributary, the Rio Piracicaba. Also AngloGold Mining and iron ore giants Vale and Samarco operate there, concentrated in the iron ore quadrangle of Minas Gerais State. Added to these large enterprises are the export of marble and granite, as well as the production of coffee and fruit pulp. Through this variety of export articles – iron ore, steel, pulp, and coffee – the Rio Doce basin has been contributing significantly to the generation of foreign exchange for the country at large.

Samarco Mineração S.A. was created in 1977 and is nowadays a joint venture between Vale S.A. and Anglo-Australian BHP Billiton, each with 50% of the company's shares. Samarco was the first mining company in Brazil to replace the railroad transportation of iron ore with tubular (pipeline) transport. In 2014, operating three pipelines, Samarco was ranked as 10th among the largest exporters in Brazil. Its nominal production capacity was 30.5 million tons at the Germano mine (in Mariana, Minas Gerais),[1] linked by three pipelines to the complex of four pelletizing plants and a port in Ponta Ubú, at the Atlantic Ocean.

[1] Revista Exame. 40 largest brazilian export companies in 2014. By Tatiana Vaz, published Jan 27, 2015; updated Sept 13, 2016. Available at http://exame.abril.com.br/negocios/as-40-maiores-exportadoras-brasileiras-de-2014/

6.3 The Disaster

On November 5, 2015, the pact linking big business with political actors, media, and public opinion was broken with the rupture of the Samarco/Vale/BHP tailings dam. The 60,000,000 cubic meters of tailings descended from the Fundão dam, located more than 900 m above sea level in the municipality of Mariana in Minas Gerais, through the Gualaxo do Norte, Carmo, and Doce rivers, reaching the ocean in the state of Espírito Santo, about 600 km away. Large-scale destruction of biodiversity, socioeconomic environments of habitats, and sociocultural and psychological basis of people and their communities lay in its wake. Life in the rivers Gualaxo do Norte and Carmo was basically destroyed and the river course drastically altered. The first place hit was the village of Bento Rodrigues, located 3 km downstream. The villagers watched from the top of the hills as the mud completely annihilated the place where their families had lived for generations and generations.

When the tailings reached the Rio Doce and the sea, the direct impacts were numerous and devastating, including immediate destruction of all life in the river, threats to species dependent on the river (ducks, herons, tapirs, jaguars, etc.), collapse of city water supply services, interruption of economic activities, prohibition of irrigation and supply of water to rural properties, suspension of river and ocean fishing, and paralysis of tourism and river and maritime leisure activities (Fig. 6.1).

Initial responses to the tragedy were provided solely by the two state governments in the devastated area (Minas Gerais and Espirito Santo) and by the municipalities affected. The Samarco managers turned out to be completely unprepared to deal with the events caused by the rupture of the dam. The company had no warning system, emergency plan, or staff to deal with the crisis situation.

The disaster had a major impact on both human and nonhuman communities. In order to make a realistic assessment of the damage and impact, an integrated analysis considering the biocultural dimension is needed which deals with issues such as the territorial context, water systems, local ecosystems, interrelations between habits and habitats that shape local community identity, as well as the homogenizing dynamics created by the mining-metallurgical complex installed in the Rio Doce basin. Such an integrated analysis has yet to be conducted or planned.

The uncertainties experienced since the November 5, 2015, dam rupture are amplified by the political, legal, and technical-scientific controversies of the different actors responsible for the disaster. Overly technical language has made communication difficult and increased suspicion about the accounts of the true extent and duration of impacts. Authorities and the media mainly invoke the scientific knowledge of "experts," not acknowledging the particular expertise of those whose lives were disrupted. The knowledge of their lives, their habitats, and their losses is barely taken into account.

The radical difference in describing or framing the disaster has far-reaching consequences. The companies Samarco/Vale/BHP call the rupture and devastation an accident; the population of the Rio Doce speaks of a crime. This reveals an important ethical-political change: those who inhabit the Rio Doce basin and have suf-

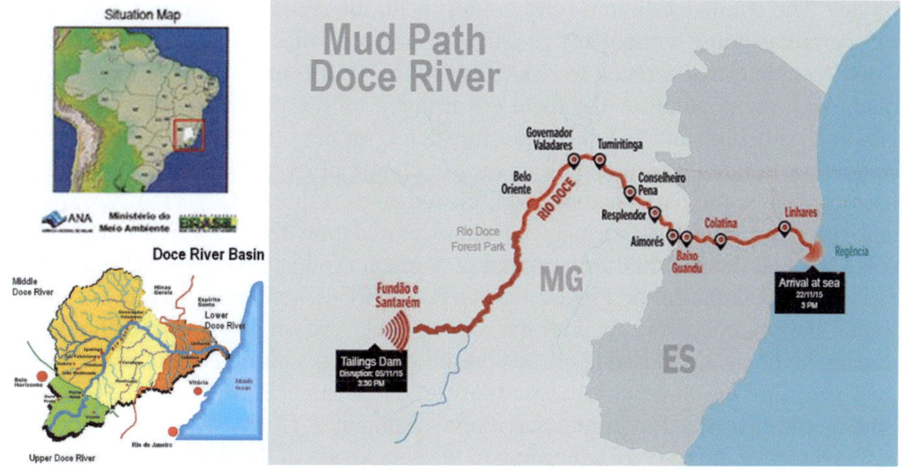

Fig. 6.1 Rio Doce basin showing its course from the disaster point at the dam. Credit: Claudio Guerra/Dulce Albarez design. (Source: Agência Nacional das Águas (ANA))

fered profoundly from the violence they were exposed to have learned to name the people and communities affected, to name the causes of the disaster, and to name those responsible for the disaster.

Until the day of the disaster, the Samarco official website affirmed absolute commitment to the environment, sustainability, and social responsibility. Samarco had received several awards as an exemplary company in the environmental and social fields. In an interview for the magazine *Ecológico*, the president of Samarco stated: "I want to draw your attention to our values. They are: respect for people, integrity and mobilization for results. It is not enough to say that we respect people and are honest. It is necessary to show, to prove in practice" (Firmino and Morais 2014).[2] The disaster and its ongoing aftermath of the reveal that the company's social and environmental ethics were little more than marketing strategies.

Five days after the rupture of the Fundão dam, the public prosecutor of Minas Gerais, Carlos Ferreira Pinto, declared that it was not an accident but clearly a mistake in the operation and negligence in monitoring.[3] November 5, 2015, marked, in fact, 2 years since the imposition of an action by the Public Prosecutor's Office (MP) of Minas Gerais against Samarco. The Company was ordered to resolve problems pointed out by the Pristino Institute's technical report, requested by the MP and presented on October 21, 2013. The problems were clearly stated: dam safety, risks for the community of Bento Rodrigues, and lack of a contingency plan required

[2] FIRMINO, H.; MORAIS, L. A mineração de valores. Revista Ecológico, Belo Horizonte, v. 71, July 2014

[3] "There was negligence," says MP on dam disruption in Mariana . Document states that there was a risk of dam rupture. G1 web Portal, published in 10/11/2015. Available at http://g1.globo.com/minas-gerais/noticia/2015/11/houve-negligencia-diz-mp-sobre-rompimento-de-barragens-em-mg.html. Accessed on 12 December 2016

by law. The company did not fully comply with the recommendations; still it kept its operations going without any constraint. Samarco's *modus operandi* clearly indicated complete indifference to risk.[4] On one hand, it took advantage of its legitimacy with the population and local government, of the positive public image and prestige with state and federal authorities; and, on the other hand, it simply side-stepped or ignored government agencies in charge of inspection. To make matters more complex, the performance of public agencies responsible for the supervision and licensing of dams, such as the State Foundation for the Environment MG (FEAM)[5] and the National Department of Mineral Production (DNPM),[6] was rather faulty. The Federal Public Prosecution Office (MPF) asked Federal Policy to investigate these public agencies with the objective of verifying "any crime committed during the licensing, possible crime committed by the omission of the State, by the negligence of the State and to ascertain the environmental damage."[7] These two bodies and also the Environmental Policy Council – COPAM-MG – are jointly responsible for the tragedy, according to the conclusion of that Federal Police investigation.[8] Till the day of this account (July 2017), none of these agencies have made any public announcement about their role in the disaster.

[4] INSTITUTO PRÍSTINO. Technical Report Parecer Único N° 257/2013 Description of the event: Technical Analysis Concerning the Revalidation of the Operational License of the Fundão tailings dam – SAMARCO MINERAÇÃO S/A. Available at http://giaia.eco.br/wp-content/uploads/2015/11/Resposta-parecer-257-2013.pdf. Accessed on 4 February 2017

[5] PARREIRAS, Mateus. FEAM admits discrepancy between actual volume of dam in Mariana and official data. The organization confirms that the register on the volume of tailings in the Samarco dams in Mariana has lapsed since 2012. The State of Minas Gerais revealed that the Fundão dam, which broke, had 20 times the volume recorded. Journal Estado de Minas, published in 2/12/2015. Available at http://www.em.com.br/app/noticia/gerais/2015/12/02/interna_gerais,713362/feam-admite-discrepancia-entre-real-volume-de-barragem-em-mariana-e-da.shtml. Accessed 2 February, 2017

[6] RODRIGUES, Léo. TCU shows problems of DNPM that contributed for the tragedy in Mariana. Agência Brasil. Available at http://agenciabrasil.ebc.com.br/geral/noticia/2016-09/tcu-aponta-falhas-do-dnpm-que-contribuiram-para-tragedia-em-mariana. Accessed on 2 February, 2017

[7] Pedro Ângelo. "Polícia Federal abre inquérito para apurar crime ambiental em Mariana. MPF pediu que Samarco, FEAM, OPAM e DNPM sejam investigados". G1 web Portal, published in 17/11/2015. Available at http://g1.globo.com/minas-gerais/noticia/2015/11/policia-federal-abre-inquerito-para-apurar-crime-ambiental-em-mariana.html

[8] Jorge Munhós de Souza, Procurador da República do Ministério Público Federal aponta falha de fiscalização e co-responsabilidade. See "MPF cobra R$ 155 bi para reparar danos do rompimento da barragem da Samarco". EBC – Agência Brasil web Portal, published in 03/05/2016. Available at https://www.escavador.com/sobre/3516567/jorge-munhos-de-souza

6.4 The Biocultural Ethics, Economic Power, and People

The disaster was not an event with an easily defined beginning and end: the negative effects of the dam disruption continue to produce consequences, and political, legal, technical, and ethical controversies continue to arise. The damage is individual and collective. With the framework of the biocultural ethic (Rozzi 2013), we identify that the disaster affects both the habitats and habits of both territorial communities and ecosystems and their nonhuman co-inhabitants. The disaster has revealed the explorationist logic that governs the technical, economic, ethical, and governance decisions of the territorial system created by large mining and steelmaking companies. Also, it revealed the threats to the sites, ecosystems, and co-inhabitants of the Rio Doce basin.

The local and global impacts caused by Samarco/Vale/BHP are of a radically different order and scale. At the local level, we have the simultaneous and interconnected losses of biological and cultural diversity. On the global scale, we have decision-makers worried about the interruption of the companies' activities and the effort they have to expend to regain "normality." Tax losses, legal disputes in the courts, commodity market, exports, foreign exchange, balance of payments, stock values, and other problems at the global scale are far from the lives of the people of affected region. A biocultural ethics requires a shift in scale to consider people and their communities in the multiple territories of the Rio Doce basin. It is at the local level that we can verify the specific impacts on each place. The river has different stretches, and each one of them and their co-inhabitants suffered from the disaster differently, in the biotic and abiotic as well as socioeconomic and cultural ways. Our analysis foregrounds the devastating effects of the global interests on the local scale. We advocate discursive practices that consider and interact with the human and nonhuman communities of each place affected by this disaster.

What happened in the stretch between the tailings dam of Fundão and the dam of the Candonga Hydroelectric Power Plant, 100 km downward, was like a "tsunami." After this stretch, the mud settled in the channel of the Rio Doce, except at points where it overflowed into the alluvial planes and some tributary areas. A year and a half later, the disaster continues to produce its effects. The violence of the uprooting suffered by the people continues. In addition to destroying Bento Rodrigues village, the mud damaged six other locations in the municipality of Mariana and the city of Barra Longa. Before reaching the Rio Doce, it destroyed cultural and historical patrimony, devastated 195 rural properties, rendered the soils unusable, ripped up the riparian forests, and dramatically altered the landscape.

The official figures show 536 injured and 19 deaths, but Priscila Monteiro, 28, fights to have the 20th death recognized. She suffered a miscarriage when she was taken and carried by the mud, for more than a kilometer. "The walls of the house began to fall on top of us, and the mud came and plucked my 2-year-old son and my niece from my arms, both of them inside the mud, I sank. I could not see anything. I felt the pain of the miscarriage, I lost my baby and I was swept away by the mud…" Priscilla stayed 13 days in the hospital and, since then, fights for the recognition: "It's not 19, it's 20 (deaths). I loved him. I had his clothes ready."[9] Seven-year-old Thiago Santos also died in the mudslide. As his mother says: "Every day I wake up crying. I dream of him saying 'mama, I will not let go of you ever again'…."

The tricentennial village of Bento Rodrigues was a community with a deep sense of belonging, so it was deeply traumatic for its people to watch their houses, stories, and ways of life destroyed in a matter of minutes. For Daniella de Souza, it was "a whole life, a whole story, everything is under the mud now." The residents return to the ruins of the village to "calm the heart and soul," when they "realize that the destruction is real." According to the former resident Francisco de Paula: "The people who come here want to kill the homesickness, to walk here, to see where their house was." On September 24, 2016, the celebrations of the patron saint feast of Our Lady of Mercy were celebrated among the ruins, and, at the same time, people said a last good-bye to the place. It would be covered by water after the construction of the S4 dike, to block the tailings still coming down from the Fundão dam. Like the other residents, Monica Santos is against the dike: "I am totally against this dam S4 because, once flooded the place, I doubt they will want to disassemble it later. Once again, they will take our story and cover up the crime that happened there."

Samarco promised to build a "new Bento Rodrigues" and deliver it in 2019, which means 4 years of waiting for a new home! There is a feeling that even if houses are built, the wounds will not be healed. The houses of Bento Rodrigues had large yards with many fruits, joys, and memories. Residents perceive the house as *the* existential place: "my home," "where the heart is," they say. The indemnity covering house rebuilding does not compensate for the loss. As Onézio de Souza, 52, said: "My little house was modest, but full of affection, had chickens, a dog, was a good place, quiet place, easy to raise the family…if I could I would have saved my dog, because he is life, he took care of my family…."

When the flood of tailings reached the Rio Doce, the biggest problem was the turbidity of the water, because it affected the aquatic life and prevented its use and consumption by human and nonhuman co-inhabitants. The turbidity was extremely high when the mud reached the ocean, damaging the fragile coastal ecosystem. More than a year after the disaster, the dam of the Candonga Hydroelectric Plant continues to release the mud it has retained in its reservoir. The flowing mud has churned the soils and riverbed, putting into suspension different chemical components and contaminated sediments from the mining processes of the past (gold). With this, a significant increase in concentrations of heavy metals occurred. With the "normalization" of water supply by water treatment services, the court suspended Samarco's obligation to provide bottled mineral water, but about 400,000 inhabitants of the riverside towns have not yet been convinced that treated water from the river is safe. Almost 2 years after the rupture of the tailings dam, a large part of the population continues to buy bottled water, though not all residents can afford this alternative to drinking the river water.

[9]Reportagem de Ricardo Senra para a BBC Brasil, published on 4/11/2016. Available at http://www.bbc.com/portuguese/brasil-37829548

Material losses and interruption of drinking water supply have gained prominence in the media, but immaterial losses are a source of continuing distress. The impacts on the environment persist – the mud has devastated aquatic biodiversity and made fishing in the river and coastal waters unacceptable – as do the prohibitions on activities necessary to the survival of family farmers, fishermen, indigenous people, and river and beach tourism workers (merchants, hostel owners, various service providers). Two years after the flood, affected people and communities continue to experience cultural uprooting and vulnerability.

We must also consider the loss of the modus vivendi of thousands of families, as social worker Bianca Pavan asserts: "We need answers because we can not freeze people." The deterritorialization of people left emotional trauma and constant symptoms of distress, aggravated by lack of attention and information. Samarco, government institutions, and universities have failed to answer basic questions such as: Can I use the river water to irrigate the plantations? Can I fish and eat the fish of the Rio Doce? Is there a health problem if children swim in the river?

Uncertainty is unsustainable as it makes life difficult for people already shocked by the tragedy that drastically suppressed their habits and habitats. Furthermore, it creates an environment of growing anxiety. The persistence of this situation has negative repercussions on health; depression has become common, as noted by public health agents in many towns. In the municipality of Mariana, for example, where the mud drastically affected about 1500 people, in three villages, there are public reports of increases in cases of alcoholism, violence, depression, and suicide caused by the abrupt process of deterritorialization (Rodrigues et al. 2016).

In the middle reach of Rio Doce basin live the Krenak indigenous people. They have been deeply impacted by the arrival of the mud. Here is the testimony of the cacique Rondon Krenak, to the Free Journalists, on the banks of the Rio Doce, in the municipality of Resplendor, MG 11/12/2015:

> *Watu* [the great river] is the source for everything in the life of the Krenak people … bathing, hunting, fishing, religious practices… We have a strong relationship with the river, we consider the river our mother. With all that happened, the Krenak community decided to take a stance [*demonstrate by sitting on the tracks of the Vale train*].[10] We cannot be in this situation. Those responsible have to come here and have to talk, to know how we are going to live from now on… We are here watching the river, since it is dead. They thought they killed the river alone, but they killed us too.[11]

The members of the Krenak tribe decided to occupy the tracks of the railroad Vitoria-Minas (Vale), which passes through their lands. When they interrupted the transportation of iron ore, Vale was forced to negotiate directly with them. There were a couple of meetings, and Vale provided food for a certain period of time and made some promises.

[10] Portal de Notícias G1/MG, published on 14/11/2015. Available at http://g1.globo.com/minas-gerais/noticia/2015/11/indios-bloqueiam-saida-de-trem-da-vale-em-estacao-de-belo-horizonte.html. Accessed on 27 January 2017.

[11] Reportagem da BBC Brasil. "Não podemos pescar, não podemos tomar banho, o rio morreu para nós." 16 November 2015. Available at https://www.youtube.com/watch?v=qfk3yiAwl-Q

The close relationship between habits and habitats is revealed by the speeches of these natives of the Rio Doce basin. "Our tradition has been swept away. We always live in these lands. Our language, our leisure, our fishing and even our religion are associated with the Rio Doce. What will the future of our children be like now? It's a global ecological tragedy. Our ancestors are certainly very sad .... We have to begin to understand how our relationship with our own culture will be from now on. This is going to be the biggest challenge facing our people. But in this process we can not fail because, if this happens, Vale will pass over us" (Djukurna Krenak, El Pais, 12/31/16).

Fishermen forbidden to carry out their activities have spoken out. "Our life was the river because the people here live on it. I only know how to fish. My life and the families here have collapsed. They killed our river" (Adroaldo Gonçales Filho, resident of the village of Mascarenhas, municipality of Baixo Guandu-ES). "We no longer use the water of the Rio Doce for anything … Now the river water is gone," said José Lemos, president of the Association of Fishermen of Maria Ortiz, in Colatina, ES.

Farmers, fishermen, sand collectors, merchants, innkeepers, various service providers linked to the tourism of the Rio Doce, and the beach of Regencia suffered a severe disruption of their activities and emotional losses. The monthly payments through the "Samarco financial aid card" do not cover the loss of the modus vivendi of people and their habitats.

If on one side there are losses, on the other there is a growing desire to resist and fight for their habitats and way of life: "They came into our house, did not ask if they could enter, did not wipe their feet, they are trying to decimate our village, our culture, our way of living, simplicity, our essence, but we will resist. As long as I have the strength, we will resist, we will not want this inheritance, the inheritance we want is at least the possibility of my children swimming in this river" (Luciana Oliveira, resident of the Mascarenhas village, in testimony to the newspaper El País).

6.5 Final Considerations

Scholars and engineers have tools of observation and analysis that have been sharpened by years of training and experience. In the midst of a crisis, those tools can be used to analyze problems and to suggest answers and to bring the crisis before the eyes of the broader community. Clarity of thinking can go hand in hand with passion. We assume our readers will understand when our passion for the people, animals, and environment of the Rio Doce is made obvious in this writing. The loss of lives, the loss of ways of life, and the destruction of a fundamentally important river basin require our passion along with our intellect.

Those responsible for the biocultural disaster used the strategy of silence: only the communication and marketing sectors were allowed to speak about the case. With this silence, Samarco sought "to gain time," seemingly believing that the legal department would find alternatives to circumvent the crisis situation. In the political field, the speeches were also framed by marketing concerns. Millions of dollars were directed to advertising on major TV networks, national newspapers, and the Internet. The use of the word "accident" was an attempt to "naturalize" the biocultural disaster and thereby to reduce Samarco's liability. The slogan created was: "To do what must be done. This is our commitment." In fact, during the first year of the disaster, instead of proactive conduct, Samarco acted, on different occasions, by court judicial decision-making power. Samarco's business strategy seriously increased uncertainty and speculations.

In March, 2016, Samarco and governments (Federal and of the states of Minas Gerais (MG) and Espirito Santo (ES)) signed an agreement called "Compromise of Adjustment of Conduct" (TTAC). To the surprise of all, the State Public Prosecutor's Office (MP) of MG and ES refused to participate in it. According to Carlos Pinto, Public Prosecutor MG, this agreement clearly favors the companies and gives them total control of post-tragedy actions. The creation of a Foundation to assume all the responsibilities and relationships with the affected people and other actors, as put forth in the agreement, will serve to shield Samarco, Vale, and BHP from their liability for the disaster, Mr. Pinto reaffirmed.

As defined by TTAC, Samarco, Vale, and BHP created the Fundação Renova, which began work in August 2016. The entity undertakes all repair activities previously conducted by Samarco and its shareholders. This new actor plays the role of shifting the focus away from the companies and allowing a "business-as-usual" scheme. The Board of Directors and the Fiscal Council of the foundation are controlled by representatives of the three mining companies.

Even with the performance of the Fundação Renova, Federal MP entered the court questioning and demanding the effective participation of civil society in the process of reparation of damages, both material and immaterial. The Federal MP claims that the affected population must have guaranteed principles, such as human rights, self-organization of the affected people and communities, respect to the collective logics of belonging and territorialities, access to information, transparency of decision-making processes, and implementation of programs and projects.

Finally, although the Federal Police investigation concluded that Samarco/Vale/BHP leaders were guilty (22 people were indicted), after more than a year and a half since the tragedy, no one has been arrested, nor has any of the fines imposed on the company been paid to federal or state environmental control agencies. The lawsuits against Samarco are paralyzed in a legal imbroglio (in August 2016, there were around 48,000 lawsuits in the Court of Justice in MG e ES).

Nowadays, Samarco presses forward with resuming its operations, which involves strong economic, social, and political interests. It receives support from the federal, states, and municipal governments, from the population of the affected municipalities, and from the business sector. Not surprisingly, Samarco already has its environmental licensing process in a well advanced stage, in the certainty that it will be operating again by the end of 2017.

The Rio Doce basin today is facing a continuing biocultural disaster caused by the world's mining giants. For this reason, we consider it a fundamental ethical obligation to appoint oversight of Samarco/Vale/BHP officers. To protect biocultural diversity, one can never simply trust companies as Samarco/Vale/BHP, since they operate on a different scale and within a frame that does not take into account the interests of local habitat and habits. Our analysis is centered on the local scale and specifically on the effects of global companies on local life. We position ourselves to further an ongoing discursive practice that considers and interacts with the human and nonhuman communities of each place affected by this disaster. For any analysis, for any understanding of the situation, the stories of the victims, humans, and nonhuman entities alike must be taken into account and the effects of the global on the local spelled out to safeguard biocultural diversity from devastating homogenization.

Acknowledgments This work was supported by undergraduate students from Univale, Iesmy Elisa Gomes Mifarreg (Law) and Natália Moreira Ferreira (Civil Engineering), from the scientific initiation program. We are grateful for the close reading and extensive editing of Irene J. Klaver and Brian C. O'Connor.

References

Firmino H, Morais L (2014) A mineração de valores. Revista Ecológico, Belo Horizonte, v. 71, julho de 2014

Guerra C, Torres F (2012) Waters and culture in the Doce River Basin of southeast Brazil: mobilization and construction of a new perspective. In: Johnston BR, Hiwasaki L, Klaver IJ, Ramos Castillo A, Strang M (eds) Water, cultural diversity & global environmental change: emerging trends, sustainable futures? UNESCO international hydrological program, Springer Press, Dordrecht, pp 21–24

Rodrigues et al (2016) Algumas análises sobre os impactos à saúde do desastre em Mariana (MG). In: Milanez, Bruno; Losekann, Cristiana (Org.). Desastre no Vale do Rio Doce. Antecedentes, impactos e ações sobre a destruição. Rio de Janeiro: Folio Digital: Letra e Imagem, pp 163–193

Rozzi R (2013) Biocultural ethics: from biocultural homogenization toward biocultural conservation. In: Rozzi R, Pickett STA, Palmer C, Armesto JJ, Callicott JB (eds) Linking ecology and ethics for a changing world: values, philosophy, and action. Ecology and ethics, vol 1. Springer, Dordrecht, pp 9–32

Chapter 7
Land Grabbing and Violence Against Environmentalists

Roy H. May Jr.

Abstract Ignoring traditional people's rights and resorting to intimidation and even murder are strategies used to grab land and impose economic models. The history of violence and intimidation against those who defend the land – its people, its beauty, and its vitality – is long and worldwide. It is history made whenever vital local and environmental interests conflict with the neoliberal capitalist economic development projects aimed at the enrichment of the few. In 2015 alone, Global Witness documented 185 killings of land and environmental defenders; since 2002, it has confirmed at least 1176 such murders across the globe. Among these is the murder of Honduran environmentalist and land defender Berta Cáceres in 2016. The constant intimidation and murder of environmentalists and defenders of land rights of indigenous and peasant people, and the imposition of economic enterprises inappropriate to ecological conditions, demonstrate the persistent coloniality that affects nature and people.

Keywords Murder · Land defenders · Indigenous people · Berta Cáceres · Coloniality · New extractivism

Berta Cáceres, an indigenous woman in Honduras, actively campaigned against the construction of the Agua Zarca dam and reservoir on the Gualcarque River in southeast Honduras, not far from the border with El Salvador. "The Gualcarque River has called upon us, as have other gravely threatened rivers," she said. "We must answer their call." For her and her Lenca people, the most important indigenous group in Honduras, the Earth, its life, and landscape – especially rivers – were to be defended. "The Lenca people are ancestral guardians of the rivers, in turn protected by the spirits of young girls, who teach that giving ourselves in various ways for the protection of the rivers is giving our lives for the well-being of humankind and of this

R. H. May Jr. (✉)
Departamento Ecuménico de Investigaciones, San José, Costa Rica

R. Rozzi et al. (eds.), *From Biocultural Homogenization to Biocultural Conservation*, Ecology and Ethics 3, https://doi.org/10.1007/978-3-319-99513-7_7

planet," she explained. So women have a special responsibility to defend them. Through her leadership of the Civic Council of Popular and Indigenous Peoples Organizations of Honduras (COPINH), from 2010 to 2014 she led a massive protest movement that effectively stopped the hydroelectric project that would have dammed the river, which is sacred for the Lenca people. The Chinese company that was constructing the dam pulled out after the International Finance Corporation of the World Bank withdrew its support. Taken over by its Honduran partner, it too found it impossible to go forward. Momentarily, at least, the river, defended by Cáceres and others, had won. "When we began the struggle against Agua Zarca I knew it was going to be hard," she said, "but I knew we would win. The river told me." For her courageous defense of the river and the culture of her people, she was awarded the Goldman Environmental Prize in 2015 (Box 7.1). On March 2, 2016, Berta Cáceres was murdered. Three others had been murdered during the long protest campaign (Martins 2016; La Nación 2016). Sadly, their murders are among many others past and present; for developers, land grabbing is more important than human lives.

Box 7.1 Berta Cáceres: Goldman Environmental Prize Acceptance Speech 2015

In our [Lenca] worldview, we are beings who come from the Earth, from the water, and from corn. The Lenca people are ancestral guardians of the rivers, in turn protected by the spirits of young girls, who teach that giving ourselves in various ways for the protection of the rivers is giving our lives for the well-being of humanity and this planet. COPINH, walking alongside of people struggling for emancipation, validates this commitment to continue protecting our waters, the rivers, our shared resources, and nature in general, as well as our rights as a people. Let us wake up! Let us wake up, humankind. We're out of time. We must shake our conscience free of the rapacious capitalism, racism… and patriarchy that will only assure our own self-destruction. The Gualcarque River has called upon us, as have other gravely threatened rivers. We must answer their call. Our Mother Earth, militarized, fenced in, poisoned, a place where basic rights are systematically violated – demands that we take action. Let us build societies that are able to coexist in a dignified way…in a way that protects life. Let us come together and remain hopeful as we defend and care…for the blood of the Earth and of its spirits. I dedicate this award to all the rebels out there, to my mother, to the Lenca people, to Rio Blanco, and to the martyrs who give their lives in the struggle to defend our natural resources. (Translation Goldman Environmental Prize) (Fig. 7.1).

http://www.goldmanprize.org/recipient/berta-caceres/

(continued)

Box 7.1 (continued)

Fig. 7.1 Berta Cáceres with COPINH and Rio Blanco community members honoring colleagues killed during the 2-year struggle. (Photo by the Goldman Environmental Prize)

Clearly powerful interests that prefer hydroelectric dams and other uses of the land over traditional culture and life habits, local communities of human and other-than-human coinhabitants, and native habitats are willing to resort to violence to force the success of their projects. This substitution of native habitats, traditional life habits, and unique local communities of coinhabitants by global models of economic development and lifestyles drives processes of biocultural homogenization worldwide and socio-environmental injustices (Rozzi 2012). The deaths of Cáceres and others illustrate a pattern common to many places of Latin America and other regions of the world: the selective elimination of those who raise their voices in defense of the land and its people. "Killing has become politically acceptable to achieve economic goals," observes Felipe Milanze (Hill 2016), a political ecologist at the Federal University of Recôncavo of Bahía in Brazil. The murder and intimidation of environmental defenders – defined by the Center for International Environmental Law (CIEL 2010) as persons who protect the environment and defend the rights of the victims of environmental degradation, whether or not they are formally affiliated with environmental organizations – facilitate processes of biocultural homogenization.

7.1 A History of Violence and Intimidation

This history of violence and intimidation against those who defend the land – its people, its beauty, and its vitality – is long and worldwide. It began with European colonial expansion and today is repeated whenever vital local and environmental interests conflict with the neoliberal capitalist economic development projects aimed at the enrichment of the few that have dominated much of the world since the 1970s. Neoliberalism, as all forms of capitalism, reduces natural resources to commodities that in turn are converted into money. The "problem," of course, is that much of the Earth is occupied by indigenous or traditional peoples or by peasant farmers that derive their subsistence from the land. These people may or may not have legal title; many are there in usufruct because that's where their people have always been, or because they have been pushed off the lands where they had previously lived. Worldviews respecting the spirits of young women residing in rivers are deemed silly. So if indigenous and other local people won't move out on their own from their ancestral territories, then powerful interests will do it for them, all too often resorting to violence to do so. "Nature, the Amazon, the Indians…they're all seen as obstacles," says Milanze (Hill 2016).

This is a long history: colonialism not only pushed native peoples off their lands in order to create many of today's nations; it also declared war against them, killing and enslaving thousands. In more recent times, development projects removed countless traditional peoples from their lands in order to establish agricultural resettlement colonies to be occupied by others who were to become modern farmers. Plantation agriculture, logging, mining, petroleum, and hydroelectric interests became hugely present in the displacement of poor and traditional peoples as countries, pushed by big internal as well as international economic interests, surged to become modern and developed. Murder was never shunned as a strategy to ensure development success. Land grabbing – committed by individuals against individuals or facilitated through occult policies decided by the corporations and governments against whole groups – is a longtime strategy used by powerful interests for their own benefit.[1]

7.2 Contemporary Violence Against Defenders of the Environment

Sadly, this is not past history. As the murder of Berta Cáceres shows, this history is present history. Multiple situations have been confirmed by human rights and environmental groups. The London-based watchdog organization, Global Witness (2016), has documented hundreds of examples of this violence due to land

[1] For historical background on conflicts over land worldwide, see Christodoulou 1990; for conflict and forest and environmental destruction, consult Colchester and Lohmann 1993.

grabbing. In 2015 alone, 185 killings of land and environmental defenders were documented; particularly violent were Brazil with 50 murders, the Philippines with 33, and Colombia 26. Since 2002, Global Witness has confirmed at least 1176 such murders across the globe; probably there are more since many are unreported. Indigenous peoples are especially vulnerable due to their often weak legal hold on traditional lands and their isolated geographical locations and because their lands often are the richest in natural resources. The drivers of violence are the same historic economic sectors: mining and extractive industries, mega hydroelectric projects, agribusiness, and logging. According to Global Witness, "Collusion between state and corporate interests shield many of those responsible for the killings." These are carried out by paramilitary groups, the army, police, or private security guards, thus "strongly implying state or company links to the killings." Continuing, Global Witness concludes, "There is little evidence that the authorities either fully investigated the crimes or took actions to bring the perpetrators to account" (2016, p. 4). The Global Witness report is backed up by a similar review of the situation in Latin America conducted by the Center for International Environmental Law (CIEL). This report also illustrates the severe human rights violations against environmental defenders, who engage in lawful activities that bring to light environmental damage and human rights abuses. The CIEL describes specific cases including violent attacks, torture, disappearances, and killings. It notes especially that indigenous people are most often targeted and represent 40% of the deaths (Article 19 2016, p. 4).

Furthermore, Global Witness (as does CIEL) shows that this violence goes hand in hand with the defamation and criminalization of environmental activism. In Latin America, according to the Inter-American Commission on Human Rights (IACHR 2011), company lawyers and government officials increasingly accuse environmentalists of crimes such as sabotage, terrorism, insurrection, unlawful association, and instigation to commit crimes, among others (IACHR 2011). These activists oppose land grabbing due to industrial exploitation of natural resources and usurpation of traditional land tenure systems wanted for economic development projects (Birss 2017). According to the Mexico-based Observatory for the Protection and Defense of Human Rights (El Observatorio 2016), this legal harassment is increasing throughout Latin America. Months before being murdered, Berta Cáceres was detained by police and accused of carrying a concealed weapon and aggression against the internal security of the State. After these charges were dropped, she was accused of usurpation, coercion, and damages against private property. All were related to her leadership protesting hydroelectric projects (El Observatorio 2016, p. 19).

Sometimes grassroots movements are infiltrated by government agents who commit crimes in order to blame the environmentalists. For example, this strategy was revealed in 2014 involving the Chilean police and the Mapuche movement defending the monkey puzzle tree (*Araucaria araucana*) forests (Miranda 2014). These types of actions, the Observatory (2016, p. 19) argues, "show the interrelationship among public authorities, judicial, and private company officials" to defame and criminalize activists.

Global Witness especially signals Africa where defenders of the environment are accused of being anti-development and are defamed publicly, sometimes even being accused of crimes. This recalls the tragic execution by the (then) Nigerian military government of writer and environmentalist Ken Saro-Wiwa. Like Berta Caceres, Saro-Wiwa was awarded the Goldman Environmental Prize (in 1995). He was a member of the Ogoni people who inhabit the delta of the Niger River in southeast Nigeria. He denounced the enormous environmental damaged caused by the petroleum industry, especially the indiscriminate dumping of toxic wastes, and criticized companies, particularly Royal Dutch Shell, as well as the government, for doing nothing about it. His leadership of a highly public protest campaign proved extremely costly for Shell. Accused on trumped-up charges of murder, Saro-Wiwa and eight other activists – the Ogoni Nine – were hanged on November 10, 1995 (Odunfa 2005; Wiwa 2005). Although it has never admitted any responsibility for the executions, in 2009 Shell agreed to an out-of-court settlement, awarding the aggrieved families $15.5 million as a gesture of "reconciliation," according to Shell (Pilkington 2009).

There are other forms of land grabbing that recur to violence against defenders of the environment, traditional lifeways, and the well-being of other-than-human coinhabitants. Criminal syndicates engage in illegal mining, logging, and poaching – either for the national bushmeat market or the lucrative illegal international trafficking of wild animals – occupying territory and terrorizing or expelling indigenous and peasant farmers. For example, in Costa Rica, a country praised for its conservation efforts and nature tourism destinations, Diego Saborío was murdered in 2014 by a would-be poacher because he would not allow hunting on his family's farm (Hernández 2016). The year previously, Jairo Mora, who patrolled the Caribbean beaches at night to protect sea turtle nests from poachers, was murdered by a gang of local thugs who made their living selling turtle eggs on the illegal market (Arguedas 2016). In the heavily forested Talamanca region in the Atlantic side of Costa Rica, drug lords have appropriated large tracts to grow marijuana, effectively expelling the indigenous people who live there (OIJ 2013; Solano 2015). This is not a local phenomenon. Especially notable are attacks on rangers who guard parks and wildlife reserves. Worldwide over the past decade, according to the International Union for Conservation of Nature (IUCN), at least 1000 rangers have been assassinated in 35 countries. The number could be three times as high since reporting is voluntary. In the Democratic Republic of the Congo (DRC) in Africa and Thailand in Southeast Asia, violence against defenders of wildlife has become commonplace as several hundred rangers have been killed in recent years. In Thailand during the last 10 years, at least 50 rangers have been murdered, 26 injured, and 23 left in critical physical condition (Global Conservation 2016). In the DRC a pathetic example occurred in April 2018 when a vehicle carrying rangers patrolling the Virunga National Park was ambushed. Five rangers and their driver were killed and another ranger was severely injured. During the past 20 years, more than 170 rangers have died defending the wildlife of the national park. According to government officials, these attacks are carried out by anti-government rebels, local bandits, and rare animal hunters that roam throughout the area (Ansede 2014, Downs 2018). Common criminals also know that the Earth is the source of wealth and don't hesitate to use violence to obtain it.

7.3 Violence and Land Grabbing Across the Global South

Conflicts due to land grabbing are widespread across the Global South.

In Southeast Asia, industrial-scale rural land acquisitions by large corporations are actively encouraged by governments, causing widespread conflict over forests and land between locals, primarily indigenous people, and governments, companies, and foreign investors. Many economic pursuits are involved, from logging to mining to shrimp farming to hydroelectric, but agribusiness, especially palm oil production, is paramount. Palm oil is a $44 billion-a-year industry, and its production has increased dramatically worldwide in recent years. Expanding by millions of hectares especially in Indonesia and Malaysia, these two countries produce 85% of the world's supply. Palm oil plantations take out huge parcels of mostly primary rain forest affecting not only wildlife such as orangutans (whose numbers have plummeted) but many indigenous peoples. As one observer writes, "palm oil is in everything and it's destroying Southeast Asia's forests" (Mosberger 2015). Although Indonesia has taken important steps to regulate the forestry sector, deforestation for establishing palm oil plantations also has fomented a highly lucrative trade in illegal timber. Numerous plantations, many of which operate outside the law, were originally fronts for illegal logging (EIE 2014). In these countries, frequently local or traditional landholders are forcibly evicted by police or the private security guards of companies so that large investors can take the land (Cherry 2013), whether it be for palm oil or other enterprises. Land grabbing deals are completed in secret; local residents, who have lived off the land for generations, are ignored. People – numbering from hundreds to millions – are displaced, livelihoods destroyed, rights abused, and communities disrupted (CSR Asia 2014; Yasmi et al. 2010). Market forces are given priority over nature and culture.

Global Witness reports few murders of land defenders and environmentalists in Indonesia, Myanmar, Cambodia, and Thailand, but the Philippines is especially dangerous for environmental defenders: 33 were killed in 2015 alone. Among them, 22 were in Mindanao where the Lumad people are defending their right to their traditional lands against large agribusiness and mining interests. The region has substantial reserves of coal, nickel, and gold and is excellently located for large-scale plantation agriculture. During recent years, more than 500,000 hectares of Mindanao land have been allotted to mining exploitation and over 700,000 hectares for agribusiness plantations (Global Witness 2016). The Lumad people are a coalition of 18 different ethnolinguistic groups that have long defended their people's culture and traditional land rights. Often accused of being Communists, the Lumad have suffered considerable violence and persecution since the Ferdinand Marcos regime (Spear 2015).

Throughout sub-Saharan Africa, in spite of increasing urbanization, much of the population is still rural and depends on access to arable land and pasture. However, growing population, internal migration, and changing economic models and land use patterns increasingly provoke conflicts. Market-driven economic models favoring urban interests and foreign investors increase the value of land, which in turn pressures for privatization of communal lands and elimination of smallholders, consequently eroding traditional cultures and forcing landlessness. As a result, land grabbing is increasingly taking place throughout the continent by economically powerful groups, including government officials and politicians (Lund et al. 2006).

The great mineral wealth of parts of Africa represents another driver of social conflicts and land grabbing as competing interests struggle to control it. The estimated $24 trillion worth of unexploited minerals in the Democratic Republic of the Congo (DRC), for example, is the source of extreme violence. During the last quarter of a century, an estimated five to seven million Congolese have died in ethnic rivalries and land disputes related to mining interests. Private militias in the employ of powerful, shady, economic interests control mining operations and mineral smuggling to nearby countries and then into the supply chains of multinational corporations (Lasker 2015).

Throughout the continent, non-African governments and corporations are investing in large-scale agricultural, hydroelectric, mining, and other economic ventures. Global Witness has identified few cases of violence against environmentalists and land defenders since 2002, but in 2015, 13 murders were confirmed, mostly park rangers. Global Witness suggests that the actual number is much higher and that murders go unreported because they occur in isolated geographical regions and because they may be bound to other causes, thus making invisible the defense of the environment and the land. Nevertheless, as the Global Witness Report (2016, p. 22) points out, all the potential drivers of violent land grabbing are "undeniable present in many African countries."

In Latin America, land grabbing takes many forms, affecting especially indigenous people.[2] Historically mining has been a prime land grabber and continues to generate much conflict. Demand for new minerals, but also "classic" minerals such as gold, and improved technology that makes it economically feasible to exploit hard-to-get-at reserves, has spurred renewed investments throughout Latin America. Many of these concessions have provoked severe, even violent, conflicts, such as in the Cajamarca region of Peru and Central America. The Center for International Environmental Law (CIEL 2010) reviews numerous cases and provides evidence of the complicity of the state in many such cases. It concludes that repression and violence are a growing trend in the region.

As energy needs grow, hydroelectric projects loom as another especially powerful driver of land dispossession and environmental destruction, as the case of Berta Cáceres illustrates. Mexico and Central America have embarked on numerous ambitious hydroelectric projects. Associated with this process, in 2015 alone, 15 people were murdered due to their opposition to hydroelectric projects in Mexico,

---

[2] For background and overview, see Borras et al. 2011 and Sexton 2012.

Honduras, and Guatemala (Global Witness 2016). Social conflict has been especially serious in Guatemala with numerous Maya activists murdered, threatened, or criminalized. Vandalism has been committed against the property of construction companies. Towns have been placed under martial law and leaders jailed (Merida 2014). In all cases, "They were opposing the threat of hydroelectric dam's displacing villages, disrupting farmers' irrigation, and drowning fertile valleys" (Global Witness 2016, p. 16). Social, sometimes violent, conflicts have occurred in Panama when the government approved the construction of hydroelectric projects on rivers crossing territories legally assigned to the Gnöbe-Buglé people in 2012 (Prieto-Valladares 2012). In Costa Rica, although free from violence, the mega hydroelectric project Diquís in the southern part of the country has stalled due to opposition from indigenous people whose lands will be inundated. No proper consultation with them was conducted before initiating construction of the huge dam and reservoir (Ramírez 2010; La Nación 2012).

In South America, the construction of dozens of large hydroelectric dams and more than a hundred smaller ones is projected over the next 20 years for the greater Amazon Basin, mostly in Brazil, Peru, and Bolivia but also Ecuador, Colombia, Venezuela, Guyana, and Surinam. The Andes Mountains and the Amazon River are geologically and ecologically linked, and so the disruption of connectivity could have severe and unpredictable environmental impacts. Some of the impact is obvious: these reservoirs will flood millions of hectares of rain forest, thus destroying prime habitat for multiple forms of life as well as forcing the removal of dozens of tribal peoples and others. Furthermore, the construction of access roads and electric power transmission lines contributes to additional forest destruction. The resulting fragmentation of the rivers and consequent loss of connectivity also have serious negative environmental as well as social impact; fish communities and migration are disrupted, and native peoples who depend on fishing for their livelihoods are effectively expelled from their riverine homelands. Furthermore, in deep time, changes in river patterns might influence rivers as drivers of speciation. Amazonia riparian systems affect hybridization in Amazonia birds; hence significant changes in their geography could have implications for future gene flow (Weir et al. 2015). Although depending on the characteristics of each, the reservoirs also will emit considerable quantities of greenhouse gases due to underwater decaying vegetation and thus contribute to global warming (Finer and Jenkins 2012; Fearnside 2014).

Hydroelectric development, however, faces severe protest by environmentalists but especially by peasant farmers and indigenous peoples whose lands and livelihoods are affected by the reservoirs. In Brazil the Movement of People Affected by Dams networks hundreds of local organizations to protest dams and loss of land rights (Muller 2017; MAB n.d.). Tribals such as the Juruna and the Kayapó have vigorously resisted the construction of dams, even refusing large financial settlements as compensation for loss of their traditional territories (Paz and Miño 2013; Watts 2014). These protests often have been violently suppressed by government and company agents. Across the Amazon region of Brazil, Bolivia, Peru, Ecuador, and Venezuela, indigenous groups have organized the Coordination of Indigenous Organizations of the Amazon Basin (COICA) in an effort to stop the construction of hydroelectric plants (Fellet 2012; Watts 2015). Sadly, however, these protests have had little success.

In addition to hydropower, oil and gas projects also are serious threats to wilderness, biodiversity, and indigenous people, especially in the western Amazon Basin where there are large, mostly unexploited, reserves. Governments are opening these strategic areas for exploration and exploitation. These regions, however, are among the most biologically diverse and geographically isolated areas – and therefore containing largely intact ecosystems – on Earth. These areas overlap the traditional lands of many groups of indigenous people with their rich linguistic diversity and traditional ecological knowledge (see Introduction to this volume). Hydrocarbon production began in the area in the 1920s, and today there are numerous large projects, notably in Colombia, Ecuador, Peru, and Brazil; Bolivia is a longtime producer of natural gas. Many of these projects have been marked by serious environmental damage due to deforestation, urbanization, and large oil spills. Likewise they have provoked severe social conflict, often marked by violence, as indigenous peoples have defended their traditional land rights and the environmental integrity of their geographical territories (Rozzi 2012, 2015). These newer projects cover nearly 700,000 km^2 and involve 35 transnational corporations (Finer et al. 2012).

International law (Indigenous and Tribal Peoples Convention No. 169 of the International Labor Organization) requires that indigenous peoples whose lands will be affected by these and other development projects be consulted and that their full and informed consent be obtained before advancing such projects. However, historical experience shows that these consultations seldom are conducted and that the consent of native peoples seldom, if ever, is obtained. Nevertheless, the projects go forward.

Land grabbing, however, is not limited to the Global South. In the United States, in 2016 the Lakota Sioux vigorously protested the construction of an oil pipeline crossing lands sacred to them and posing a threat to their people's water (Global Witness 2017, pp. 36–38). The Sioux charged that they never were fully consulted and argued that adequate environmental and archaeological investigation was not done before the project received governmental approval. In a court proceeding, a federal judge ruled against the Sioux and in favor of the pipeline company, allowing construction to continue. Nevertheless, agencies of the federal government were able to suspend the construction temporarily, and the White House called for a full review of how indigenous people are consulted and involved when similar projects affect their lands (Kennel-Shank 2016; Medina and Rafols-Nuñez 2016). However a new US president, intimately involved with huge business interests including the pipeline and wholly unsympathetic to Native Americans and social justice, ordered that the pipeline be continued.

This is not the only contemporary example of grabbing land from Native Americans. Months previously, in Arizona, a gigantic copper mine was approved on lands sacred to the Apaches (Dokoupil 2015; Estrada 2015). This project has been temporarily delayed pending judicial review, environmental impact, and consultations with the Apache nation (Quiroz 2017), but the land will never again be available to the Apache. Especially in the Western United States, from Alaska to Arizona, private interests, particularly mining and hydraulic fracturing or "fracking" for natural gas, vigorously lobby to privatize public lands and to have access to national parks, forests, and wildlife refuges. At the same time, they are pressuring Native Americans to accept economic projects, such as "fracking" on Navajo land in New Mexico or tourist development at the confluence of the Little Colorado River and the Grand Canyon in Arizona, which will effectively marginalize them from their own land (Riggs 2017). For these interests, "sacred land" is a meaningless concept. Such an idea is "silly." Land is deemed valuable only if it is privately owned and functions commercially.

7.4 Land Grabbing, Persistent Coloniality, and the "New Extractivism"

Land grabbing is inextricably linked to colonialism. Historically, colonialism was shaped by European nations imposing their political and economic hegemony over countries of the Global South for the purpose of gaining access to vast mineral wealth and rich agricultural lands, as well as new markets. Both people (culture) and land (nature) were turned into subalterns in their own homes. All kinds of power strategies were utilized to subjugate them, but violence was always among them. Those who resisted were eliminated, and production processes ecologically inappropriate were imposed on nature. Although they tended strongly toward sustainable conviviality with nature, ancient traditions and environmental practices not only were deemed primitive; they too were eliminated.

Although formal colonialism has passed with the independence of the countries, its legacy remains as the former colonial worldwide structure of center-periphery largely remains intact. Today, former colonies still exhibit many aspects of the old colonial ways in their social structures and thought patterns. Contemporary neoliberal capitalism replicates historic colonialism as it views nature and people as subalterns in function of economic enterprises and markets dominated by relatively few nations and economic elites. As Latin American political ecologist Hector Alimonda has highlighted (2011, p. 22), the "persistent coloniality that affects nature," is pertinent far beyond the geographical confines of the Americas. Alimonda argues that "[b]iophysical reality (flora, fauna, human inhabitants, ecosystem biodiversity) as well as territorial configuration (the sociocultural dynamic that significantly articulates those ecosystems and landscapes) … are viewed by global hegemonic thought and regional elites as subaltern space that can be exploited, obliterated, or reconfigured, according to the needs of the current accumulation regime." The constant intimidation and murder of environmentalists and defenders of land rights of indigenous and peasant people, and the imposition of economic enterprises inappropriate to ecological conditions, confirm Alimonda's assertion.

In recent years "new extractivism" – extraction of raw materials through public-private joint ventures in which the state is the dominate partner – (or its variant, "market economy with a preferential option for the poor") has become the development model for Latin American countries endeavoring to throw off vestiges of colonialism (Burchardt and Dietz 2014), yet it replicates it as it also is defined by the center-periphery structure of worldwide political economy. Eduardo Gudynas (2010, pp. 5–7), the Uruguayan environmentalist and development theorist and who coined the term, argues that while the government's control and share of income through joint ventures with private companies increases, enabling it to fund more social programs and exercise control over environmental and social justice concerns, since the new extractivism is functional for commercial globalization, "it will finance and maintain the international subordination of South America." Furthermore, Gudynas explains that even "under progressive governments … except for the ownership of the resources, the rules and functions of productive processes [that are] oriented to enhanced competition [are repeated], [and] increase profits according to classic criteria of efficiency including the externalization of social and environment impacts." For these reasons, he concludes, the "social and environmental impacts [of the new extractivism] have increased, [and] actions to confront them and to deal with them are still ineffective and even, on occasion have been weakened." The upshot is that the new extractivism also grabs land from traditional people and peasant farmers, affects forests with their great biodiversity, and contributes to social conflict (Burchardt and Dietz 2014, pp. 478–480; Walters 2017, p.10; Becker 2017, p. 130). Control of nature, even by progressives, is profoundly conflictive.

7.5 Conclusion

This violence tells us that ignoring traditional people's rights and resorting to intimidation and even murder are prime strategies used to grab land and impose economic models. Time is not on the side of the Earth nor the majority of its people. The persistent violence and environmental destruction wrought by land grabbing accelerate a bleak environmental and social future of biocultural homogenization that has little room for the richness that diversity brings. "Let us wake up, let us wake up, humankind," Berta Cáceres insisted in her acceptance speech for the Goldman Environmental Prize. "We're out of time … Let us come together and remain hopeful as we defend and care… for the blood of the Earth and its spirits."

References

Alimonda H (2011) La colonialidad de la naturaleza. Una aproximación a la ecología política latinoamericana. In Alimonda H, coord. (2011) La naturaleza colonizada. Ecología política y minería en América Latina. CLACSO y Ciccus, Buenos Aires

Ansede M (2014) Más de 1.000 guardabosques han sido asesinados en la última década. Materia. Published online June 30. Available via: http://esmateria.com/2014/07/30/mas-de-1-000-guardabosques-han-sido-asesinados-en-la-ultima-decada/
Arguedas C (2016) Jairo Mora sufrió brutal tortura antes de morir. La Nación [San José, Costa Rica] 3 April. Available via: http://www.nacion.com/sucesos/juicios/Jairo-Mora-sufrio-brutal-tortura_0_1552444779.html
Article 19 (2016) A deadly shade of Green. Threats to environmental human rights defenders in Latin America. Article 19 and CIEL, London. Available via: http://www.ciel.org/wp-content/uploads/2016/08/Deadly_shade_of_green_English_Aug2016.pdf
Becker M (2017) Ecuador: slowing a conservative restoration. NACLA Rep Am 49(2):127–131
Birss M (2017) Criminalizing Environmental Activism. NACLA Report on the Americas 49(3):316–322 Published online 14 September. Available via: http://www.tandfonline.com/doi/full/10.1080/10714839.2017.1373958
Borras Jr SM, Franco JC, Kay C, Spoor M (2011) Land grabbing in Latin America and the Caribbean viewed from broader international perspectives. Paper presented to the Latin America and Caribbean seminar: Dinámicas en el mercado de la tierra en América Latina y el Caribe, 14–15 November. Food and Agriculture Organization (FAO), Santiago, Chile. Available via: file:///C:/Users/Personal/Downloads/borras_franco_kay__spoor_land_grabs_in_latam__caribbean_nov_2011.pdf
Burchardt J-H, Dietz K (2014) Neo-extractivism – a new challenge for development from Latin America. Third World Q 35(3):468–486
Cherry J (2013) The great southeast Asian land grab. The Diplomat. Available via: http://thediplomat.com/2013/08/the-great-southeast-asian-land-grab/
Christodoulou D (1990) The unpromised land, agrarian reform and conflict worldwide. Zed Books, London
CIEL (2010) Environmental defenders in danger. The situation in Mexico and Central America in the context of the mining industry. Center for International Environmental Law, Washington, DC/Geneva
Colchester M, Lohmann L (eds) (1993) The struggle for land and the fate of the forests. Zed Books, London
CSR Asia (2014) Conflicts over land. A role for responsible and inclusive business. RIB Asia. Available via: http://www.csr-asia.com/report/RIB_Asia_Land_report.pdf
Dokoupil T (2015) A battle over land sacred for the Apache and lucrative for a mining company. MSNBC (07/31). Available via: http://www.msnbc.com/msnbc/battle-over-land-sacred-apache-and-lucrative-mining-company
Downs R (2018) 6 park rangers killed in Democratic Republic of the Congos's Virunga National Park. UPI. Published online April 9. Available via: https://www.upi.com/Top_News/World-News/2018/04/09/6-park-rangers-killed-in-DRCs-Virunga-National-Park/8021523315142/
EIE (2014) Permitting crime: how palm oil expansion drives illegal logging in Indonesia. In: Environmental Investigation Agency, London
El Observatorio para la protección de los defensores de derechos humanos (2016) Criminalización de defensores de derechos humanos en el contexto de proyectos industriales: un fenómeno regional en América Latina. FIDH, Paris. Available via: https://www.fidh.org/IMG/pdf/criminalisationobsangocto2015bassdef.pdf; https://www.fidh.org/en/region/americas/criminalization-of-human-rights-defenders-an-alarming-phenomenon-in
Estrada S (2015). Stealing Indian lands for a copper mine. DiveristyInc.com. Available via: http://www.diversityinc.com/news/stealing-indian-land-copper-mine/
Fearnside PM (2014) Análisis de los principales proyectos hidro-energéticos en la región amazónica. DAR-CLAES, Lima
Fellet J (2012) Latin American indigenous groups join to fight dams. BBC News (26 April). Available via: http://www.bbc.com/news/world-latin-america-17827131

Finer M, Jenkins CN (2012) Proliferation of hydroelectric dams in the Andean Amazon and implications for the Andes-Amazon connectivity. PLoS One 7(4):e35126. https://doi.org/10.1371/journal.pone.0035126
Finer M, Jenkins CN, Primm SL, Keane B, Ross C (2012) Oil and gas projects in the Western Amazon: threats to wilderness, biodiversity, and indigenous peoples. PLoS One 3(8):e2932. https://doi.org/10.1371/journal.pone.0002932
Global Conservation (2016) Park Rangers on the Frontline of Being Killed at an Astonishing Rate from India to Thailand to Africa. Global Conservation. Published online March 31. Available via: http://globalconservation.org/news/park-rangers-frontline-being-killed-astonishing-rate-new-solutio/
Global Witness (2016) On dangerous ground. Global Witness, London. Available via: https://www.globalwitness.org/en/reports/dangerous-ground/
Global Witness (2017) Defenders of the earth. Global killings of land and environmental defenders in 2016. Global Witness, London. Available via: https://www.globalwitness.org/en/campaigns/environmental-activists/defenders-earth/
Gudynas E (2010) The new extractivism of the 21st century: ten urgent theses about extractivism in relation to current South American progressivism. Americas Program Report, Center for International Policy, Washington, DC
Hernández C (2016) Autopsia confirma muerte asesino de ambientalista. La Nación [San José, Costa Rica] 4 September Available via: http://www.nacion.com/sucesos/crimenes-asaltos/Autopsia-confirma-muerte-asesino-ambientalista_0_1583241709.html
Hill D (2016) Never seen it so bad: violence and impunity in Brazil's Amazon. The Guardian. Available via: https://www.theguardian.com/environment/andes-to-the-amazon/2016/feb/16/never-seen-it-so-bad-violence-and-impunity-in-brazils-amazon
IACHR Inter-American Commission for Human Rights (2011) Second report on the situation of human rights defenders in the Americas. Organization of American States, Washington, DC
Kennel-Shank C (2016) Indigenous people gather to block oil pipeline. Christian Century (September 28): 14–15
La Nación [San José, Costa Rica] (2012) 3 May
La Nación [San José, Costa Rica] (2016) 4, 6, 13 March, 3 May
Lasker J (2015) Finding and tracking conflict minerals in the heart of darkness. Earth (December) 60: 40–49
Lund C, Odgaard R, Sjaastad E (2006) Land rights and land conflicts in Africa: a review of issues and experiences. Danish Institute for International Studies, Copenhagen
MAB (n.d.) Dams-affected peoples movement. Movimento dos Atingidos por Barragens. Available via: http://www.mabnacional.org.br/search/english%20index
Martins A (2016) Matan a Berta Cáceres, la activista que le torció la mano al Banco Mundial y a China. BBC Mundo. Available via: http://www.bbc.com/mundo/noticias/2015/04/150423_honduras_berta_caceres_am
Medina D, Rafols-Nuñez D (2016) Protesters vow to stop oil pipeline from traversing Sioux land. NCC News. 9 September. Avaialble via: http://www.nbcnews.com/news/us-news/north-dakota-protesters-vow-stop-oil-pipeline-traversing-sioux-land-n644861
Merida H (2014) La conflictividad social detrás de los proyectos hidroeléctricos. República GT. Available via: http://www.republicagt.com/seguridad-y-justicia/la-conflictividad-social-detras-de-los-proyectos-hidroelectricos-_c72efc/
Miranda O (2014) Agente infiltrado de carabineros confiesa autoría de atentados incendiarios en la Araucania. Diario Uchile. Available via: http://radio.uchile.cl/2014/02/12/agente-infiltrado-de-carabineros-confiesa-autoria-de-atentados-incendiarios-en-la-araucania/
Mosberger D (2015) Palm oil is in everything—and it's destroying Southeast Asia's forests. The World Post. Available via: http://www.huffingtonpost.com/entry/palm-oil-impacts_us_55a4c391e4b0b8145f737dd5
Muller L (2017) Em jornada de lutas, atingidos por barragens se mobilizam en todo Brasil. Luíz Muller Blog. 15 March. Available via: https://luizmuller.com/2017/03/15/em-jornada-de-lutas-atingidos-por-barragens-se-mobilizam-em-todo-o-brasil/

Odunfa S (2005) Ogonis still mourning 10 years on. BBC News. Available via: http://news.bbc.co.uk/2/hi/africa/4424238.stm

OIJ (2013) Reporte de situación, tráfico de drogas y amenazas del crimen organizado en Costa Rica. Organismo de Investigación Judicial, San José

PazyMiñoP(2013)Brazil'sKayapórejectAmazondams,refusemillions.AmazonWatch.Availablevia: http://amazonwatch.org/news/2013/0313-brazils-kayapo-reject-amazon-dams-refuse-millions

Pilkington E (2009) Shell pays out $15.5 m over Saro-Wiwa killing. The Guardian. Available via: https://www.theguardian.com/world/2009/jun/08/nigeria-usa

Prieto-Valladares JA (2012) Indignación del pueblo Gnöbe-Buglé en sus luchas por la vida y el agua (Panamá). La Voz 56:10–13

Quiroz D (2017) Copper mine on sacred Apache land in Arizona complains about delays. Indianz.com. March 22. Available via: https://www.indianz.com/News/2017/03/22/cronkite-news-copper-mine-on-sacred-apac.asp

Ramírez E (2010) Represa Diquís del ICE genera divergencias: indígenas y comunidades combaten proyecto de electricidad. Semanario Universidad, San José (23 June)

Riggs S (2017) Is Grand Canyon Escalade the next "Savior"? Advocate (Fall/Winter): 4–9

Rozzi R (2012) Biocultural ethics: recovering the vital links between the inhabitants, their habits, and habitats. Environ Ethics 43:27–50

Rozzi R (2015) Earth stewardship and biocultural ethics: Latin American perspectives. In: Rozzi R, Chapin FS III, Callicott JB, Pickett STA, Power ME, Armesto JJ, May RH Jr (eds) Earth stewardship: linking ecology and ethics in theory and practice. Ecology and ethics, vol 2. Springer, Dordrecht, pp 87–112

Sexton R (2012) Environment and conflict in Latin America. Humanitarian Aid on the Move 9:10–13 Available via: http://www.urd.org/Environment-and-conflict-in-Latin?artpage=2-3

Solano H (2015) Erradicación de cultivos de marihuana en Talamanca persiste pero sin capturas. La Nación, San José, 20 May. Available via http://www.nacion.com/sucesos/narcotrafico/Erradicacion-marihuana-Talamanca-persisten-capturas_0_1488651245.html

Spear L (2015) A civil war is being waged against indigenous tribes in the southern Philippines, rights groups say. Time. Available via: http://time.com/4028811/philippines-lumad-mindanao-indigenous-military-war-killings/

Walters J (2017) Capitalism a la Sandinista. NACLA Rep Am 49(1):7–10

Watts J (2014) Belo Monte, Brazil: the tribes living in the shadow of a megadam. The Guardian (16 December). Available via: https://www.theguardian.com/environment/2014/dec/16/belo-monte-brazil-tribes-living-in-shadow-megadam

Watts J (2015) Amazonian tribes unite to demand Brazil stop hydroelectric dams. The Guardian (30 April) Available via: https://www.theguardian.com/world/2015/apr/30/amazonian-tribes-demand-brazil-stop-hydroelectric-dams

Weir T, Faccio MS, Pulido-Santracruz P, Barrera-Guzmán AO, Alexio A (2015) Hybridization in headwater regions, and the role of rivers as drivers of speciation in Amazonian birds. Evolution (July) 69:1823–1834. https://doi.org/10.1111/evo.12696

Wiwa K (2005) Ken Wiwa: committed to remembering. BBC News. Available via: http://news.bbc.co.uk/2/hi/africa/4421188.stm

Yasmi Y, Kelley L, Enters T (2010) Conflict over forests and land in Asia. Impacts, causes, and management. The Center for People and Forests (RECOFT), Bangkok

Chapter 8
The Changing Role of Europe in Past and Future Alien Species Displacement

Bernd Lenzner, Franz Essl, and Hanno Seebens

Abstract Human activity has resulted in a massive reshuffling of the world's biota by introducing species into regions outside their native range worldwide. Alien species introduction leads to the breakdown of biogeographic barriers, thereby promoting a homogenization of the world's biota. The observed pattern of alien species distributions today is a result of past connectivity of different regions of the world through, for example, trade, human migration, and political affiliation (e.g. historic empires), all of which have changed in time. A historical perspective on human activity is therefore essential to understand the processes underlying biotic homogenization. During the fifteenth to nineteenth century, growing processes of global trade of commodities occurred between Europe and North America and Europe and Southeast Asia. The colonization of the North American continent led to a strongly directed introduction pattern of species with European origin; however, towards the end of this period, many colonies in temperate regions (e.g. Australia, New Zealand) received increasing numbers of alien species. In the twentieth century, the world markets and societies moved closer together due to a further increase in the global connectivity of countries, which resulted in an intensification of biotic homogenization worldwide. Changing political and economic situations of countries caused continuous changes in trade and migration networks. These changes in connectivity have profound effects on the displacement of species around the globe. In this chapter, we investigate the role of Europe in historical times as a central agent in global species displacement. Subsequently, we discuss how this role changed recently and will likely change in the future due to dynamics in the global economy and changes in the importance of countries as key players in a globally interconnected world.

B. Lenzner (✉) · F. Essl
Division of Conservation Biology, Landscape & Vegetation Ecology, University of Vienna, Vienna, Austria
e-mail: Bernd.lenzner@univie.ac.at; franz.essl@univie.ac.at

H. Seebens
Senckenberg Biodiversity and Climate Research Centre (SBiK-F), Frankfurt am Main, Germany
e-mail: hanno.seebens@senckenberg.de

R. Rozzi et al. (eds.), *From Biocultural Homogenization to Biocultural Conservation*, Ecology and Ethics 3, https://doi.org/10.1007/978-3-319-99513-7_8

Keywords Alien species · Biological homogenization · Global trade · Empires · Future trends · Global connectivity

8.1 Introduction

The anthropogenic transport and subsequent spread of alien species into regions beyond the reach of natural colonization contributes strongly to biological homogenization (Olden and Rooney 2006; Winter et al. 2009). A considerable share of the world's biota has already established in regions outside their native ranges due to human-mediated transportation (e.g. 4% of all vascular plant species, 6% of birds, 2% of reptiles and 1.1% of amphibians worldwide) (Cassey et al. 2015; van Kleunen et al. 2015; Capinha et al. 2017). The introduction of alien species is driven by various factors associated with global trade and transport and intentional introductions of species for ornamental and utilitarian purposes being among the most important ones (Levine and D'Antonio 2003; Hulme 2009; Seebens et al. 2015). Once introduced, establishment of alien species is facilitated by habitat transformation, infrastructure development and biodiversity degradation in the recipient region (Kennedy et al. 2002; Theoharides and Dukes 2007; Hulme et al. 2008). All these processes have constantly changed during the last centuries, and it is likely that they continue to change in the future.

Europe has been assumed to be one of the major donor regions for species that have become alien elsewhere during the last centuries particularly due to its politically and economically dominant role in colonial history (Di Castri 1989). This assumption has recently been supported for global established plants, showing that a disproportionately large amount of alien plant species worldwide originated from Europe: Europe's contribution to the global pool of alien plant species is 288% higher than expected compared to 57% for North America and 52% for temperate Asia (van Kleunen et al. 2015). However, in an era of increasing socio-economic globalization, shifting international trade relations and transport routes, this pattern will likely change in the future with emerging economies becoming more important for the global displacement of species. In this chapter, we discuss the contribution of Europe to global biological homogenization driven by the anthropogenic exchange of species. We also discuss changes in the major global socio-economic drivers that underlie alien species displacement and how they might affect the role of Europe as a donor of alien species in the future.

8.2 The Role of Europe in Historic Species Exchanges

The sun never sets on the empire of the dandelion. Edgar Anderson (1954, p. 19)

Early species introductions were particularly prominent in regions with long-standing agricultural activity such as Europe and Western and Central Asia (Di

Castri 1989). In these regions, large ancient empires such as the Persian Empire at around 500 BCE were among the first to establish far-reaching trade networks, which increased regional connectivity, but also the exchange of species between central Asia and the Mediterranean region. Besides trade, human migration emerged as a vector for alien species spread with the migration of people often accompanied by species of practical use (e.g. crops, livestock), their pests, or with strong associations with humans (e.g. house mouse *Mus musculus* sp., Sax and Brown 2000; Cucchi et al. 2005). However, travel and transportation at that time were cumbersome and slow, resulting in a low rate of species exchanges. The subsequent rise of the Roman Empire led to an intensification of human migration and associated species introductions particularly towards Central and Northern Europe (Di Castri 1989). Indeed, some of the most widely distributed alien species in Europe (e.g. sweet chestnut or common pheasant) were introduced by the Romans (Zeuner 1963; Conedera et al. 2004).

The European discovery of the Americas in 1492 and the emergence of a global trade network mark the onset of a new era: the era of globalization. In subsequent centuries, European countries expanded politically and economically into new regions, often accompanied by discoverers and scientists who systematically explored the world. The establishment of livestock in these regions was essential for European explorers even if they did not intend to immediately form settlements. For instance, on islands livestock species such as goats and sheep were introduced to establish viable populations to ensure food supply for later expeditions (Crosby 1972). Once a region was occupied by Europeans, the cultivation of European crops in that region was crucial for sustaining the first settlements. In addition to the introduction of domestic species, Europeans unintentionally introduced pathogens to the new environment causing severe epidemic outbreaks, which often played an important role for the success of European colonization (Diamond 2003). In essence, the introduction of alien livestock, crops and their associated pathogens was crucial for the success of European expansion (Beinart and Middleton 2004). The legacy of these past introductions of species is still apparent in the wide distribution of European (domestic) species globally.

The successive intensification of trade relations worldwide led to the establishment of a truly global trade network and marked a transition from regionally connected societies to a globally interconnected world. The increasing connectivity of continental regions resulted in a breakdown of natural dispersal barriers for many species. Colonial empires played a crucial role in global species exchanges by extending their administrative area through integration of their newly acquired regions into this global trade network (Fig. 8.1).

Regions, which were part of a colonial empire between 1870 and 1913, experienced an increase in trade by up to 270% compared to non-colonized regions (Mitchner and Weidenmier 2008). Likewise, the colonies received disproportionately many species via intentional (i.e. crops or livestock) and unintentional (i.e. via seed contamination or as stowaways) introduction (Beinart and Middleton 2004). To ensure market stability, trade agreements led to preferential trade within the boundaries of the empire (Mitchner and Weidenmier 2008), which further fostered

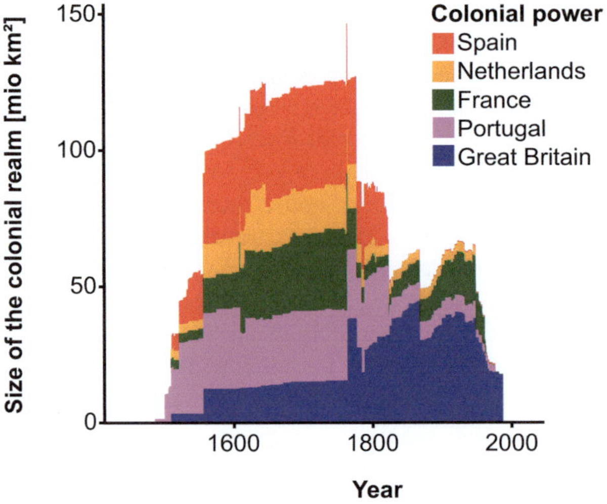

Fig. 8.1 Illustration of the temporal change in the extent of the large European colonial empires between 1444 and 1997. Bars are stacked to provide the total amount of area of global terrestrial surface under European occupation (unpublished data). Regions have been assigned as belonging to an empire once they had been politically annexed or when first permanent settlements by colonial empires were established

the spread of alien species within the range of an empire. Further major global migration movements were directed from Europe to North America resulting in more alien species transported in that direction (Jeschke and Strayer 2007). All of this led to a continuous intensification of global exchanges of alien species during 1500–1800 ("Columbian exchange"; Crosby 1972; Nunn and Qian 2010) with the majority of species introduced to the New World. Consequently, the dominance of European empires, coupled with scientific curiosity and a utilitarian self-perception resulted in a preferential expansion of native European species across the world (Crosby 1972). At that time, Europe constituted by far the most important donor region for naturalized alien species (Di Castri 1989).

In the nineteenth and early twentieth centuries, the global exchange of alien species further intensified also due to the foundation of European societies and institutions, such as botanical gardens and Acclimatization Societies. These institutions were formed for the purpose to collect and describe global biodiversity and to exchange plants, plant material, and animals for scientific inquiry, aesthetic pleasure and economic purpose (e.g. crop development and medicinal usage; Brockway 1979). These institutions played a central role for the intentional global exchange of plant materials and life specimens (Brockway 1979). Among the most prominent institutions were the Royal Botanical Gardens Kew in London, the Buitenzorg Botanical Garden in Java and the Société Zoologique d'Acclimatation in Paris. The British Botanical Garden Network counted over 100 botanical gardens around 1900, and at roughly the same time, the number of Acclimatization Societies peaked with around 50 societies in the British Empire alone (Osborne 2000). In addition, Acclimatization Societies aimed to improve agriculture and to "enrich" native communities for the leisure of European settlers in overseas territories by acclimatizing European species to the new and generally perceived as less valuable ecosystems

(Osborne 2000). At the same time, Europeans became more and more fascinated by exotic species, particularly plants. This resulted in a boom in horticulture with many new plantations of alien species in parks and botanical gardens. The increasing demand of new species for the European market led to the emergence of new companies and so-called "plant hunters" seeking for attractive plants worldwide (Musgrave et al. 1999; Stoner and Hummer 2007). This was further complemented by the increasing role of North America in the world. Reduced immigration from Europe and increasing immigration from North America to mainly Europe added to the changing pattern of human migration-mediated alien species distribution (Jeschke and Strayer 2007). In summary, the nineteenth century represents the onset of a shift from Europe being mostly an exporter of alien species to a more balanced global exchange of alien species (Seebens et al. 2015).

8.3 Shifting Trade Patterns in Times of Modern Globalization

With the end of Europe-centred empires after World War II, the dominance of non-European political and economic superpowers and the advancing economic globalization of trade have dramatically increased as has connectivity between regions. Vital for the reconfiguration of global trade was the formation of international economic institutions (e.g. the World Bank) and the establishment of multilateral political and economic agreements (e.g. General Agreement on Tariffs and Trade; World Trade Organization 2013). Preferential trade agreements to ensure market increase and stability were put in place at an increasing pace over the past from 70 in 1990 to around 300 by 2010 (World Trade Organization 2011, 2013). Consequently, global trade value increased by 8.2% between 1980 and 2011. However, international trade developed unequally across countries with an increase from 8% to 21% in developing and a decrease from 56% to 36% in developed countries between 1990 and 2011 (World Trade Organization 2011, 2013).

This restructuring of the global economy, the decentralization of global trade and the establishment of bilateral agreements have led to a changing importance of regions as donors and recipients in the global displacement of alien species and will continue to do so. Especially emerging economies in Asia and South America are projected to gain more established plant species per trade value compared to developed countries in the coming decades as a result of their increasing importance in global markets (Fig. 8.2, Seebens et al. 2015). Hence, in the light of changes in international trade networks and related effects like infrastructure development, the exploitation of new markets (e.g. in the global pet trade) and the transnational acquisition of land for agriculture and forestry goods production, donor and recipient regions for alien species will likely change in the future.

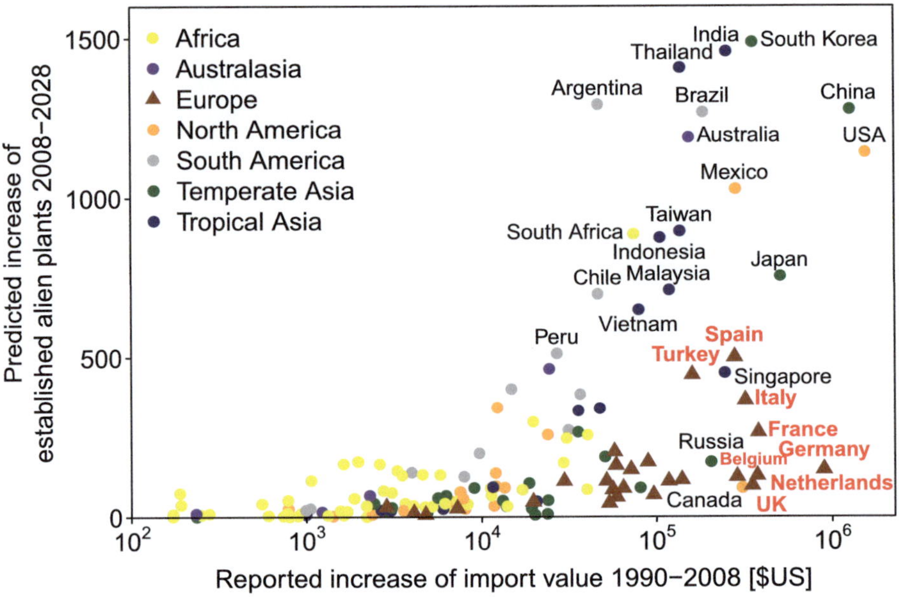

Fig. 8.2 Predicted increase in established plant species numbers between 2008 and 2028. Accounting for an estimated time lag of around 20 years between the introduction of a species and its record as established in the wild, the increase in trade value during 1990–2008 is assumed to result in different increases in alien species numbers across countries in the future. The highest established plant species numbers are predicted for developing economies with largest increases in import values. In contrast, European countries (large triangles and red-coloured names) with similar or higher increases in trade values are expected to receive fewer numbers of established plant species. (The figure is adopted and customized from Seebens et al. 2015)

8.4 Effects of Changing Economic Patterns on Alien Species Displacement

Growth in economic importance of developing regions is accompanied by a variety of developments affecting alien species establishment. Economic prosperity changes regional consumption patterns as wealth and well-being grow. The rise in consumption leads to increased pressure on local lands to meet rising demands, especially for agricultural products but also for housing and infrastructure. This increase in land transformation elevates the amount of entry hubs for alien species as disturbed areas are generally more prone to invasion compared to intact nature. Additionally, the number of introduced individuals and species in these systems is often high due to intense human activity, thereby increasing the probability of new species establishment (Sax and Brown 2000).

While this is important within a country, more recently land acquisitions outside the territories of each nation have become crucial to cover the needs for agricultural and forestry goods. Generally, this practice was pursued by developed countries located on temperate continents (mainly Europe with 287 investments encompassing

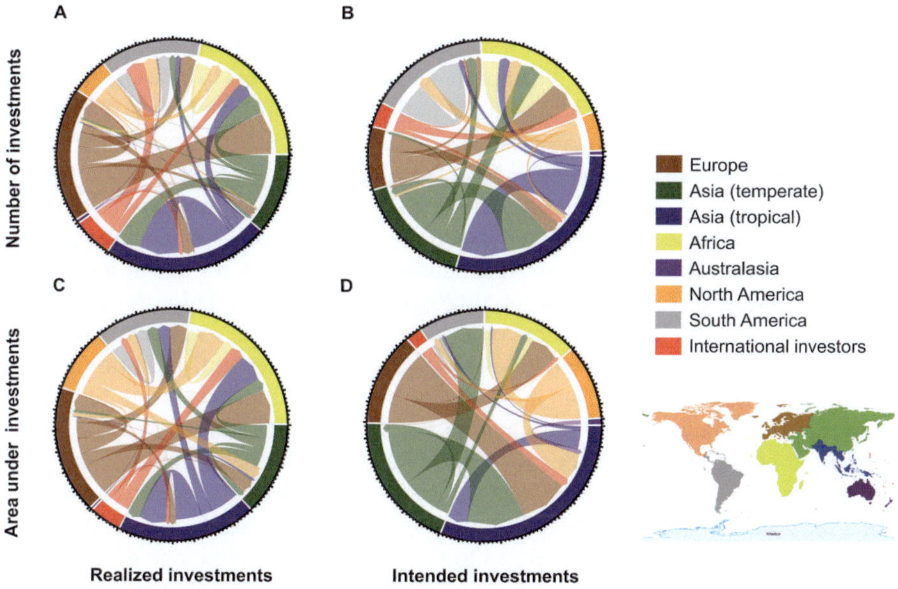

Fig. 8.3 (**a**, **b**) Numbers of transnational investment agreements and (**d**, **e**) amount of area [ha] under investment with distinction between realized investments (**a**, **c**) and future intended investment only (**b**, **d**). Regions were grouped following the TDWG continental classification (see insert) according to the origin of the investor country (or countries if multiple investors all originate from the same region). When investments are made by multiple investors from different continents, they are classified as "international investors". The arrowheads and the white space between the outer circle and the flow indicate the direction of investment. (Data are obtained from the Land Matrix project www.landmatrix.org)

9,444,298 ha, Fig. 8.3a, c) that predominantly invested in countries located in tropical continents. However, more recently developing economies (mainly temperate Asian countries with 34 planned investments encompassing 2,499,054 ha, Fig. 8.3b, d) have started to preferentially invest outside their own territories, resulting in land-use intensification and land cover change (Messerli et al. 2014).

Transnational land purchase has thus emerged as an important new feature in globalized land use, with an estimated 70.2 million ha of agricultural land area (this is four times the agricultural area of Germany) bought by companies outside their home region since 2000 (Scherer 2012). These large-scale land acquisitions have been shown to be mainly directed towards the global south (Messerli et al. 2014); the contracted area will likely increase in the future with roughly 450 million ha of land being estimated as potentially suitable for foreign investment (Deininger and Byerlee 2011). Consequently, the likely increase of bilateral exchange of agricultural commodities between country of investment and investor country will have intensifying ramifications for alien species exchanges.

The changing powers in the global economy will inevitably alter trade networks due to preferential trade agreements, the exploitation of new production areas, and an increased globalization of product supply chains (World Trade Organization 2013). In the last decades, global trade has already shifted more strongly towards Asia, thereby reducing the relative importance of Europe (World Trade Organization 2013). Changes in trade relations will affect the importance of existing sea and air traffic routes in terms of quantity and frequency of travels (e.g. extension of the Panama Canal or construction of the Nicaragua Canal; Muirhead et al. 2015; Haerer et al. 2017). Further, under future climate change, new trade routes will emerge. With the estimated sea ice decline in the Arctic over the twenty-first century, trans-Arctic shipping routes will open up, reducing travel times for Asian and North American shipping routes (Melia et al. 2016). Faster transportation times, higher trade frequency, and higher volume of exchanged commodities will severely affect trade-related introduction of alien species as more species will be introduced at higher quantities with a reduced mortality during transportation (Hulme 2009).

Besides the unintentional introduction via exchanged commodities, one of the main reasons for species introductions worldwide is for aesthetic, cultural, and utilitarian purposes (e.g. horticulture and pet trade). Facilitated by international trade and global market accessibility (e.g. Internet trade; Bush et al. 2014), wildlife trade increased in importance and is related to the level of socio-economic development of a region (Baker et al. 2013; Bush et al. 2014). In addition, illegal wildlife trade is flourishing with roughly 200,000 live animal species under trade embargo by the Convention of International Trade in Endangered Species of Flora and Fauna (CITES) being seized between 1996 and 2008 (D'Cruze and Macdonald 2016). Global trade patterns in the pet, wildlife, and horticultural trade are more ambiguous and demand driven. While mammals are mainly exported from African countries, most reptiles originate from North America and Europe, and international bird trade flourishes mainly between South America and Asia (Bush et al. 2014). In animals, deliberate introductions and releases (e.g. in birds for ceremonial purposes, Su et al. 2015) lead to higher establishment probabilities, whereas plants often naturalize as escapees from private or botanical gardens (Hulme 2011).

8.5 Final Remarks

Historically, European institutions have dominated the global spread of biota. However, macroeconomic and geopolitical transformations have strongly changed the role of different continents as donor and recipient regions for alien species and continue to do so for the foreseeable future. Globalization of trade and transport lead to a shift in the relative importance of countries in the displacement of species towards developing economies in South America and Asia. However, as the world becomes more interconnected and international relations become more complex, it will become less clear which regions will be the major donor and recipient regions of alien species. Lower predicted increases of alien species in developed countries might be a result of more strict regulations in species import and exports as well as the enforcement of laws and other legal requirements regulating live species and plant material trade. Developing countries often lack such regulations, or, if in place, they are not adequately enforced by regional authorities (Bush et al. 2014; Essl et al. 2015; Hulme 2015). However, at the same time, developing nations provide different opportunities for effective alien species management. For example, an increase in societal awareness, as peoples' livelihood often directly depends on natural resources, combined with lower labour costs likely increases future implementation effectiveness for management strategies (Nuñez and Pauchard 2010). Overall, to reduce uncontrolled species introductions and an advancing homogenization of the world biota, it becomes even more important to foster global collaboration and knowledge exchange among nations. It is critical to simultaneously and jointly develop and enforce globally binding standards and protocols to prevent introductions and to develop detection and early warning mechanisms to prevent species establishment after successful introduction.

References

Anderson E (1954) Plants, man and life. Andrew Melrose, London

Baker SE, Cain R, Kesteren FV et al (2013) Rough trade. Bioscience 63:928–938. https://doi.org/10.1525/bio.2013.63.12.6

Beinart W, Middleton K (2004) Plant transfers in historical perspective: a review article. Environ Hist Camb 10:3–29. https://doi.org/10.2307/20723465

Brockway LH (1979) Science and colonial expansion: the role of the British Royal Botanic Gardens. Am Ethnol 6:449–465. https://doi.org/10.1016/0378-8741(81)90039-8

Bush ER, Baker SE, Macdonald DW (2014) Global trade in exotic pets 2006–2012. Conserv Biol 28:663–676. https://doi.org/10.1111/cobi.12240

Capinha C, Seebens H, Cassey P et al (2017) Diversity, biogeography and the global flows of alien amphibians and reptiles. (in press)

Cassey P, Vall-Llosera Camps M, Dyer E, Blackburn TM (2015) The biogeography of avian invasions: history, accident and market trade. In: Biological invasions in changing ecosystems: vectors, ecological impacts, management and predictions. De Gruyter Open, Warsaw/Berlin, pp 37–54. https://doi.org/10.1515/9783110438666-006

Conedera M, Krebs P, Tinner W et al (2004) The cultivation of Castanea sativa (Mill.) in Europe, from its origin to its diffusion on a continental scale. Veg Hist Archaeobotany 13:161–179. https://doi.org/10.1007/s00334-004-0038-7

Crosby AW (1972) The Columbian exchange: biological and cultural consequences of 1492. Recording for the Blind & Dyslexic, Princeton

Cucchi T, Vigne JD, Auffray JC (2005) First occurrence of the house mouse (Mus musculus domesticus Schwarz & Schwarz, 1943) in the Western Mediterranean: a zooarchaeological revision of subfossil occurrences. Biol J Linn Soc 84:429–445. https://doi.org/10.1111/j.1095-8312.2005.00445.x

D'Cruze N, Macdonald DW (2016) A review of global trends in CITES live wildlife confiscations. Nat Conserv 15:47–63. https://doi.org/10.3897/natureconservation.15.10005

Deininger K, Byerlee D (2011) Rising global interest in farmland. World Bank, Washington, DC

Di Castri F (1989) History of biological invasions with special emphasis on the old world. In: Drake JA, Mooney HA, di Castri F, Groves RH, Kruger F, Rejmánek M et al (eds) Biological invasions: a global perspective. Wiley, Chichester, pp 1–30

Diamond J (2003) Guns, germs, and steel in 2003. W. W. Norton & Company, New York

Essl F, Bacher S, Blackburn TM et al (2015) Crossing frontiers in tackling pathways of biological invasions. Bioscience 65:769–782. https://doi.org/10.1093/biosci/biv082

Haerer A, Torres-Dowdall J, Meyer A (2017) The imperiled fish fauna in the Nicaragua Canal zone. Conserv Biol 31:86–95. https://doi.org/10.1111/cobi.12768

Hulme PE (2009) Trade, transport and trouble: managing invasive species pathways in an era of globalization. J Appl Ecol 46:10–18. https://doi.org/10.1111/j.1365-2664.2008.01600.x

Hulme PE (2011) Addressing the threat to biodiversity from botanic gardens. Trends Ecol Evol 26:168–174. https://doi.org/10.1016/j.tree.2011.01.005

Hulme PE (2015) Invasion pathways at a crossroad: policy and research challenges for managing alien species introductions. J Appl Ecol 52:1418–1424. https://doi.org/10.1111/1365-2664.12470

Hulme PE, Bacher S, Kenis M et al (2008) Grasping at the routes of biological invasions: a framework for integrating pathways into policy. J Appl Ecol 45:403–414. https://doi.org/10.1111/j.1365-2664.2007.01442.x

Jeschke JM, Strayer DL (2007) Invasion success of vertebrates in Europe and North America. Proc Natl Acad Sci U S A 102:7198–7202. https://doi.org/10.1073/pnas.0501271102

Kennedy TA, Naeem S, Howe KM et al (2002) Biodiversity as a barrier to ecological invasion. Nature 417:636–638. https://doi.org/10.1038/nature00776

Levine JM, D'Antonio CM (2003) Forecasting biological invasions with increasing international trade. Conserv Biol 17:322–326. https://doi.org/10.1046/j.1523-1739.2003.02038.x

Melia N, Haines K, Hawkins E (2016) Sea ice decline and 21st century trans-Arctic shipping routes. Geophys Res Lett 43:9720–9728. https://doi.org/10.1002/2016GL069315

Messerli P, Giger M, Dwyer MB et al (2014) The geography of large-scale land acquisitions: analysing socio-ecological patterns of target contexts in the global South. Appl Geogr 53:449–459. https://doi.org/10.1016/j.apgeog.2014.07.005

Mitchner KJ, Weidenmier M (2008) Trade and empire. Econ J 118:1805–1834

Muirhead JR, Minton MS, Miller WA, Ruiz GM (2015) Projected effects of the Panama Canal expansion on shipping traffic and biological invasions. Divers Distrib 21:75–87. https://doi.org/10.1111/ddi.12260

Musgrave T, Gardner C, Musgrave W (1999) The plant hunters two hundred years of adventure and discovery. Seven Dials, London

Nuñez MA, Pauchard A (2010) Biological invasions in developing and developed countries: does one model fit all. Biol Invasions 12:707–714. https://doi.org/10.1007/s10530-009-9517-1

Nunn N, Qian N (2010) The Columbian exchange: a history of disease, food, and ideas. J Econ Perspect 24:163–188. https://doi.org/10.1257/jep.24.2.163

Olden JD, Rooney TP (2006) On defining and quantifying biotic homogenization. Glob Ecol Biogeogr 15:113–120. https://doi.org/10.1111/j.1466-822X.2006.00214.x

Osborne MA (2000) Acclimatizing the world: a history of the paradigmatic colonial science. Osiris 15:135–151. https://doi.org/10.1086/649323

Sax DF, Brown JH (2000) The paradox of invasion. Glob Ecol Biogeogr 9:363–371. https://doi.org/10.1046/j.1365-2699.2000.00217.x

Scherer C (2012) Foreign investment in agricultural land down from 2009 peak. Vital Signs 20:74–77

Seebens H, Essl F, Dawson W et al (2015) Global trade will accelerate plant invasions in emerging economies under climate change. Glob Chang Biol 21:4128–4140. https://doi.org/10.1111/gcb.13021

Stoner A, Hummer K (2007) 19th and 20th century plant hunters. Hortscience 42:197–199

Su S, Cassey P, Vall-Llosera M, Blackburn TM (2015) Going cheap: determinants of bird price in the Taiwanese pet market. PLoS One 10:1–17. https://doi.org/10.1371/journal.pone.0127482

Theoharides KA, Dukes JS (2007) Plant invasion across space and time: factors affecting nonindigenous species success during four stages of invasion. New Phytol 176:256–273. https://doi.org/10.1111/j.1469-8137.2007.02207.x

van Kleunen M, Dawson W, Essl F et al (2015) Global exchange and accumulation of non-native plants. Nature 525:100–103. https://doi.org/10.1038/nature14910

Winter M, Schweiger O, Klotz S et al (2009) Plant extinctions and introductions lead to phylogenetic and taxonomic homogenization of the European flora. Proc Natl Acad Sci U S A 106:21721–21725. https://doi.org/10.1073/pnas.0907088106

World Trade Organization (2011) World trade report 2011 – the WTO and preferential trade agreements: from co-existence to coherence

World Trade Organization (2013) World trade report 2013 – dactors shaping the future of world trade

Zeuner FE (1963) A history of domesticated animals. Hutchinson & Co. (Publishers) Ltd, London

Chapter 9
Dürer's Rhinoceros: Biocultural Homogenization of the Visual Construction of Nature

José Miguel Esteban

Abstract In this paper I will try to show that the printing press was forming a visual culture that uniformized the construction of images of the fauna discovered in Africa, Asia, and the Americas during the Renaissance. Isolated from its background or ecological context, the figure of the animal, unattached and floating, absorbed a symbolic load that assimilates it to other images constructed according to that visual culture. In this chapter we will see how the figure of Dürer's rhino absorbed the symbolic load of the visual culture of Renaissance colonialism. The warlike attributes of the printed image incorporated in the Indian rhinoceros the epic and military dimension of the colonial adventure. The visual construction of Dürer could very well represent the cultural homogenization of the biotas east and west of Europe. The pictorial construction of the otherness of exotic animals reaffirmed the beneficial exceptionalism of Europe and, consequently, reinforced the legitimacy of Western colonization of a wild and alien nature, waiting to be reduced and converted into merchandise. The history of the numerous reprints of Dürer's rhino reproduces the biocultural consequences of positive feedback between processes such as representing, conquering, and commodifying nature. Finally, I present Dalí's rhinoceros as a reference to the quantitative homogenization of the images of nature and culture. To conclude, I conclude that one of the challenges of biocultural conservation is to denounce the construction of homogeneous biocultural habitats based on habits such as visual production and the consumption of images.

Keywords Art · Biocultural ethics · Biocultural conservation · Colonialism · Eurocentrism

---

J. Miguel Esteban (✉)
Facultad de Filosofía, Universidad Autónoma del Estado de Querétaro, Santiago de Querétaro, Mexico

R. Rozzi et al. (eds.), *From Biocultural Homogenization to Biocultural Conservation*, Ecology and Ethics 3, https://doi.org/10.1007/978-3-319-99513-7_9

9.1 Introduction

In 1958, Charles Elton, a father of the ecology of invasive species, predicted that in the long run, the biota of all continents would become increasingly similar (Elton 1958: p. 51). In 2016, Elizabeth Kolbert warned that this would happen sooner rather than later. Today, global processes seem to drive us backward through geological history at *full speed*, operating as a kind of inverse tectonics that impoverish and homogenize the biological diversity of the planet (Kolbert 2016: p. 22). The reports of the Anthropocene Working Group of the International Union of Geological Sciences reinforce that warning. The stratigraphy of the Anthropocene will be characterized by biocultural homogeneity, with an overwhelming predominance of plastic residues and chicken bones (Carrington 2016).

As with many other systemic processes, planetary biocultural homogenization (*sensu* Rozzi 2013) is not growing in lineal but rather exponential progression. In this I address one of its possible points of inflection, the Renaissance. In this historical period, biological homogenization received positive feedback from ascendant cultural processes such as economic monetarization, expanding interurban commerce, concentration of wealth and the incipient creation of financial services, strengthened military power of monarchs and nobles, and the technological renewal of the sciences, arts, and trades. The biocultural colonization of the world also was propelled by the printing press, which unleashed a new visual culture that revolved around the graphic icon of the globe.

The British anthropologist Tim Ingold argues that the move from the notion of a spherical cosmos, characteristic of non-modern societies, to that of a global cosmos, had deep cultural implications. "[T]he lifeworld, imagined from an experiential centre, is spherical in form, whereas a world divorced from life, that is yet complete in in itself, is imagined as a globe," he explains (Ingold 2000: p. 210). This external perspective reduces the properties of places to those that allow its cylindrical projection onto a Euclidean plane. The Mercator chart is a uniform cylindrical projection that retains the angles and shapes but deforms and sacrifices distances and surfaces. Mercator's map is not a neutral reflection of nature, but projects the world's availability for a linear, homogeneous, and constant navigation that maximizes the arrivals to port in a process of colonial expansion. Spheres are experienced from within; globes can be perceived only from without. For Ingold, the image of the earthly globe supposes a kind of cosmic exile, in which some parts of humanity imagine themselves as separated from the world in order to *contemplate* it from without, putting in parenthesis the worldly relationships of interdependence that the spherical notion required. The image of the globe leads to the objectification of the world as homogeneous and universally available extension that can be controlled by meridians and parallels.

At the beginning of the sixteenth century, two woodcuts by Albrecht Dürer, prepared at the request of Stebius, the official geographer of Maximilian I of Habsburg, seem to corroborate this thesis. Dürer located in this cosmic exile a muse and four great figures in the history of astronomy. "Urania the Muse of Astronomy"

(c. 1502) (Fig. 9.1) portrays Urania, a daughter of Zeus and the Titaness Mnemosyne, who was said to have inspired curiosity in people about the firmament and usually was represented with a light blue cloak, a diadem of stars, a globe, and a compass. Dürer simplifies these aesthetic attributes, presenting her as a powerful, nude woman who holds in her hands a circumference of the zodiacal globe that seems to rotate

Fig. 9.1 Albrecht Dürer, "The Muse Urania with the Zodiac" (c. 1502). Woodcut. (Current location: Staatliche Graphische Sammlung München/Public Domain {PD-art})

thanks to the muse herself. Along the equator is a band containing twelve astrological symbols. In the interior of the zodiacal globe, the terrestrial globe appears as segmented by parallels and meridians. Together they are similar to an armillary sphere, a celestial sphere that was employed from the time of Ptolemy to determine star coordinates and to show their apparent movement in relation to the Earth and the Sun. The armillary sphere is part of the coat of arms adorning the flag of Portugal. The idea of an external imperial power over the globe is reinforced by three bundles of fine lines that represent the supposed influence of the Zodiac on our planet.

The strict relationship between Renaissance astrology and astronomy characterizes this historical period as an interregnum or "in-between" stage. "Map of the Northern Sky" (Fig. 9.2) (c 1515) shows the celestial globe of the northern

Fig. 9.2 Albrecht Dürer, "Map of the Northern Sky" (c. 1515). Woodcut. (Source: National Gallery of Art/Public Domain {PD-art})

hemisphere with the positions of the stars around 1499–1500. The hemisphere is surrounded by Zodiac signs. Each of the four corners of Dürer's woodcut is occupied: an astronomer who, although outside the celestial globe, holds in his hands a smaller sphere that represents the celestial globe.

Like Urania and the astronomers, Europeans saw themselves as "outside the globe" although holding it in their hands. Like Urania, they could manipulate it and like the astronomers contemplate it from outside, thus, as Ingold says, separating themselves from ecological interdependencies. The world understood as a globe could be universally homogenized.

Maps and muses are not the only images of the period that are marked by cultural biases. Renaissance images of animals share these cultural biases and represent what the US art historian Janice Neri (2011) calls the *logic of the specimen.* According to this logic, the figure of a biological organism appears isolated on a uniform background, empty or reduced to schematic representations, which permit geometric techniques of perspective. Two artistic renderings, both following Dürer's previous works, illustrate this logic of the specimen. In "The Small Horse" by Franz Isaac Brun (1550–1610) (Fig. 9.3), all socioecological context is removed, and grids and geometric techniques are emphasized to create the optical illusion of depth and tridimensionality. This mathematical and geometrical vision might symbolize the imposition of rational control over animal behavior. In the "Beetle" by Hans Hoffmann (1574) (Fig. 9.4), the insect is illustrated as though in motion over a cream-colored oval superimposed on a gray background. The beetle is completely removed from any ecological context.

I endeavor to show that, thanks to the printing press, the logic of the specimen facilitated a uniform visual culture that governed the construction of homogeneous images of colonial biota. In the first place, the graphic reproduction of exotic animal *figures* on a neutral and homogenous *background* favored the mental construction of habitats as undifferentiated receptacles, functionally empty, that could be occupied by any organism, *ad libitum.* On the other hand, by isolating it from its context, the image of the animal, unanchored and floating, made it possible to join it to other images constructed according to that same visual logic. Consequently, this Renaissance artistic custom homogenizes not only the representation of habitats but also the inhabitants of the distinct regions of the planet. In this way emerged the pervasive root of biocultural homogenization, conceptualized by Rozzi (2012) as the homogenization of life habits, habitats, and identities of coinhabitants. Note that, also, following the logic of the specimen, individuals are represented as isolated inhabitants, not as coinhabitants that interact and coinhabit their respective regions (cfr. Rozzi 2013, 2015).

This interpretation can be illustrated by a notable example of the logic of the specimen and the root of Renaissance biocultural homogenization: Dürer's "Rhinoceros" (Fig. 9.5). This woodcut by the German painter and printmaker symbolically represents the visual culture of European colonialism. This image was printed on all types of backgrounds, from flat and empty planes to Chinese, European, or tropical landscapes. For example, the Dutch tapestry artist Pedro Van Elst (1549) inserted it into a forest capable of accommodating the fauna of any

Fig. 9.3 Franz Isaac Brun (after Albrecht Dürer), "The Small Horse" (1550–1610). Engraving. © The Trustees of the British Museum

continent, from Asiatic elephants to African camels, even including a marine crustacean (Fig. 9.6). The Dutch artist Willem Goeree (1689) shows it disembarking the biblical ark, together with unicorns and South American armadillos (Fig. 9.7). Both animals are armored mammals, with external defenses made of bony plates – the reason Plinio, Dürer, and De Huerta related them with the turtles. Goeree's inclusion of the armadillo apparently was inspired by the German Jesuit Athanasius

Fig. 9.4 Hans Hoffmann (after Dürer), "Beetle" (1574). Watercolor on paper. Public Domain {PD-art}

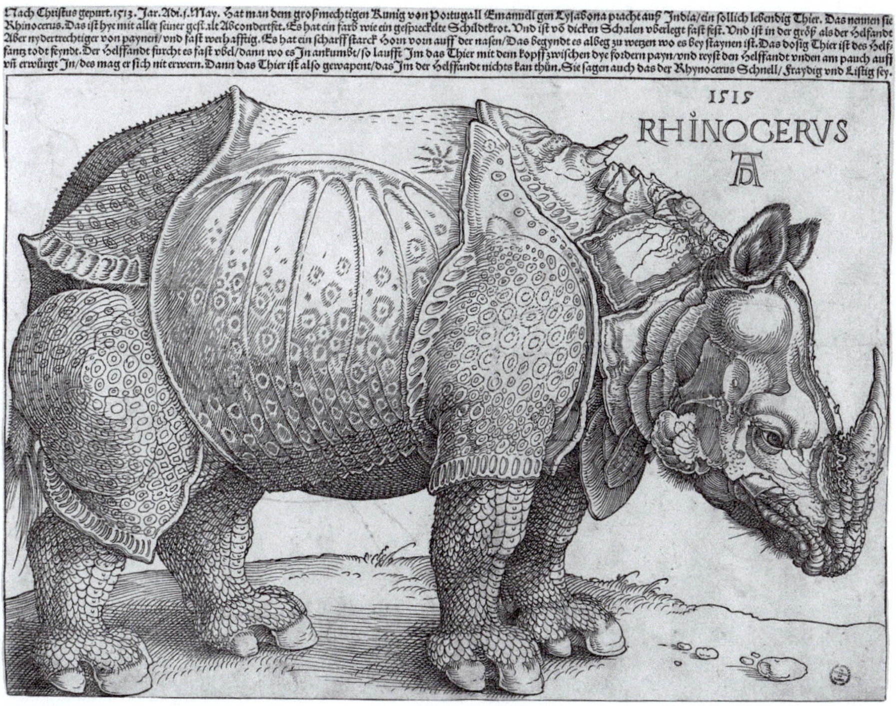

Fig. 9.5 Albrecht Dürer, "Rhinoceros" (1515). Xylograph engraving. Collection National Gallery of Art/Public Domain {PD-art}

Fig. 9.6 Peter van Elst, "Animals in a wood, including an elephant, rhinoceros, monkeys, camels and a lobster" (1549). Pen and brown ink and gray wash, squared for transfer. Tapestry. © The Trustees of the British Museum

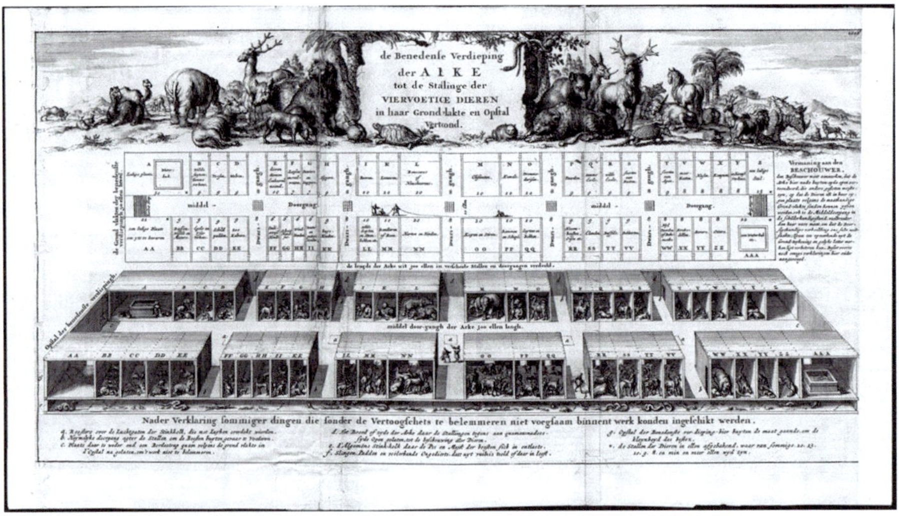

Fig. 9.7 Willem Goeree, "Noah's Ark" (1689). Engraving on paper. The engraving formed part of a series dedicated to the ark and its construction in *Introduction into Biblical Wisdom and the Use of the Holy and Ecclesiastical Histories* by Willem Goeree and Jan Luyken, his Dutch engraver (Amsterdam 1689), about the presence of New World animals in the ark. (Courtesy of the Amsterdam Museum/Public Domain {PD-Art})

Kircher who put both Dürer's rhinoceros and the armadillo into Noah's ark in a 1675 book. His doing so, however, betrays deep cultural Eurocentrism. Kircher believed that after the biblical deluge, life forms entered into a process of irreversible decomposition. Animals such as the armadillo were the degenerate products of hybridization that occurred as the animals that had been on the ark immigrated into various parts of the world. According to Kircher, the animals of the New Indies were products of corruptions and aberrations of the pristine fauna of the biblical earth, so the Mexican armadillo is the result of the cross between turtle and hedgehog. The German Jesuit never touched the American continent, of course. His image of the Mexican armadillo actually comes from the representation of one of the Brazilian species that another German naturalist, Georg Marcgraf, who explored Brazil in the seventeenth century, captured in his *Historia Naturalis Brasiliae* (1648). We know little of Willem Goeree, but it seems obvious that his inclusion of the armadillo among the passengers of the ark implicitly defies the degenerationist theory of Kircher. Visual representations of the logic of the specimen inevitably were culturally conditioned.

These representations are not limited to the Renaissance. In the final part of this chapter, I analyze a contemporary version of Dürer's "Rhinoceros," the sculpture by Salvador Dalí, another graphic representation of the rationalization – homogenization – of nature. Dürer's "Rhinoceros" and its many reproductions illustrate the ecological consequences of the cultural processes of representing, conquering, and commodifying nature. This rhinoceros condenses visually what Canadian philosopher Herbert Marshall McLuhan called in the mid-twentieth century "The Gutenberg Galaxy," a period crucial in the gestation of biocultural homogenizing processes that have affected all continents.

Although it is difficult to classify a scholar such as McLuhan, the truth is that his work is much broader than the slogan that made him popular in the 1960s and 1970s: "the medium is the message." In my opinion, McLuhan is one of the first humanists of the twentieth century who practiced an interdisciplinarity without being ashamed. He started as an engineer, studied classical literature, and was able to build his arguments nourished by philosophy, literature, history of art, history of technology, ethnology, psychiatry, or psychology of perception. Those who accused him of being a dilettante did not know how to understand the guiding thread of his work. Witnessing the various technological revolutions of the twentieth century, McLuhan devoted much effort to support the thesis that the so-called media, from the alphabet and printed materials (*The Gutenberg Galaxy*) to the telephone, radio, and television, were something more than mere means of expression. With a historical perspective, McLuhan sought to prove that these means are extensions of the contemporary human perceptual and cognitive structure. In addition to enhancing the transmission of information, these media profoundly alter our mental and social life; therefore, they operate as causal factors in the development of human societies. For McLuhan, remaining ignorant about these causal mechanisms limits the possibilities of reflective thinking, and this ignorance submits us to the forces of technological determinism. In the *Global Village*, McLuhan was able to successfully anticipate many of the scenarios that characterize digital globalization of our time. Following McLuhan, I argue that today the logic of the specimen and the biocultural homogenization it implies is carried on aggressively by digital technologies that reduce reality to quick, simplistic, and manipulable visual images easily transported throughout the globalized world.

9.2 The Rhinoceros that Drowned in the Sea

Historians of Renaissance art consider it likely that history's first *best seller* was not a book, but a woodcut that Albrecht Dürer signed with his celebrated logo, a kind of copyright for future reproductions. The truly original animal represented in the engraving was a rhinoceros that the German artist never even saw. Nevertheless, the representation of the animal amazed the European bourgeoise of the time, so fond of spectacles of exotic specimens.

The rhinoceros had been captured in the Portuguese Indies, loaded in Goa and unloaded in Lisbon on May 20, 1515. The animal disappointed the expectations of King Manuel I of Portugal, who decided to give it as a gift to Pope Leon X as a kind of bribe to assure his mediation in the king's colonial disputes in Asia. Unfortunately, the boat transporting the animal to Rome sank off the coast of Liguria on January 24, 1516. Bound in chains, the rhinoceros perished in the sinking ship. But the cadaver was brought up from the bottom of the sea, dried out and filled with straw, displayed in various European cities, and finally being given a fixed residence in Rome. In the end, the stuffed carcass of the rhinoceros disappeared following the sacking of Rome by Emperor Charles V in 1527.

What Dürer had at hand was a sketch of King Manuel's rhinoceros (Fig. 9.8) made by a merchant living in Lisbon who had sent it to colleagues in Nuremberg, Dürer's birthplace. Today we know that the German painter forged armor for military horses for the armorer's guild. Not having all the descriptive details he needed, Dürer decided to fashion the rhinoceros of Manuel I as a horse with armor.

Easily reproducible, thanks to readily available printing formats, along with accessible prices, the woodcut was widely distributed among the European bourgeoise, becoming a kind of prototype for many other artists to rework in their own way. During the sixteenth and seventeenth centuries, there were other more realistic representations of a rhinoceros, but Dürer's woodcut was the most imitated. It figured in the *Cosmographia* of the German cartographer Sebastian Münster (1544), in the *Historiae Animalium* by the Swiss naturalist and bibliographer Conrad Gessner (1551), and in the bestiary *Historie of Foure-footed Beasts* of the English naturalist and cleric Edward Topsell (1607), among others, thus becoming recognized as *the* canonical visual representation of the rhinoceros for zoological treatises up to the mid-eighteenth century. The inclusion of Dürer's woodcut by

Fig. 9.8 Gianni [Giacomo] Penni, "Sketch of rhinoceros" (1515). Above the sketch, Penni wrote in verse, "Form, nature and customs of the rhinoceros taken to Portugal by the captain of the King's fleet together with other things worthy of admiration coming from these recently found islands." Following the death of the animal, the sketch was not printed but a copy was placed in the Biblioteca Colombina de Sevilla/Public Domain {PD-old}

Topsell was especially influential because it reaffirmed many of the myths about the animal kingdom prevalent among classical authors such a Pliny, and it imprinted fantastic ideas about wild animals in people's imaginations (e.g., the idea that elephants worship the stars). Among the fantastic animals that inhabit the bestiary of Topsell, we can find carnivorous hippos that devour crocodiles, an image also present in the graphic index of the translation of *De Huerta of the Historia Naturalis* by Pliny. An example of the profound cultural impact that Topsell's bestiary had is demonstrated by the appearance of his animals in Shakespeare's *Macbeth* where reference is made to "The arm'd rhinoceros, o th'Hyrcan tiger," allusions to Dürer's works that appear in Topsell (Macbeth Act 3, Scene 4; Jackson 2013). In 1708, the

Fig. 9.9 François Leguat, "Divers kinds of rhinoceros" (1708). Florón del libro, *A new voyage to the East-Indies by Francis Leguat and his companions. Containing their adventures in two desert lands, and an account of the most remarkable things in Maurice Island, Batavia, at the Cape of Good Hope, the Island of St. Helena, and other places in their way to and from the Desert Isles.* Adorned with maps and figures (London, 1708). Courtesy of the Rhino Resource Center/Public Domain {PD-old}. The image of Dürer's rhinoceros with variations appears on page 297. The book contains many easily recognizable images of species of insects, fish, and marine animals observed during the trip

French explorer[1] and naturalist François Leguat used Dürer's rhinoceros for imagining various distinct species of rhinoceros (Fig. 9.9). His influence also was retroactive as he illustrated classical works by Pliny and Aristoteles using Dürer's rhinoceros (Fig. 9.10).

[1]Years earlier, Dutch traveler Caspar Schmalkalden (1618–1668) included a drawing of Dürer's rhinoceros carbon as an image of the Java rhino among the 111 illustrations of his manuscript Description of *Travel to the West and East Indies* (currently in the library of the University of Erfurt University Chart B 533). The rhino of Schmalkalden also serves to illustrate part of the biocultural itinerary of the engraving of Dürer. Supposedly, the Dutch author compiled these illustrations from his travels as an officer of the Dutch Companies of the West Indies and East Indies. It appears that the material in which the rhinoceros of Java is printed differs markedly from the rest of the manuscript, suggesting that it belongs to a later addition (Somers 2005, 166). According to the Dutch author, the drawing was made live by a Chinese painter from a rhinoceros who was in Batavia, present-day Jakarta. According to Krauss (2005) this shows the influence of Dürer in Asian culture through illustrations in zoological works such as the Johannes Jonstonus *Historiae naturalis de quadrupedibus libri,* cum *aeneis figuris, Johannes Jonstonus...concinnavit* (J.J. Schipperi, Amsterdam, 1657). On the other hand, Somers emphasizes that Schmalkalden reports having seen skins and horns of rhinoceros during his trips in Asia, which may have led him to accept the representation received from these data and from the confluence of his own biocultural load with that of the Chinese painter. The Schmalkalden rhino can be seen at https://commons.wikimedia.org/wiki/Caspar_Schmalkalden#/media/File:CasparSchmalkalden_Rhinoceros.jpg

Fig. 9.10 Page from Aristotelis *Meteorologicarum libri* IV (Valencia, 1555). (Courtesy of the Biblioteca Valenciana Nicolau Primitiu Biblioteca Valenciana Digital (BIVALDI) http://bivaldi.gva.es/es/consulta/registro.cmd?id=3973/Public Domain {PD-old})

9.3 Rhinoceros and Elephants

In his woodcut of the Rhinoceros (Fig. 9.5), Dürer significantly fills the upper part of the image with a written text (that also is part of the image's visual text), situating the animal in the Lisbon of Manuel I. Dürer wrote, in German, the following text:

> The first of May in the year 1513 [sic] [1515], the powerful King of Portugal, Manuel of Lisbon, brought from India a live animal called Rhinoceros. This is a faithful representation. It has the color of a mottled turtle and is almost completely covered by thick scales. It is the size of an elephant, but has shorter feet and is almost invulnerable. It has a powerful and sharply pointed horn on the tip of its nose that it files on rocks. The stupid animal is the mortal enemy of the elephant. The elephant is frightened by the rhinoceros because when they meet, the rhinoceros charges head-first between the hind legs of the elephant and tears out the stomach. Against this the elephant has no defense. The rhinoceros is so finely plated with armor that the elephant cannot injure it. It is said that the rhinoceros is rapid, impetuous, and astute.

Dürer took this description of the rhinoceros as the brutal enemy of the elephant from chapter XX of book VIII of the *Historia Naturalis* of Pliny the Elder. The biocultural history of the West from Roman times to the nineteenth century cannot be understood without knowing this work of Gaius Plinius Secundus (23–79 CE). It greatly influenced explorers such as Marco Polo and Hernán Cortés, and during many centuries zoology was no more than commentary on *Historia Naturalis*. In this work, Pliny the Elder dedicated the first 11 chapters to elephants and their exemplary behavior. According to Pliny, the elephant understands what it is ordered to do and can remember tasks that it learns, so it is really part of a domesticated nature and therefore capable of being incorporated into the Roman biocultural imagination as something suitable for themselves. For Pliny, the rhinoceros, to the contrary, embodied the residues of savage nature yet to be dominated: the wild beast, as stupid as dangerous, could not be domesticated, at best only reduced. In Lisbon, it seems the good King Manuel I wanted to prove with his own eyes what Pliny spoke of and so organized a duel between a rhinoceros and an elephant in the palace gardens. Both animals disappointed the king by their gentleness, but the elephant fled terrorized. Interestingly, Pliny's description of the bellicosity of both animals seems to have been taken literally from book II of the *Library of History* of the first-century Greek historian Diodorus of Sicily. However, Diodorus' rhinoceros easily can be defeated by the elephant if it can use its trunk and tusks to ward off the rhinoceros' attack (Diodorus 1935). As had been the chiefs of defeated tribes, the rhinoceros already had been paraded in the triumphal marches of Pompey, and in time, the images of the rhinoceros even appeared on some Roman coins.

It is known that Pliny's *Historia Naturalis* fed the zoological expectations of Christopher Columbus who carried a copy with him on several of his voyages. It is also known that Pliny's book circulated widely in colonial America. In 1624, Gerónimo de Huerta, physician of the Inquisition, translated it into Spanish and added an index of images. Dürer's rhinoceros opens chapter 20 (Fig. 9.11). So more than a century after drowning in the Mediterranean Sea, the rhinoceros of Manuel I reappeared in Spain and America. The extensive annotations made by De Huerta on chapter 20 of Pliny's book recall the celebrated, heroic animal brought to Lisbon in 1515. Apparently the king wanted to see with his own eyes the combat of which Pliny spoke, organizing a real duel between the rhinoceros and an elephant in the palace gardens. Both animals disappointed the king by his meekness, but the elephant fled in terror. The Inquisitor corroborates what Pliny says and adds that only the faithful horse is capable of driving the rhinoceros away. The rhinoceros

Fig. 9.11 Graphic index of *Historia Naturalis* by Pliny the Elder, translated and annotated by Gerónimo de Huerta (1624). Note Dürer's rhinoceros in the upper part of Chapter XX. (Courtesy of the Bibliothèque Sainte-Geneviève; https://ia800803.us.archive.org/20/items/OEXV10R/BSG_OEXV10_02_000009.jpg/Public Domain {PD-old})

sharpens the horn against the stones before the combat in which, according to Pliny, invariably defeats the elephant.

Some other of De Huerta's annotations show Dürer's rhinoceros as an image that encompasses wild nature, dangerous and useless, that had to be replaced in the colonies by species useful to Christian colonists. The translator assumes Augustine's classification in which animals could be harmful, innocuous, or useful. In order to accent the absolute uselessness of the rhinoceros for human endeavors that the horse and elephants fulfilled perfectly, De Huerta appeals to Job. This biblical figure contrasts the willing disposition of horses to go into battle with soldiers and to do so covered with armor (Job 39:19–25), with the stupidity of other beasts, whom God "made forget wisdom" and to whom had "given no share in understanding" (Job 39:17).

The struggle between elephants and rhinoceros also formed part of the biocultural imaginary that colonial Europe inherited from Rome. Inspired by its supposed bellicosity toward elephants, Alexander of Medici elected Dürer's rhinoceros as his emblem, with the slogan, "I will not return without victory." Based on Dürer, the father of modern surgery, Ambroise Paré, reproduced a battle between the two species (Fig. 9.12). In 1608, the Italian painter and engraver Antonio Tempesta also

Fig. 9.12 Ambroise Paré, "Combat between rhinoceros and elephant" (1589). *Les œuvres d'Ambroise Paré*, P. Rigaud, Lyon 1652. (Courtesy of BIU Santé, Paris/Public Domain {PD-art})

Fig. 9.13 Jan Griffier (after Barlow), "Elephant struggling against the rhinoceros" (1684). Engraving on paper. Current location: British Museum of London. This is not the only engraving that Jan Griffier (1652–1718) made of animals fighting following Francis Barlow. The British Museum has an engraving of a monkey forcing a cat to put its paw into a fiery furnace in order to reach some nuts. In a 1686 engraving, a vulture and a monkey guard the leftovers of a hare that had been caught by a pair of raptors. (These images can be viewed in http://www.britishmuseum.org/research/collection_online/search.aspx?searchText=Jan+Griffier. © The Trustees of the British Museum)

depicted this combat. The English painter and prolific book illustrator Francis Barlow illustrated in 1604 the battle that Dürer's rhinoceros never had. That same year the Dutch painter Jan Griffier, who was living in London, reproduced the same battle Barlow imagined (Fig. 9.13), thus reinforcing a long-standing cultural *meme.*

Dürer was named the official painter of the Sacred Empire of Maximilian I. His rhinoceros figured in a graphic image of the coronation of his successor, Maximilian II (Fig. 9.14). Dürer placed his rhinoceros at one of the foundations of the arch, face-to-face with an elephant located at the other foundation base. Both animals symbolize the robustness of the empire, capable of supporting on their backs the personifications of six feminine provinces. Above them is the imperial lion flanked by prudence and justice. These animal figures visually reaffirmed the colonial power of the imperial dynasty. Decades before, the rhinoceros of Dürer easily could have been the visual symbol of the Treaty of Tordesillas, in which the Catholic kings of Spain and John II of Portugal, father of Manuel I, decided to divide the world between them along a meridian. The Indian rhinoceros pertained to the biota east of

Fig. 9.14 Caspar Steinhofer, "The entry of Emperor Maximilian II in Vienna after his coronation at Frankfort" (1563). From a series of six woodcuts. 1563–1566. (Current location: British Museum of London. © The Trustees of the British Museum)

Fig. 9.15 Wenceslaus Hollar, "Map of Africa" (1670). Engraved on p.1 of John Ogilby, *Africa* (London 1670). Current location: British Museum of London. The map of Africa appears at the center of the image, drawn over the taunt and stretched skin of a lion. An admirer of Dürer, Hollar, used animals as hunting trophies as the theme of several of his works. (Courtesy of the University of Toronto Wenceslaus Hollar Digital Collection/Public Domain {PD-Art})

the line, reserved for the king of Portugal, but Dürer's visual construction, large and hyperbolic, could very well represent the cultural homogenization of the biotas to the east and west of Europe. In fact, in his map of Africa, the Dutch artist Wenceslaus Hollar elected Dürer's rhinoceros from India and the elephant as the emblematic species of African fauna (Fig. 9.15).[2] An admirer of Dürer, Hollar, often portrayed the rhinoceros along with elephants.

9.4 Rhinoceros, Dragons, and Behemoth

Dürer's rhinoceros has the scaly texture of Saint George's legendary dragon (Fig. 9.16). Furrowed by nerves, the lateral outer shell is similar to the winged extremities of dragons. These warlike attributes given to the Indian rhinoceros effectively incorporated the animal into the European colonial and military epic. This imaginary would last until the end of the nineteenth century, as illustrated by the 1890s engraving "Behemoth" by Reginald Savage (Fig. 9.17). In this engraving, Dürer's rhinoceros is represented as Behemoth, who, in the biblical tradition, has an immensely destructive power that blindly charges anything that moves. The image comes from Job 40:15–24 where Behemoth is described as the primeval creation of God, an implacable monster that only God is capable of taming. In the biblical text, Behemoth is parallel with Leviathan (Job 41: p. 1–34) who also is a terrible,

[2]Wenceslaus Hollar (1607–1667) was a Dutch artist who lived on horseback between the Netherlands and Great Britain, where he enjoyed the protection of the English nobility. His accession to the royalist camp in the English Civil War (1642–1651) costs him some time in jail, before escaping to Antwerp. His vision of the so-called English revolution is embodied in a work titled *Civil Sedition*, in which a snake is torn by the thrust of its two heads, one at each end. He was able to live on the spot of the battle of the ship St. Rose Mary in front of seven Algerian ships, represented in an engraving of 1643. It is said that he charged per hour of his works, counted with a clepsydra. Apparently he passed away ruined. Most of Hollar's works can be viewed at http://www.britishmuseum.org/research/collection_online/search.aspx?searchText=+Hollar&page=1

Fig. 9.16 Albrecht Dürer, "Saint George slays the dragon" (1501–1504). Xylograph engraving. Current location: British Museum of London. The dragon's body armor is identical to that of the devil of Saint Michael or the apocalyptic beast of Babylon. (Courtesy of the National Gallery of Art/Public Domain {PD-art})

Fig. 9.17 Reginald Savage, "Behemoth" (1890s). (Current location: British Museum of London {PD-Art})

primeval monster. Michael Coogan believes that it is probable that Behemoth and Leviathan were understood as primitive animals that God had to tame at the beginning of time and which, following the final judgment, will be food for the just (Coogan and Metzger 2001, p. 33). In Savage's engraving, the rhinoceros, the most powerful of beasts, lies defeated or at least sleeping, suggesting that Europeans, like God in the Bible, have tamed the forces of evil, thus making possible civilization.

9.5 Animals and Otherness

One by one, the rhinoceros' traits were sufficiently recognizable to have been assimilated by European visual culture, although, seen all together, nothing like this had been seen on the continent before. Europe craved a cynegetic component that promised much more fame than the game animals so benignly engraved also by Dürer: the elk eating from the huntress Diana's hand (Fig. 9.18) or the elk above

Fig. 9.18 Albrecht Dürer, "Apollo and Diana" (c.1502). Current location: British Museum of London. (Courtesy of the National Gallery of Art/Public Domain {PD-art})

Fig. 9.19 Unknown, "Ghent 1767" (1767) Rhino Resource Center

whose head the figure Christ on the cross appears and provoked the conversion of Saint Eustace.

The artistic construction of the otherness of exotic animals reaffirmed the beneficent exceptionality of Europe and, consequently, reinforced the legitimacy of Western colonialization of the wild and alien nature of Asia, Africa, and America in order to convert it into exotic and lucrative commodities. An anonymous engraving of the celebrated parade in Ghent in 1767 shows Dürer's rhinoceros with other wild animals, mounted and tamed (Fig. 9.19), as a propitiatory offering to Saint Macario who protected the city against the plague. In 1591, the supposed miracle powers of the rhinoceros horn seduced the physicians of Pope Gregory XVI, and today's price on the black market of the horn is equal to that of cocaine (Carrington 2017).

Dürer's rhinoceros represents a decisive phase in biocultural homogenization associated with mercantilism that reduces diverse forms of life to exchange value subject to the ups and downs of the market. The environmental economy continues being a global market: its objective is to internalize biodiversity into a global price system in order to unquestionably determine its value as *monetary value*, independent of any other biocultural, ecological, or ethical consideration.

Ironically, an event that occurred in an ex-colony reveals the biocultural inadequacy of the market economy and at the same time shows the falsity of the narratives by Pliny, Dürer, and De Huerta. After thousands of years of cohabitation among rhinoceros and elephants, African aboriginal people have known that, when they confront each other, it is the rhinoceros that invariably is defeated by the elephant. Westernized elites in the former Asiatic and African colonies have inherited a series of mercantile habits that alter the ecological equilibriums of the habitats of large mammals. Without any apparent motive, in July of 2003, three large, male elephants killed 63 rhinoceros in a park in Pilanesberg, South Africa (Siebert 2006). The three were part of a group of young elephants that had been uprooted and transferred to the Pilanesberg National Park in order to increase the park's economic potential as a place where Western tourists could encounter elephants during photography safaris. Traumatic separation from their elder family members and being transported to a strange place explain the elephant's aggressive behavior. This is a good example of the devastating ecological effects that some human habits can have on habitats and coinhabitants exclusively governed by economic rationality, homogenized by the law of supply and demand.

9.6 Dalí's Rhinoceros: Rationalization of Nature and Culture

In 1956, Salvador Dalí added another version to the history of Dürer's rhinoceros. The sculpture "Rhinoceros Dressed in Lace" (Fig. 9.20) can be interpreted as a symbol of self-referential Western art. Dalí's composition viewed as Western art self-reference, turned inward toward oneself, reinforces a constructive rationality common to Western natural history and fine arts. The rhinoceros of Dalí is not alone.

Fig. 9.20 Salvador Dalí, "Rhinoceros Dressed in Lace" (1956). This sculpture is located in Puerto Banús, Marbella, Spain. (Photo by Manuel González Olaechea used under Creative Commons License)

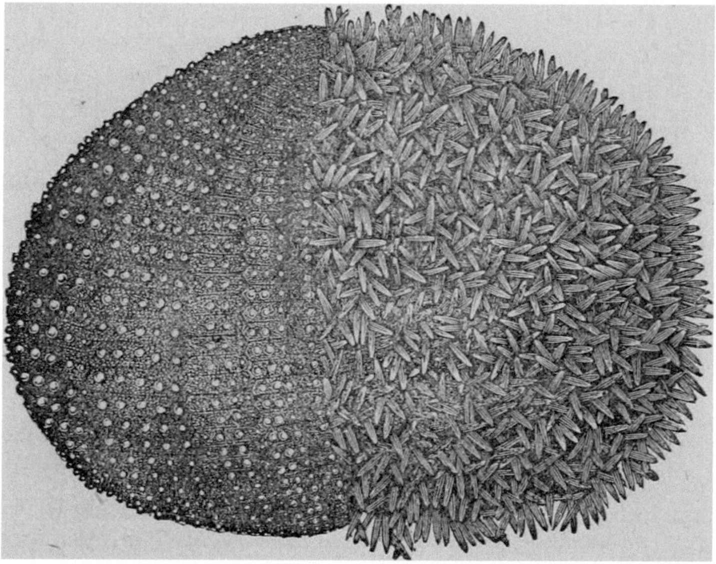

Fig. 9.21 Louis Agassiz et al., "Comparative anatomy of Echinodermata" (1870). https://archive.org/stream/outlinesofcompar00agas/outlinesofcompar00agas#page/n130/mode/1up/Public Domain {PD-old}

On its back and next to the horn, Dalí has displayed sea urchin skeletons, seemingly a reminder of the shipwreck that Dürer's "model" suffered.

Abundant among his works, the rhinoceros incarnated for Dalí animality and irrationality as the irreducible conditions of life (Dalí 2003, cited by Moure 2011). Sea urchins, with or without spines, also are an important element in Dalí's iconography. With spines, they appear numerous times in his works as icons of tangled and disordered natural parts of the female anatomy, such as axillary hair or that of the pubic triangle: nature in its organic and corruptible character (Fig. 9.21). Without spines, the urchin skeleton seems to incarnate the abstract idea of underlying rational order, a lifeless, concave figure ordered in convergent points: nature rationalized in a mathematical formula or in genetic sequence A-C-T… (Moure 2011). The connection of sea urchin skeletons with Dürer's rhinoceros in the sculpture "Rhinoceros Dressed in Lace" can be understood as a critique of the cultural rationalization of nature: the rhinoceros represents nature drowned by the cultural enterprises of merchants, kings, and popes, an animal that reemerges from the sea after death, sacked of its vital internal parts by taxidermists, and pulling with it from the sea floor calcified organic remains. These are skeletons of sea urchins, of whose rotten spines only stubs remain, now aligned in a series of convergent successions as a synopsis of its nature in exquisite proportionality to its *ratio* or the beautiful exactness of a mathematical equation. Cloth work such as crochet and lace also requires ordered, numerable, and recurrent patterns like algorithms and genetic sequences (Fig. 9.22).

Fig. 9.22 Thérèse de Dillmont, "Illustration of crochet techniques" (1922). Courtesy of Smithsonian Libraries. (Source: https://archive.org/details/motifspourbroder01dill/Public Domain {PD-old-70-1923})

In this way, we can understand Dalí's rhinoceros "dressed in lace" as a complex of biocultural and interdependent metaphors. To dress is to culturally cover the nakedness of nature. By dressing Dürer's rhinoceros, covering the bulges of the skin and its tridimensional projection with the "spiny points" of sea urchin skeletons (calcitic structures shaped in lines converging in a small pentagon that are geometric symbols of the pentaradial symmetry of echinoderms such as starfish and urchins), Dalí covers nature's nakedness. Perhaps with some sarcasm, Dalí summarizes in this lace dress the reduction of nature to a homogeneous and quantifiable visual extension, shaped by the abstract rationality that is shared by the plastic arts and sciences at least since the Renaissance.

9.7 Final Comments

Following the ideas of British ethno-psychiatrist J.C. Carothers (1959), in the 1960s Marshall McLuhan attributed this logic of visual abstraction to the perceptive and conceptual revolution that the invention of the printing press presupposed. McLuhan popularized during the 1960s and 1970s the idea that "the medium is the message." Witness to numerous technological revolutions, he argued that the so-called means of communication, from the alphabet and the printing press to the telephone, radio, and television, were more than just means of expression. From a decidedly historical perspective, McLuhan endeavored to show that these means are extensions of the perceptive and cognitive structure of humans that, besides optimizing the circulation of information, profoundly altered our mental and social life and, consequently, were causal factors of the development of societies. For McLuhan, to remain ignorant of these mechanisms limits the possibilities of reflexive thought and subjects us to the forces of technological determinism. Both texts and printed images, he argues, pertain to the abstract logic of visual space, thus permitting subjective distancing from the emotional constrictions of orality, of the word spoken and heard (McLuhan 2011, p. 133). The abstract character of visual images in some way mutes the emotions and determines them in semantic and behavioral terms. Their visual construction supposes the suspension of peremptory emotionality that gives the act of speaking an obligatoriness difficult to bribe away. The emotional neutralization of recurrent and identical visual sensations, all precedent from the same graphic impression, opens the possibility of representing the image as an abstraction that, for the subject that contemplates, can open up a repertory of available alternatives.

The graphic image of Dürer's rhinoceros pertains to what Heidegger called "the epoch of the image of the world" (Heidegeer 1995) that in reality was the epoch in which the world began to be reduced to a visual extension, homogeneous and abstract, fragmented into floating, assimilable and available, and reproducible and interchangeable parts. The US environmental thinker Paul Shepard warned in the 1970s that, in order to confront the non-differentiation of biotas and cultures, the human species would have to go far beyond abstractions to the time of perceiving nature: "The substitution of places by images was the first step in the construction of places similar to the images" (Shepard 2003, p. 36).

Today the image of the world has become the tactile screen, a technological device that reinforces more than ever the manipulable condition of images that can be expanded, multiplied, interspersed, or hyperlinked. One might say that the biocultural diversity of the world of life is buried under a digital surface, flooded by uncountable number of *menus* superimposed as visual layers that release images on the screen and that end up materialized as market offerings. The new technologies even allow incorporating into our texts graphics and icons that are printed on crystal lenses with lightweight touches of support or optical devices impressed directly on contact lenses that "enrich" the visual stimuli of the retina surfaces, therefore constructing an "enlarged" reality in view of maximizing the information supposedly useful for the preceptor.

Marshall McLuhan feared that visual technologies would end up producing a kind of psychological death by separating us from the ecological order through a narcissistic turning inward on ourselves (McLuhan and Powers 1990, p. 17). Heidegger's phrase "there toward where man looks, he finds only himself" perhaps diagnoses the narcissistic craziness that biocultural homogenization of the planet means. Under the never-ending offer of visual products, digital technologies are gestating a globally uniformized culture, whose inhabitants occupy anthropogenic habitats that are increasingly homogeneous (Ellis and Ramankutty 2008; Meyer 2006) and, following the logic of the specimen, that are populated by biological species best accommodated to the habits of humans.

Paraphrasing Ricardo Rozzi, we can conclude that, in times of tablets, smart phones, Facebook, and Twitter, our minds acquire mental habits globally homogeneous and construct globally homogeneous habitats (Rozzi 2013, p. 14), hence the urgency for biocultural conservation of "specific ecosystemic units of habitats-habits-inhabitants," permeated by an ethos that reintegrates the identity of inhabitants with their own way of inhabiting their local habitats (Rozzi and Arango 2008, p. 117). This ethos is fundamentally distinct from that of *Homo economicus*, whose global habits are ruled by the subjective preferences of individuals, whose inhabiting is reduced to consuming, and whose habits are built on the absolute preponderance of economic relations over ecological interdependencies. The ethos of biocultural conservation is incapable of constructing habitat only to be contemplated or as a place only to produce and to consume. As Heidegger said, we build habitat, but to build, in the sense of protecting and caring, is not to produce. Caring is something more than abstaining from harm. The fundamental characteristic of inhabiting is caring as surrounding protection, not *looking to* but rather *looking at* the world that encompasses us (Heidegger 1994, p. 128–131).

References

Carothers JC (1959) Culture, psychiatry, and the written word. Psychiatry 22:307–320

Carrington D (2016) How domestic chicken rose to define the Anthropocene. https://www.theguardian.com/environment/2016/aug/31/domestic-chicken-anthropocene-humanity-influenced-epoch. Accessed 15 Jan 2017

Carrington D (2017) High price of rhino horn trade leaves bloody trail across the globe. https://www.theguardian.com/environment/2017/mar/10/high-price-of-rhino-horn-leaves-bloody-trail-across-the-globe. Accessed 13 June 2017

Coogan M, Metzger B (2001) The Oxford guide to people and places of the Bible. Oxford University Press, New York
Dalí S (2003) Diario de un genio. In: Obra Completa de Salvador Dalí, vol I. Destino, Fundación Gala-Salvador Dalí, Sociedad Estatal de Conmemoraciones Culturales, Barcelona
Diodorus Siculus (1935) The library of history. The Loeb classical library. http://penelope.uchicago.edu/Thayer/E/Roman/Texts/Diodorus_Siculus/3C*.html. Accessed 11 June 2017
Ellis EC y Ramankutty N (2008) Putting people in the map. Anthropogenic biomass of the world. http://ecotope.org/people/ellis/papers/ellis_2008.pdf. Accessed 14 Jan 2017
Elton C (1958) The ecology of invasions by animals and plants. Chapman and Hall, London
Heiddeger M (1994) Construir, habitar, pensar. In: Conferencias y Artículos. Del Serbal, Barcelona
Heiddeger M (1995) La época de la imagen del mundo. In: Caminos de Bosque. Alianza, Madrid
Ingold T (2000) The perception of the environment, essays on livelihood, dwelling, and skill. Routledge, London/New York
Jackson S (2013) Shakespeare's world in 100 objects. http://findingshakespeare.co.uk/shakespeares-world-in-100-objects-number-91-a-book-of-beasts. Accessed 14 June 2017
Kolbert E (2016) La sexta extinción. Crítica, México
Krause W (2005) Chinese influence on early modern Indonesian art ? Archipel, volume 69, 2005. Autour de la peinture à Java. Volume I. pp 61–86
McLuhan M (2011) The gutenberg galaxy. University of Toronto Press, Toronto/Buffalo/London
McLuhan M, Powers BR (1990) La aldea global. Gedisa, Barcelona
Meyer S (2006) The end of the wild. MIT Press, Cambridge, MA
Moure I (2011) Apunte paranóico-crítico sobre Dalí y el rinoceronte, Asociación Aragonesa de Críticos de Arte, número 5. http://www.aacadigital.com/contenido.php?idarticulo=495. Accessed 2 June 2017
Neri J (2011) The insect and the image. The University of Minnesota Press, Minneapolis
Plinio Segundo C (1624) Historia naturalis, Gerónimo de Huerta ed., Luis Sánchez impresor, Madrid. https://books.google.com.ar/books?id=s5CpZ6f9gpkC. Accessed 3 June 2017
Rozzi R (2012) Biocultural ethics: the vital links between the inhabitants, their habits and regional habitats. Environ Ethics 34:27–50
Rozzi R (2013) Biocultural ethics: from biocultural homogenization toward biocultural conservation. In: Rozzi R, Pickett STA, Palmer C, Armesto JJ, Callicott JB (eds) Linking ecology and ethics for a changing world: values, philosophy, and action. Ecology and ethics, vol 1. Springer, Dordrecht, pp 9–32
Rozzi R (2015) Earth stewardship and the biocultural ethic: Latin American perspectives. In: Rozzi R, Chapin FS III, Callicott JB, Pickett STA, Power ME, Armesto JJ, May RH Jr (eds) Earth stewardship: linking ecology and ethics in theory and practice. Ecology and ethics, vol 2. Springer, Dordrecht, pp 87–112
Rozzi R, Arango X, Massardo F et al (2008) Field environmental philosophy and biocultural conservation: the Omora Ethnobotanical Park educational program. Environ Ethics 30:325–336
Shepard P (2003) Where we belong. Beyond abstraction in perceiving nature. University of Georgia Press, Athens
Siebert Ch (2006) An elephant crack up? New York Times, 8 October 2006. http://www.nytimes.com/2006/10/08/magazine/08elephant.html?pagewanted=all&_r=02003? Accessed 20 Dec 2013
Somers M (2005) An early traveler's compendium: Caspar Schmalkalden's images of asia. Archipel 70:145–184

Chapter 10
Biocultural Exoticism in the Feminine Landscape of Latin America

Angelina Paredes-Castellanos and Ricardo Rozzi

Abstract The colonial language "of the imperial eye" about Latin America is summarized in an exotic myth about the landscape, which today combines with Western globalization at its ecological crossroads. Latin America offers a key hermeneutic for analysis, because it exposes aesthetic ideological roots that accompany biocultural homogenization. In the twentieth century, orientalism came to be criticized for perpetuating a false, Western colonial image of the East. Here, we focus on Latin American exoticism by critically examining themes of feminized geographic landscapes. We review epistemological, aesthetic, and ethical alternatives to the exoticism of Latin American landscapes and bodies as seen by the "imperial eye." Latin American artistic, literary, and political concepts and movements have been taking seriously landscapes and bodies, enabling a critique of the exotic. These expressions also offer an alternative, post-exotic, eco-epistemology, and aesthetic hermeneutic to the negative hermeneutic or the curse that resulted from the colonial expansion of Europe and the new postmodern biocultural homogenization or biocultural exoticization imposed by Western globalization. We emphasize intimate associations between ecological contexts and social practices to revalue the geographies and co-inhabitants, including the political subjects, their identities, and their cultures. A major contribution offered by Latin American environmental arts, thoughts, and movements is the understanding of nature as a great body in which we co-inhabit.

A. Paredes-Castellanos (✉)
Facultad de Filosofía, Universidad Michoacana San Nicolás de Hidalgo, Morelia, México

R. Rozzi
Department of Philosophy and Religion and Department of Biological Sciences, University of North Texas, Denton, TX, USA

Sub-Antarctic Biocultural Conservation Program, University of North Texas, Denton, TX, USA

Instituto de Ecología y Biodiversidad and Universidad de Magallanes, Punta Arenas, Chile
e-mail: Ricardo.Rozzi@unt.edu

R. Rozzi et al. (eds.), *From Biocultural Homogenization to Biocultural Conservation*, Ecology and Ethics 3, https://doi.org/10.1007/978-3-319-99513-7_10

Keywords Biocultural homogenization · Biocultural conservation · Colonialism · Ecofeminism · Globalization

10.1 Introduction

At the beginning of the twentieth century, the French ethnographer and writer Victor Segalen initiated a critique of exoticism, the Western tendency to define the non-Western world as "other." Notably in his *Essay on Exoticism: An Aesthetics of Diversity* (*Ensayo sobre el exotismo* 1989, p. 62), he says he will not speak of the palm trees, women, indigenous people, camels, or paradises that figure prominently in Western exoticism but rather about diversity and difference. Yet he does not say what is diverse or different nor how he understands exoticism; he supposes this is obvious when it is not obvious. This critique continued throughout the century (Paredes-Castellanos 2017). In the decade of the 1970s, the literary critic and Palestinian-American activist Edward Said, following the same tendency as Segalen, explains in *Orientalism* that he will not talk about mysteries, enchantments, nor love affairs among natives and foreigners. Instead, he says he will focus only on how, in the realm of knowledge and Western culture (literature, science, politics, universities, etc.), a false image of the East is perpetuated (Said 2016, p. 447).

In contrast, we will speak about the themes these authors did not want to mention: feminized geographic landscapes, indigenous people, romantic encounters between Europeans and Native Americans, aesthetic experiences, the mysterious, the aesthetic enchantment of landscapes, and the nude bodies. Segalen and Said consider these to be of minor importance. Hence, while they stimulated a critical discussion about discrimination and xenophobia directed to "the other," the "Southern foreigner," about the supposed "third world" as defined by the "first world," they did not provide epistemological or aesthetic alternatives.[1] In our text, however, these themes are taken seriously and are given prominence in ways that take us closer to seeing hermeneutic alternatives to the exoticism of Latin American landscapes and bodies as seen by the "imperial eye" and its romantic and commodified idea of "exotic America."

We will look especially at Latin American artistic and literary expressions, which by taking seriously landscapes and bodies reveal "exotic America" as having important cultural elements that permit a *critique of exotic reason*. These expressions also

[1] In *Orientalism*, Said criticized the constellation of false prejudices that Western imperial culture uses to interpret Eastern cultures and denounced Eurocentric prejudices against the Arab-Islamic peoples and their culture and the longtime modern Western tradition of false images about Asia and the Middle East. He denounced these images as having served as an implicit justification for the colonial and imperial ambitions of the West over the East (Said, 2016, pp. 339–431). However, by not providing epistemological nor aesthetic alternatives, to some extent, Said unintentionally cooperated with the continuity of *Orientalism*.

offer an alternative, post-exotic, eco-epistemology, and aesthetic hermeneutic to the negative hermeneutic or the curse that resulted from the colonial expansion of Europe and the new postmodern imperialism represented by biocultural homogenization or biocultural exoticization of Western globalization.

10.2 Biocultural Homogenization and Exoticism

Biocultural homogenization refers to the interrelated uniformization of habitats and life habits (Rozzi 2013). It alludes to the Western tendency to universalize or to make everything uniform that today, with hegemonic globalization, imposes its own vision uniformizing hermeneutic on all localities, affecting both the eco-systemic biodiversity and the cultural diversity of every corner of the planet (Rozzi 2012a). Rich biological and cultural diversity is replaced by a uniformized biota and culture that have been defined by Western colonialism and modernity since the sixteenth century.

This process also involves biocultural exoticism: the non-Western world envisioned, defined, and imposed by the West. In this manner, biocultural exoticism feeds the neocolonialist tendency to impose the West over the regional. This imposition of the foreign over the local reduces or even eliminates rich biotas and local cultures. Consequently, biocultural homogenization implies oppressive forms of losses of local knowledge, values, and life habits that represent atrocious forms of socio-environmental injustice (Rozzi 2012b, 2015a). Paradoxically, a greater valuation of local interrelated ecosystems and cultures (that are being rapidly eliminated) could offer sustainable alternatives for an Anthropocene that could overcome the current global socio-ecological crisis by focusing on the diverse biocultural mosaic that characterizes the different regions of the planet (Rozzi 2015b).

The exoticism of the "imperial eye" rejects differences and otherness. Rather, these are redefined, and a foreign aesthetic and epistemological model acceptable to the West is imposed. This process clearly is evident in the globalized market that is little interested in regional ecological, cultural, and social questions, nor in customs and local relationships, except for commercial ends and political influence (Paredes-Castellanos 2017). Rather than bridges of respect and exchange, walls both physical and nonphysical are put up that negate "the other." By putting up walls instead of bridges, exoticism loses the horizon of a broad, human community, a cosmopolitan and multicultural horizon where all learn from each other. Preserving and maintaining biocultural integrity are a matter of socio-environmental justice and of the sustainability of multiple species and multiple cultures in each of the different regions of the planet. In order to achieve this goal, as Rozzi et al. (2008) argue, biocultural environmental education centered on that which is local is necessary to eradicate epistemological, ecological, and cultural colonialism and to foment appreciation for the immediate surroundings of original peoples and local communities.

10.2.1 On Palm Trees, Dinner Plates, and Reptiles

The waiter and the prostitute constantly are cited in the artistic work of the Mexican-Costa Rican painter Joaquín Rodríguez del Paso (Chavarría 2013, p. 13). Through his paintings, Rodríguez del Paso denounces the supposed "servile and weak" character of the falsely stereotyped Latin American man and woman. In *Biodiversity*, he criticizes these stereotypical and false images by portraying a woman sitting on a crocodile, surrounded by palms and served on a dinner plate (Fig. 10.1). The exotic

Fig. 10.1 Joaquín Rodríguez del Paso, *Biodiversidad* (*Biodiversity*), 2008. (Chavarría-Zamora 2013, p. 13). Reproduced with kind permission of the family

feminine of the continent and a Latin American body served for pleasure and economic exploitation. The woman seated on a reptile references the exotic image of the "wild woman" of the New World, which is distortedly considered to be wild and servile at the same time. It is not by chance that Latin America is a sex tourism destination in which a particular way of seeing the female body is revealed: the feminine body as an object of pleasure. In the same image, the male figure is a waiter, the one who serves and takes orders. He makes no decisions of his own. Rodríguez del Paso vividly illustrates biocultural exoticism by portraying the female body as an exotic oversexualized fetish and the masculine body as servile, both located in the laid-back and wild tropics. This work speaks of exotic beauty as part of the exotic myth of Latin America, an aesthetic interpretation formed by the "imperial eye" about the feminine landscape.

10.2.2 "Imperial Eyes" View of the American Geography

During the so-called discovery (or invention) of America until the eighteenth century, the romantic idea of "exotic America" was recorded by diverse European travelers and others: conquerors, religious workers, naturalists, scientists, and painters, among others. For these creators of colonialist and exotic representations, the enormous environmental and social wealth of Latin America was to be exploited. At the end of the nineteenth century and the beginning of the twentieth, other European travelers with longtime residence in South America reenforced the colonialist idea of "exotic Latin America" through graphic representations of exoticism: "the savage landscape," "the virginal primeval nature," "the noble savage," and even the "cannibal." The exoticization of South America can be called *the exotic myth of Latin America* in which nature is portrayed as an untouched virgin, wild, servile, exhibited, and a sleeping machine.

10.3 The Exotic Encounter Between Natives and Foreigners

Canadian linguist Mary Louise Pratt (2010) approaches the problematic nature of contacts and relations between local regions and the American continent as a whole, with Europeans. She proposes that the American landscapes receive their primary semantic interpretation as an aesthetic-negative impression by European colonizers. The encounter of the European with the recently conquered landscape of the New World, later called America, determined its prevailing image in the Modern Era (see also Mignolo 2007). For the first "imperial eyes"—masculine and patriarchal—the vision of American nature was always that of possession and constant exploitation. The virginal myth of a feminine nature, youthful, untouched, flowered, wild, exuberant, rebellious, and Amazonian transformed into "America dead and mute" (Pratt 2010), provided the ideological justification and arguments for the expanding limitless intervention of Western masculinity. The aesthetic side was the image of a deformed, feminine geography redrawn as a wild "montage" to be controlled and ravished.

The Uruguayan thinker Eduardo Gudynas (2009a, b) proposes that ever since the arrival of the first conquerors, for all colonialist and modernist actors (external and internal), Latin American nature has been viewed as a "great basket of natural and bodily resources." I note, as especially important, the body that is within this "basket" (Paredes-Castellanos 2017). No obstacle impeded the conquerors past nor present; the exploitative rape of the American Earth justified by the idea that nature is only an object and thing to be used. The Uruguayan journalist and writer Eduardo Galeano (2011) recalls how Latin American nature and its inhabitants were interpreted as a "great treasure" that offered immense wealth.

Pratt and ecofeminism identify a coincident and constant idea among both European foreigners and natives about the association of nature with the feminine. This association suggests feminine geography as a woman map (Fig. 10.2). Both the conquerors and native peoples understood this. The first forged an image of the feminine geographic nature based on the concept of control and domination. In contrast, even today, the "other gaze" of native men and women considers the same geography as feminine, rooted in a hermeneutic of Mother Earth—not in the notion of domination.

10.3.1 On "Love Affairs" Between Native Women and Foreign Men

Not only was the American natural landscape the object of domination, so too was the body landscape of the native woman (along with mammals, birds, trees, vegetable bodies, etc.). Consequently, the theme of the seminude body of the female Native American is key for understanding the wider American geography that is projected as a female body to be dominated. The geographies of both the environment and the body are viewed through the same patriarchal "imperial eyes" of Western domination. We can understand how a negative aesthetic of a whole corporal geographic zone would impact for posterity the image of the native woman of the New World. This image is crystalized in the female body stereotyped as part of the exoticism of the feminine South: the tendentious image of the Latin American woman as "exotic."

The hermeneutic of the curse that is part and parcel of the dualist Christian priestly masculinist vision of the feminine was imposed semantically on both native women and nature. Latin American woman and Latin American landscape are interchangeable for understanding the violent history of the drama of American nature. The encounter of the European male with the Native American female reveals this deeply. The activist and decolonial feminist thinker Yuderkys Espinosa Miñoso argues:

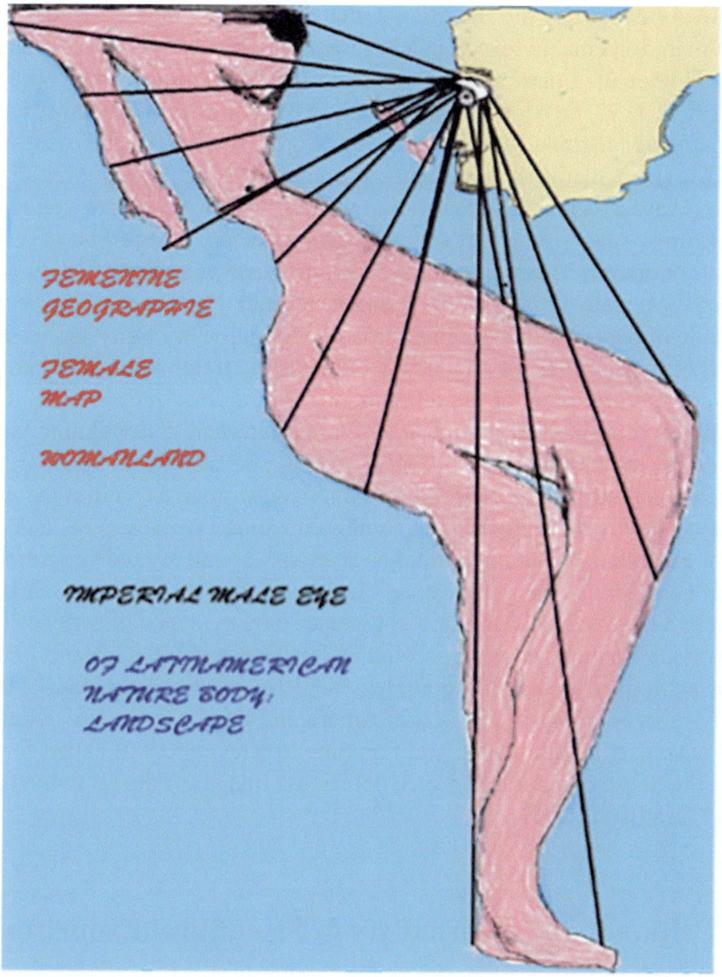

Fig. 10.2 "Ojos imperiales sobre un mapa femenino" (the imperial eyes on a female map), drawing by Angelina Paredes Castellanos

> The story about the white man "in love" with the indigenous or African slave, hides the truth about the obligatory sexual encounter, about the creation of a feminine body in the service of the colonial and patriarchal enterprise. The naturalization of the native or slave woman as part of the conquered landscape is in effect not only the colonial logic but the patriarchal and heteronormative logic also. It is that both logics, more than just being articulated, have been part of the same conspiracy of domination. It is not possible to think of one without the other: the history of the European invasion of these lands has also been the invasion of the violable body of the aboriginal women. (Espinosa 2009)

Exoticism imposed by "imperial eyes" defines native women as sensual bodies, pleasurable for foreign males, just as nature is viewed as a body to be exploited. Both are to be conquered.

The novel *Iracema* by the Brazilian writer José Martiniano de Alencar is about a native woman, Iracema, in love with a Portuguese man. She has been considered the founding mother of a new mestiza or mixed heritage cultural future (de Alencar 2000). Born in 1892, de Alencar participated in the politics and literature movement that inaugurated a national vision of the New World. Toward the end to the twentieth century, Mexican historian and writer Fernanda Núñez Becerra criticized how Malinche (Marina)—the Mexican Indian woman who was advisor, translator, and lover of Cortes—has been portrayed as a "whore" and a traitor who gave in to the European conqueror. This negative and exoticist interpretation of Malinche again underlines the negative image of the feminine (Núñez-Becerra 2002). This exoticist hermeneutic of the curse and domination of the feminine in Latin America is persistently part of the colonialist hermeneutic of insult and domination. The exoticism of the Mexican cultural view of the feminine is seldom criticized; instead, it is problematically incorporated as part of a continuing internal, national, and patriarchal exoticist Western colonialism.

Exoticism has stereotyped the Native American woman as well as the American geography: The first as a prostitute, a thingified woman body, servile, and enslaved for sexual exploitation and the second as an enslaved and sacked geography—"the American continent as the great whore." Uruguayan writer Eduardo Galeano refers to the "open veins of Latin America," thus describing the raped and sacked womb of Latin American nature as a deep colonial imperial wound (Galeano 2011). The image of ecological damage is complemented by another, particularly feminine, metaphor: "the open legs of the Latin American body." Given this dominant and negative narrative, an alternative and feminine hermeneutic will called for a narrative from the South counteracting such deforming images of the feminine landscape (Paredes-Castellanos 2017).

10.4 Critiques and Alternatives to Biocultural Exoticism

We can now appreciate how, during the twentieth century, four South American countries—Brazil, Chile, Bolivia, and Ecuador—began to gestate, to weave, or to draw an image different from the deformed "exotic America." This reevaluation of Latin American nature and societies relies on a hermeneutic of ecological recognition and post-exotic alterity to forge new attitudes toward environmental and social identity.

10.4.1 On Mysteries: "The Cannibal"

Brazil, with a biodiversity among the greatest on the planet (Primack et al. 2006), has produced the artistic and intellectual movement of *Antropofagia*. One of its mottos is "*tupí* or not *tupí*" (from the Tupí people indigenous to Brazil) with an ironical intercultural reference to the Western philosophical canon "to be or not to be" (Giraudo 2016). In the 1920s, this movement began to sketch a new image of the region. Today we can appreciate how *Antropofagia* offered biocultural alternatives; at the same time, it questioned the European aesthetic cultural canon and its supposed cultural superiority. This Brazilian aesthetic movement takes on the accusation that cannibalism by the Tupí was part of the original culture but subverts the idea, intellectualizing it in such a way that it becomes a part of the place's identity, returning it to its roots, to its tradition, and to its original cultural myths.

Two paradigmatic paintings described this new identity formulation: *Abaporu* and *Antropofagia* (Fig. 10.3). *Abaporu* is by Brazilian artist Tarsila do Amaral, who painted it in 1928 for the poet Oswald de Andrade, her life partner. De Andrade was one of the founders of Brazilian modernism, and *Abaporu* inspired him to write the *Manifesto Antropófago* (Cannibalist Manifesto), published in the first issue of the primitivist, modernist magazine *Antropofagia*, in the same year 1928. *Abaporu* is a Tupi Indian name that means "man (*abá*) that eats (*u*) people (*poro*)." Interpreting the painting, we discover that the predatory root of nature is modern humanity itself. It preys on nature that is almost finished off as a green *inferno*, completely alone with an enormous body and heavy feet; this is the "modern cannibal." Rooted in the predation of their own nature, the modern Western subject has become capable of causing their own self-destruction. The second painting, *Antropofagia* (1929), is emblematic of the homonymous artistic movement and magazine. The central feature is a large feminine figure whose great breast is lifted up, along with a smaller masculine figure. Both are enormous corporal presences with diminished heads, wrapped in the green splendor of tropical nature that brings to memory the ancient sacred visions of the feminine and its strict relation to the Earth. The theme of the body is underlined in both canvases. This painting allows us to understand a non-exoticist view of the corporal for the study of nature in Latin America.

Fig. 10.3 Emblematic oils on canvas by the Brazilian artist Tarsila do Amaral: Abaporu (left), and Antropofagia (right). Reproduced with kind permission of the family

During the twentieth century, by centering on the importance of nature for Brazilian identity, the philosophical environmental ideas of Brazilian thinkers became part of this new spirit. The Brazilian philosopher Mauro Grün uses a hermeneutic of nature taken from the ecological themes of the German philosopher Hans-Georg Gadamer. Both point to the danger of anthropocentrism, involving an interpretation that insists on the superiority of (Western and modern) humanity over (European and American) nature. This human is the master who ought to dominate nature for its own good. Grün calls this "anthropofascism" (Grün 2012). As an alternative view, Grün adds the concern for the Latin American identity based on the ancestral feminine vision of the Earth itself. However, he cautions against the problematic character of a potential "ecofascism" derived from supposing nature as a great feminine subject that could lead us to diminish humanity. Therefore, this could lead to objectifying humans and eliminating them, as subjects with rights. Concurring with Gadamer, Grün recommends thinking of nature from the ethic of "the other," with whom we coexist and respect it in its otherness.

10.4.2 Overcoming Exoticism: Kinship Between Humans and Birds in Southern South America

In southern Chile, where one of the world's last remaining extensive wilderness areas is protected by the Cape Horn Biosphere Reserve (Rozzi et al. 2012), we are challenged to overcome exotic representations of the indigenous Fuegians of Tierra del Fuego made by the nineteenth-century photographers. These photographs were distributed throughout Europe as "human zoos" that were invented to exhibit and "to scientifically study" the supposedly exotic customs of native peoples from different parts of the world (Báez and Mason 2006). In their book *Human zoos: Photographs of Fuegians and Mapuche in the Jardin d'Acclimatation in Paris, XIX century*, Chilean historian Christian Báez and English anthropologist Peter Mason describe in detail the sad trajectory of the Fuegians and Mapuches of central Chile who were taken to Europe to be exhibited at international fairs, variety theaters, and anthropozoological exhibitions. Groups of native people were even abducted and caged, in order to be exhibited like monkeys in zoos and circuses, in European cities. Such devaluing of people and nature resulted from viewing the bodies and landscapes of the southern New World as exotic.

In an ongoing biocultural conservation program whose purpose is to overcome the exotic distortion of cultural traditions and colonialist oppression in southern Chile, we have been working with a post-exotic epistemological and ethical perspective centered on indigenous Fuegian cultures and their ancestral knowledge of nature (Rozzi 2001). By affirming this knowledge, a new identity emerges based on intercultural dialogues, education, and collaborative research. Through this intercultural partnership, we have found remarkable convergences between the forms of ancestral ecological knowledge and that of contemporary ecological and evolutionary science. Traditional life habits allow them to co-inhabit sustainably with a great diversity of biological species in their shared habitats, by establishing respectful ways of representing and relating to nonhuman co-inhabitants (Rozzi 2015b). These life habits offer keys for biocultural ethics and environmental education in which ancestral knowledge is revalued and the evolutionary kinship of humans to nature is expressed in different ways through myths, traditions, songs, and poetry (Rozzi 2004).

Working together, the Mapuche poet Lorenzo Aillapan and the Chilean philosopher and ecologist Ricardo Rozzi have composed a selection of *Twenty Winged Poems* (*Veinte Poemas Alados*, Aillapan and Rozzi 2001). Extending beyond the human community of Pablo Neruda's *Twenty Love Poems*, the *Twenty Winged Poems* include stories of affection, play, and care between birds and humans, as well as onomatopoeic verses that allude to interspecies dialogues. Bilingual poems written in Mapudungun and Spanish suggest a common genealogy shared by birds and humans. For example, regarding the raptor *Caracara* or *Traru* (*Caracara plancus*), Aillapan writes:

This Caracara is a millenary bird. Of eagle its variety pertains, Carnivorous bird of raptor legs and talons. Today it is on its way to extinction.* Few are yet left due to absence of native forest, as the only reason. For its reproduction there is no appropriate tree. *¡Trarrrr trarrrr traruuuu traruuuu!*	*Kuyfi ünüm may tachi traru* *Feyti ñamku pingey ñi küpal kuyfideuma* *Weñefe wili namun niey ka itro fill ilotufe* *Newe ngewelay ella müfü mütem miyaukey* *Mauwidantu kuyfi aliwen newe ngewelay* *feymeu aftuy reke ñi rumel chauken.* *¡Trarrrr trarrrr traruuuu traruuuu!*
Its form makes him known as the police bird, of police hat with an ancestral hat-band, figure and form equal to an armed police officer. In addition, its speed of flight with contagious thinking, its onomatopoeic, rough call resounding in the territory announce war, when they noisily appear in flocks. *¡Trarrrr trarrrr traruuuu traruuuu!*	*Miyaukechi tralkatufe ka che pingekey üytungen* *Nienmeu allangechi trarilonko kuyfi chetuley* *Pidpid meñkituley tralkatufe lonkotraru üytungey* *Kimfaluwi ngenikayauwün rakiduam reke pünerpuy* *Fütrarume wakeñ/ülkantun wepümniey lofmeu* *Muñkupüle nomnom ñi wefün müleyaludauma aukan.* *¡Trarrrr trarrrr traruuuu traruuuu*!
For people in the territory it announces misfortune. Several Caracaras detect visions among the people, then the people of the community comment and shout; "Like that the outsiders come to cause war in the land." A young man named Traru, incorporates its fleet spirit. A novel figure, his first name becomes Lautaro. *¡Trarrrr trarrrr traruuuu traruuuu!*	*Wefünmeu weda inautun dungu tachi Lofmeu* *Femngechi wüne kimfaluwi rupayawiyum wera traru* *kiñeina may pi pu che mülerkeyay tachi fütra* *Aukan akulu deuma ka tripaye che Lofmeu* *Fey kiñe weche tuy traru ünüm ñi püllü pepilkawi* *Wüne Leftraru che wefrumey, feychi üy tukuy.* *¡Trarrrr trarrrr traruuuu traruuuu*!

(Aillapan and Rozzi 2004, p. 426)

This poem about Traru (Caracara) alludes to the ancestral warrior that belonged to the same genealogy as eagles and who became the greatest of Mapuche warriors: Lautaro. His name signifies swift (*lef*) falcon (*Traru*). That is, "Traru transforms into Lautaro" in allusion to the relationship of humans to eagles and their admiration for the swift flight of these birds (Aillapan and Rozzi 2004, p. 430). In this manner, the genealogical histories of birds are interwoven with the genealogies of human beings. The Mapuche worldview and the scientific theory of biological evolution converge by affirming the common origin of all living beings. Both worldviews, each in its own way, imply a degree of genealogical kinship between birds and humans. The poetry of Aillapan transmits this kinship relationship involving a multiple species of common family through creative onomatopoeic and not technical language (Rozzi 2004).

10.4.3 On "Sacred" American Nature

In Bolivia, a country with a rich cultural heritage, we find the philosophy (and theology) of the *Pachamama* or *pachasophy* (Estermann 2011). This philosophy complements the theory of ecofeminism in order to approach the local question of American nature as the greater ecological question of the Earth or Pachamama as the Universe. The concept of "exotic Latin America" as a "raped feminine continental body" is associated with the distorted and exotic image of the body geography of the Americas, which pretends to impose Western consciousness through a biocultural homogenization of the Earth. In response to this reifying, commodifying, and destructive trend, an alternative concept of body earth or body world is key for a post-exotic, ecological comprehension for our times. The Bolivian ecofeminist Vicenta Mamani proposes a theological view of the Pachamama in which the Mother Earth is effectively a Great Sacred Body. The Pachamama, Mamani explains:

> … is like a material body that is born, lives and finally goes to another mysterious life … She is Pachamama, like a mother who provides us everything to nourish us abundantly and to live fraternally … She is complementarity through reciprocity … she is all that exists, and therefore all pertains to her and is fulfilled in her because everything is previously ordered to the rhythm of life … she is time, space, spirit and flesh; light and darkness; above and below. You, me and we in her plenitude … she is the placenta of a mother, where there is fullness of life for even a little living being waiting to come into the world from the maternal womb. (Mamani and Quispe 2007, 11–50)

The Swiss philosopher and theologian who resided many years in Bolivia, Josef Estermann, proposes a pachasophy:

> "*Pacha* means earth, world, 'planet,' 'space for living' but also 'universe, cosmic stratification' … When dealing with the 'earth' as the foundation of life, one uses both the expression Pachamama ('mother earth') or simply *Pacha* … Philosophically *Pacha* signifies the universe ordered in special-temporal categories, but not simply something physical and astronomical. *Pacha* also could be an equivalent homeomorphic of the Latin word *ese* ('to be'); *Pacha* is 'that which is,' universal existence, 'reality' … *Pacha* as 'interrelated cosmos' or 'cosmic relationality.'" (Estermann 2011, pp. 156, 157, 158)

For Estermann, pachasophy is not a philosophy that speaks exclusively of a particular or regional worldview but rather is a universal philosophy. It is the philosophy of the *Pacha* with global implications because pachasophy is "integral reflection about cosmic relationality, as the manifestation of the collective Andean experience of 'reality'" (Estermann 2011, p. 158). It is a way of thinking about reality from *Pacha*, that is, the whole of existence. The Andean philosophy of *Pacha* thinks of the human community as a great family that lives beneath a single roof for its home. The Universe is a House. The Pachamama is Mama Pacha, a living place or body home, the Mother Earth.

The Andean gaze of indigenous people looking upon nature is a non-exotic view of feminine geography. This Andean gaze generates a great semantic change: from the disdain and commodification viewed though "imperial eyes" to the great love and respect for Mother Earth or Pachamama by native peoples. Rising from the local, the idea of Pachamama counters the curse of "exotic America" and its synonym "great American whore." From a universal perspective, Latin America can serve as a new eco-spiritual feminine center of the Earth and make great contributions in areas such as judicial, ethical, aesthetic, and spiritual (Paredes-Castellanos 2017).

10.4.4 On a "Reserve of Life" to be Defended

In the case of Ecuador, the first country to recognize Mother Earth or Pachamama as having legal rights by incorporating them into the nation's constitution, Gudynas explains:

> nature or Pachamama is given rights recognized by the Constitution and among them are 'integral respect for her existence'; the 'maintenance of regeneration of her vital cycles, structural functions and evolutionary processes'; the 'restoration and mitigation of harmful environmental consequences'; and the 'integrity of national genetic patrimony and the permanence of biological diversity and natural cycles.' (2009a, p. 195)

Recognizing nature as subject to rights in no way counters humans as having rights. Rather it affirms the rights of the men and women of the South by making them even more visible and their rights legally protected. Latin American biocentrism asserts that the defense of nature as a subject does not somehow objectify human beings. As the phenomenology of the body (Merleau-Ponty 1975) reminds us, people are bodies and, therefore, are part of the landscape. This phenomenological perspective points to a new ontological formulation of the body-landscape relation, which has been undertaken by the proposal of *Florestania* (Guydnas 2009a, p. 175).

"Florestania" is a term in Portuguese that results from combining the words *floresta* (jungle) and *ciudadanía* (citizenship). Its origin is not an academic analysis but emerged among environmental activists, journalists, and politicians from the Amazonian state of Acre in Brazil (Gudynas 2009b). In the heart of the Amazon, Acre hosts a unique political and environmental context, being the state where the leader *siringueiro* Chico Mendes defended the Amazon rain forest by proposing a combination of traditional and innovative sustainable practices.[2] At the end of the

[2] Francisco Alves "Chico" Mendes Filho coined and implemented the concept of "extractive reserves" as a way to defend the Amazonian forests and the rights of sustainable use practices by rubber tapper communities in the 1980s (see Gross 1989).

1980s, Mendes' practices and demands entailed the defense of ecosystems and respect for local communities and their traditional forms of life. Likewise, his militancy as a union and social leader invoked a citizen practice. The arrival of the Brazilian Workers' Party to Acre in the late 1990s led to the rapid adoption of the term *florestania*, as one of the slogans of state management. From there, the term was transferred to other instances such as productive enterprises, the promotion of a university center, and even a distinction: the Chico Mendes Prize of Florestania awarded annually by the government of the State of Acre since 2004.

Unlike the proposal of environmental citizenship by British political author Andrew Dobson, the proposal of *Forestania* is more diffuse, and it arises from a close link to the native habitat, implying an ethical demand. Firstly, it involves a precise ecological context: the Amazon rain forest ecosystems. Secondly, these ecosystems exhibit a great heterogeneity, including little modified patches along with others where human intervention exists, either by indigenous groups or by the *siringueiros*. Thirdly, *Florestania* emphasizes an intimate association between this ecological context and social practices; one is not possible without the other. This perspective coincides with the core statement of the biocultural ethic: the continuity of life habits is only possible if habitats and communities of multiple-species co-inhabitants are conserved (Rozzi 2013). Extractive reserves and rubber gathering are only possible if there is an Amazon jungle; *siringueiros* can only exist in the rubber fields. Therefore, the concept of *Florestania* involves a demand for socio-environmental justice (Rozzi 2015a) and resistance to the drastic modification of these ecosystems by processes such as modern agriculture or extensive livestock farming (Gudynas 2009b). These changes lead to the disappearance of both the original environments and the local indigenous and mestizo communities. Therefore, the existence of these "citizens of the forest" requires the permanence of that Amazonian forest. Fourthly, "forestry" is not a citizenship of strangers (as in Dobson's proposal 2003) but involves communities of individuals who share common histories, identities, and direct links between them. In this way, the concept of *Florestania* allows us to revalue the ecological space of the forest for its environmental as well as social importance, as well as to the political subjects, their identities, and their culture.

In the *Florestania* formulation, the classic modern division between society and nature is blurred. Philosophically it is understood that the same human body is part of the geographic corporal landscape just because it is the body. The *Florestania* concept implies that *to be* is *to be landscape*. This philosophy is the foundation for another, the philosophy of Living Well or *buen vivir* in Spanish, *sumac kawsay* in Quechua, and *suma qamaña* in Aymara. This philosophy refers to a life of dignity as part of the Pachamama (Acosta and Martínez 2009). *Buen vivir*, *sumac kawsay*, or *suma qamaña* is a philosophy for an alternative to Western-style development, one that respects, and always is aware of, the limits of Pachamama. From the South, a new way of life and human self-understanding are revealed in this new eco-ontological, post-exotic wisdom.

10.5 Concluding Remarks from the South

Today, biocultural exoticism is resurfacing. The myth of landscapes and bodies of "exotic America" again affects Latin America. But at the same time, in Latin America, post-exoticism constantly pushes efforts to overcome the exoticism of environment and society. An appreciation, without prejudices, stereotypes, deception, or deformed images, emerges for a whole realm of cultural and biological knowledge. This renewed appreciation is contributing to building a new future for human and other-than-human life in the face of climate change, global warming, and worldwide pollution, as well as of growing technological impositions that tend to supplant the bodies of nature (humans included) by machines and robots; the new "foreigner" with new machines, the new technological exoticism.

The greatest contribution that we can take away from our concise review of Latin American environmental arts, thoughts, and movements, especially from the "eco-feminism of the South," is the understanding of nature as a great body in which we co-inhabit. The feminine geographic body has been penetrated violently, and this geo-corporal damage has made it gravely ill. The ecological-corporal health of the Earth body of the Pachamama needs an "eco-medicine" that takes into account biological-ecological, anatomic-geographic, cultural, philosophical, and spiritual dimensions in complementary ways.

For global society, the philosophy of the Pachamama is a great contribution. The international ecological situation could be nourished by local environmental philosophies with their respective and diverse ecological knowledge that offer novel understandings about respectful and responsible relationships between humans and the Earth (Callicott 1997). To overcome the colonial wounds in the great body of Mother Earth—Pachamama—we must take again an eco-loving ethics toward nature, a Southern eco-eroticism.[3]

Acknowledgments We thank the valuable, and many editions by Roy May, and the critical comments by Francisca Massardo and Irene Klaver who helped improving this chapter. Angelina thanks also the Universidad Michoacana de San Nicolás de Hidalgo, Mexico, and the Institute of Ecology and Biodiversity (IEB-Chile; grant Financiamiento Basal CONICYT AFB170008) and Universidad de Magallanes for the support that allowed her to conduct research at the Omora Ethnobotanical Park in the Cape Horn Biosphere Reserve, in Chile.

References

Acosta A, Martínez E (eds) (2009) El Buen Vivir. Una vía para el desarrollo. Ediciones Abya Yala, Quito

Aillapan L, Rozzi R (2001) Veinte Poemas Alados de los Bosques Nativos del Sur de Chile. Editorial Plaza y Valdés, México City

[3]We borrow, and adapt, here the creative analysis of J.B. Callicott on "South American Eco-Eroticism" in his book *Earth's Insights* (1997).

Aillapan L, Rozzi R (2004) Una etno-ornitología mapuche contemporánea: Poemas alados de los bosques nativos de Chile. Ornitología Neotropical 15(Suppl):419–434
Báez C, Mason P (2006) Zoológicos Humanos, fotografías de fueguinos y mapuche en el jardín d'acclimatation de París, siglo XIX. Pehuen, Santiago
Chavarría-Zamora MJ (2013) Construcciones/Invenciones. De la Suiza Centroamericana al país más feliz del mundo. Museo de Arte y Diseño Contemporáneo, MADC, San José
Chavarría MJ (ed) (2016) Joaquin Rodriguez del Paso, Super Moderno. San José, Costa Rica, Diseño Contemporáneo
Callicott JB (1997) Earth's insights: a multicultural survey of ecological ethics from the Mediterranean Basin to the Australian outback. University of California Press, Berkeley
de Alencar MJ (2000) Iracema. Barcelona, Ediciones Obelisco
Dobson A (2003) Citizenship and the environment. Oxford University Press, Oxford
Espinosa-Miñoso Y (2009) 12 de octubre: conmemorar la violación originaria. Suplemento Pan y Rosas 12(September)
Estermann J (2011) Filosofía andina. Sabiduría indígena para un nuevo mundo. Plural, La Paz
Galeano E (2011) Las venas abiertas de América Latina. México DF, Siglo XXI
Giraudo V (2016) Antropofagia y modernidad, Arte brasileño en la colección Fadel. MALBA/Museo de Arte Nacional, Buenos Aires/México
Gross T (ed) (1989) Fight for the forest—Chico Mendes in his own words. Latin America Bureau, London
Grün M (2012) Ética E Educação Ambiental: a Conexão Necessária. Papirus, Campinas/Sao Paulo
Gudynas E (2009a) El mandato ecológico, Derechos de la Naturaleza y políticas ambientales en la nueva Constitución. Ediciones Abya Yala, Ecuador
Gudynas E (2009b) Ciudadania ambiental y meta-ciudadanias ecológicas:revision y alternativas en America Latina. Desenvolvimento e Meio Ambiente 19:53–72
Mamani-Bernabe V, Quispe-Huanca C (2007) Pacha. Editorial Verbo Divino SRL – Bolivia, La Paz
Merleau-Ponty M (1975) Fenomenología de la percepción, tr. Jem Cabanes. Ediciones Península, Barcelona
Mignolo WD (2007) La idea de América Latina, la herida colonial y la opción decolonial. Gedisa, Barcelona
Núñez-Becerra F (2002) La Malinche, de la historia al mito. Divulgación serie Historia, INAH, México DF
Paredes-Castellanos A (2017) "América Latina Exótica": Hermenéutica Ecoestética de la Naturaleza y el Cuerpo. Una Aproximación al tema. Doctoral Dissertation in Philosophy, Instituto de Investigaciones Filosóficas "Luis Villoro", Universidad Michoacana de San Nicolás de Hidalgo, Morelia, Michoacán, Mexico
Pratt ML (2010) Ojos imperiales. Literatura de Viajes y Transculturación, tr. Ofelia Castillo. Fondo de Cultura Económica, México DF
Primack R, Rozzi R, Feinsinger P, Dirzo R, Massardo F (2006) Fundamentos de Conservación Biológica: Perspectivas Latinoamericanas. Fondo de Cultura Económica, México D.F
Rozzi R (2004) Implicaciones éticas de narrativas yaganes y mapuches sobre las aves de los bosques templados de Sudamérica austral. Ornitología Neotropical 15:435–444
Rozzi R (2001) Éticas ambientales latinoamericanas: raíces y ramas. In: Primack R, Rozzi R, Feinsinger P, Dirzo R, Massardo F (eds) Fundamentos de Conservación Biológica: Perspectivas Latinoamericanas. Fondo de Cultura Económica, Mexico DF, pp 311–362
Rozzi R (2012a) Biocultural ethics: the vital links between the inhabitants, their habits and regional habitats. Environ Ethics 34:27–50
Rozzi R (2012b) South American environmental philosophy: ancestral Amerindian roots and emergent academic branches. Environ Ethics 34:343–365
Rozzi R (2013) Biocultural ethics: from biocultural homogenization toward biocultural conservation. In: Rozzi R, STA P, Palmer C, Armesto JJ, Callicott JB (eds) Linking ecology and ethics

for a changing world: values, philosophy, and action. Ecology and ethics, vol 1. Springer, Dordrecht, pp 9–32

Rozzi R (2015a) Implications of the biocultural ethic for earth stewardship. In: Rozzi R, Chapin FS, Callicott JB, Pickett STA, Power ME, Armesto JJ, May RH Jr (eds) Earth stewardship: linking ecology and ethics in theory and practice. Ecology and ethics, vol 2. Springer, Dordrecht, pp 113–136

Rozzi R (2015b) Earth stewardship and the biocultural ethic: Latin American perspectives. In: Rozzi R, Chapin FS, Callicott JB, Pickett STA, Power ME, Armesto JJ, May RH Jr (eds) Earth stewardship: linking ecology and ethics in theory and practice. Ecology and ethics, vol 2. Springer, Dordrecht, pp 87–112

Rozzi R, Arango X, Massardo F, Anderson C, Heidinger K, Moses K (2008) Field environmental philosophy and biocultural conservation: the Omora Ethnobotanical Park educational program. Environ Ethics 30:325–336

Rozzi R, Armesto JJ, Gutiérrez J, Massardo F, Likens G, Anderson CB, Poole A, Moses K, Hargrove G, Mansilla A, Kennedy JH, Willson M, Jax K, Jones C, Callicott JB, Kalin MT (2012) Integrating ecology and environmental ethics: earth stewardship in the southern end of the Americas. BioScience 62(3):226–236

Said WE (2016) Orientalismo, tr. María Luisa Fuentes, Penguin Random House, Debolsillo, México DF

Segalen V (1989) Ensayo sobre el exotismo. Una estética de lo diverso (y textos sobre Gauguin y Oceanía. Fondo de Cultura Económica, México DF

Chapter 11
Overcoming Biocultural Homogenization in Modern Philosophy: Hume's Noble Oyster

Ricardo Rozzi

Abstract The great influence that the eighteenth-century philosopher David Hume had on Darwin's conception of his evolutionary theory offers today a paradigmatic case for advancing an interdisciplinary integration between philosophical and scientific ideas. This interdisciplinary integration offers novel approaches to address some of the complex indirect drivers of current socio-environmental problems, such as biocultural homogenization. The identification of philosophical factors linked to losses of biological and cultural diversity adds to the concept of indirect drivers used by the Millennium Ecosystem Assessment. In this chapter, I undertake three interrelated goals. The first is to expose philosophical concepts and methods that are helpful to understand some complex indirect drivers of biocultural homogenization. The second is to investigate in Hume's work philosophical foundations to overcome the prevailing taxonomic bias that favors only a few vertebrates and to contribute overcoming the exclusion of moral consideration for the most diverse groups of animals inhabiting our planet. My third, and the most general, goal is to demonstrate that it is possible to *de-homogenize* a prevailing negative view about European modern philosophy and to invite readers to discover, instead, some environmental values in Western thinkers and schools of thought that can be key for overcoming taxonomic biases and their associated impact on biocultural homogenization.

Keywords Animals · Darwin · Ethics · Sentient · Taxonomic bias

---

R. Rozzi (✉)
Department of Philosophy and Religion and Department of Biological Sciences, University of North Texas, Denton, TX, USA

Sub-Antarctic Biocultural Conservation Program, University of North Texas, Denton, TX, USA

Instituto de Ecología y Biodiversidad and Universidad de Magallanes, Punta Arenas, Chile
e-mail: Ricardo.Rozzi@unt.edu

R. Rozzi et al. (eds.), *From Biocultural Homogenization to Biocultural Conservation*, Ecology and Ethics 3, https://doi.org/10.1007/978-3-319-99513-7_11

11.1 A Potential Philosophical Driver of Biocultural Homogenization

Biocultural homogenization involves a severe reduction in biological and cultural diversity and their interrelations. It causes a reduction in the diversity and abundance of native co-inhabitants (humans and other-than-humans) and their replacement by a few cosmopolitan co-inhabitants and a few global life habits. This reduction involves complex processes of elimination of co-inhabitants from the mind-sets of human colonizers (and/or colonized populations), as much as from their native biophysical habitats.

The identification of cultural-linguistic and biophysical factors linked to losses of biological and cultural diversity echoes the concepts of direct and indirect drivers used by the Millennium Ecosystem Assessment to identify causal agents of anthropogenic change (MEA 2005). Direct drivers are those agents of change that are physical or mechanical processes, whereas indirect drivers are cultural, ethical, and socially related agents of change. As highlighted by Alexandria Poole (2018, p. 317), "understanding the inter-dynamics among culture, values, and lifestyles within the indirect drivers is a complex task that is often demarcated to the humanities or social sciences, fragmenting causal discussions regarding management decisions and ecological knowledge across the disciplines." According to Stephen Carpenter and collaborators (Carpenter et al. 2006), focusing on indirect drivers remains a pressing research need in the process of linking social to ecosystem change, because most research related to ecosystem management focuses on direct drivers, such as land use change or invasive species. However, effective management requires more attention to indirect drivers such as cultural factors.

Carpenter's viewpoint is applicable to one of the central goals of our *Biocultural Homogenization* book: to contribute to reorienting this process toward the conservation of biological and cultural diversity. Toward this aim, this chapter analyzes the diversity of all animals included in the work of one of the most influential modern philosophers: David Hume. This eighteenth-century Scottish empiricist had a seminal influence on Charles Darwin's evolutionary thinking (Huntley 1972). The historical links between Hume and Darwin's ideas offer a paradigmatic case for the type of integration between philosophical and scientific ideas that we need today to address complex socio-environmental problems (Rozzi 1999). My purpose in this chapter is to find in Hume's work modern philosophical foundations to overcome the exclusion of moral consideration for the most diverse groups of animals inhabiting our planet.

When the field of environmental philosophy was established in the 1970s, several philosophers concerned with the current biodiversity crisis and the widespread abuse of nonhuman animals criticized and even rejected "conventional Western moral philosophy" as a viable option to address these pressing problems in contemporary society (cf. Passmore 1974; Sessions 1994; Zimmerman 1994). In my view, these criticisms stated in such general terms limit opportunities for finding contrasting environmental values and attitudes, as well as philosophical traditions in Western civilization; moreover, a given philosophical school or author admits diverse readings and is susceptible to more interpretations than the critics tend to admit (cf. Gare 1995, 1998).

To support this thesis, which is relevant to *de-homogenize modern philosophy*, I will inquire to what extent the work of Scottish philosopher David Hume does, or does not, provide a philosophical foundation to broaden the boundaries of moral consideration beyond humans toward nonhuman animals. First, I will investigate which animals are included in Hume's work and to what extent he proposes that human and nonhuman animals share the attributes of reason and sentiency. I chose to focus on animals and these two attributes, because for the two main schools of modern philosophy, any being that deserves moral consideration must have the faculties of reason and/or sentiency (cf. Passmore 1975; Midgley 1978; Sorabji 1985; Singer 1993; Palmer 2013). In the second part, I will discuss how Hume's work could provide foundations for a moral consideration of all kinds of animals and more broadly for environmental ethics.

11.2 Hume's Inclusion of Sections on Animals in His Main Works About Human Nature

In his philosophical enterprise to understand human nature, Hume directs his attention to animals in the search for common natural processes.[1] In his main works on human nature, he dedicates an independent section to the "reason of animals": Section XVI of the *Treatise of Human Nature* (T) (Hume 1978), which he expanded a decade after the *Enquiry Concerning Human Understanding* (EHU) (Hume 1976a). Hume proposed his approach in contrast to Cartesian rationalist philosophy that attributed reason only to humans and provocatively opened his Section XVI *Of the Reason of Animals* by writing that:

> No truth appears to be more evident, than that beasts are endowed with thought and reason as well as men. The arguments are in this case so obvious, that they never escape the most stupid and ignorant. (T 176)

Later, in Book II of the *Treatise*, Hume extends the resemblance between human and animal nature to the origin of the passions. He dedicates two particular sections to it: "Of the pride and humility of animals" (in Part I) and "Of the love and hatred of animals" (in Part II). In this manner, Hume provides a bridge between human and animal reason and emotions, which would leave behind Cartesian dualistic distinctions. However, which animals does Hume consider in his analogical thinking?

[1] Because Hume dedicates separate sections *in* THN and EHU to humans and [other] animals, I will maintain Hume's language in some passages of this paper, by referring to humans and animals. However, for my own analyses, I understand humans as another animal species.

11.3 In Hume's Work: Which Animals Share with Humans Attributes of Reason and Sentiency?

Rephrasing from a contemporary scientific evolutionary perspective, to what extent can Hume's analogy between humans and other animals be projected along the evolutionary tree? Which kind of animals are mentioned by Hume? Two basic branches of animal types associated with the scientific evolutionary tree correspond to the vertebrate and invertebrate divisions. Hence, a basic first question is: Do Hume's analogies apply to both vertebrates and invertebrates?

In the sections of the *Treatise* and the *Enquiry* dedicated to animal reason, Hume provides only a few examples of animals and only vertebrates. However, when he makes general statements about reason and sentience, Hume expresses them in such a way that enables us to extend them to the whole animal kingdom. For example, when Hume introduces the topic of animal reason in Book I of the *Treatise*, he writes:

> We are conscious, that we ourselves, in adapting means to ends, are guided by reason and design, and that 'tis not ignorantly nor casually we perform those actions, which tend to self-preservation, to obtaining pleasure, and avoiding pain. When therefore we see *other creatures, in millions of instances*, perform like actions, and direct to like ends, all our principles of reason and probability carry us with an invisible force to believe the existence of a like cause.' *Tis needless in my opinion to illustrate this argument by the enumeration of particulars*. The smallest attention will supply us with more than are requisite. The resemblance betwixt the actions of animals and those of men is so entire in this respect, that *the very first action of the first animal we shall please to pitch on*, will afford us an incontestable argument for the present doctrine. (T176) [emphasis added]

Hume affirms here that examples are unnecessary; it is needless to enumerate particular animals to prove that animals share "a like cause" with humans. The analogy between human and animal reason seems to Hume such a general phenomenon that the "very first action of the first animal" we may look at would confirm the resemblance between animal and human actions. Therefore, we can conclude that when Hume suggests that "a like cause" that "guide[s] reason and design," his analogical thinking could be applied to all three: human, vertebrate, and invertebrate animals.

This interpretation is, however, in marked contrast with the actual examples of animals that Hume provides to develop his analogy. For example, at the beginning of the section "Of the Reason of Animals" in the *Treatise*, Hume gives only two examples, and both correspond to vertebrates:

> Here we must make a *distinction betwixt* those actions of *animals*, which are *of a vulgar nature*, and seem to be on a level with their own capacities, and those more extraordinary instances of *sagacity*, which they sometimes discover for their own preservation, and the propagation of their species. *A dog*, that avoids fire and precipices, that shuns strangers, and caresses his master, affords us an instance of the first kind. *A bird*, that chooses with such care and nicety the place and materials of her nest, and sits upon her eggs for a due time, and in a suitable season, with all the precaution that a chymist is capable of in the most delicate projection, furnishes us with a lively instance of the second. (T177) [emphasis added]

This passage seems to contradict Hume's previous statement about the futility of examples. First, he defines a hierarchy among animals: vulgar and sagacious. Second, he illustrates the analogy between human and animal reason only with a mammal and a bird (the most "evolved" or "sagacious" vertebrates). Hume maintains the trend of mentioning only vertebrates, particularly mammals with the general term of "beasts,"[2] throughout the section on "Animal Reason," concluding that:

> [L]et any philosopher make a trial, and endeavor to explain that act of mind, which we call belief, and give an account of the principles, from which it is derived, independent of the influence of custom on imagination, and *let his hypothesis be equally applicable to beasts as to the human species*; and after he has done this, I promise to embrace his opinion. (T178) [emphasis added]

In Book II of the *Treatise*, in the sections dedicated to the passions of animals, Hume provides again general statements that allude, at least potentially, to all animal species (invertebrates included), but his specific examples are restricted to birds and mammals. For instance, in the section "Of the Pride and Humility of Animals," he writes:

> It is plain, that almost in *every species of creatures*, but especially of the nobler kind, there are many evident marks of pride and humility. The very port and gait of a *swan*, or *turkey*, or *peacock* show the high idea he has entertained of himself, … The vanity and emulation of *nightingales* in singing have been commonly remarked; as likewise that of *horses* in swiftness, of *hounds* in sagacity and smell, of the *bull* and *cock* in strength, and of *every animal* in his particular excellency. (T326) [emphasis added]

However, at the end of the section that includes the former paragraph, after mentioning only examples of birds and mammals, Hume seems to privilege his general statement alluding to all animal species. He concludes categorically that: "All these are evident proofs, that pride and humility are *not merely human passions, but extend themselves over the whole animal creation.*" [emphasis added].

To clarify Hume's puzzle about which animals can be considered in his analogies, I prepared a synthesis of the specific examples of animals that Hume gives throughout the *Treatise* and the *Enquiry*, as well in his other writings. Table 11.1 includes all references to animals I found in Hume's complete works by using the search tools of the Past-Masters CD (Hume 1948, 1976a, b, 1978, 1995). In Table 11.1, a conspicuous feature strikes us immediately: vertebrate animals fill most of the table. Among vertebrates, in turn, the bias toward birds and mammals is evident. This conspicuous inclination toward vertebrates is consistent throughout the different works: the *Treatise*, both *Enquiries* and Hume's letters, and other published works.

Table 11.2 presents a quantitative analysis of the animals detailed in Table 11.1, providing the total numbers and relative percentages for each species or kind of animals. These "species or kinds" of animals do not necessarily correspond with the concept of biological species, because they include different taxonomic categories. If we use our contemporary biological taxonomic criteria, we find that Hume refers

[2] The eighteenth-century *Samuel Johnson's Dictionary* defines beast as "an animal distinguished from birds, insects, fishes, and man" (see Reddick 1996).

Table 11.1 Animals found in Hume's *Treatise of Human Nature* (T), *An Enquiry Concerning Human Understanding* (EHU), *An Enquiry Concerning the Principles of Morals* (EPM), other published works, and letters

		Animals mentioned by David Hume			
Animal taxa	Animal taxa	T[a]	EHU[a]	EPM[a]	Other works and letters[b]
Invertebratesws	**Mollusca**	Oyster (634)	–	Snail (footnote 24)	Cockle [1], oyster [1]
	Arthropoda	Insect (28), mite (28)	–		Bee [1], butterfly [1], drone [1], flea [1], fly [2], insect [2], silkworm [2]
Vertebrates	**Fishes**	Fish (303)	Fish (104)	–	Eel [1], fish [12], herring [2], lampreys [3], salmon [1], shad [1], sprat [1]
	Amphibia	–	Frog (104)	–	Frog [1]
	Reptiles	–	–	Serpent (179, 271), tortoise (273)	Chameleon [1], crocodile [2], serpent [2], snake [1], turtle [1]
	Birds	Bird (177, 284), cock (326), crow (451), magpies (451), nightingale (326, 485), partridge (451), peacock (326), pheasant (451), plover (452), swan (326), Turkey (326), woodcock (452)'	Bird (108), dove (271)	Pigeon (198)	Bird [3], chicken [5], cock [4], crow [2], dove [1], duck [1], eagle [1], gamecock [1], goose [6], fowl [1], hawk [1], hen [6], ibis [1], kite [1], magpie [1], nightingale [2], partridge [6], peacock [2], pheasant [4], pigeon [6], plover [1], pullet [2], swan [2], Turkey [2], woodcock [2]

	Mammals	Bull (326), cat (398), dog (177, 279, 327, 397, 398, 660), horse (10, 279, 310, 326, 378, 385, 397, 398, 529, 557, 576, 660), hound (310, 326), ion (398, 484), ox (397, 398, 484, 485), sheep (484, 485), tiger (398)	Dog (105), greyhound (105), hare (105), horse (19, 47, 48, 105, 129, 179), lion (28), tiger (28)	Hare (237), horse (244, 310), wolf (179, 271)	Ape [2], ass [5], bear [2], bitch [1], boar [4], buck [1], bulldog [2], bull [8], camel [1], capon [1], cat [11], cow [7], deer [7], dog [27], elephant [1], fox [2], hare [4], heifer [1], hog [5], horse [199], hound [2], lamb [4], lion [4], monkey [3], mouse [2], nag [1], ox [6], pig [2], rabbit [1], ram [2], rat [1], rhinoceros [1], sheep [13], sow [1], stag [1], swine [1], tiger [2], veal [1], wether [2], whale [1], wolf [5]

[a]Numbers within round parenthesis indicate the page numbers for the occurrence of each kind of animal species indicate the number of paragraphs in Hume's works (including his letters) in which they are mentioned

[b]Numbers within square parenthesis indicate the total number of paragraphs in Hume's work in which each kind of animal occurs. The list and sums of kinds of animals found in Hume's written documents was elaborated on the basis of the text and computational tools provided by the Complete Works of David Hume in "Past Masters". (CD-ROM, Charlottesville VA: InteLex Corporation, Hume 1995)

Table 11.2 Total numbers of occurrences of vertebrate and invertebrate groups of animals in Hume's texts. The columns "kind of animal" provide an indicator for the diversity of animals mentioned by Hume. The columns "total occurrences" provide the total number of paragraphs within Hume's work in which a member of each animal group is mentioned and is proposed as an indicator of the "abundance" of each group of animals. The percentage (%) is calculated as the number kinds of animals or occurrences within each animal groups divided by the total number of kinds of animals or occurrences, respectively. On the left part of the table, the values for the *Treatise* (T), the *Enquiry concerning Human Understanding* (EHU), and *An Enquiry concerning the Principles of Morals* (EPM) are indicated

	Numbers and relative percentages of animals mentioned by David Hume							
	T + EHU + EPM				Hume's complete works			
Group of animals	Number of kinds	(%)	Total occurrences	(%)	Kinds	(%)	Total occurrences	(%)
Invertebrates								
Mollusca	2	5.9	2	2.7	3	3.2	4	0.8
Arthropoda	2	5.9	2	2.7	9	9.7	12	2.3
Subtotal	**4**	**11.8**	**4**	**5.4**	**12**	**12.9**	**16**	**3.0**
Vertebrates								
Fishes	1	2.9	2	2.7	7	7.5	23	4.4
Amphibian	1	2.9	1	1.4	1	1.1	2	0.4
Reptiles	2	5.9	3	4.1	6	6.5	10	1.9
Birds	14	41.2	17	23.0	25	26.9	81	15.4
Mammals	12	35.3	47	63.5	42	45.2	394	74.9
Subtotal	**30**	**88.2**	**70**	**94.6**	**81**	**87.1**	**510**	**97.0**
Total	**34**	**100.0**	**74**	**100.0**	**93**	**100.0**	**526**	**100.0**

sometimes to "supra-specific taxonomic categories" – such as families (i.e., kite, eagle), orders (i.e., butterfly, frog), and classes (bird, insect) – or "infra-specific taxonomic categories," such as races (i.e., bulldog and greyhound), gender (i.e., cock and hen), development stages (i.e., lamb and sheep), fertility condition (i.e., bull and ox), or linguistic synonyms (i.e., dog and hound) that belong to the same biological species. (To minimize confusion about Hume's taxonomic categories, I unify them under the term "kinds" of animals.) The disparity between Hume's and contemporary biological taxonomic categories does not obscure the impressive inclination in favor of vertebrates: they constitute between 87% and 97% of the examples given by Hume in different works.

In his complete writings, including his letters, Hume refers to 93 kinds of animals (Table 11.2). Among these, approximately 90% correspond to vertebrates. Among vertebrates, mammals and birds are markedly dominant: they alone account for more than 70% of the animal kinds mentioned by Hume. When we focus on total number of occurrences (considering repetitions of the same kind of animal), then the bias toward vertebrates, particularly mammals and birds, is even more marked. Of the 526 occurrences of animals found in Hume's complete work, 97% correspond to vertebrates. Moreover, among the 510 mentions of vertebrates, two animals account for more than 50%: the horse (43%) and the dog (8%).

Table 11.3 World's total numbers of animal species that have been described and scientifically accepted (Chapman 2009)

Group of animals	Number of species	(%)
Invertebrates		
Mollusca (snails, oysters, etc.)	~85,000	6.0
Arthropoda (insects, arachnids, etc.)	1,175,873	82.6
Other invertebrates (sponges, worms, etc.)	101,465	7.1
Subtotal	***1,362,338***	***95.6***
Vertebrates		
Fishes	31,269	2.2
Amphibia	6515	0.5
Reptiles	8734	0.6
Birds	9990	0.7
Mammals	5487	0.4
Subtotal	***61,995***	***4.4***
Total	***1,424,333***	***100.0***

This analysis of animals that Hume mentions in his works shows that his mindset is not only vertebrate-centric but also mammal-centric, with an unbalanced focus on horses and dogs. Hume's taxonomic bias could be a symptom, as well as a driver, of a general trend in modern European culture that tends to focus on, and favor, a few mammals. This narrow focus on mammals collides with the actual diversity of animals. Among animals, less than 5% are vertebrates (Table 11.3). Moreover, invertebrates not only represent more than 95% of all known species, but they also include 30 phyla, while vertebrates include only one phylum: Chordata (Mora et al. 2011).[3]

Conservation biologists have criticized that more than 95% of the animal species correspond to a group almost completely neglected by modern philosophy, as well as by conservation biology literature: the invertebrates (Wilson 1988; Clark and May 2002). Only 1% of the described animal species correspond to "higher" mammals or birds. The "rest" represent 99% of animal diversity and includes cold-blooded vertebrates (or ectotherms – fishes, amphibians, and reptiles) and invertebrates, which contribute slightly more than 3% and 95% of animal species, respectively (Table 11.3). This philosophical bias toward "higher vertebrates" seems to be more marked when we introduce the evolutionary temporal dimension, because compared to the 700-million-year-old invertebrates, "higher mammals" are recent newcomers (Rozzi 2001).

Hume could be condemned for promoting biocultural homogenization, due to the drastic reduction in the spectrum of kinds of animals that he mentions in his work. He is one of the most influential philosophers in modern schools of thought, and the fact that most of the animals inhabiting the planet are not considered in his

[3] Of the 30 phyla of invertebrates that are known to science, only 2 are included in Hume's examples of invertebrates: Mollusca (oysters, cockles, snails) and Arthropoda (insects [bees, drones [male bees], butterflies, fleas, flies, silkworms], mites) (see Table 11.1).

philosophical arguments might have influenced a reduction in the spectrum of animals included in modern culture. Through colonialism, European culture, including Hume's philosophy, has been disseminated worldwide. Under this influence, students and citizens remain blind about the philosophical and particularly ethical values of most animal species on planet Earth.

If the previous reasoning would exhaust Hume's work, then the work of this philosopher should be "thrown into the flames"[4] for two reasons. First, there is an obvious dissociation between the empirical evidence about the diversity of animals and the representation of this diversity in his examples. Second, Hume's taxonomic bias, his vertebrate-centrism, might have dire consequences for the relationship between human beings and most animals and lead to processes of drastic biocultural homogenization.

The purpose of our book on *Biocultural Homogenization* is, however, not only to criticize this process but also to explore options to reorient it toward processes of conservation of biological and cultural diversity. In that vein we ask: Could we find in Hume's work some fracture in his vertebrate-centrism? Could we find in Hume's work a broader philosophical foundation that would justify ethical consideration of a wide spectrum of vertebrate and invertebrate animals?

11.4 Hume's Noble Oyster

The results presented in the former section seem to threaten the thesis that Hume's analogy between animal and human reason would be applicable to the "whole animal kingdom." However, let's give him a historical release for the taxonomic bias in his examples of animals. Hume's bias could be associated with historical circumstances, such as the prevalence of vertebrates in eighteenth-century sciences (see Bowler 1993). Hence, before rejecting the thesis that Hume's analogy between human and animal nature is applicable to both invertebrates and vertebrates, we should analyze his few examples of invertebrates.

In the *Treatise*, Hume refers to invertebrates only three times. Two of them (an insect and a mite) appear in a section on *the infinite divisibility of our space and time*, which says nothing about their capacity for reason and sentience. However, Hume's third invertebrate example appears at the end of the *Treatise* and corresponds to an oyster used to illustrate the case of an animal provided with the lowest potential for having a mind. Hume writes:

[4]Hume concluded his *Enquiry Concerning Human Understanding* (p 166) by stating that: "If we take in our hand any volume; of divinity or school metaphysics, for instance; let us ask, Does it contain any abstract reasoning concerning quantity and number? No. Does it contain any experimental reasoning concerning matter of fact and existence? No. *Commit it then to the flames,* for it can contain nothing but sophistry and illusion" (emphasis added). This passage has played a crucial role in the appropriation of Hume's philosophical legacy by twentieth-century positivist philosophers (see Frasca-Spada 1996).

> We can conceive a thinking being to have either many or few perceptions. Suppose the mind to be reduced even below the life of an oyster. Suppose it to have only one perception, as of thirst or hunger. (T634)

This passage represents a radical statement by Hume: the oyster, an animal that lacks any obvious resemblance to the human body, much less to human behavior, is still considered by Hume as deserving a mind. The thinking or sentient oyster: What a shock for British society accustomed to knowing about this animal only on their dinner plates! Hume is radical because nothing stops him in his empiricist project, and therefore he is able to conceive a thread that links all animals, from the highest to the lowest, even below the life of an oyster. In connecting the processes of perception and behavior among diverse animals, Hume affirms:

> There are also instances of relation of impressions, sufficient to convince us, that there is a union of certain affections with each other in the inferior species of creatures as well as in the superior, and that their minds are frequently conveyed through a series of connected emotions. (T327)

Although Hume conceives a hierarchy among animals that places humans at the top, he proposes that even the few perceptions of the lowest living animals – such as the oyster's thirst or hunger – rely on processes that operate in kindred ways among higher animals, such as vertebrates, humans included. The elegance of the argument and the radical departure of Hume's imagination from his social environment incline me to accept his previously quoted claim that it is "needless to illustrate this argument by the enumeration of particulars.... The resemblance betwixt the actions of animals and those of men is so entire in this respect, that the very first action of the first animal we shall please to pitch on, will afford us an incontestable argument for the present doctrine." In Hume's view an oyster adapts "means to ends" struggling for self-preservation, to obtain pleasure and avoid pain, "guided by reason and design" in a way comparable to birds (such as swans, turkeys, peacocks, nightingales) or mammals (such as horses, hounds, or bulls).

The disproportion of examples in favor of the better known vertebrates can lead to misleading interpretations concerning the extent of Hume's analogy. For example, the ethologist William Homan Thorpe (1979) writes:

> David Hume held it be self-evident that the *beasts*, as brothers of men, were endowed with thought and reason.... Naturalists had by that time [the last quarter of 19th century] generally accepted the conclusion of David Hume's *A Treatise of Human Nature* that *higher animals* use the same principles of intel.

Thorpe addresses the relevance of Hume for the origins of ethology. However, Thorpe's interpretation restricts the extension of reason to beasts or higher animals. This represents a serious mistake because it overlooks the refinement of Hume's argument tracing the analogy throughout the "whole animal creation." Moreover, it reduces the scope of animals actually or potentially considered by Hume. Blinded by the abundance of references by Hume to higher vertebrates, Thorpe cuts the invertebrates' limb off the ethological "tree of animals." This amputation is not only problematic for its scientific consequences but also for our moral consideration for invertebrates. In Thorpe's work, oysters, butterflies, or silkworms are not conceived as being capable of suffering pain and pleasure, even less as "thinking beings having either many or few perceptions," as Hume sagaciously did two centuries earlier.

Hume's perspective dissolves the conceptual discontinuities between human and animal nature, and he highlights that "a theory about the operation of human understanding will acquire further authority if proved for nonhuman animals." In this way, Hume stimulated the search for common biological processes, including the investigation of the biological origin of all living beings – humans included.

11.5 The Influence of Hume on Darwin: The Common Evolutionary Origin of Human and Nonhuman Animals

To propose that Hume's conception of a common biological nature laid the groundwork for the development of the Darwinian evolutionary theory may seem merely speculative. However, we can find historical evidence for its support. In Charles Darwin's personal notebooks, David Hume is the most quoted philosopher.[5]

The Humean influence on the conception of Darwin's evolutionary theory can be further tracked back to Charles' grandfather, Erasmus Darwin. For his physiological psychology, Erasmus followed Hume's distinction of "three classes of associated ideas, based on contiguity, causation, and resemblance" (Logano 1972, p 43). Essential to my argument is that Charles Darwin's grandfather quoted Hume's posthumous book *Dialogues Concerning Natural Religion* when providing his first unequivocal evolutionary pronouncement (cf. Harrison 1971). Erasmus Darwin (1794) wrote in his main work *Zoonimia* that:

> [T]he late Mr. David Hume… concludes that the world itself might have been generated rather than created; that is, it might have been gradually produced from very small beginnings, increasing by the activity of its inherent principles, rather than by a sudden evolution of the whole by the Almighty fiat. (pp 245–246)

Erasmus Darwin's statement resembles closely the thoughts of Philo, Hume's skeptical character in the *Dialogues Concerning Natural Religion*. Part VII of that book begins with the following "discovery" by Philo:

> In examining the ancient system of the soul of the world, there strikes me, all on a sudden, new idea, which, if just, must go near to subvert all your reasoning…. If the universe bears a greater likeness to animal bodies and to vegetables, than to the works of human art, it is more probable that its cause resembles the cause of the former than that of the latter, and its

[5] Major historian on the development of evolutionary theories, Robert Richards (1989), has described David Hume as "Darwin's favorite philosophical author." Philosopher Edward Manier (1978) attempted a quantitative analysis of the incidence of different philosophers on Darwin's thought by composing a table, which provides the numbers of quotes for each philosopher mentioned by Darwin in his notebooks. In Manier's table, Hume ranks first with nine quotes. Five other philosophers appear below Hume in Darwin's notebooks ranking of frequency: Auguste Comte (eight quotes), David Hartley (six quotes), Dugald Stewart (six quotes), William Paley (two quotes), and Immanuel Kant (one quote).

origin ought rather to be ascribed to generation or vegetation than to reason or design. (DCNR, p 176)

Via Philo the "late Mr. Hume" planted the germ for the evolutionary theory in Darwin's family. The grandson, Charles Darwin, was however reluctant to quote Hume in his published work. A footnote in *The Descent of Man* constitutes Darwin's single public citation of Hume. This scarcity of references to Hume could be associated with a well-established aspect of Darwin's personality. He avoided publicizing his controversial considerations on human evolution and delayed as long as possible his publication of *The Descent of Man.*[6] In fact, the publication of that book revolted Victorian society far more than the *Origin of Species*. Darwin was aware of the theological implications of his theory and wanted to elude further controversies, which could have occurred by referring to the atheist philosopher par excellence. As historian William Huntley (1972, p 465) surmised "Darwin (and Lyell) did not wish to introduce a host of unwanted implications that the mentioning of the name Hume would suggest to some readers."

In contrast to his public writing, Charles Darwin did not hesitate to quote Hume in his personal notebooks, which contain insightful records. In August 1838, just 2 weeks before coming up with his fundamental evolutionary mechanism of natural selection, Darwin wrote in his personal notebooks (1836–1844) that "Hume's essay on the Human Understanding [is] well worth reading." The temporal contiguity between Charles Darwin's readings of Hume and his conception of the evolutionary mechanisms is astonishing. Later, during the spring of 1839, when Darwin continued to work on the formulation of his evolutionary theory, he added several notes on Hume in his notebooks M and N dedicated to the topics "Man, Mind and Materialism." In Table 11.4, I summarize all quotes that Charles Darwin made of Hume in his personal notebooks. These quotes provide ample evidence for how the father of the modern theory of biological evolution was inspired by Hume's *Enquiries*, *A Treatise*, *Natural History of Religion, Dialogues*, *The Dissertation on the Passions*, *The Skeptic*, and other essays, at the moment he was conceiving his evolutionary theory.

Darwin's entries on January 1839 refer to the section *Of the Reason of Animals* in Hume's *Enquiry* that I discussed above and to Sect. XV of Hume's *Dialogues Concerning Natural Religion*. The latter corresponds to the same work in which Philo says that "the world might have generated rather than created," inspiring Charles' grandfather's idea of evolution. Thus, Hume's skeptical character touched the imagination of both Darwin's grandfather and grandson. But the grandson went further and collected the empirical data required by Philo.

Part II of the *Dialogue Concerning Natural Religion* concludes with Philo's questions to Cleanthes: "Can you pretend to show any (such) similarity between the fabric of a house, and the generation of a universe? Have you ever seen nature in any such situation as resembles the first arrangement of the elements? Have worlds ever

[6] John T. Bonner and Sir Robert M. May (Bonner and May 1981) elegantly develop this point in their introduction to *The Descent of Man* by Charles Darwin.

Table 11.4 Quotes of Hume made by Charles Darwin in his personal notebooks and published works. In parenthesis the page numbers for the edition of *Charles Darwin's Notebooks* by Barret et al. 1987* and for *The Descent of Man* by Princeton University Press (1981), which reproduces the first edition of the work in 1871. The dates for the entries in the notebooks were defined as the month mentioned in the actual entry or in the first previous entry found. The date for *The Descent of Man* is the publication date, and not the date when Darwin wrote his quote on Hume, as in the dates given for the entries of the notebooks

Source	Date	Darwin's Quotations of Hume
Notebook M, entry 104 (p.545)	August 1838	"As some impressions 'Hume' become unconscious. So may some ideas.- ie habits, which must require idea to order muscles to do 'certain' the actions."
Notebook M, entry 155 (p.559)	September 1838	"Hume's essay on the Human Understanding well worth reading"
Notebook C, entry 270 (p.321)	October 1838	"Hume's do, with correspond. With Rousseau"
Notebook C, entry 267 (p.325)	January 1839	"Hume's essay on H. Understanding (some time)"
Notebook O, entry 52b (p.627)	May 1839	"Hume's Inquiry—good abstract of Butler & arguments of beneficial tendency of affections."
Notebook N, entry 101 (p.591)	July 1839	"Hume has section (IX) on the Reason of Animals. Essays Vol. 2"
Notebook N, entry 101 (p.591)	July 1839	"Also on origin of religion or polytheism, at p. 424 Vol.II "Sect XV Dialogues on Natural Religion. " however, he seems to allow it as an instinct."
Notebook N, entry 101 (p.592)	July 1839	"I suspect the endless round of doubts and skepticism might be solved by considering the origin of reason. As gradually developed. See Hume on Skeptical Philosophy
		"[Of the Skeptical and Other Systems of Philosophy]"
Notebook N, entry 101 (p.592)	July 1839	"Hume has written 'Natural Hist. Of Religion' and its origin in the Human mind."
Notebook N, entry 184 (p.595)	April 1840	"'Adam Smith Moral Sentiments' much on life & Character"
		"'Hume's Dissertation on the Passions' [A Dissertation on the Passions]"
The Descent of Man (p.85)	February 1871	"Hume remarks ('An Enquiry concerning the Principles of Morals,' edit 1751, p.132), 'there seems a necessity for confessing that the happiness and misery of others are not spectacles altogether indifferent to us, but that the view of the former … communicates a secret joy; the appearance of the latter … throws a melancholy damp over the imagination.'"

*I performed the search of quotes on Hume in Darwin's work by looking at the available concordances for *The Origin of Species*, *The Descent of Man*, *The Expression of the Emotions in Man and Animals*, and most importantly the personal notebook that Darwin kept between the years 1836 and 1844 (edited by Barret et al. 1987). I also reviewed *The Voyage of the Beagle* and *The Autobiography* of Charles Darwin. In total, I found 11 quotes. Only one of these was found in a published work (*The Descent of Man*). All the remaining quotes were found in Darwin's personal notebooks, most of them in the M and N notebooks. Gruber (1974) described the subjects of these notebooks as M: "Metaphysics-Morals and Speculation on Expression" and N: "Metaphysics and Expressions Selected for Species and Theory." Three other quotes are found in his notebooks C and O, which deal with evolution in general.

been formed under your eyes?... If you have, then cite your experience, and deliver your theory" (DCNR p151). Cleanthes' creationist argument surrenders when confronted with Philo's challenge. Cleanthes could not refer to any experience to answer Philo's questions, and in consequence could not prove his analogy between the creation of the universe and the creation of a house. Furthermore, the dissimilarity between the two events is too striking to be acceptable for Philo, who says:

> If we see a house, CLEANTHES, we conclude, with the greatest certainty, that it had an architect or builder; because precisely that species of effect, which we have experienced to proceed from that species of causes. But surely you will not affirm, that the universe bears such a resemblance to a house that we can with the same certainty infer a similar cause. (DCNR 144)

Philo continues developing an argument based on Hume's central principle "like effects arise from like causes" and arrives in Part VII to a more tenable analogy. In the passage that inspired Darwin's grandfather, Philo affirms that:

> [I]f the universe bears a greater likeness to animal bodies and to vegetables, than to the works of human art, it is more probable that its cause resembles the cause of the former than that of the latter, and its origin ought rather to be ascribed to generation or vegetation than to reason or design. (DCNR, VII, 78)

Another character of the *Dialogues*, Demea, challenged Philo's conclusion that the universe was generated rather than created, by asking "What *data* have you for such extraordinary conclusions?" (DCNR, p 80, emphasis added). Philo concedes here saying that "I have still asserted, that we have no *data* to establish any system of cosmogony" (DCNR, p 80, emphasis added). This is the task that Hume's character Philo left for Charles Darwin: amassing the data through which the process of generation may be deduced.

Sixty years after Philo's words were published, Charles Darwin wrote in his personal notebook that "we can allow 'satellites,' planets, suns, universes, nay whole systems of universes 'of man' to be governed by laws, but the smallest insect we wish to be created at once by special act, provided with its instincts its place in nature" (N36). We can interpret Darwin's statement as a development of Philo's statement: if the universe bears a likeness to animal bodies and to vegetables, then its origin ought rather to be ascribed to generation or vegetation than to reason or design.

In order to elaborate his own evolutionary theory based on a natural mechanism for the generation of the diversity of living beings, Darwin needed – as much as Philo did – to supplant the prevailing explanation based on creation by design. Darwin's entry on November 27, 1838, expresses that:

> *Arguing from man to animals* is philosophical, viz.; man is not a cause like a deity...because if so *ourang outang* [sic], *oyster & zoophyte*. (N49, emphasis added)

In this note Darwin affirms conclusively that human beings are not created by God, when he says "man is not a cause like a deity." Darwin sees no reason to attribute a different origin to nonhuman and human animals. He also knows that nonhuman animals are naturally generated – not specially created by God. Therefore, humans are naturally generated too. Like Hume, Darwin concludes that the same process of generation "applies to the whole animal kingdom." The origin of humans, orang-utans, oysters, and zoophytes – i.e., living beings that are placed in the evolutionary tree below the oyster – is explained by the mechanism.

The oyster that seemed to be an isolated case in Hume's work acquires a notable role in Darwin's reflections on the extent to which human and nonhuman animal's properties are shared. In his Notebook N, Darwin wrote:

> Origin of cause & effect being a necessary notion … connect[s] …'our' willing [with that] of the simplest animal, as hydra toward light…The Cyanocephalus [a monkey] when fondling the keeper, clasping '& rubbed' his arms & show signs of affecting something *like man. Has an oyster necessary notion of space*. (N12–13, emphasis added)

In this note, Darwin searches and finds again, based on the Humean principle "like causes, like effects," common properties concerning perception and behavior shared by humans, monkeys, oysters, and animals below the oyster, such as the hydra. In this entry, Darwin seems to examine the possibility of extending one of the two Kantian *a priori* forms of sensibility – space – beyond humans to the oyster and hence to the whole animal kingdom. Darwin not only extends one of the Kantian necessary transcendental forms of human sensuous intuition to the oyster, but he proceeds to explore links between free will, the highest faculty of humans. Free will is examined by Darwin in nonhuman bizarre aquatic animals, when he writes in Notebook M:

> With respect to free will, seeing a puppy playing cannot doubt that they have free will, if so *all animals*, then an *oyster* has & a *polype…* now *free will of oyster,…*[is the] *direct effect of organization, by the capacities its senses give it of pain or pleasure*. (M72, emphasis added)

Darwin associates here free will with the capacity of feeling pain or pleasure, which results from an organization shared by all living organisms (in other entries of Darwin's notebooks, plants are included as well). In this manner, Darwin definitively extends sentiency to all animals. Like Hume, the naturalist includes in his note a dog, an oyster, and a living being that is evolutionary placed "below an oyster," the polyp (or coral). In the hand of Darwin, Hume's analogy between human and animal nature is undoubtedly applied to all animals: vertebrates and invertebrates, terrestrials and aquatics.

Through the development of the theory of biological evolution, Charles Darwin and numerous other biologists furnished the *data* requested by Demea of Philo, in Hume's *Dialogues Concerning Natural Religion.* The Darwinian theory of a common origin for all living forms, including human, was inspired and stimulated by Hume's philosophy. In turn, Darwinian theory provided the empirical support requested by Hume's philosophy. Abundant evidence based on cellular theory, comparative anatomy and physiology, and development biology offered the necessary support for the understanding of humans as one biological species among many. *Homo sapiens* was classified as an animal which shares its origin, embryological development, physiological functions, and anatomical structures to various degrees, with the rest of biological species placed in diverse branches of the evolutionary "tree of life."

11.6 Concluding Remarks

The proposition of a common nature shared by all animals developed by Hume and Darwin unsettled not only Victorian society; today it still challenges ethical deliberations concerning our relations with animals. When Peter Singer (1975) published his influential book *Animal Liberation*, he introduced it by stating that his attempt was "to think through, carefully and consistently, the question of how we ought to treat non-human animals.... A liberation movement is a demand for an end to prejudice and discriminations based on an arbitrary characteristics like race or sex" (pp xii-xiii). Later in the book, when Singer argues for vegetarianism, he asks: "How far down the evolutionary scale shall we go? Shall we eat fish? What about shrimps? Oysters?" (Singer 1975, p 176). And Singer's answer is: "Oysters, ... and [other] mollusks are in general very primitive organisms... Most mollusks are such rudimentary beings that it is difficult to imagine them feeling pain, or having other mental states" (Singer 1975, p 179).

The answers that Hume or Darwin might have given to the oyster question are more radical than Singer's. For both the modern philosopher and the founder of the modern evolutionary theory, oysters and organisms "even below the oyster" do have some mental states and sentiency. Hume's principle of "like causes, like effects" permits a closer relation between oysters and humans than Singer's argument. Darwin and other evolutionary biologists supported Hume's perspective by providing vast empirical evidence. Humans and oysters are sentient organisms that share a common evolutionary origin. Based on comparative anatomy and other observation of the oyster's organization, Darwin not only attributed to these organisms the capacity of feeling pain or pleasure, but he even speculated about the oyster's perception of space.

Intimately, Hume links not only the nature of human and nonhuman animals but also equals the value of human and nonhuman existences. In the essay "Of Suicide" that the Scottish philosopher refused to publish in life (see Gaskin 1995), he argues that suicide is neither immoral nor irreligious and arrives to the limits of his argument linking human nature to the "whole animal kingdom." Then he arrives to his most radical conclusion:

> The lives of men depend upon the same laws as the lives of all other animals... the life of a man is of no greater importance to the universe than that of an oyster. (Of Suicide, p 371)

Two allusions to the oyster in his complete written works suffice Hume to disolve the barriers between the natures of human and nonhuman animals, and between the values of their existences. What counts for humans counts for all other animals too!

The analysis developed here of the "traditional Western" philosopher David Hume not only invites us to reconsider the answer about the oyster given by contemporary utilitarian philosopher Peter Singer. It also underscores the relevance of re-reading and reinterpreting the implications of Western historical philosophers for contemporary projects that rethink and transform our ethical attitudes toward non-human living beings.

David Hume was himself inspired by much earlier Western philosophical traditions, namely the Skeptics – particularly, Pyrrho[7] and the Cynics (Clark 1985). On the one hand, these ancient Greek philosophical traditions exerted a decisive influence on Hume's view of animals. On the other hand, the modern Scottish philosopher brought these traditions to new horizons. Likewise, today the work of Hume pushes contemporary philosophers and biologists to consider animal understanding and sentiency even beyond their current boundaries. In turn, we can develop implications of philosophical traditions beyond the limits explored by their authors. For example, the ethical notion of co-inhabitants (Rozzi 2015) and the demands for reorienting processes of biocultural homogenization go beyond Hume's moral considerations for nonhuman animals (Rozzi 2018).

Instead of a dichotomy between "traditional" and "radical" modern philosophies, the analysis of this suggests an evolutionary metaphor of dynamic branching and flowering of philosophical ideas, nourished by diverse roots. A more fluid communication with the philosophical roots of Western civilization and modern thought, like the Cynics or Hume, may supply contemporary ecologists, ethicists, and more broadly biocultural conservationists with stronger perspectives when confronted with monolithic views about modern anthropocentric philosophy as being irremediably indifferent toward invertebrates and other less conspicuous nonhuman living beings. Hume stated in his own words that the values of the life of an oyster and a human being are equal, thereby fracturing the anthropocentrism of Victorian society and providing a philosophical foundation for overcoming taxonomic biases in the type of contemporary environmental ethics we need to coinhabit the planet in the Anthropocene.

Acknowledgments I thank Donald Baxter, Kurt Heidinger, Scott Lehmann, Francisca Massardo, Peter D'Alesandre, and Roy May for their insightful comments on the manuscript and zoologist Sacha Spector for his valuable help identifying animal kinds in Hume's work.

References

Bonner JT, May RM (1981) Introduction. In: The Descent of Man (Charles Darwin). Princeton University Press, Princeton

Bowler P (1993) The Norton history of the environmental sciences. WW Norton & Company, New York

Carpenter SR, De Fries R, Dietz T, Mooney HA, Polasky S, Reid WV, Scholes RJ (2006) Millennium Ecosystem Assessment: research needs. Science 314(5797):257–258

Chapman AD (2009) Numbers of living species in Australia and the world. Report for the Australian biological resource study, Department of the Environment, Water, Heritage and the Arts, Australian Government. Canberra: Australia

Clark SRL (1985) Hume, animals and the objectivity of morals. Philos Q 35:117–133

[7] See Popkin (1980), which includes "David Hume: His Pyrrhonism and his critique of Pyrrhonism" (pp 103–132), "David Hume and the Pyrrhonian controversy" (pp 133–148), and "Bayle and Hume" (pp 149–160).

Clark JA, May RM (2002) Taxonomic bias in conservation research. Science 297:191–192

Darwin E (1794) Zoonimia; or the laws of organic life. Republished in 1974. AMS Press, New York

Darwin C (1987) Charles Darwin's notebooks, 1836–1844. Transcribed and edited by Barrett PH, Gautrey PJ, Herbert S, Kohn D, Smith S. British Museum Natural History/Cornell University Press, Ithaca

Frasca-Spada M (1996) David Hume, the Caliph Omar and the burning issue of metaphysics. Trans Camb Bibliogr Soc 11(1):49–58

Gare AE (1995) Postmodernism and the environmental crisis. Routledge, New York

Gare AE (1998) MacIntyre, narratives, and environmental ethics. Environ Ethics 20:3–21

Gaskin JCA (1995) Hume on religion. In: Norton DF (ed) The Cambridge companion to Hume. Cambridge University Press, New York, pp 313–344

Gruber HE (1974) Darwin on man: a psychological study of scientific creativity. EP Dutton, New York

Harrison J (1971) Erasmus Darwin's view on evolution. J Hist Ideas XXXII:247–264

Hume D (1948) In: Smith NK (ed) Dialogues concerning natural religion. Social Sciences Publishers, New York

Hume D (1976a) In: Selby-Bigge LA, Nidditch PH (eds) An enquiry of human understanding (EHU). Oxford University Press, Oxford

Hume D (1976b) In: Selby-Bigge LA, Nidditch PH (eds) An enquiry concerning the principles of morals (EPM). Oxford University Press, Oxford

Hume D (1978) In: Selby-Bigge LA, Nidditch PH (eds) A treatise of human nature (THN). Oxford University Press, Oxford

Hume D (1995) The complete works of David Hume, "past masters" CD-roam. InteLex Corporation, Charlottesville

Huntley WB (1972) David Hume and Charles Darwin Journal of the History of Ideas 33:457–470

Logano JV (1972) The poetry and aesthetics of Erasmus Darwin. Princeton University Press, Princeton

Manier E (1978) The young Darwin and his cultural circle: a study of influences which helped shape the language and logic of the first drafts of the theory of natural selection. D. Reidel Pub. Co., Dordrecht

MEA (2005) Millennium Ecosystem Assessment. Ecosystems and human well-being: synthesis. Island Press, Washington, DC

Midgley M (1978) Beast and man. The roots of human nature. Cornell University Press, Ithaca

Mora C, Tittensor DP, Simpson AG, Worm B (2011) How many species are there on earth and in the ocean? PLoS Biol 9(8):e1001127

Palmer C (2013) Introduction to environmental philosophy: ethics, epistemology, justice. In: Rozzi R, SF Chapin, JB Callicott, STA Pickett, M Power JJ Armesto, RH May Jr (eds) Earth stewardship: linking ecology and ethics in theory and praxis. Ecology and ethics, vol 1. Springer, Dordrecht, pp 187–190

Passmore J (1974) Man's responsibility for nature: ecological problems and western traditions. Charles Scribner's Sons, New York

Passmore J (1975) The treatment of animals. J Hist Ideas 36:195–218

Poole AK (2018) The UN sustainable development goals and the biocultural heritage lacuna: where is goal number 18? In: Rozzi R, May RH Jr, Chapin FS III, Massardo F, Gavin M, Klaver I, Pauchard A, Nuñez MA, Simberloff D (eds) From biocultural homogenization to biocultural conservation. Ecology and ethics, vol 3. Springer, Dordrecht, pp 315–331

Popkin RH (1980) The high road to Pyrrhonism. Austin Hill Press, SanDiego

Reddick A (1996) The making of Johnson's dictionary 1746–1773. Cambridge studies in publishing and printing history. Cambridge University Press, Cambridge

Richards RJ (1989) Darwin and the emergence of evolution theories of mind and behavior. The University of Chicago Press, Chicago

Rozzi R (1999) The reciprocal links between evolutionary-ecological sciences and environmental ethics. Bioscience 49:911–921

Rozzi R (2001) Ética ambiental: raíces y ramas latinoamericanas. In: Primack R, Rozzi R, Feinsinger P, Dirzo R, Massardo F (eds) Fundamentos de Conservación Biológica: Perspectivas Latinoamericanas. Fondo de Cultura Económica, Mexico, pp 311–359
Rozzi R (2015) Implications of biocultural ethics for earth stewardship. In: Rozzi R, Chapin FS III, Callicott JB, Pickett STA, Power ME, Armesto JJ, May RH Jr (eds) Earth stewardship: linking ecology and ethics in theory and practice. Ecology and ethics, vol 2. Springer, Dordrecht, pp 113–136
Rozzi R (2018) Biocultural homogenization: a wicked problem in the Anthropocene. In: Rozzi R, May RH Jr, Chapin FS III, Massardo F, Gavin M, Klaver I, Pauchard A, Nuñez MA, Simberloff D (eds) From biocultural homogenization to biocultural conservation. Ecology and ethics, vol 3. Springer, Dordrecht, pp 21–47
Sessions G (1994) Deep ecology for the twenty-first century. Sambahala, San Francisco
Singer P (1975) Animal liberation. A new ethics for our treatments of animals. Avon Books, New York
Singer P (1993) Practical ethics. Cambridge University Press, New York
Sorabji R (1985) Animal minds & human morals: the origins of the western debate. Cornell University Press, Ithaca
Thorpe WH (1979) Origins and rise of ethology. Heinemann Educational Books, London
Wilson EO (1988) Biodiversity. National Academic Press, Washington, DC
Zimmerman ME (1994) Contesting Earth's future: radical ecology and postmodernity. University of California Press, Berkeley

Part II
Biotic Homogenization

Chapter 12
Nature, Culture, and Natureculture: The Role of Nonnative Species in Biocultures

Daniel Simberloff

Abstract "Nature" and "culture" are defined variously. The dualist end of a spectrum sees "culture" as consisting of humans and their products and "nature" as comprising whatever is unaffected by humans. The other end sees but one "natureculture" integrating humans, other animals, and plants, and considers "nature" an artifact that humans construct in their observations and discussions. A bioculture is a local collection of humans, other species, and their interactions. Nonnative species can thus be parts of biocultures and may even replace native species to form new biocultures. To the extent that interactions among species arise through long processes of evolution and coevolution, a new bioculture thus formed will likely differ substantially from the previous one. Whether there are characteristic types of differences between such new biocultures and ones unaffected by nonnative species is an unexplored topic, just as is the question of whether novel ecosystems dominated by nonnative species characteristically differ from long-standing ones barely affected by humans. Whether we lament the loss of traditional biocultures probably depends largely on our perception and deep appreciation of the long sweeps of time underpinning much evolution and coevolution. A similar difference probably underlies conflicts between those who deplore ecological changes and local species extinctions wrought by nonnative species and those who accept, welcome, or even advocate producing them.

Keywords Artifact · Coevolution · Ecological fitting · Globalization · Novel ecosystem

---

D. Simberloff (✉)
Department of Ecology and Evolutionary Biology, University of Tennessee, Knoxville, TN, USA
e-mail: dsimberloff@utk.edu

R. Rozzi et al. (eds.), *From Biocultural Homogenization to Biocultural Conservation*, Ecology and Ethics 3, https://doi.org/10.1007/978-3-319-99513-7_12

12.1 Conceptions of Nature and Its Relation to Culture

"Biocultural" is a word that appeared in the fields of evolution, ecology, ethnobiology, and anthropology in the 1960s to refer to interactions among humans and co-occurring species in particular habitats or regions (Rozzi 2013a). One might then use the term "bioculture" to refer to the human populations, other species, habitats, and the "totality of interacting biological and cultural relationships" (Bennett et al. 1975) occupying a particular site or region. Much of the literature on biocultures refers to age-old customs of local human residents that entail interactions with co-occurring species. However, Rozzi (2013a, b, 2015a, b) and Simberloff (2013) have noted with concern the apparent integration of recently arrived nonnative species and their incorporation into biocultures, speculating on the possible threat of this process to existing biocultures. Assessing the extent of this threat requires first examining the relationship between nature and culture and, in fact, what "nature" and "culture" actually mean. Attitudes about nonnative species and attempts to manage or eradicate them align tightly with how people define "nature" and "culture."

Philosophers have long debated the relationship between nature and culture. The ancient Greeks first articulated one extreme, postulating a strict dichotomy between nature and culture, with culture being the products of humans (Tassin 2014). This conception of nature, as an independent counterpoint to human contrivance, was common throughout ancient and early modern times (Merchant 1980; Coates 1998). In the most rigorous version of this conception of nature, nature would be free of all human intervention and influence. Among contemporary philosophers, this view of nature is probably most closely associated with Taylor (1986). Philosopher Eric Katz (1997, 2000) notably articulated this stance, seeing humans as a species consisting of evolved biological entities that are clearly part of nature and depend on a "functioning biosphere" to survive. But human artifacts, products of human intentions, are not part of nature in this view and are relegated to the category of "culture." This is why Katz, following Elliot (1997), largely rejects the enterprise of ecological restoration: no matter how closely a reconstructed ecosystem approximates some historical reference system, humans planned and constructed it. Notably, many environmentalists, though they would not reject traditional ecological restoration, characterize nature as being free of human influence. In fact, they almost seem to assume this definition of "nature" is obvious, with justification unnecessary, and lament what they perceive as its decline. Environmental historian Carolyn Merchant's *The Death of Nature: Women, Ecology, and the Scientific Revolution* (1980) and journalist Bill McKibben's *The End of Nature* (1989) are in this category.

In many ways, "nature" in the eyes of dualists – that is, the part of the environment not influenced by human productions and artifacts – is closely akin to "wilderness," defined as "a tract or region uncultivated and uninhabited by human beings" (Merriam-Webster 1961). Many authors (e.g., McKibben 1989) treat these two terms as synonyms. Historian William Cronon (1983, 1996) does not deny the possibility that nature may exist somewhere separate from culture but cites abundant evidence of pre-Columbian human impact throughout North America to question whether any part of North America qualifies as wilderness.

The other extreme in the debate over the relationship between nature and culture arose prominently in the 1960s, led by the French poststructuralist philosopher Michel Foucault (1966), who argued that the barrier between nature and culture is an artificial, invalid construct (Tassin 2014). A key proponent of this view is Donna Haraway (1999), a postmodernist historian of science and technology, who coined the widely cited term "natureculture," seeing no distinction between the two categories. Although she recognizes the dwindling of nature in the dualist sense of "nature," Haraway – unlike Merchant and McKibben – is enthusiastic about the various new possibilities for both humans and other species that this fusion of nature and culture affords (Haraway 1997, Huggan and Tiffin 2010).

Haraway (1999) erected the notion of "natureculture" in the belief that humans, other animals, and plants have inextricably intertwined histories and that humans "construct" nature in the act of observing and discussing it. This is part of a postmodern, poststructuralist effort to erase what are seen as artificial, essentialist dualisms – not only between nature and culture but between animate entities and machines (cf. cyborgs [Haraway 1991]), between humans and other animals, and among species in general (e.g., Cardozo and Subramaniam 2013). Cultural theorist and philosopher Jean Baudrillard (1972) went so far as to contend that there really is no such thing as "nature," but rather a multitude of signs in the environment to be interpreted by humans as they construct their cultures. Political scientists William Chaloupka and R. McGreggor Cawley (1993) see "nature" as primarily an "artifact of language," a collection of terms for human relationships with their environment.

Authors at the dualist end of the spectrum may appear to be uncompromising advocates of an essential, ontological, stark difference between nature and culture. In fact, many would concede the existence of a spectrum but view intermediate entities as less important than the clear endpoints. Thus Katz (2000, p. 40) says, "This dualism is not absolute, for naturalness and artifactuality exist along a spectrum of various kinds of entities. A wooden chair is more natural than a plastic chair, because it is more closely related to the naturally produced material that forms its basic structure." Similarly for Merchant (1996, p. 153): "Nor are nature and culture, men and women binary opposites with universal or essential meanings. Nature, wilderness, and civilization are socially constructed concepts that change over time and serve as stage settings in the progressive narrative."

Although he does not define "nature" explicitly, writer Fred Pearce (2015) treats nature as whatever nonhuman species are present at a site, whether the site is what might be seen as "wilderness" or a city center. Thus, this conception of nature, rather than placing it at the end of a nature-culture spectrum, locates it within culture as a constitutive component. Perhaps some such conception of nature – whatever species happen to be around – is common in the public at large, even among those who consider themselves conservationists. The mission statement of one of the largest conservation NGOs, The Nature Conservancy, features the opening statement, "Protecting nature, for people today and future generations," below a photograph of a young boy netting butterflies in an old field (<https://www.nature.org/about-us/vision mission/index.htm?intc=nature.tnav.about>, accessed 7/18/17), the antithesis of a location free from human influence. Tim Low (2002), biologist and environmental author, takes a similar tack in *The New Nature*: "Forget about wilderness… nature lives here in our cities and gardens, exploiting everything we do, forging new connections with us" (back cover). He invites readers to revel in the species that turn up in cities and industrial zones.

12.2 How Nature Is Integrated into Culture

However, modern anthropological and sociological research (e.g., Maffi and Woodley 2010) suggests that, even with nature conceived as a constitutive part of culture, nature for many is not just whatever species happen to be around. Rather, the complement of species present is often tightly and intricately integrated into local and regional cultures through language and tradition. Anthropologist Anna Tsing (2005) calls such integrated complexes of people and other species "social-natural landscapes," exemplified by the Meratus Dayaks, inhabitants of a mountainous region of Borneo. Rejecting the notion that this region should be strictly divided between zones of intensive agriculture and zones of pristine nature, Tsing details the myriad ways in which the Meratus Dayaks use many species for food and other purposes and how these extractive uses maintain much of the landscape in a species-rich ecological disclimax state, sustaining populations of some species that would be at best uncommon if the entire region succeeded to forest.

Rozzi (2013a, b, 2015a, b) and Simberloff (2013) have called attention to how globalization poses threats, including nonnative species, to such complexes formed from the interactive entwinement of human populations and the communities of species that constitute their environment – that is, "biocultures." One's conception of nature is integral to one's concern about this threat or even whether one perceives the threat at all. If one believes, as does Pearce (2015), that nature is whatever species happen to be around, then nonnative species are no worse for nature than any others. This rationalizes the otherwise paradoxical part of the title of Pearce's book: *Why Invasive Species Will Be Nature's Salvation*. If, as is often the case, invasive nonnative species thrive where native species do not – even if this stems from human disturbance that transforms the environment into a site where nonnative species can outcompete or otherwise prevail over native species – then one might view a "nature" composed wholly or largely of nonnative species as just as acceptable as one without them, and certainly better than no species at all.

Several enthusiasts of Haraway's concept of natureculture extol novel collections of species, including nonnatives, living in anthropogenic environments. Kirksey (2015) describes "emergent ecologies" whose constituent species are "lively," "transformative," "cosmopolitical," "flexible," "hearty," "symbiotic," and "riotous" combined in "convivial" assemblages, worlds, communities, associations, and alliances. Invasive nonnative species are "rambunctious" (Marris 2011) and "go-getters" (Pearce 2015). Those who lament species, communities, and ecosystems that have been replaced by emergent ecologies are indulging themselves in "anachronistic depictions of past environments" (Kirksey 2015, p. 3). Similarly, Kirksey and Helmreich (2010) hail the recognition of collections of humans, animals, plants, and microbes, no matter the species, as "biocultures" that are the proper subject of a new "multispecies ethnography." Subramaniam (2014) sees no inherent primacy for native species in naturecultures that she believes humans should assemble according to their collective desires.

12.3 And How Do Nonnative Species Fit in Nature, or Do They?

One of the main dichotomies that postmodernists and poststructuralists strive to break down, in addition to that between nature and culture, is the dichotomy between native and nonnative (Simberloff 2012a); critics of invasion biology also do this. Some critics attempt this by pointing to unusual cases of species that do not have the normal history of having evolved somewhere and then, over time and with or without human assistance, spreading to other areas. Thus geographer Charles Warren (2011) points to capercaillie in Scotland, reintroduced in 1837 from Scandinavia after having been eliminated in the mid-eighteenth century, largely through habitat destruction. Is the current population native or nonnative? Similarly, Warren questions the status in Scotland of hybrids between native red deer and introduced Asian Sika deer: are they native or nonnative? However, cases such as these are a small minority and rarely subjects of controversies over management; they simply remind us that any classification scheme is somewhat arbitrary and will have to rely on certain conventions (Simberloff 2012a). Others attempt to invalidate the native/nonnative dichotomy by pointing to the arbitrariness of the cutoff date of arrival, before which a species is considered native and after which it is considered introduced (e.g., Thompson 2014). After all, if most species have dispersed from their region of origin to their current geographic range, would they not all be nonnative in much of that current range? It is true that we do not call a species that somehow arrived many millennia ago in a discontinuous, distant region without human assistance a nonnative. For instance, the hemlock woolly adelgid is considered a native of the Pacific Northwest even though we now know that it somehow arrived there from Asia without human assistance ca. 20,000 years ago (Havill et al. 2016), whereas the species is considered nonnative in eastern North America, having arrived in the 1940s or 1950s. For the great majority of populations recognized as nonnative, documentary, molecular, or other evidence strongly implies an arrival after the Columbian Exchange.

Other authors do not question the legitimacy of a native/nonnative dichotomy but seek criteria by which a species that would routinely be classified as nonnative could achieve a sort of "honorary native" status. For philosopher Ned Hettinger

(2001), it is not length of time since arrival that qualifies a species as native but rather that it coadapt with the native biota and physical environment and persist without human assistance. The criteria of political scientist John Rodman (1993) also demand that the species develop "mutual dependence and mutual controls" with the native species, though Rodman recognizes that increasing length of time since arrival increases the likelihood of such interactions. Philosopher Clare Palmer and sociologist Brendon M.H. Larson (2014, p. 651) advance a similar argument to support "assisted migration" of whitebark pine, plagued by white pine blister rust in its native range, to a region in which it is not native: "The whitebark pine might be able to forge valuable new ecological connections; if so, this would give us reason to support assisted migration not solely based on the preservation of the species, but on its contributory functions in a broader ecological community." "Valuable" and "contributory" are undefined. Ecologists Alexandra Carthey and Peter Banks (2012), examining the response of native Australian mammals to the dingo, introduced ca. 4000 years ago, suggest that possibly native prey's behavioral modification to avoid a long-term nonnative species would be one criterion for declaring the latter native.

We see in the previous paragraph the notion that introducing a nonnative species may be acceptable, despite risks of undesired possible outcomes, so long as the species integrates in some usually unspecified way(s) into the existing community, or perhaps a new community. This notion leads to a series of questions about the structure of communities in general and what it means for nonnative species to constitute a new "normal" community or to integrate seamlessly into an existing one. Janzen (1985), observing communities of mixed native and nonnative species in a Costa Rican national park, noted that it was not obvious how, if at all, these communities, constituted primarily of species that had evolved elsewhere over long periods amidst other groups of species, differed from communities of species that had long coexisted and had coevolved. He termed "ecological fitting" the process by which such groups of species coexisted, even though for the most part they had not coevolved with one another. However, he did not specify exactly what he meant by fitting together or what features showed that the species fit together. One can imagine comparing community-level traits such as the shape of the species-abundance curve or species-to-genus ratios, or ecosystem-level traits such as aspects of nutrient pools and cycles. Pearce (2013), pointing to a plant community on Ascension Island composed wholly of species relatively recently introduced from various other places (Wilkinson 2004), stated that this community functioned perfectly normally. He maintained that it exemplifies ecological fitting and contradicts what he saw as the widespread belief that well-functioning communities need long amounts of time to coevolve and that nonnative species disrupt this process. In fact, as both Sax et al. (2007) and Simberloff and Strong (2013) pointed out, referring to this same Ascension Island forest, there is as yet no general answer to Janzen's question of whether ecologically fitted communities differ in some characteristic way(s) from highly coevolved ones, and certainly no research has been conducted on the Ascension Island forest on candidate ecosystem and community traits that might differ. However, recalling Pearce's conception of nature as whatever species happen to be present at a site, we should not be surprised that he did not recognize the need for various sorts of ecological data that would be required to assess how the Ascension Island forest compares to coevolved communities lacking nonnative species. Perhaps he is not seeing the forest for the trees.

Authors who have applauded the newly formed communities and naturecultures, and those who have sought routes to legitimize the presence of nonnative species in such assemblages, are for the most part not biologists, but scholars in the social sciences and humanities, or journalists. It seems possible that they deemphasize the enormously long sweeps of time through which ecological communities are typically formed in nature simply because they have not focused on the general tenets of evolution and historical biogeography. McKibben (1989) begins his eulogy for what he sees as a dying nature with an extended discussion of the sheer spans of time during which life on earth has been evolving and the great difficulty for humans to grasp and to internalize these spans and thus the length of the evolutionary process. The recent demonstration of rapid evolution and coevolution in certain circumstances (e.g., Hendry 2017) does not negate the fact that many of the communities we see in "nature" (in the dualistic sense) and that are disappearing are the product of hundreds of thousands if not millions of years of coevolution plus ecological fitting (however the latter comes about). This fact, and an understanding of the eons that characterized their development, accounts for many scientists' great concern over the various forces, including the increasing wave of nonnative species, threatening their existence. To return to the example of the hemlock woolly adelgid, in its native range (China and Japan) it is not a common insect, and the native hemlocks are in no way devastated by it (McClure and Cheah 1999). In the Pacific Northwest, where it arrived ca. 20,000 years ago, it has no substantial impact on the native hemlock, *Tsuga heterophylla* (McClure and Cheah 1999). In eastern North America, where it arrived less than a century ago, it has so devastated populations of the native hemlocks (*T. canadensis* and *T. caroliniana*) as to cast doubt on the continued existence of these trees as common forest constituents (Crabtree 2014; Eschtruth et al. 2013). This example gives some sense of the time scale over which coevolution often operates.

Aldo Leopold's growing antipathy toward nonnative species (Simberloff 2012b) developed apace with his development of a "land aesthetic," a "systematic theory of natural beauty and the criteria for its appreciation" (Callicott 1994, p. 179). The "picturesque" aesthetic tradition that long dominated western paintings of nature adhered to formal requirements of painting (perspective, balance, etc.) but might be considered a "surface aesthetic," whereas Leopold's land aesthetic is partially analytic and reflects a profound knowledge of the ecological workings and context and evolutionary history of a natural entity (Callicott 1994, 2008). An excellent example of Leopold's acute sensitivity to the vast sweep of evolutionary time in his land aesthetic is his essay "Marshland Elegy" in *A Sand County Almanac* (Leopold 1970), in which the tremendous span of time during which the marsh species and their complex interactions evolved is represented by the detailed evolutionary history and ecology of the sandhill crane (Callicott 1994, 2008). This evolutionary history of the marsh confers "a paleontological patent of nobility, won in the march of eons" (Leopold 1970, p. 103). In fact, neither the crane nor the marsh is inherently beautiful in the traditional formal sense, but "evolutionary literacy can alter and deepen perception" (Callicott 1994, p. 175).

It is possible that those who venerate biocultures laden with nonnative species do so based on an aesthetic or ethical perspective that does not particularly value evolution and coevolution. More generally, most critics of invasion science and management come from the social sciences and humanities (Simberloff 2003, 2012a), as with enthusiasts over newly assembled naturecultures described above, and seem uninterested in the evolutionary trajectories of the components of either biocultures or ecological communities invaded by nonnative species. This attitude contrasts strikingly not only with Leopold's land aesthetic but also with the views of Elliot (1997) and Katz (1997, 2000), who even deplore restoration projects that rigorously exclude nonnative species simply because such projects are conceived and implemented by modern humans.

The "novel ecosystems" that currently enthrall some restoration ecologists (Hobbs et al. 2013) are, of course, highly germane to a discussion of the role of nonnative species in nature and in biocultures. They consist of new combinations of species, including particularly nonnative ones, that arise spontaneously in response to climate and land use change (or that can possibly be engineered to accommodate these changes), that persist autonomously, and that fulfill human needs for ecosystem services such as flood control or provision of food. As they are of interest particularly because of their potential benefits to humans and are envisioned as being associated with and exploited by humans, they will obviously affect existing biocultures and perhaps generate new ones. Novel ecosystems are a controversial concept (e.g., Hobbs et al. 2014; Murcia et al. 2014), and the controversy is beyond the scope of this essay. Suffice it to say that the same questions Janzen (1985) posed about the nature of relatively recent, ecologically fitted communities pertain to novel ecosystems and have not been answered: in what ways do they differ from other ecosystems, exactly? I have treated the specific role and impacts of nonnative species in novel ecosystems in Simberloff (2015).

12.4 Biocultures and Nonnative Species

It is apparent that, in certain circumstances, nonnative species can quickly become embedded in local and regional cultures (Nuñez and Simberloff 2005; Nuñez et al. 2018). In many instances most local people do not even realize that a culturally significant species is not native, but in other cases the nonnative origin is widely recognized and sometimes even publicized, as for the South American jacaranda trees in Pretoria, South Africa (Nuñez et al. 2018). Although Nuñez et al. (2018) list several factors (e.g., time since introduction, aesthetic appeal) that may affect the likelihood that a bioculture will incorporate a particular nonnative species, there is as yet no systematic exploration of these factors (or others) and their role in producing biocultures. Nor is there any systematic examination of whether recently assembled biocultures differ in any characteristic ways from those that have developed over millennia. This situation is quite analogous to the still unanswered key question of Janzen (1985) regarding ecological communities: do coevolved communities differ characteristically from those assembled wholly or primarily by ecological fitting, however ecological fitting comes to be defined and measured? Whatever our attitude about these new biocultures, the destruction of old biocultures noted by Rozzi (2013a, b, 2015a, b) and Simberloff (2013) by biotic homogenization may be lamentable on aesthetic or ethical grounds, much as many people deplore the McDonaldization of cultures (Ritzer 1993) partly for these reasons. However, just as McDonaldization has other characteristic fundamental impacts on societies (e.g., socioeconomic ones), it may be that biotic homogenization produces characteristically distinct biocultures. This possibility has simply not been explored; it is not even obvious what traits or characteristics could be measured or at least assessed for such a comparison between ancient and newer biocultures.

It is also worth observing that the very fact of mourning the passing of traditional biocultures and their replacement by new ones including nonnative species need not be a manifestation of xenophobia, as is suggested by many critics of invasion science and management (e.g., Raffles 2011; Subramaniam 2014; Kirksey 2015). Perhaps, because these critics seem unconcerned with evolutionary history and tend to deemphasize ecological impacts, they are driven to focus on the possibility of psychological predisposition against species from elsewhere as a motivating factor. Poststructuralists and postmodernists in general seek patterns and causal explanations in aspects of the psychology of the human participants of whatever phenomenon they are exploring, and this approach leads to social constructions of science (cf. Huggan and Tiffin [2010] on Merchant and Haraway). A general assessment of the validity of social constructivist interpretations of science (Brown 2001) is well beyond the purview of this paper. Suffice it to say that there is nothing necessarily xenophobic about a desire to preserve traditional biocultures, which is motivated by the same general principles supporting the right of human societies to maintain their cultural distinctness (e.g., Council of Europe 2000; UNESCO 2001).

Acknowledgments Mary Tebo, Christy Leppanen, and Tony Ricciardi contributed many thoughtful comments on an early draft of this manuscript.

References

Baudrillard J (1972) Pour une critique de l'économie politique du signe. Gallimard, Paris

Bennett KA, Osborne RH, Miller RJ (1975) Biocultural ecology. Annu Rev Anthropol 4:163–181

Brown JR (2001) Who rules in science? An opinionated guide to the wars. Harvard University Press, Cambridge, MA

Callicott JB (1994) The land aesthetic. In: Chapple C (ed) Ecological prospects: scientific, religious, and aesthetic perspectives. State University of New York Press, Albany, pp 169–183

Callicott JB (2008) Leopold's land aesthetic. In: Carlson A, Lintott S (eds) Nature, aesthetics, and environmentalism. From beauty to duty. Columbia University Press, New York, pp 105–118

Cardozo K, Subramaniam B (2013) Assembling Asian/American naturecultures. Orientalism and invited invasions. J Asian Am Stud 16:1–23
Carthey AJR, Banks PB (2012) When does an alien become a native species? A vulnerable native mammal recognizes and responds to its long-term alien predator. PLOS ONE 7(2):e31804
Chaloupka W, Cawley RM (1993) The great wild hope: nature, environmentalism, and the open secret. In: Bennett J, Chaloupka W (eds) In the nature of things. Language, politics, and the environment. University of Minnesota Press, Minneapolis, pp 3–23
Coates P (1998) Nature: western attitudes since ancient times. University of California Press, Berkeley
Council of Europe (2000) Declaration on Cultural Diversity, Council of Europe. https://search.coe.int/cm/Pages/result_details.aspx?ObjectID=09000016804bfc0b
Crabtree T (2014) Tennessee Natural Heritage Program Rare Plant List. Division of Natural Areas, Tennessee Department of Environmental and Conservation. Nashville. http://www.state.tn.us/environment//natural-areas/docs/plant_list.pdf
Cronon W (1983) Changes in the land: Indians, colonists, and the ecology of New England. Hill and Wang, New York
Cronon W (ed) (1996) Uncommon ground. Norton, New York
Elliot R (1997) Faking nature: the ethics of environmental restoration. Routledge, London/New York
Eschtruth AK, Evans RA, Battles JJ (2013) Patterns and predictors of survival in *Tsuga canadensis* populations infested by the exotic pest *Adelges tsugae*: 20 years of monitoring. For Ecol Manage 305:195–203
Foucault M (1966) Les mots et les choses. Une archéologie des sciences humaines. Gallimard, Paris
Haraway DJ (1991) Simians, cyborgs, and women: the reinvention of nature. Routledge, New York
Haraway DJ (1997) Modest_Witness@Second_Millennium.FemaleMan_Meets_OncoMouse: feminism and technoscience. Routledge, New York
Haraway DJ (1999) How like a leaf: an interview with Thyrza Nichols Goodeve. Routledge, New York
Havill NP, Shiyake S, Galloway AL, Foottit RG, Yu G, Paradis A, Elkinton J, Montgomery ME, Sano M, Caccone A (2016) Ancient and modern colonization of North America by hemlock woolly adelgid, *Adelges tsugae* (Hemiptera: Adelgidae), an invasive insect from East Asia. Mol Ecol 25:2065–2080
Hendry AP (2017) Eco-evolutionary dynamics. Princeton University Press, Princeton
Hettinger N (2001) Exotic species, naturalisation, and biological nativism. Environ Values 10:193–224
Hobbs RJ, Higgs ES, Hall CM (eds) (2013) Novel ecosystems: intervening in the new ecological world order. Wiley-Blackwell, Oxford
Hobbs RJ, Higgs ES, Harris JA (2014) Novel ecosystems: concept or inconvenient reality? A response to Murcia et al. Trends Ecol Evol 29:645–646
Huggan G, Tiffin H (2010) Postcolonial criticism. Literature, animals, environment. Routledge, Abingdon
Janzen D (1985) On ecological fitting. Oikos 45:308–310
Katz E (1997) Nature as subject: human obligation and natural community. Rowman and Littlefield, Lanham
Katz E (2000) Another look at restoration: technology and artificial nature. In: Gobster PH, Hull RB (eds) Restoring nature. Perspectives from the social sciences and humanities. Island Press, Washington, DC, pp 37–48
Kirksey E (2015) Emergent ecologies. Duke University Press, Durham
Kirksey E, Helmreich S (2010) The emergence of multispecies ethnography. Cult Anthropol 25:545–576
Leopold A (1970), A Sand County almanac, with essays on conservation from Round River. Ballantine Books, New York (originally published as Leopold A [1949] A Sand County almanac and sketches here and there, Oxford University Press, New York).

Low T (2002) The new nature. Penguin, Camberwell

Maffi L, Woodley E (2010) Biocultural diversity conservation: a global sourcebook. Earthscan, London

Marris E (2011) Rambunctious garden: saving nature in a post-wild world. Bloomsbury Press, New York

McClure MS, Cheah CAS-J (1999) Reshaping the ecology of invading populations of hemlock woolly adelgid *Adelges tsugae* (Homoptera: Adelgidae), in eastern North America. Biol Invasions 1:247–254

McKibben B (1989) The end of nature. Random House, New York

Merchant C (1980) The death of nature: women, ecology, and the scientific revolution. Harper & Row, San Francisco

Merchant C (1996) Reinventing Eden: western culture as a recovery narrative. In: Cronon W (ed) Uncommon ground. Norton, New York, pp 132–159

Merriam-Webster (1961) Webster's new collegiate dictionary. G. and C. Merriam, Springfield

Murcia C, Aronson J, Kattan GH, Moreno-Mateos D, Dixon K, Simberloff D (2014) A critique of the 'novel ecosystem' concept. Trends Ecol Evol 29:548–553

Nuñez MA, Simberloff D (2005) Invasive species and the cultural keystone species concept. Ecol Soc 10(1)

Nuñez MA, Dimarco RD, Simberloff D (2018) Why some exotic species are deeply integrated into local cultures while others are reviled. In: Rozzi R, May R, Chapin T, Massardo F, Gavin M, Klaver I, Pauchard A, Nuñez MA, Simberloff D (eds) From biocultural homogenization to biocultural conservation, Ecology and ethics, vol 3. Springer, Cham, pp 219–231

Palmer C, Larson BMH (2014) Should we move the whitebark pine? Assisted migration, ethics, and global environmental change. Environ Values 23:641–662

Pearce F (2013) On a remote island, lessons in how ecosystems function. Yale Environment 360. http://e360.yale.edu/feature/on_a_remote_island_lessons__in_how_ecosystems_function/2683/

Pearce F (2015) The new wild – why invasive species will be nature's salvation. Beacon, Boston

Raffles H (2011) Mother nature's melting pot. New York Times, April 3, p WK12

Ritzer G (1993) The McDonaldization of society. Sage, London

Rodman J (1993) Restoring nature: natives and exotics. In: Bennett J, Chaloupka W (eds) In the nature of things. Language, politics, and the environment. University of Minnesota Press, Minneapolis, pp 139–153

Rozzi R (2013a) Introduction to integrating philosophy and ecology: biocultural interfaces. In: Rozzi R, Pickett STA, Palmer C, Armesto JJ, Callicott JB (eds) Linking ecology and ethics for a changing world. Ecology and ethics, vol 1. Springer, Dordrecht, pp 3–7

Rozzi R (2013b) Biocultural ethics: from biocultural homogenization toward biocultural conservation. In: Rozzi R, Pickett STA, Palmer C, Armesto JJ, Callicott JB (eds) Linking ecology and ethics for a changing world. Ecology and ethics, vol 1. Springer, Dordrecht, pp 9–32

Rozzi R (2015a) Earth stewardship and the biocultural ethic: Latin American perspectives. In: Rozzi R, Chapin FS III, Callicott JB, Pickett STA, Power ME, Armesto JJ, May RH Jr (eds) Earth stewardship. Linking ecology and ethics in theory and practice. Ecology and ethics, vol 2. Springer, Dordrecht, pp 87–112

Rozzi R (2015b) Implications of the biocultural ethic for earth stewardship. In: Rozzi R, Chapin FS III, Callicott JB, Pickett STA, Power ME, Armesto JJ, May RH Jr (eds) Earth stewardship. Linking ecology and ethics in theory and practice. Ecology and ethics, vol 2. Springer, Dordrecht, pp 113–136

Sax DF, Stachowicz JJ, Brown JJ, Bruno JF, Dawson MN, Gaines SD, Grosberg RK, Hastings A, Holt RD, Mayfield MM, O'Connor MI, Rice WR (2007) Ecological and evolutionary insights from species invasions. Trend Ecol Evol 22:465–471

Simberloff D (2003) Confronting introduced species: a form of xenophobia? Biol Invasions 5:179–192

Simberloff D (2012a) Nature, natives, nativism, and management: worldviews underlying controversies in invasion biology. Environ Ethics 34:5–25

Simberloff D (2012b) Integrity, stability, and beauty: Aldo Leopold's evolving view of non-native species. Environ Hist 17:487–511

Simberloff D (2013) Introduced species, homogenizing biotas and cultures. In: Rozzi R, Pickett STA, Palmer C, Armesto JJ, Callicott JB (eds) Linking ecology and ethics for a changing world. Ecology and ethics, vol 1. Springer, Dordrecht, pp 33–48

Simberloff D (2015) Invasive species and novel ecosystems. F1000Prime Rep 7:47. https://doi.org/10.12703/P7-47

Simberloff D, Strong DR (2013) Counterpoint: scientists offer a dissenting view on Ascension Island. Yale Environment 360. http://e360.yale.edu/Counterpoint_Scientists_Offer_Dissenting_View_on_Ascension_Island.msp

Subramaniam B (2014) The aliens in our midst: managing our ecosystems. In: Kleinman DL, Cloud-Hansen KA, Handelsman J (eds) Controversies in science and technology, from sustainability to surveillance, vol 4. Oxford University Press, New York, pp 228–240

Tassin J (2014) La grande invasion. Qui a peur des espèces invasives? Odile Jacob, Paris

Taylor P (1986) Respect for nature: a theory of environmental ethics. Princeton University Press, Princeton

Thompson K (2014) Where do camels belong? The story and science of invasive species. Profile Books, London

Tsing AL (2005) Friction. An ethnography of global connection. Princeton University Press, Princeton

UNESCO (2001) UNESCO Universal Declaration on Cultural Diversity, UNESCO. http://portal.unesco.org/en/ev.php-URL_ID=13179&URL_DO=DO_TOPIC&URL_SECTION=201.html

Warren C (2011) Nativeness and nationhood: what species 'belong' in post-devolution Scotland? In: Rotherham ID, Lambert RA (eds) Invasive and introduced plants and animals. Human perceptions, attitudes and approaches to management. Earthscan from Routledge, Abingdon, pp 67–79

Wilkinson DM (2004) The parable of Green Mountain: Ascension Island, ecosystem construction and ecological fitting. J Biogeogr 31:1–4

Chapter 13
Why Some Exotic Species Are Deeply Integrated into Local Cultures While Others Are Reviled

Martin A. Nuñez, Romina D. Dimarco, and Daniel Simberloff

Abstract One key challenge for invasive species management is finding support from local communities. Without local support, management plans can be severely compromised. What makes people support or reject management of invasive species can be linked to their perception of the target invasive species. In other words, it can be harder to control invasive species that are assimilated into the local community. We identify five factors associated with how quickly invasive species can become culturally assimilated. These factors are arrival time, economic impact, aesthetic value, effect on human health, and origin of nonnative species and of human immigrants. We suggest that understanding how these factors contribute to the incorporation of nonnative species into local cultures is important in determining effective control measures. In this vein, publicly accessible educational programs explaining the problems that invasive species produce will be required to implement effective invasive species management.

Keywords Invasive species · Management · Culture impact · Nonnative species

13.1 Introduction

Management of invasive species presents several challenges. Among these are the technical challenges of learning how to manage or eradicate a given species. Because invasive populations of these species are prolific colonizers, various new

---

M. A. Nuñez (✉)
Grupo de Ecología de Invasiones, INIBIOMA, CONICET-Universidad Nacional del Comahue, Bariloche, Argentina

R. D. Dimarco
Grupo de Ecología de Poblaciones de Insectos, CONICET, INTA, Bariloche, Argentina

D. Simberloff
Department of Ecology and Evolutionary Biology, University of Tennessee, Knoxville, TN, USA
e-mail: tebo@utk.edu

R. Rozzi et al. (eds.), *From Biocultural Homogenization to Biocultural Conservation*, Ecology and Ethics 3, https://doi.org/10.1007/978-3-319-99513-7_13

methods of manual, mechanical, biological, and chemical control are constantly being developed. However, a main challenge often comes from local communities that, for various reasons, oppose managing the target invasive (Dickie et al. 2014; Crowley et al. 2017). Some of these invasive species become incorporated into local cultures, increasing the difficulties in obtaining local support to control them. On the other hand, other species are fiercely hated and local populaces welcome management efforts. For example, the Burmese python in Florida is universally reviled, while some nonnative ornamental trees that become invasive are nevertheless defended by locals (e.g., the callery pear, *Pyrus calleryana*; Culley et al. 2011).

Some nonnatives species, including some problematic invasive species, are cultural icons in different areas of the world. Examples include tomatoes (*Lycopersicon esculentum*) in Italy, bananas (*Musa paradisiaca*) in Ecuador, coffee (*Coffea* spp.) in Colombia, *Cannabis sativa* in Jamaica, and in the United States bluegrass (*Poa pratensis*) in Kentucky, peaches (*Prunus persica*) in Georgia, *Eucalyptus* spp. in California, horses (*Equus caballus*) in the West, and kudzu (*Pueraria montana*) in the Southeast (cf. Nuñez and Simberloff 2005). In many cases, residents are not even aware that a species integral to their culture is not native.

A striking instance of a nonnative species deeply incorporated into a local culture comes from *Eucalyptus* in California. Beginning in the nineteenth century, many species of *Eucalyptus* trees introduced from Australia were planted in California. Some species formed massive groves and became a characteristic feature in the region. Among artists of the Eucalyptus school, a major school of art that originated in the early twentieth century, paintings of landscapes featured these invasive species (Moure 1982). Some introduced *Eucalyptus* populations negatively affect native species, in part because they are fire-tolerant and promoters, and in certain circumstances they propagate far better than native species (Williams 2002). In addition, they consume massive amounts of water. Despite these problems, many attempts to remove these trees for ecological restoration have suffered fierce opposition from passionate defenders. A group called "POET" (Preserve Our Eucalyptus Trees) greatly delayed the attempt to restore Angel Island, in the Golden Gate National Recreation Area, calling National Park Service employees "plant Nazis" for the plan to remove *Eucalyptus* (Williams 2002). Ironically, native bird, plant, and insect biodiversity is greatly reduced in *Eucalyptus* forest established in California and can be restored only if these trees are replaced by native species (Williams 2002).

Yet cases such as these in which a nonnatives species is welcomed into a culture, sometimes to the point of becoming a cultural icon, constitute a small minority of nonnative populations. Under what circumstances does such cultural assimilation occur (cf. Cardozo and Subramaniam 2013)? Why are some species welcomed and others deplored?

Nonnative species arrive for various reasons. Immigrants introduce some of them from their own native areas for aesthetic or cultural reasons, as many Asians,

North Americans, and Europeans did in the Pacific islands and Europeans also did when they immigrated to the Americas (Crosby 2003; Dillingham 1936; Guild 1938; Lever 1992). Also, nonnative species are purposely introduced for economic reasons. These species can adapt to the new environment and become an important part of the local culture and ecosystems, owing simply to their use by local people (e.g., for medicine) or just because of their abundance. Some inadvertent introductions are also able to become hyperabundant or useful in their new homes. Nonnative populations can become so firmly integrated into local cultures as to become cultural icons. Examples include potatoes in Europe or apples for the native Mapuche in Northern Patagonia, who call their land "the country of the apples" (Capella 1998).

Although other factors can also play a role, we suggest that five key factors – arrival time, economic impact, aesthetic preferences and phobias, effect on human health, and origin of nonnative species and origin of human immigrants – can profoundly affect whether and when a species is reviled or prized by the people of its new residence (Table 13.1). It is important to note that these five factors are not intrinsically independent. For example, economic impact (e.g., by using the species) can be higher for species with a long history in the region than for recent arrivals. Below we describe the five factors identified as important to understand species assimilation into local cultures.

Table 13.1 Factors that can accelerate or retard incorporation of a nonnative species into the local culture

Factor	How it operates	Example
Arrival time	Over time, people perceive that the nonnative species "belongs" there as much as native species	Horses (*Equus caballus*) brought by the Spaniards to the Americas
Economic impact	The use of species can accelerate its incorporation into the local culture and economy. By contrast, species that generate economic losses are difficult to incorporate into local cultures	Bananas (*Musa paradisiaca*) in Ecuador. Lodgepole pine (*Pinus contorta*) in New Zealand (Nuñez et al. 2017)
Aesthetic preferences and phobias	A species perceived as ornamental may be easier to be incorporated in the local culture than species perceived as ugly	Lupines (*Lupinus polyphyllus*) in New Zealand. Exotic weeds in gardens
Effect on human health	Species inimical to human health will be reviled; medicinal species will be incorporated	Poisonous mushrooms (Trim et al. 1999; Pringle et al. 2009)
Origins of human immigrants and nonnative species	Large groups of immigrants can introduce species from their homelands and incorporate them into the new culture	Hydrangeas in the Azores

13.2 Arrival Time

Time since introduction into a region can affect how a nonnative species is perceived. For example, species that arrive during the life span of the residents are clearly recognized as nonnative, while species that have been present for generations may require a taxonomist to identify the original area of distribution and may be perceived by locals as natives (Speziale et al. 2012). Among the cultural icons listed above, tomatoes reached Italy, peaches reached Georgia, and horses reached North America in the sixteenth century, bluegrass was well-established in Kentucky by the eighteenth century, and by the nineteenth century bananas were a major crop in Ecuador, coffee was the dominant export crop in Colombia, and kudzu, cannabis, and eucalyptus were established in their new homes. People were used to them, and probably in many cases only experts realized they were not native.

Also, in relation to other factors that can affect assimilation, with time chances are increased that a species will be used and therefore more incorporated into the local culture, as is the case for all these icons. Further, a nonnative species long present in a region is likely already to have exerted its most evident impact on the native biotic community, long before current residents would have noticed it; thus, the species is less likely to be perceived as invasive, except perhaps by scientists who perform experiments (e.g., exclosures of grazers or predators) and find major ecological changes. Finally, coevolution between the nonnative and coexisting native species will occur and, over time, facilitate accommodation between the newcomer and residents (see Crego et al. 2018). All of these factors will lessen the likelihood that a species will be perceived as inimical and may increase the probability that it will be valued.

13.3 Economic Impact

The use of a species can help to incorporate it quickly into a culture. A good example is the use of species as food; if a species becomes popular in local cuisine, it can be perceived as being an intrinsic part of the local culture (Nuñez et al. 2012), as with tomatoes in Italy and peaches in Georgia. Heated opposition to attempts to manage or eliminate environmentally damaging nonnative boar (*Sus scrofa*) and strawberry guava (*Psidium cattleianum)* in Hawaii was probably generated primarily by their use as food, although other factors (hunting for boar, ornamental status for strawberry guava) contributed to the fact that each species had a constituency (Simberloff 2010). Other uses for economic reasons, such as timber, fur, or sportfishing, can also help a species to become part of the local economy and hinder an efficient control campaign. Persistent controversy surrounding attempts to manage ecologically damaging nonnative salmonids in South Africa stems from concerted opposition by the trout angling industry (Zengeya et al. 2017). Similarly, sportfishing for introduced salmonids in Lago Nahuel Huapi, the hub of Argentina's largest national park, is such a major economic pillar of the region that no one would suggest removal, despite the fact that the salmonids are widely known to be introduced and to have detrimentally affected native fishes (Vigliano et al. 2007). The salmonids have even become a cultural icon, with t-shirts depicting them and touting angling in the lake.

By contrast, weeds and pests are by definition species that have negative effects on human economies (Lockwood et al. 2013). Although the definition does not consider whether the species is native or exotic, many nonnative species are problematic weeds and pests. These species are not incorporated into local cultures, and the goal is to manage or eradicate them. However, management or eradication of "environmental weeds or pests" that have negative ecological impact but are used by some stakeholders, especially if the use is economically beneficial, can also be resisted by local communities. A good example is the opposition by hunters to the control of introduced pigs in Hawaii, where extensive research demonstrates their harmful ecological impacts (Nogueira-Filho et al. 2009).

13.4 Aesthetics and Phobias

Perceived beauty is an important source in the valuation of nature (Millennium Ecosystem Assessment 2005). Species perceived as being beautiful or ugly can be valued differently in local communities. Nonnative species of Asian ornamental *Hydrangea* in the Azores archipelago have become a national symbol for the archipelago, with tourists timing their visits to coincide with maximum *Hydrangea* flowering and the island of Faial known as the "Blue Island" because of the density of blue *Hydrangea* flowers. Remarkably, California regions heavily settled by three waves of Azorean immigrants are known as "new Blue Islands," and the immigrants have planted *Hydrangea* in their yards and decorated homes, carts, and household items with paintings of *Hydrangea* flowers as a cultural marker (Helzer and Machado 2011).

Removal of invasive ornamental species with high aesthetic appeal tends to be resisted by the local community (Dickie et al. 2014). The controversy surrounding removal of *Eucalyptus* trees as part of the ecological restoration of Angel Island, cited above, is but one of several disputes based on aesthetics that have arisen over planned removals of *Eucalyptus*. Similarly, although nonnative European common buckthorn (*Rhamnus cathartica*) and glossy buckthorn (*R. frangula*) have replaced prairie and oak savanna in parts of the US Midwest, a plan to restore the native community in suburban Chicago elicited such heated opposition that it was canceled (Simberloff 2012). Many citizens valued the appearance of the large, leafy buckthorn shrubs far more than that of the prairie plants they have replaced. In Florida, Australian "pine" (*Casuarina* spp.) causes major ecological damage to native plants and nesting sea turtles, but passionate admirers of the frequently tall tree with sweeping, slender scale leaves have impeded several removal projects (Simberloff 2012). Pretoria, the capital of South Africa, is known as the "Jacaranda City" because of the tens of thousands of south-central South American jacaranda trees (mostly *Jacaranda mimosifolia*) lining the streets beginning in 1888. It came to be viewed as "almost…South Africa's national tree" (Moll and Moll 1994), and President Nelson Mandela, in his 1994 inaugural speech, proclaimed "To my compatriots, I have no hesitation in saying that each one of us is as intimately attached to the soil of this beautiful country as are the famous jacaranda trees of Pretoria and the mimosa trees of the bushveld" (Nur 2015). However, in 2001, jacarandas were classified as a category 3 invader, based on their heavy use of water and inhibition of ground cover. This classification, which allows existing trees to be kept but prevents sale, movement, or planting of new individuals, set off a persistent controversy among residents unwilling to lose a beloved part of their heritage (<http://www.bbc.com/news/av/world-africa-20317884/jacaranda-tree-debate-in-pretoria>(12/13/2012), accessed 6/23/17).

A striking case concerns the raccoon (*Procyon lotor minor*) present on the island of Guadeloupe in the eastern Caribbean Sea, long believed to be an endemic species or subspecies resident prehistorically. Recent research shows that the type specimen was a juvenile, thus incorrectly classified, and that the Guadeloupe raccoon is almost certainly an introduction from mainland North America within the last 200 years (Helgen and Wilson 2003). In addition, Guadeloupe, as a French overseas territory, is subject to European Union regulations, including the 2016 Regulation on invasive alien species, which requires member states to take measures for early detection and rapid eradication of these species and to manage species that are already widely distributed. The raccoon is on the initial list of 37 species. However, islanders, even those suffering substantial agricultural losses from raccoons, are loath to kill them, and, in fact, residents have long been proud of the raccoon as an "endemic" unique to their island and one of very few mammals in the Lesser Antilles. Even the new knowledge that it is a relatively recent invader has not led many islanders to forswear its iconic status (https://www.acast.com/radiolab/stanger-in-paradise, accessed 6/23/17).

By contrast, the public applauds and encourages efforts to manage or eradicate nonnative species that elicit a sort of instinctive revulsion or fear. Thus the brown tree snake (*Boiga irregularis*) in Guam and the Burmese python (*Python bivittatus*) in Florida have become not only locally reviled targets of massive (and so far futile) control campaigns but internationally recognized symbols of the threat posed by biological invasions (e.g., Anthony 2017). In these two instances, the ecological damage caused by the invaders is staggering (Perry and Rodda 2011; Dorcas et al. 2012). On the other hand, the northern snakehead (*Channa argus*), an eel-shaped predatory Asian fish, was predicted to have staggering ecosystem impacts (Dolan 2003). Dubbed as "Frankenfish" it elicited national concern and garnered numerous newspaper headlines in the United States after its discovery near Washington, D.C., in 2002. However, 15 years later, it is recognized to have been introduced multiple times into the Eastern United States (Wegleitner et al. 2016), and no evidence of ecological impact has yet been documented (Fuller et al. 2017). Nevertheless, the snakehead continues to be a target of public fear and odium (e.g., Zauzmer 2015). There is little likelihood that the snakehead fish or any other nonnative snakelike animal will ever become a beloved, protected cultural icon owing to a widespread generalized phobia regarding snakes (cf. Öhman and Mineka 2003).

13.5 Effect on Human Health

Nonnative species that threaten human health are not assimilated in local cultures and often elicit quick support for their management or eradication. Mosquitoes transmitting the Zika virus are the latest example (e.g., Servick 2016). Giant hogweed (*Heracleum mantegazzianum*), a tall, striking, invasive shrub from the Caucasus Mountains whose sap, under certain conditions of moisture and sunlight, can cause severe skin and eye irritation, blistering, permanent scarring, and even blindness, is a widely targeted and despised invasive in the United States, Great Britain, and continental Europe, despite appeals by the garden architect Gilles Clément (2002) that its aesthetic qualities should nevertheless earn it a place in gardens. Even nonnative species that do not threaten human health can be tarred by association with human immigrants believed to be unclean reservoirs of human disease. The Old World house sparrow (*Passer domesticus*), introduced to North America in the mid-nineteenth century, was long reviled as an inhabitant of urban slums inhabited by despised immigrants such as Italians and Eastern European Jews thought to spread germs to the general population (Tomes 1998; Coates 2006).

On the other hand, nonnative species are widely used for medical purposes (including in traditional medicine) and are often prized in local cultures, even though they are recognized as introduced (e.g., Bennett and Prance 2000; Palmer 2004; Alencar et al. 2014). Even highly invasive plants can be valued by local populations for medical uses. For instance, one of the most ecologically damaging invaders in Africa, Siam weed (*Chromolaena odorata*), is prized as a cosmetic (Holou et al. 2013).

13.6 Origin of Nonnative Species and Origin of Human Immigrants

Immigrants have been a main source of biological invasions. Some groups of immigrants have introduced species from their original homelands into their new homes as sources of psychological support and aesthetic pleasure. These species are by definition part of the culture of the immigrant group and therefore part of the local culture where they have established, if immigration is substantial. Thus, in Hawaii, competing acclimatization societies, the Hui Manu and the Honolulu Mejiro Club, brought North American and Japanese birds, respectively, to the Hawaiian Islands, thinking the native avifauna was drab and boring (Simberloff 2010). Similarly, nostalgia for the flora and fauna of home was a major factor leading members of

Australian and New Zealand acclimatization societies to import massive numbers of European plants and animals (Thomson 1922; Crosby 1986; Lever 1992).

By the same token, antipathy toward particular peoples can cause natives or earlier immigrants to a region to scorn and stigmatize species associated with subsequent immigrants. Above we noted the early opprobrium attached by nineteenth-century Americans of English and German extraction to the introduced house sparrow because of its association, in their minds, with Italian and Eastern European Jewish immigrants. Several authors, reviewed by Simberloff (2003) and Cardozo and Subramaniam (2013), have argued that Asian species of plants and animals introduced to the United States are feared and loathed as extensions of the "yellow peril," while Tsing (1995) sees concern about the spread of Africanized honey bees (*Apis mellifera scutellata*) as partly due simply to their African origin rather than to the real threat they pose to human health and apiarists' livelihoods. Certainly the Danish-American garden architect Jens Jensen despised plants native to Latin America because of his antipathy toward Latin-American immigrants (Simberloff 2003).

13.7 How Quickly Can Invasive Species Become Culturally Assimilated?

Although we noted above that time since arrival contributes to the likelihood that a nonnative species will not be reviled and may even become culturally assimilated, it is important to note that nonnative species, even highly invasive ones, can be incorporated into cultures very quickly and be perceived as intrinsic parts of local ecosystems. Speed of incorporation can depend heavily on the local culture's ecological understanding of their surroundings, which can vary among different societies and can depend on where they live (e.g., rural vs. urban) among other factors. A key factor is that memory of a culture is not static: it changes through time, and it is impacted by various forces (Olick and Robbins 1998); therefore, customs, mythology, and history of a people cannot be trusted as precise delineations of old or even recent ecosystems. A case in point is found in the Nuer tribe from East Africa. The tree under which the Nuer believed that humanity appeared was still alive in the early 1900s (Evans-Pritchard 1940), which exemplifies how traditional beliefs can be gained in societies without written records.

Another distinctive example can be found in the impacts of the Spanish conquest of the Americas, which entailed introducing many nonnative species, some of them now attached to various local cultures as if they have always been present in the region. Such is the situation with horses (*Equus caballus*), which are so integrated into local cultures that it can appear to be unconceivable that they were not present before the way of life emerged (Musters 1964), and there are even laws to protect them. The "Wild Free-Roaming Horses and Burros" Act states:

> That Congress finds and declares that wild free-roaming horses and burros are living symbols of the historic and pioneer spirit of the West; that they contribute to the diversity of life forms within the Nation and enrich the lives of the American people.... (U.S.A. Public Law 92–195, 1971)

Despite a US National Park Service policy that states "nonnative species of plants and animals will be eliminated where it is possible to do so," feral horses are protected in some units, such as Assateague Island National Seashore and Cumberland Island National Seashore, where they are favorites of visitors despite strong evidence of their harmful impacts (Simberloff 2017).

In Maine, USA, the European periwinkle (*Littorina littorea*) is likely the most unmistakable gastropod in the state (Norton 1993), and in Maine and Nova Scotia it supports a large fishery. However, it arrived less than two centuries ago and has massively changed local biological communities, transforming most of the coastline from mud flats, marshes, and algal-covered rocks to the stark, rockbound vista depicted by innumerable landscape painters (Bertness 1984; Chapman et al. 2008). Even recently introduced plants perceived to be of medicinal value have been incorporated in the native Hawaiian pharmacopeia, with nearly 50% of medicinal plants recent introductions, all now with Hawaiian names, and the majority native to Europe, Africa, the Neotropics, and Australia (Palmer 2004).

13.8 Concluding Remarks and Implications for Management

It has become a commonplace in ecology and environmental science that effective resource management requires engagement and support from various stakeholders, particularly the local community and government. As part of this recognition, it is now widely accepted that management of invasive species must often involve interaction with people with different views of whether a particular nonnative species should be a target for management and, if so, what appropriate management goals and approaches should be (Rotherham and Lambert 2011; Simberloff et al. 2013; Van Driesche et al. 2016; Crowley et al. 2017; Nuñez et al. 2017; Zengeya et al. 2017). The lack of support by key groups has slowed management and eradications or completely stopped those programs. An eradication program to eliminate the North American gray squirrel in Italy was terminated because of a lawsuit by advocates of animal rights (Bertolino and Genovesi 2003), and the species has now spread nearly to the French and Swiss borders. In Argentina and Chile, governments restrict hunting of the exotic red deer (*Cervus elaphus*), which have become important commercially (Lambertucci and Speziale 2011) (Fig. 13.1). These deer have become highly ecologically problematic in Patagonia (Relva et al. 2010). The similar conflicting views of introduced salmonids in South Africa are cited above. In Hawaii, controversy over management of highly damaging strawberry guava, noted above, delayed a biological control program by several years (Johnson 2016).

Understanding factors that contribute to the incorporation of a nonnative species into local cultures is clearly important to determining effective control measures.

Fig. 13.1 Monument to the red deer (*Cervus elaphus*) in San Martin de los Andes, Patagonia, Argentina. This deer is a European invader that is highly problematic for regenerating native forest. However, the species is a prized target for hunting, which is an important industry in the region. (Photo Martin A. Nuñez)

For instance, the frequent advocacy of using invasive species as food items to control them can become problematic if this leads to incorporation of the species into the local diet, economy, and culture (Nuñez et al. 2012). The matter of how exactly to engage stakeholders and mediate controversies is beyond the scope of this chapter, but it is apparent that integration of invasive nonnative species into local cultures will increase the management challenge. Certainly, substantial, publicly accessible educational programs explaining the problems that the invasive species produce (economic and/or ecological) and programs of appreciation of native species will be required.

References

Alencar NL, Santoro FR, Albuquerque UP (2014) What is the role of exotic medicinal plants in local medical systems? A study from the perspective of utilitarian redundancy. Rev Bras 24:506–515

Anthony L (2017) The aliens among us. Yale University Press, New Haven

Bennett BC, Prance GT (2000) Introduced plants in the indigenous pharmacopeia of South America. Econ Bot 54:90–102

Bertness MD (1984) Habitat and community modification by an introduced herbivorous snail. Ecology 65:370–381

Bertolino S, Genovesi P (2003) Spread and attempted eradication of the grey squirrel (*Sciurus carolinensis*) in Italy, and consequences for the red squirrel (*Sciurus vulgaris*) in Eurasia. Biol Conserv 109:351–358

Capella WD (1998) Pasos cordilleranos, grupos sociales y procesos hegemónicos en norpatagonia hacia fines del siglo XIX. Revista de Historia – Universidad de Concepción 7:215–228

Cardozo K, Subramaniam B (2013) Assembling Asian/American naturecultures. J Asian Am Stud 16:1–23

Chapman JW, Blakeslee AMH, Carlton JT, Bellinger MR (2008) Parsimony dictates a human introduction: on the use of genetic and other data to distinguish between the natural and human-mediated invasion of the European snail *Littorina littorea* in North America. Biol Invasions 10:131–133

Clément G (2002) Éloge des vagabondes. Herbes, arbres et fleurs à la conquête du monde. Nil éditions, Paris

Coates P (2006) Strangers on the land. American perceptions of immigrant and invasive species. University of California Press, Berkeley

Crego R, Rozzi R, Jiménez JE (2018) Fur trade and the biotic homogenization of sub-polar ecosystems. In: Rozzi R, May R, Chapin SF, Massardo F, Gavin M, Klaver I, Pauchard A, Nuñez MA, Simberloff D (eds) From biocultural homogenization to biocultural conservation. Ecology and ethics, vol 3. Springer, Dordrecht, pp 233–243

Crosby AW (1986) Ecological imperialism: the biological expansion of Europe. Cambridge University Press, Cambridge, pp 900–1900

Crosby AW (2003) The Columbian exchange: biological and cultural consequences of 1492. Greenwood Publishing Group, Westport

Crowley SL, Hinchliffe S, Redpath SM, McDonald RA (2017) Disagreement about invasive species does not equate to denialism: a response to Russell and Blackburn. Trends Ecol Evol 32:228–229

Culley TM, Hardiman NA, Hawkes J (2011) The role of horticulture in plant invasions: how grafting in cultivars of Callery pear (*Pyrus calleryana*) can facilitate spread into natural areas. Biol Invasions 13:739–746

Dickie IA, Bennett BM, Burrows LE, Nuñez MA, Peltzer DA, Porté A, Richardson DM, Rejmánek M, Rundel PW, van Wilgen BW (2014) Conflicting values: ecosystem services and invasive tree management. Biol Invasions 16(3):705–719

Dillingham WF (1936) Hui Manu: friend of birds. Parad Pac 48:20–21

Dolan EJ (2003) Snakehead. A fish out of water. Smithsonian Books, Washington DC

Dorcas ME, Willson JD, Reed RN, Snow RW, Rochford MR, Miller MA, Meshaka WE Jr et al (2012) Severe mammal declines coincide with proliferation of invasive Burmese pythons in Everglades National Park. Proc Natl Acad Sci USA 109:2418–2422

Evans-Pritchard EE (1940) The Nuer, a description of the modes of livelihood and political institutions of a Nilotic people. Clarendon Press, Oxford

Fuller PL, Benson AJ, Nunez G, Fusaro A, Neilson M (2017) Channa argus. USGS nonindigenous aquatic species database. https://nas.er.usgs.gov/queries/factsheet.sspx?speciesid=2265

Guild E (1938) Tahitian aviculture: acclimatization of foreign birds. Avic Mag 3:8–11

Helgen KM, Wilson DE (2003) Taxonomic status and conservation relevance of the raccoons (*Procyon* spp.) of the West Indies. J Zool 259(1):69–76

Helzer J, Machado E (2011) The new blue islands: Azorean immigration, settlement, and cultural landscapes in California's San Joaquin Valley. Calif Geogr 51(6), on-line URL: http://hdl.handle.net/10211.2/2817

Holou RAY, Achigan–Dako EG, Sinsin B (2013) Ecology and management of invasive plants in Africa. In: Jose S, Singh HP, Batish DR, Kohli RK (eds) Invasive plant ecology. CRC Press, Boca Raton, pp 161–174

Johnson MT (2016) Managing conflict over biological control: the case of strawberry guava in Hawaii. In: Van Driesche R, Simberloff D, Blossey B, Causton C, Hoddle M, Marks C, Heinz K, Wagner D, Warner K (eds) Integrating biological control into conservation practice. Wiley Blackwell, Chichester, pp 264–276

Lambertucci SA, Speziale KL (2011) Protecting invaders for profit. Science 332(6025):35–35

Lever C (1992) They dined on eland: the story of the acclimatisation societies. Quiller, London

Lockwood JL, Hoopes MF, Marchetti MP (2013) Invasion ecology. Wiley, West Sussex

Millennium Ecosystem Assessment (2005) Ecosystems and human well-being: synthesis. Island Press, Washington, DC

Moll E, Moll G (1994) Common trees of South Africa. Struik, Cape Town

Moure N (1982) Impressionism, post-impressionism, and the eucalyptus school in Southern California. In: Westphal R (ed) Plein air painters of California: the Southland. 1, Irvine, pp 4–15

Musters GC (1964) Vida Entre los Patagones. Ediciones Solar, Buenos Aires

Nogueira-Filho SLG, Nogueira SSC, Fragoso JMV (2009) Ecological impacts of feral pigs in the Hawaiian Islands. Biodivers Conserv 18:3677–3683

Norton PT (1993) Fossils of the Maine state capitol. Maine Nat 1(4):193–204

Nuñez MA, Simberloff D (2005) Invasive species and the cultural keystone species concept. Ecol Soc 10(1):r4

Nuñez MA, Kuebbing S, Dimarco RD, Simberloff D (2012) Invasive species: to eat or not to eat, that is the question. Conserv Lett 5(5):334–341

Nuñez MA, Chiuffo MC, Torres A, Paul T, Dimarco RD, Raal P, Policelli N, Moyano J, García RA, van Wilgen BW, Pauchard A, Richardson DM (2017) Ecology and management of invasive Pinaceae around the world: progress and challenges. Biol Invasions 19(11):3099–3120

Nur S (2015) Analysis of interpersonal metafunction in public speeches: a case study of Nelson Mandela's presidential inauguration speech. Int J Soc Sci Humanit Stud 30(1):52–63

Öhman A, Mineka S (2003) The malicious serpent: snakes as a prototypical stimulus for an evolved module of fear. Curr Dir Psychol Sci 12:5–9

Olick JK, Robbins J (1998) Social memory studies: from "collective memory" to the historical sociology of mnemonic practices. Annu Rev Sociol 24:105–140

Palmer CT (2004) The inclusion of recently introduced plants in the Hawaiian ethnopharmacopeia. Econ Bot 58(Suppl):S280–S293

Perry G, Rodda GH (2011) Brown treesnake. In: Simberloff D, Rejmánek M (eds) Encyclopedia of biological invasions. University of California Press, Berkeley, pp 78–81

Pringle A, Adams RI, Cross HB, Bruns TD (2009) The ectomycorrhizal fungus *Amanita phalloides* was introduced and is expanding its range on the west coast of North America. Mol Ecol 18(5):817–833. https://doi.org/10.1111/j.1365-294X.2008.04030.x

Relva MA, Nuñez MA, Simberloff D (2010) Introduced deer affect native plant communities and facilitate non-native tree species in a temperate forests in Patagonia, Argentina. Biol Invasions 12(2):303–311

Rotherham ID, Lambert RA (eds) (2011) Invasive and introduced plants and animals. Human perceptions, attitudes and approaches to management. Routledge, Abingdon

Servick K (2016) Winged warriors. Science 354:164–167

Simberloff D (2003) Confronting introduced species: a form of xenophobia? Biol Invasions 5(3):179–192

Simberloff D (2010) Hawaii: paradise lost. In: Weber K (ed) Cane toads and other rogue species. Public Affairs, New York, pp 81–97
Simberloff D (2012) Nature, natives, nativism, and management: worldviews underlying controversies in invasion biology. Environ Ethics 34:5–25
Simberloff D (2017) Biological invasions in the national parks and *Park Science*. In: Beissinger SR, Ackerly DD, Doremus H, Machlis GE (eds) Science, conservation, and national Parks. University of Chicago Press, Chicago, pp 161–188
Simberloff D, Martin J-L, Genovesi P, Maris V, Wardle DA, Aronson J, Courchamp F, Galil B, García-Berthou E, Pascal M, Pyšek P, Sousa R, Tabacchi E, Vilà M (2013) Impacts of biological invasions: what's what and the way forward. Trends Ecol Evol 28:58–66
Speziale KL, Lambertucci SA, Carrete M, Tella JL (2012) Dealing with non-native species: what makes the difference in South America? Biol Invasions 14(8):1609–1621
Thomson GM (1922) The naturalisation of animals and plants in New Zealand. Cambridge University Press, Cambridge
Tomes N (1998) The gospel of germs. Men, women, and the microbe in American life. Harvard University Press, Cambridge
Trim G, Lepp H, Hall M, McKeown R, McCaughan G, Duggin G (1999) Poisoning by *Amanita phalloides* ("deathcap") mushrooms in the Australian capital territory. Med J Aust 171:247–249
Tsing AL (1995) Empowering nature, or: some gleanings in bee culture. In: Yanagisako S, Delaney C (eds) Naturalizing power. Essays in feminist cultural analysis. Routledge, New York, pp 113–143
Van Driesche R, Simberloff D, Blossey B, Causton C, Hoddle M, Marks C, Heinz K, Wagner D, Warner K (eds) (2016) Integrating biological control into conservation practice. Wiley Blackwell, Chichester
Vigliano PH, Alonso MF, Aquaculture M (2007) Salmonid introductions in Patagonia: a mixed blessing. In: Bert TM (ed) Ecological and genetic implications of aquaculture activities. Springer, Dordrecht, pp 315–331
Wegleitner BJ, Tucker A, Chadderton WL, Mahon AR (2016) Identifying the genetic structure of introduced populations of northern snakehead (*Channa argus*) in Eastern USA. Aquat Invasions 11:199–208
Williams T (2002) America's largest weed. Audubon 104(1):24–31
Zauzmer J (2015) Spread of snakehead fish in Potomac River concerns area officials. Washington Post, August 5. https://www.washingtonpost.com/local/invasive-fish-expands-territory-this-is-just-absolutely-something-we-didnt-want-to-happen/2015/08/05/75e5fe3c-3afc-11e5-b3ac-8a79bc44e5e2_story.html?utm_term=.34f643ef87b8
Zengeya T, Ivey P, Woodford DJ, Weyl O, Novoa A, Shackleton R, Richardson D, van Wilgen B (2017) Managing conflict-generating invasive species in South Africa: challenges and trade-offs. Bothalia Afr Biodivers Conserv 47(2): online URL: https://doi.org/10.4102/abc.v47i2.2160

Chapter 14
Fur Trade and the Biotic Homogenization of Subpolar Ecosystems

Ramiro D. Crego, Ricardo Rozzi, and Jaime E. Jiménez

Abstract At the southern end of the Americas exist one of the last pristine ecosystems in the world, the sub-Antarctic Magellanic forests ecoregion, protected by the Cape Horn Biosphere Reserve (CHBR). Despite its remote location, the CHBR has been subject to the growing influences of globalization, a process that has driven cultural, biotic, and economic transformations in the region since the mid-twentieth century. One of the most important threats to these unique ecosystems is the increase of biological invasions. Motivated by the expanding fur industry that responded to the globalization process, American beavers (*Castor canadensis*), muskrats (*Ondatra zibethicus*), and American minks (*Neovison vison*) were introduced, independently, to the southern tip of South America. Research has shown that these three North American species have reassembled their native interactions to affect negatively the invaded ecosystems of the CHBR. Beavers affect river flow and

R. D. Crego (✉)
Department of Biological Sciences, University of North Texas, Denton, TX, USA

Instituto de Ecología y Biodiversidad, Santiago, Chile

Sub-Antarctic Biocultural Conservation Program, University of North Texas, Denton, TX, USA

R. Rozzi
Department of Philosophy and Religion and Department of Biological Sciences, University of North Texas, Denton, TX, USA

Sub-Antarctic Biocultural Conservation Program, University of North Texas, Denton, TX, USA

Instituto de Ecología y Biodiversidad and Universidad de Magallanes, Punta Arenas, Chile

J. E. Jiménez
Department of Biological Sciences, University of North Texas, Denton, TX, USA

Instituto de Ecología y Biodiversidad, Santiago, Chile

Sub-Antarctic Biocultural Conservation Program, University of North Texas, Denton, TX, USA

Department of Philosophy and Religion, University of North Texas, Denton, TX, USA

Universidad de Magallanes, Punta Arenas, Chile

R. Rozzi et al. (eds.), *From Biocultural Homogenization to Biocultural Conservation*, Ecology and Ethics 3, https://doi.org/10.1007/978-3-319-99513-7_14

native vegetation, changing forests into wetlands, creating suitable habitats for muskrats. Muskrats, in turn, are the main prey of inland mink populations. The latter has major impacts by preying opportunistically on the native biota, especially native birds and small rodents. In this chapter, we explore this multi-species invasive system as an example of biotic homogenization, in which the introduction of these species and their subsequent reassembling of their interactions, together with the ecosystem impacts, offer a novel example of complex processes of biotic homogenization involving both biological and sociocultural dimensions.

Keywords Cape Horn Biosphere Reserve · Biocultural homogenization · Invasive meltdown · Invasive species

14.1 Introduction

South American temperate forests represent a unique biome with extremely high endemism. Close to 90% of the woody plants, about 60% of the bryophytes, 50% of fish, 80% of amphibian, 36% of reptile, 30% of land bird, and 33% of mammal species are endemic to the forest biome (Armesto et al. 1998; Rozzi et al. 2008). This high endemism is associated with the isolation of the austral South American forest biome from the nearest tropical forests by 1500–2000 km (Armesto et al. 1998). Topographic and climatic barriers include the high Andes along with the vast dry steppe of Argentina, the hyper-arid Atacama Desert, and the southern Pacific Ocean. The biome extends for approximately 3000 km along southwestern South America from central Chile (35°S) to the southern tip of the continent in Cape Horn (56°S). At its southern end, the temperate forest biome includes the sub-Antarctic Magellanic forests ecoregion, which is dominated by trees of the genus *Nothofagus* (southern beech), one evergreen, *N. betuloides*, and two deciduous, *N. pumilio* and *N. antarctica*. The forest is embedded in the Magellanic moorland complex, composed of a matrix of peatlands, bogs, and meadows (Rozzi et al. 2006). Given its remote location, low human density, large geographical extent, and high percentage of remaining non-fragmented native forests, this area has been identified as 1 of the last 24 wildernesses areas of the world (Mittermeier et al. 2003). To protect these remote and unique ecosystems together with their cultural heritage, in 2005 a team of scientists, the local community, and the Chilean government successfully proposed to UNESCO the creation of the Cape Horn Biosphere Reserve (CHBR) (Rozzi et al. 2006).

Despite being isolated and located in a remote region, the CHBR has been subject to the growing influences of globalization, a process that has driven cultural, biotic, and economic transformations in the area since the mid-twentieth century (Rozzi et al. 2006, 2012; Berghöfer et al. 2008). One of the salient agents of change within this reserve is the introduction of invasive mammalian species, particularly those associated with the fur trade. This process has involved the relocation of species from North to South America and provides an exemplar case of human-mediated translocation of species that contributes to a pervasive phenomenon known as biotic homogenization (McKinney and Lockwood 1999), in which distant ecosystems end up dominated by similar groups of plants and animals (Simberloff 2013).

The process of biotic homogenization often involves parallel and interrelated processes of cultural homogenization. To characterize the feedback between these two processes, Rozzi (2012) coined the term "biocultural homogenization." The habitats and biotas as well as the life habits (customs and worldviews) of local people are replaced by those of global culture, a process that has been intensified during the Great Acceleration since the mid-twentieth century (Rozzi 2015).

One of the industries that expanded with the globalization process during the twentieth century was the fur industry. Humans have historically used hides as clothes for protection from the environment, including the Native American cultures that inhabited the hostile climates of subpolar regions of the Americas. Hides were an essential component of the clothing of Koyukon, Inuit, Algonquian, and Iroquois people inhabiting the temperate and sub-Arctic zones of North America and the Fuegian people inhabiting terrestrial and coastal ecosystems in the sub-Antarctic region of South America (Martin 1982; McEwan et al. 2014). However, during the nineteenth and twentieth centuries, the fur industry prospered as a response to more than just using clothes for protection. It was a response to an increasingly profitable clothing industry motivated by western ideas of aesthetics and fashion in the United States and Europe, which started to expand in the twentieth century with the growing free market and global economic system (Skov 2005; IFTF 2017). The fur industry settled at the south end of South America between the 1930s and 1950s.

Among the 12 documented non-native mammals introduced into the CHBR, 3 are furbearing species brought as a response to the promotion of fur industry by the Argentinean National Direction of Wildlife (Lizarralde 1993). These are the American beaver (*Castor canadensis*), the muskrat (*Ondatra zibethicus*), and the American mink (*Neovison vison*). These three species that interact in their native range in North America (Viljugrein et al. 2001; Shier and Boyce 2009; Mott et al. 2013) have reassembled those native interactions in the non-native ecosystems of the CHBR (Fig. 14.1; Crego et al. 2016). In this chapter, we explore the reassemblage of these three North American species and potential ecological impacts on the ecosystems of the sub-Antarctic Magellanic ecoregion, which represents a unique example of biotic homogenization between two subpolar ecosystems.

14.2 The Trio and a Trans-Hemispheric Journey

Beavers are important components of North American ecosystems as keystone and ecosystem engineer species (Baker and Hill 2003). With their dam- and den-building activities, together with foraging, they can alter the hydrology, geomorphology, and chemistry of freshwater ecosystems, having large effects on plant and animal species composition, density, and distribution (Baker and Hill 2003; Rosell et al. 2005). There is one species that benefits from beaver behavior, the muskrat, which frequently occurs in habitat associated with beaver dams and dens (Mott et al. 2013). Muskrats are also known to affect invertebrate and plant abundance and nutrient flows of aquatic habitats (Connors et al. 2000; de Szalay and Cassidy 2001). Additionally, a third North American mammal can benefit from the presence of muskrats: the mink.

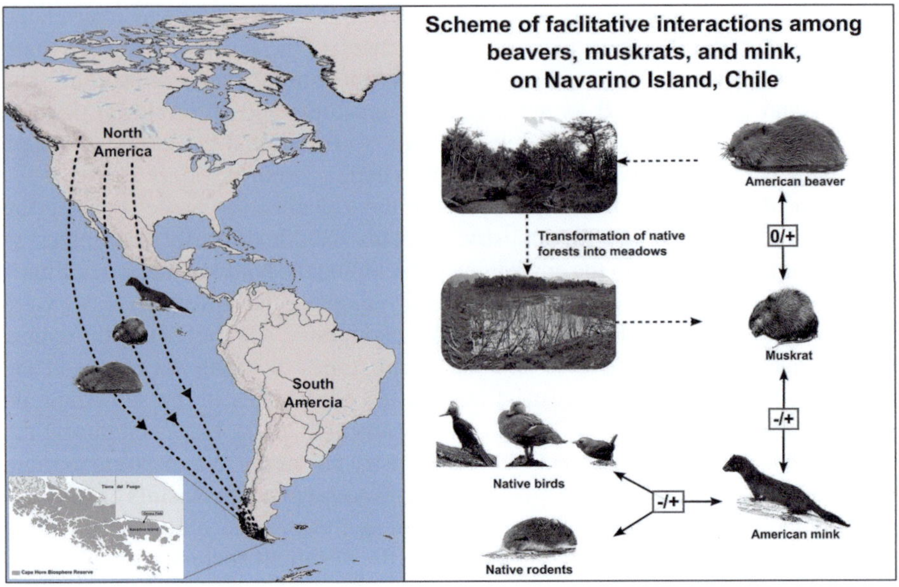

Fig. 14.1 This figure represents on the left the long-distance translocation of American beavers, muskrats, and American mink led by the fur industry from northern areas of North America to the southern regions of South America within the Cape Horn Biosphere Reserve, which promoted the biotic homogenization of both regions. On the right, the facilitative interactions among the three invasive species that were described by Crego et al. (2016) for Navarino Island are represented. American beavers modify native forests along fast water streams into meadows with calm water, favoring the establishment of muskrats (0/+ interaction type). Muskrats are the main prey of mink that inhabit in the forests and meadows of the island, facilitating mink survival, especially during the winters (−/+ interaction type). In turn, mink affect native species of rodents and birds through predation (−/+ interaction type). These interactions resulted in the reassemblage of the interactions that take place in the native habitats of North America

In many areas of their native range, muskrats are the preferred prey of the mink, a carnivore mustelid of semi-aquatic habits (Larivière 1999). Correlated population fluctuations in a predator-prey interaction between mink and muskrats have been documented in Canada indicating a tight ecological linkage between the two (Viljugrein et al. 2001; Shier and Boyce 2009). These three mammalian species form an assemblage that coevolved in a variety of habitats in North America.

In the mid-twentieth century, this trio of North American furbearing mammals was introduced in southern South America (Fig. 14.1). In 1946, 20 beavers were brought from Canada and released in the Fagnano Lake in the Argentinean side of Tierra del Fuego Island (Jaksic et al. 2002). At a similar time and in a similar location, several muskrats were also released in Tierra del Fuego freshwater systems (Jaksic et al. 2002). Since the introduction, beavers and muskrats have colonized almost every environment on the island of Tierra del Fuego and have dispersed onto other islands and the mainland of the South American continent (Jaksic et al. 2002; Graells et al. 2015). Beavers, particularly, have impacted large areas of native ecosystems, altering stream nutrient cycles (Ulloa et al. 2012), stream food webs

(Anderson and Rosemond 2007), and changing large areas from closed *Nothofagus* forest to grass and rush-dominated meadows (Lizarralde et al. 2004; Anderson et al. 2006a). These habitat alterations are landscape-wide and long-lasting (Henn et al. 2016).

The American mink was introduced from fur farms. The first mink individuals were brought from North America to farms installed in Punta Arenas, Chile, during the years 1934 and 1936 (Rozzi and Sherrifs 2003). Additionally, fur farms began operations in southern Argentina in the 1930s (Jaksic et al. 2002). By 1959 there were more than 60 mink breeding centers in southern Argentina, including Tierra del Fuego (Rozzi and Sherriffs 2003). They ultimately led to the establishment of wild populations due to escapes and intentional releases (Lizarralde and Escobar 2000). After 1990, sightings of mink have become more common in several areas of Tierra del Fuego Island (Lizarralde and Escobar 2000), and approximately at the end of that decade, mink crossed the Beagle Channel reaching Navarino Island, in southern Chile (Rozzi and Sherriffs 2003). Since then, mink have quickly dispersed throughout the CHBR (Anderson et al. 2006b; Valenzuela et al. 2014; Crego et al. 2015). Today, the populations of mink represent a major threat for native mammals, birds, and fishes that are part of the mink's diet in the CHBR (Schüttler et al. 2008, 2009; Ibarra et al. 2009).

14.3 Biotic Homogenization of Two Poles in the Americas

The CHBR is composed of many different islands. Navarino Island, located south of Tierra del Fuego, is where the town of Puerto Williams and the Omora Ethnobotanical Park are situated. This location is the scientific center of the CHBR where the majority of the studies on the reserve take place. A recent study suggests that the ecological interactions that occur among American beavers, muskrats, and mink in the sub-Arctic region of North America have been reassembled on Navarino Island in the sub-Antarctic region of South America (Crego et al. 2016).

Beavers and muskrats crossed the Beagle Channel from Tierra del Fuego into Navarino Island around 1962 (Jaksic et al. 2002). The mink was first detected on this island more recently, in 2001 (Rozzi and Sherriffs 2003), where it became the top terrestrial predator. In contrast with Tierra del Fuego and Hoste Islands where the native Culpeo fox (*Lycalopex culpaeus*) is the top predator (Sielfeld 1977), Navarino Island had no mammalian terrestrial predators. Additionally, on Navarino Island freshwater ecosystems are mainly formed on steep mountain slopes with rivers having narrow beds and high water velocities (running waters). These are not the habitat conditions that muskrats prefer; they inhabit water systems that present low variations in water levels, found in calm waters such as ponds, and with abundant aquatic vegetation (Artimo 1960).

Beavers may then benefit muskrats in Navarino Island by transforming riparian forests into meadows, changing running waters into calm waters where vegetation grows. To investigate this interaction, Crego et al. (2016) sampled for muskrat signs in four different types of aquatic habitats: active beaver dams with a pond system, inactive beaver dams with a pond system, inactive beaver dams with a running water system (i.e., old beaver dams with recovered stream flow), and streams with no beaver activity. Muskrat's signs were more abundant in habitats that had been transformed by beavers into calm water systems. Moreover, muskrat's signs were almost absent from naturally occurring fast-flowing streams. These findings revealed a novel facilitative interaction between beavers and muskrats, where beavers create suitable habitat for muskrats (Fig. 14.1; Crego et al. 2016).

Crego et al. (2016) also examined the trophic interactions of mink with muskrats and beavers. They studied the mink's diet and found that mink populations inhabiting marine coastal habitats feed mainly on fish, while populations inhabiting inland forests and meadows feed mainly on muskrats. Today, in inland habitats of Navarino Island, muskrats represent over 50% of the total biomass intake of the mink. This predator-prey interaction closely resembles trophic interactions described for mink populations in their native habitats of North America (Viljugrein et al. 2001; Shier and Boyce 2009). Additionally, previous research has shown that mink consumption of muskrats is more important during harsh winters when other prey items decrease in abundance (Schüttler et al. 2008; Ibarra et al. 2009). Therefore, muskrats may play a pivotal role in allowing part of the mink population to survive winters with low prey abundance. Mink have, in turn, major impacts on the native fauna, especially after the breeding season, when their population increases. This is particularly relevant for Navarino Island because native species of small rodents and birds evolved in the absence of terrestrial mammalian predators, making them potentially naïve to mink predation risk (Crego et al. 2016). High predation rates on native species of rodents and birds (Schüttler et al. 2008, 2009; Ibarra et al. 2009; Jiménez et al. 2014) may have important consequences on population dynamics (Fig. 14.1; Crego et al. 2014).

In summary, on Navarino Island the presence of muskrat seems to be facilitated by beaver's habitat modifications, which in turn facilitate mink survival in inland habitats (Fig. 14.2). These three species appear to be synergistically interacting to invade and transform the terrestrial biotic community of the sub-Antarctic Magellanic ecoregion in the CHBR. The habitat modifications and trophic interactions among a trio of mammals that interact in the native habitats of North America offer a novel example of complex processes of biotic homogenization involving both biological and sociocultural dimensions. As a result, today the sub-Antarctic and sub-Arctic ecosystems share an assemblage of species that make these two distant regions more similar than they were a century ago.

14.4 The Homogenization Includes Eurasia

The effect on biotic homogenization driven by American beavers, muskrats, and American mink is not limited to the subpolar regions of the Americas. With the prosperity of the fur industry in the Northern Hemisphere, these species were also

Fig. 14.2 These photographs represent the interaction of American beavers, muskrats, and American mink on Navarino Island, Chile. The photograph on the left shows the habitat transformation that results from the activity of American beavers. Trees die and are replaced by grasses and rushes at the time that the river is transformed into a pond. This new habitat is preferred by muskrats. The photograph on the right shows the remains of fur (inside the circle) left by a mink after preying on a muskrat on the shore of the same beaver pond showed on the left

introduced in Europe and Asia. Currently, American beavers are restricted to Finland and northwestern Russia (Parker et al. 2012); however, muskrats and mink inhabit great extensions of Europe and Russia, with populations still expanding (Andow et al. 1990; Bonesi and Palazon 2007; Iordan et al. 2012). In these vast territories, American beavers and American mink have a direct effect on the native Eurasian beaver (*C. fiber*) and European mink (*Mustela lutreola*), respectively, by direct competition (Andow et al. 1990; Bonesi and Palazon 2007).

American beavers, American mink, and muskrats are only sympatric in Finland and northwestern Russia; however, no study has investigated if synergistic interactions among these three species occur. Some studies have documented pairwise interactions that could suggest facilitative interactions similar to those we have described on Navarino Island in Chile. For instance, population trends of mink and muskrats in Poland and Sweden suggest tightly coupled predator-prey relationships, with mink populations affecting muskrats (Hjalten 1991; Brzeziński et al. 2010). Minks, in turn, have well-documented impacts on the local native biota (Macdonald and Harrington 2003; Bonesi and Palazon 2007). More interestingly, on the border between Mongolia and Russia, the European beaver, the muskrat, and the American mink coexist in the same habitat (Saveljev et al. 2015). Although there is no clear relationship between European beavers and muskrats, it seems that mink population expansion occurred only after the muskrat established a self-sustaining population and increased its population size, suggesting a facilitative interaction (Saveljev et al. 2015).

The replacement of native species with North American ecological equivalents, in addition to the degrading habitats by mink predation, geographically extends the area impacted by biotic homogenization processes described above for southern South America. The three furbearing North American species and their large impacts on habitats and ecosystem are contributing to increasing biotic and ecological similarities among the subpolar ecosystems of North America, Europe, some parts of Asia, and South America.

14.5 Conclusion

Large-scale spatial and temporal processes have generated marked contrasts between the subpolar regions of the Northern and Southern hemispheres (Axelrod et al. 1991; Lawford et al. 2012; Rozzi et al. 2012). At the same time, these subpolar regions present ecological similarities and a degree of connectivity between their biotic communities. For instance, some migratory birds travel each year from one pole to the other between the breeding and the wintering grounds connecting the two terrestrial ecosystems. These species are potentially capable of transporting propagules attached to the plumage, perhaps explaining the bipolar and disjoint distribution of some bryophyte species (Lewis et al. 2014) and vascular plants (Popp et al. 2011). Nevertheless, distributions including both poles are more common in marine organisms than terrestrial organisms (Stepanjants et al. 2006); and among animals, only a few are mammals. Orcas (*Orcinus orca*) can be seen swimming in waters of both subpolar regions. Yet, while orcas are present in all oceans, they present physiological and cultural differences that make populations unique and very distinct between different regions, including northern and southern subpolar populations (Foote et al. 2016).

Human populations at high latitudes also developed distinct cultures in each of the subpolar regions. North and South American Native American people remained quite isolated until the nineteenth century (McEwan et al. 2014). Today, economic and cultural globalization is bringing increasing development pressures that are contributing to homogenizing human cultures even in remote subpolar regions. The expansion of the fur industry stands as a prime of these pressures. In less than a century, the introduction of American beavers, muskrats, and American mink into the southern end of South America has reconfigured the structure and evolutionary course of biotitic communities, habitats, and cultures of the Magellanic sub-Antarctic ecoregion.

This trio of North American mammals represents a powerful example of the biological homogenization process that is occurring at a planetary scale in association with economic globalization and cultural homogenization. In short, today we could state that sub-Antarctic and sub-Artic regions are becoming bioculturally homogenized. With the fur industry prospering again, especially for the mink (IFTF 2017), it is crucial to take precautions to avoid further introductions and to prevent the expansions of already introduced species. Control of exotic mammal populations is critical for effective long-term conservation programs that protect the uniqueness of distinct biota and cultures of subpolar ecosystems in the Northern and Southern hemispheres.

Acknowledgments We want to thank Julie Lockwood, Alexandrea Safiq, Roy May, Brian C. O'Connor, Giovanni Frigo, and Nora Ward for their valuable comments and contributions that helped to improve earlier drafts of this essay.

References

Anderson CB, Rosemond AD (2007) Ecosystem engineering by invasive exotic beavers reduces in-stream diversity and enhances ecosystem function in Cape Horn, Chile. Oecologia 154:141–153

Anderson CB, Griffith CR, Rosemond AD et al (2006a) The effects of invasive North American beavers on riparian plant communities in Cape Horn, Chile: do exotic beavers engineer differently in sub-Antarctic ecosystems? Biol Conserv 128:467–474

Anderson CB, Rozzi R, Torres-Mura JC et al (2006b) Exotic vertebrate fauna in the remote and pristine sub-Antarctic Cape Horn Archipelago, Chile. Biodivers Conserv 15:3295–3313

Andow DA, Kareiva PM, Levin SA, Okubo A (1990) Spread of invading organisms. Landsc Ecol 4:177–188

Armesto JJ, Rozzi R, Smith-Ramirez C, Arroyo MTK (1998) Conservation targets in South American temperate forests. Science 282:1271–1272

Artimo A (1960) The dispersal and acclimatization of the muskrat, *Ondatra zibethicus* (L.), in Finland. Pap Game Res 21:1–101

Axelrod DI, Arroyo MTK, Raven PH (1991) Historical development of temperate vegetation in the Americas. Rev Chil Hist Nat 64:413–446

Baker BW, Hill EP (2003) Beaver (*Castor canadensis*). In: Feldhamer GA, Thompson BC, Chapman JA (eds) Wild mammals of North America: biology, management, and conservation, Second edn. Johns Hopkins University Press, Baltimore, pp 288–310

Berghöfer U, Rozzi R, Jax K (2008) Local versus global knowledge: diverse perspectives on nature in the Cape Horn biosphere reserve. Environ Ethics 30:273–294

Bonesi L, Palazon S (2007) The American mink in Europe: status, impacts, and control. Biol Conserv 134:470–483

Brzeziński M, Romanowski J, Zmihorski M et al (2010) Muskrat (*Ondatra zibethicus*) decline after the expansion of American mink (*Neovison vison*) in Poland. Eur J Wildl Res 56:341–348

Connors LM, Groffman PM, Ostfeld RS (2000) Muskrat (*Ondatra zibethicus*) disturbance to vegetation and potential net nitrogen mineralization and nitrification rates in a freshwater tidal marsh. Am Midl Nat 143:53–63

Crego RD, Jiménez JE, Soto C, Barroso O, Rozzi R (2014) Tendencias poblacionales del visón norteamericano invasor (*Neovison vison*) y sus principales presas nativas desde su arribo a isla Navarino, Chile. Boletín de la Red Latinoamericana para el Estudio de Especies Invasoras 4:4–18

Crego RD, Jiménez JE, Rozzi R (2015) Expansión de la invasión del Visón Norteamericano (*Neovison vison*) en la Reserva de la Biósfera de Cabo de Hornos, Chile. An del Inst la Patagon 43:157–162

Crego RD, Jiménez JE, Rozzi R (2016) A synergistic trio of invasive mammals? Facilitative interactions among beavers, muskrats, and mink at the southern end of the Americas. Biol Invasions 18:1923–1938

de Szalay FA, Cassidy W (2001) Effects of muskrat (*Ondatra zibethicus*) lodge construction on invertebrate communities in a Great Lakes coastal wetland. Am Midl Nat 146:300–310

Foote A, Vijay N, Avila-Arcos M et al (2016) Genome-culture coevolution promotes rapid divergence in the killer whale. Nat Commun 7:11693. https://doi.org/10.1038/ncomms11693

Graells G, Corcoran D, Aravena JC (2015) Invasion of North American beaver (*Castor canadensis*) in the province of Magallanes, Southern Chile: comparison between dating sites through interviews with the local community and dendrochronology. Rev Chil Hist Nat 88:3

Henn JJ, Anderson CB, Martínez Pastur G (2016) Landscape-level impact and habitat factors associated with invasive beaver distribution in Tierra del Fuego. Biol Invasions 18:1679–1688

Hjalten J (1991) Muskrat (*Ondatra zibethica*) territoriality, and the impact of territorial choice on reproduction and predation risk. Ann Zool Fenn 28:15–21

Ibarra JT, Fasola L, Macdonald DW, Rozzi R, Bonacic C (2009) Invasive American mink *Mustela vison* in wetlands of the Cape Horn Biosphere Reserve, southern Chile: what are they eating? Oryx 43:87–90

International Fur Trade Federation (IFTF) (2017) The socio-economic impact of international fur farming. https://web.archive.org/web/20110713004402/http://www.iftf.com/publctns/4849Intls_eEng.pd. Accessed 20 Apr 2017

Iordan F, Rushton SP, Macdonald DW, Bonesi L (2012) Predicting the spread of feral populations of the American mink in Italy: is it too late for eradication? Biol Invasions 14:1895–1908

Jaksic FM, Iriarte JA, Jiménez JE, Martínez DR (2002) Invaders without frontiers: cross-border invasions of exotic mammals. Biol Invasions 4:157–173

Jiménez JE, Crego RD, Soto GE et al (2014) Potential impact of the alien American mink (*Neovison vison*) on Magellanic woodpeckers (*Campephilus magellanicus*) in Navarino Island, southern Chile. Biol Invasions 16:961–966

Larivière S (1999) Mustela vison. Mamm Species 608:1–9

Lawford RG, Alaback P, Fuentes E (eds) (2012) High-latitude rainforests and associated ecosystems of the west coast of the Americas: climate, hydrology, ecology, and conservation, vol 116. Springer, New York

Lewis LR, Behling E, Gousse H et al (2014) First evidence of bryophyte diaspores in the plumage of transequatorial migrant birds. PeerJ 2:1–13

Lizarralde MS (1993) Current status of the introduced beaver (*Castor canadensis*) population in Tierra del Fuego, Argentina. Ambio 22:351–358

Lizarralde MS, Escobar JM (2000) Mamíferos exóticos en la Tierra del Fuego. Ciencia Hoy 10:52–63

Lizarralde MS, Escobar J, Deferrari G (2004) Invader species in Argentina: a review about the beaver (*Castor canadensis*) population situation on Tierra del Fuego ecosystem. Interciencia 29:532–536

Macdonald D, Harrington L (2003) The American mink: the triumph and tragedy of adaptation out of context. N Z J Zool 30:421–441

Martin C (1982) Keepers of the game: Indian-animal relationships and the fur trade. University of California Press, California

McEwan C, Borrero LA, Prieto A (eds) (2014) Patagonia: natural history, prehistory, and ethnography at the uttermost end of the earth. Princeton University Press, Princeton

McKinney ML, Lockwood JL (1999) Biotic homogenization: a few winners replacing many losers in the next mass extinction. Trends Ecol Evol 14:450–445

Mittermeier RA, Mittermeier CG, Brooks TM et al (2003) Wilderness and biodiversity conservation. PNAS 100:10309–10313

Mott CL, Nielsen CK, Bloomquist CK (2013) Within-lodge interactions between two ecosystem engineers, beavers (*Castor canadensis*) and muskrats (*Ondatra zibethicus*). Behaviour 150:1325–1344

Parker H, Nummi P, Hartman G, Rosell F (2012) Invasive North American beaver *Castor canadensis* in Eurasia: a review of potential consequences and a strategy for eradication. Wildl Biol 18:354–365

Popp M, Mirré V, Brochmann C (2011) A single Mid-Pleistocene long-distance dispersal by a bird can explain the extreme bipolar disjunction in crowberries (*Empetrum*). PNAS 108:6520–6525

Rosell F, Bozsér O, Collen P, Parker H (2005) Ecological impact of beavers *Castor fiber* and *Castor canadensis* and their ability to modify ecosystems. Mamm Rev 35:248–276

Rozzi R (2012) Biocultural ethics: recovering the vital links between the inhabitants, their habits, and habitats. Environ Ethics 43:27–50

Rozzi R (2015) Earth stewardship and the biocultural ethic: Latin American perspectives. In: Rozzi R, Chapin FS III, Callicott JB, Pickett STA, Power ME, Armesto JJ, May RH Jr (eds) Earth stewardship: linking ecology and ethics in theory and practice. Ecology and ethics, vol 2. Springer, Dordrecht, pp 87–112

Rozzi R, Sherriffs MF (2003) El visón (*Mustela vison* Schreber, Carnivora: Mustelidae), un nuevo mamífero exótico para la Isla Navarino. An del Inst la Patagon 31:97–104

Rozzi R, Massardo F, Anderson CB et al (2006) Ten principles for biocultural conservation at the southern tip of the Americas: the approach of the Omora Ethnobotanical Park. Ecol Soc 11:43

Rozzi R, Armesto JJ, Goffinet B et al (2008) Changing lenses to assess biodiversity: patterns of species richness in sub-Antarctic plants and implications for global conservation. Front Ecol Environ 6:131–137

Rozzi R, Armesto JJ, Gutiérrez JR et al (2012) Integrating ecology and environmental ethics: earth stewardship in the southern end of the Americas. Bioscience 62:226–236

Saveljev AP, Shar S, Scopin AE et al (2015) Introduced semiaquatic mammals in the Uvs Nuur Hollow (Current distribution and ecological vectors of naturalization). Russ J Biol Invasions 6:37–50

Schüttler E, Cárcamo J, Rozzi R (2008) Diet of the American mink *Mustela vison* and its potential impact on the native fauna of Navarino Island, Cape Horn Biosphere Reserve, Chile. Rev Chil Hist Nat 81:585–598

Schüttler E, Klenke R, McGehee S, Rozzi R, Jax K (2009) Vulnerability of ground-nesting waterbirds to predation by invasive American mink in the Cape Horn Biosphere Reserve, Chile. Biol Conserv 142:1450–1460

Shier C, Boyce M (2009) Mink prey diversity correlates with mink-muskrat dynamics. J Mammal 90:897–905

Sielfeld WH (1977) Reconocimiento macrofaunístico terrestre en el area de Seno Ponsomby (Isla Hoste). An del Inst la Patagon 8:275–295

Simberloff D (2013) Introduced species, homogenizing biotas and cultures. In: Rozzi R, Pickett STA, Palmer C, Armesto JJ, Callicott JB (eds) Linking ecology and ethics for a changing world: values, philosophy, and action. Ecology and ethics, vol 1. Springer, Dordrecht, pp 33–48

Skov L (2005) The return of the fur coat: a commodity chain perspective. Curr Sociol 53:9–32

Stepanjants SD, Cortese G, Kruglikova SB, Bjørklund KR (2006) A review of bipolarity concepts: history and examples from Radiolaria and Medusozoa (Cnidaria). Mar Biol Res 2:200–241

Ulloa E, Anderson CB, Ardón M et al (2012) Organic matter characterization and decomposition dynamics in sub Antarctic streams impacted by invasive beavers. Lat Am J Aquat Res 40:881–892

Valenzuela AEJ, Anderson CB, Fasola L, Cabello JL (2014) Linking invasive exotic vertebrates and their ecosystem impacts in Tierra del Fuego to test theory and determine action. Acta Oecol 54:110–118

Viljugrein H, Lingjærde OC, Stenseth NC, Boyce MS (2001) Spatio-temporal patterns of mink and muskrat in Canada during a quarter century. J Anim Ecol 70:671–682

Chapter 15
Non-native Pines Are Homogenizing the Ecosystems of South America

Rafael A. García, Jorgelina Franzese, Nahuel Policelli, Yamila Sasal, Rafael D. Zenni, Martin A. Nuñez, Kimberley Taylor, and Aníbal Pauchard

Abstract A large area previously dominated by native ecosystems in South America is now covered by monocultures of non-native tree species, mainly of the genus *Pinus*. Currently, pine plantations and the invasions that have been generated from these are causing a homogenization process at the landscape, stand, and even micro-site scales. The continuous and extensive areas covered by pine plantations have replaced the native ecosystem heterogeneity in many landscapes of South America. Within these plantations, the diversity of plants and animals is lower than

---

R. A. García (✉) · A. Pauchard
Facultad de Ciencias Forestales, Universidad de Concepción, Concepción, Chile

Instituto de Ecología y Biodiversidad (IEB), Santiago, Chile
e-mail: ragarcia@udec.cl; pauchard@udec.cl

J. Franzese · Y. Sasal
Laboratorio Ecotono, Instituto de Investigaciones en Biodiversidad y Medioambiente (Universidad Nacional del Comahue – CONICET), S. C. Bariloche, Argentina
e-mail: franzesej@comahue-conicet.gob.ar; yamila.sasal@comahue-conicet.gob.ar

N. Policelli
Grupo de Ecología de Invasiones, Instituto de Investigaciones en Biodiversidad y Medioambiente (Universidad Nacional del Comahue – CONICET), S. C. Bariloche, Argentina
e-mail: npolicelli@comahue-conicet.gob.ar

R. D. Zenni
Setor de Ecologia, Departamento de Biologia, Universidade Federal de Lavras, Lavras, MG, Brazil
e-mail: rafael.zenni@dbi.ufla.br

M. A. Nuñez
Grupo de Ecología de Invasiones, INIBIOMA, CONICET-Universidad Nacional del Comahue, Bariloche, Argentina
e-mail: martin.nunez@crub.uncoma.edu.ar

K. Taylor
Department of Ecosystem and Conservation Sciences, University of Montana, Missoula, MT, USA
e-mail: kimberley.taylor@umontana.edu

R. Rozzi et al. (eds.), *From Biocultural Homogenization to Biocultural Conservation*, Ecology and Ethics 3, https://doi.org/10.1007/978-3-319-99513-7_15

that of the nearest remnant native ecosystems. These plantations can also act as a barrier to the movement of species across the landscape. In addition, in most places where pine plantations have been established, invasions have occurred into the surrounding ecosystems. Overall, pine invasions are more evident in open ecosystems (e.g., grasslands, steppes, and degraded native forest), but they can also occur in denser vegetation (e.g., temperate forests). Native species loss as a consequence of pine invasions has been recorded in tropical, mediterranean, and temperate ecosystems. Increased pine abundance and the resultant native species loss bring changes to all levels of organization within the ecosystem, from soil microorganisms to invertebrates, plants, and vertebrates. These changes reduce the ecosystem's spatial heterogeneity and thus cause biotic homogenization. These biodiversity losses can affect the stability of ecosystems by decreasing their resilience to environmental change and disturbances. To mitigate the impacts caused by pines, it is important to implement comprehensive landscape planning, understanding that pine plantations coexist and interact with other land uses in a complex ecological and social setting.

Keywords Homogenization · *Pinaceae* · Invasive trees · Plant invasions · Impacts

15.1 Introduction

Pine plantations are growing in South America, and their direct and indirect effects on biodiversity remain uncertain. Currently, 68% of the tree plantation area added annually in the Southern Hemisphere (estimated in 750,000 ha year^{-1}) occurs in South America (Food and Agriculture Organization 2010). On this continent, almost all tree plantations are based on of introduced species, and conifers, particularly *Pinus* spp., are one of the most common choices (Food and Agriculture Organization 2010). The accelerated growth of forest plantations in South America raises concerns about the risks and the potential impacts of pine plantations on biodiversity and ecosystem services, especially in those areas of high conservation value (Armesto et al. 2010). Much of the research on the biotic consequences of pine plantations has been conducted at landscape or even larger scales (e.g., Echeverria et al. 2006), where plantations have been blamed for landscape homogenization and replacement of native forests (Rozzi et al. 1994), as well as for changes in hydrological regimes (e.g., Farley et al. 2005; Little et al. 2008). However, biotic impacts at smaller scales have been less studied, with local reduction of understory plant diversity as one of the most reported problems (Paritsis and Aizen 2008; Simonetti et al. 2013; Heinrichs and Pauchard 2015).

In addition to the direct impacts of pine plantations, conifers have long been recognized as one of the most invasive plant taxon (e.g., Richardson et al. 1994; Rejmánek and Richardson 1996; Ledgard 2001; Buckley et al. 2005; Essl et al. 2011; Gundale et al. 2014; Nuñez et al. 2017). Pine invasive potential emerges partially from their widespread ornamental use and mainly from their extensive use for forest plantations (Richardson 2006; Simberloff et al. 2010; Essl et al. 2010). In addition, this taxon has biological attributes that increase its invasive potential such as high reproductive rate, fast growth, and long-distance dispersal (Rejmánek and Richardson 1996). Conifer invasions can have severe impacts on ecosystem processes causing changes in water and fire regimes and reductions in local diversity (Simberloff et al. 2010). South America has been increasingly affected by conifer invasions, especially in the case of *Pinus* spp. (Pauchard et al. 2015), due to the high invasibility of the ecosystems and the novelty of the *Pinaceae* south of the equator (Lusk 2008). Nevertheless, it is only recently that researchers have started to pay close attention to pine invasions in South America and their effects on local biota (see Richardson et al. 2008).

Although the impacts of invasive pines are highly variable and depend upon the habitats they invade (Pauchard et al. 2015; Nuñez et al. 2017), biotic homogenization is one of the most noticeable and consistent impacts across latitudes and continents. Two mechanisms largely explain the process of homogenization caused by pine invasions.

1. An increase in the similarity between different invaded ecosystems caused by the dominance of a single non-native species (i.e., *Pinus* spp.), which usually has very different functional traits than the native plant communities (Pauchard et al. 2016).
2. Homogenization occurs as pine cover and biomass increase, and competition for resources causes a decrease in the abundance and diversity of the native species (Franzese et al. 2017; Taylor et al. 2016).

This process can develop quickly (in less than a year) in burned habitats invaded by pines, but in less-disturbed ecosystems, it can take multiple years or decades from the beginning of the invasion until the impacts on the native biota become evident (Franzese and Raffaele 2017). Pine invasions can have legacy impacts on plant communities, even after they are removed, resulting in increases in other exotic species (Dickie et al. 2014). Furthermore, there is growing evidence that plant community homogenization is usually preceded by a biotic homogenization of animal and fungal communities, either through increases in new invasive species or simplification of the native ecosystem (Simberloff and Von Holle 1999; Nuñez et al. 2013; Dickie et al. 2017).

The impacts of pine plantations and invasions in South America, and the consequent biotic homogenization, are expected to increase because of continuing expansion of the afforested areas (Food and Agriculture Organization 2010), as well as increases in invasion drivers, such as fire (Franzese and Raffele 2017), habitat degradation (Echeverría et al. 2007), and grazing (Loidy et al. 2010; de Villalobos et al. 2011). In this context, the aim of this chapter is to review the evidence of biotic homogenization caused by pine plantations and pine invasions on plant and fungal communities of South America to better understand the causes and consequences of this ongoing conservation problem.

15.2 Pine Plantations Dual Effect: Homogenization and Invasion

Forest plantations, especially pine plantations, generate a series of economic, social, and environmental goods and services (Sutton 1995; Gerrand et al. 2003; Vihervaara et al. 2012). On the other hand, the lack of management or inadequate management in plantations causes negative impacts that may be more noticeable to society than the supposed benefits they generate (Spellerberg 1996; Aber et al. 2000; Hartmann et al. 2010; Salas et al. 2016). Although pine invasions and pine plantations are two different phenomena, it is important to understand the ecological, social, and economic links between them, especially when addressing biotic homogenization. Commercial plantations are the most important promoter of pine invasions, particularly in South America, and also a fundamental component of the process of biotic homogenization, with strong implications at landscape and local scales.

15.2.1 Landscape Homogenization

The landscape is a mosaic of different biophysical elements (natural and/or anthropic), and the distribution of species within the landscape is determined by the diversity and spatial distribution of resources in the landscape (Debinski et al. 2001; Hartmann et al. 2010). A simplification of the landscape will, therefore, have a direct effect on the composition of animal and plant communities, as well the ecosystem services they provide (Carnus et al. 2006; Hartmann et al. 2010). Thus, the transformation of natural forest habitats into productive and homogenous systems, as a consequence of the establishment of large-scale plantations of non-native tree species, is one of the most direct threats to biodiversity conservation (Potton 1994; Larsson and Danell 2001).

In South America, a massive expansion of commercial pine plantations has occurred in tropical, mediterranean, temperate, and alpine ecosystems (Cubbage et al. 2007; Pauchard et al. 2015). This expansion has been particularly remarkable in the Coastal Range in central Chile, where native vegetation is composed of deciduous forest, with dominance by different *Nothofagus* species (i.e., *Nothofagus glauca* (Phil.) Krasser, *N. obliqua* (Mirb.) Oerst., and *N. alessandrii* Espinosa). This forest has been continuously degraded and replaced by extensive area of timber plantation (mostly *Pinus radiata* D. Don) (Bustamante and Castor 1998; Smith-Ramirez 2004; Echeverria et al. 2006). In this area, 41.5% of new plantations in the 1975–1990 period and 22.8% in the 1990–2007 period were established by clearing secondary native forests, which confirms that plantation expansion in Chile has been a direct cause of deforestation and biodiversity loss (Nahuelhual et al. 2012). Nevertheless, in recent years (2001–2011 period), the rate of forest conversion has decreased due to tighter regulations, which suggests that pressure on remaining native forests is beginning to ease (Heilmayr et al. 2016).

15.2.2 Local Biodiversity Homogenization

In addition to an increase in homogenization at the landscape scale, within pine plantations, the diversity of species is usually much lower than in adjacent natural or seminatural habitats. The change of land use (direct or indirect) from forests or other native ecosystems to commercial plantations necessarily implies a simplification of the structure and composition of species (Potton 1994; Spellerberg 1996; Freedman et al. 1996; Gjerde and Saetersdal 1997; Hartley 2002; Braun et al. 2017).

In the central zone and Chilean Patagonia, Braun et al. (2017) found a strong negative impact of extensive plantation forestry on plant biodiversity. The extensive areas of pine plantations have negative impacts on α (local), β (species turnover), and γ (landscape) biodiversity, where plant communities are predominantly native and endemic in natural forests and predominantly non-native in plantations (Braun et al. 2017). Furthermore, these pine crops do not serve as an alternative habitat for native species, leaving many species threatened by extinction (Braun et al. 2017).

A study carried out in Patagonia, which compares the structure and composition of *Nothofagus dombeyi* forests and small pine forest plantations (<5 ha), shows that even at the stand level the same pattern of habitat structure homogenization is observed (Paritsis and Aizen 2008). The largest impact was detected on understory plants, followed by the beetle and bird assemblages, with a reduction in evenness in plants and beetles, an increase of non-native plants and birds, and a loss of rare and specialist species in all three assemblages (Paritsis and Aizen 2008). Additionally, pine plantations with little or minimally developed understories contained fewer species of medium-sized mammals than plantations with more understory vegetation (Simonetti et al. 2013). When understory plants are present in commercial plantations, they can enhance the quality of plantations as habitat for native fauna (mammals, birds, and insects) and even for some vulnerable species (Briones and Jerez 2007; Tomasevic and Estades 2008; Nájera and Simonetti 2010; Simonetti et al. 2013). In more tropical areas, the effects of pine plantations depend largely on the intensity and frequency of management actions. For instance, *P. elliottii* plantations under low-intensity management were shown to have similar understory species richness and diversity as native woody Cerrado formations in Brazil (Abreu et al. 2011), but more intense management techniques (e.g., shorter rotation, higher herbicide applications) would likely reduce local biodiversity. Thus, choosing the right management options may help to increase or at least reduce the losses of local biodiversity. For example, a management scheme that allows the establishment of well-developed understories (e.g., intensive pruning and thinning) would not only minimize the impact on plant diversity by providing substitute habitats for native species but also mitigate the effects on wildlife (Simonetti et al. 2013).

15.2.3 Plantations as the Main Source of Pine Invasions

The different species of pines cultivated in South America have a variable level of invasiveness (Pauchard et al. 2015), but overall it is higher than other tree taxa. High invasive potential and high propagule pressure transform the pine plantations into the main source of invasions into native ecosystems. In fact, forest plantations release a large amount of seeds each year into the landscape, which increases the likelihood of invasion by a mass effect (Richardson and Brown 1986; Kruger et al. 1989; Richardson and Higgins 1998). This high propagule pressure overwhelms microsite-scale interactions and independently explains invasion success (Pauchard et al. 2016). Depending on the species of pine planted (invasiveness) and the characteristics of the invaded ecosystem (invasibility), invasion from the plantations can follow two basic patterns: continuous invasion and mosaic invasion.

Continuous invasion is frequent in open, less competitive ecosystems, as is the case of *Pinus contorta* Douglas ex Loudon in the Patagonian steppe (e.g., Langdon et al. 2010; Pauchard et al. 2016). The initial population growth is usually characterized by a dispersal kernel, where dense regeneration is located next to the seed source (short distance dispersal), while medium- and long-distance dispersal generates scattered outlier pines (Higgins and Richardson 1999; Richardson 2001; Ledgard 2003; Langdon et al. 2010). In this first stage, intraspecific competition is low, and therefore wildings can establish at any distance from the seed source within the first "invasion wave." After some of the pines reach maturity, propagules originate from the original source, the invasion front, and the outlier trees. Once the second wave of invasion has started, the opportunities for successful control are greatly reduced, increasing the overall impacts of the invasion in the landscape (Fig. 15.1).

Mosaic invasion occurs in forests or other competitive vegetation (i.e., shrubs or grassland) with relatively stable dense vegetation cover but some spatial heterogeneity that includes suitable habitat for the invader. In this case, the invasion process is slower due to the low light availability and absence of bare soil for seedling establishment. This process is frequent in remnants of Maulino Forest in the Coastal Range of central Chile (Bustamante and Simonetti 2005). Today this forest covers only a few percent of the original area, persisting as small remnants. These remnants are surrounded by extensive *P. radiata* plantations (Bustamante and Castor 1998). Although seeds of pines can arrive in the interior of well-developed forest fragments (Bustamante and Simonetti 2005), this shaded habitat generates abiotic conditions which impose constraints to germination and establishment that reduce seedling recruitment (Bustamante et al. 2003). The permeability of these forests to the invasion not only depends on the maintenance (or degradation) of the canopy layer but also on the size of the fragments. The fragment size seems to be a good indicator of susceptibility to being invaded by pines: small fragments are more susceptible to invasion while large fragments are more resistant (Gómez et al. 2011).

Fig. 15.1 Example of a continuous invasion of pines into grassland environments in Patagonian steppe, in Coyhaique Alto, Aysen Region, Chile. Top panel: *invasion core* of *P. contorta* from 2007 (**a**) to 2011 (**b**), 2015 (**c**), and 2017 (**d**). Bottom panel: *invasion front* from 2007 (**e**) to 2011 (**f**), 2015 (**g**), and 2017 (**h**). The high dispersal capacity and rapid growth and development of pines can generate invasion processes with a high impact on the invaded community. In short periods of time (less than 10 years), complete ecosystem transformations can occur. The change from a steppe to a monospecific pine forest results in important changes in microclimatic conditions and availability of resources, reduction in the richness and abundance of native plants, modification of the soil biota, and alteration of the trophic networks of the invaded ecosystems

15.3 Pine Invasions and the Aboveground Biotic Homogenization

Most of the invasion impact studies in South American habitats are focused on the aboveground components of the biota (e.g., Urrutia et al. 2013; Cóbar-Carranza et al. 2014; Franzese et al. 2017), with few of them focused on the belowground components (e.g., Chapela et al. 2001; Dickie et al. 2011; de Oliveira et al. 2014). This section describes work quantifying the direct impacts on aboveground biodiversity caused by pine invasions in different South American ecosystems.

Three pine species are currently recognized as invasive in the tropical ecosystems in South America: *Pinus elliottii*, *P. caribaea*, and *P. oocarpa* (Braga et al. 2014; Zenni 2015). The tropical invasive range of pines includes the Brazilian central savanna (Cerrado) and open and degraded areas of the Atlantic Forest Biome (Zenni and Ziller 2011; Zenni 2015). Pine plantations in tropical regions are usually associated with declines in species richness and abundance in native plant regeneration (Valduga et al. 2016).

In open habitats, such as grassland Cerrado and Cerrado *sensu stricto*, pine invasions can result in massive decreases in native species richness and abundance. One study comparing invaded and non-invaded grassland savanna found a tenfold decrease in plant density (non-invaded site = 12,656 plants ha^{-1}; invaded site = 1210 plants ha^{-1}) and a twofold decrease in species richness in the pine-invaded site ($H' = 2.82$) in comparison to a non-invaded reference site ($H' = 1.53$). Furthermore, the pine invasion completely excluded the herbaceous layer (Abreu and Durigan 2011).

Most pines used in forestry have evolved in more temperate or colder environments, and, therefore, the southern part of South America could be more suitable for pine invasions. In fact, much of the Andean temperate forests of Argentina and Chile are characterized by a tree layer that is relatively permeable to light (e.g., open *Araucaria araucana* forest, deciduous forest of *Nothofagus* species); for this reason, the probability of being invaded by pines is greater than in more closed forests (Peña et al. 2008; Simberloff et al. 2010). However, currently closed forests with evergreen species are also being invaded by pines, mainly by *Pseudotsuga menziesii* (Sarasola et al. 2006; Pauchard et al. 2008).

Plant diversity in open temperate forests is negatively affected by *P. contorta* invasions. Richness and cover of plants beneath pine canopies decreased with increasing pine size (i.e., height and canopy area) (Franzese et al. 2017). This decrease did not affect all species equally, generating changes in the relative cover of different life-forms between invaded and non-invaded areas (Urrutia et al. 2013). These impacts began in the early stages of the invasion, before canopy closure, when the pines had a height of less than 10 m (Franzese et al. 2017). With increasing pine canopy closure in the invaded area, the diversity of species declined considerably as a result of the decreased light availability under the tree crowns (Rodríguez-Calcerrada et al. 2011; Taylor et al. 2016). This evidence suggests that some impacts go beyond the reduction of plant diversity and are related to a change in the ecosystem functioning. In this modified ecosystem, pines not only compete for resources but also modify resource availability by altering microenvironmental conditions. Invaded sectors have a dense and deep layer of needles (Taylor et al. 2016), less light availability (Fig. 15.2), and more moderate temperatures (García unpublished data). The moderation of extreme conditions in these high-mountain forests could facilitate the arrival of new species that previously were excluded due to the harsh environment, thus promoting the homogenization of these unique ecosystems currently dominated by species adapted to these relatively harsh conditions.

Fig. 15.2 Araucaria forests in the Andes of south-central Chile with low (**a**, **c**) and high pine invasion (**b**, **d**). At the ground level, the accumulation of pine needles decreases the amount of bare soil limiting the germination and establishment of native plants. Ground view (1 m^2 plot) (top panel: **a**, **b**). Hemispheric canopy view (bottom panel: **c**, **d**). The pine invasions generate a more closed canopy with less light availability

Temperate steppes in Patagonia can be extremely susceptible to pine invasion in comparison to forested habitats (Franzese et al. 2017). In a steppe of the southern Chilean Patagonia invaded by *P. contorta*, a significant reduction in plant richness and cover of the invaded community was registered even at early pine invasion stages (i.e., pines of low height and small canopy areas) (Franzese et al. 2017). As pine canopy cover increased, there was a strong decline in native plant richness and cover (Taylor et al. 2016). The abrupt modification of the habitat conditions produced by pine invasion, along with the addition of novel functional traits, could be acting as an ecological filter on plant biodiversity of steppe ecosystems (Bravo-Monasterio et al. 2016). The differential level of impact among distinct habitat types (forest and treeless ecosystems) could be related to how adapted the invaded community is to tree cover. This highlights the importance of context-dependency as a key factor in determining the overall impact of pines on diverse natural plant communities (Fig. 15.3).

Although conifer invasions are regarded as a serious threat to biodiversity, information on their impacts beyond plant communities is very limited. In fact, little is known about changes in invertebrate assemblages. Replacement of open native treeless vegetation with dense, closed, even-aged forests is by far the most striking impact of pine invasions (Richardson et al. 1994; Richardson and Higgins 1998). These changes in vegetation composition and structure result in new abiotic conditions that could have cascading effects on invertebrates. Many invertebrates depend on particular plant species or structures for food or reproduction sites. Losses of these plants, or structural and/or compositional changes to the natural vegetation communities, may be especially detrimental for invertebrates.

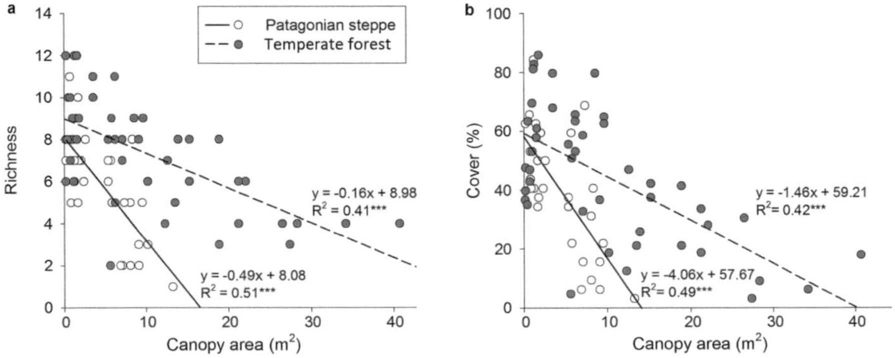

Fig. 15.3 Plant species richness and abundance under invading pine trees decreases significantly as pine canopy area increases. Species richness (**a**) and cover (**b**) in two different habitats, temperate forest (dark dots) and Patagonian steppe (white dots). The magnitude of this decrease will depend on the degree to which the native species are adapted to conditions similar to those created by the pines (e.g., level of shade tolerance). The impact of the pines is most pronounced in those ecosystems where shrub and grass life-forms are dominant (e.g., steppes). (Figure modified from Franzese et al. 2017)

In *Araucaria angustifolia* moist forest of Brazil, the average richness and abundance of land planarians were lower in areas invaded by pine than in uninvaded forests. This reduction was a consequence of the alterations in microhabitat caused by the accumulation of pine litter (de Oliveira et al. 2014). Pawson et al. (2010) examined the impact of non-native conifer density on native invertebrate assemblages in *Pinus nigra* invasions in New Zealand. They found that the effects of non-native conifer invasion on grassland invertebrate assemblages were strongly dependent on conifer density after 14 years of invasion. The relative abundance of major classes and orders of invertebrates was largely unaffected by conifer invasion at densities below 800 trees per hectare but differed in higher-density conifer stands (canopy cover >50%). At the species level, beetle species composition was highly sensitive to conifer invasion at densities as low as 400 trees ha^{-1}. Changes in beetle species composition were correlated with reduced soil moisture, increased canopy cover, and increasing trap distance from the nearest seminatural grassland. The effects of conifer invasions on invertebrates may have strong ecological consequences because invertebrates influence ecosystems as important links in the food web but also as pollinators, decomposers, and predators of pest insects (Losey and Vaughan 2006).

15.4 Biotic Homogenization Belowground

The potential replacement of native belowground biota by non-native invasive microorganisms is probably one of the most unexplored aspects of *Pinaceae* invasions. *Pinaceae* trees are only able to invade if their belowground mutualists are present in the invaded range (Nuñez et al. 2009; Hayward et al. 2015a); therefore they co-invade with a group of species of ectomycorrhizal fungi (EMF) (Dickie et al. 2010). Once introduced, ectomycorrhizal fungi can disperse via water, wind, mammals, soil movement, and intentional or accidental human transport (Nuñez et al. 2013; Dickie et al. 2016).

Since pines were introduced to South America for forestry purposes, there have also been large-scale introductions of non-native ectomycorrhizal fungi (Rivera et al. 2015; Hayward et al. 2015b), which may cause severe ecological impacts. Ectomycorrhizal fungi invasion produces loss of soil carbon, movement of phosphorus into labile pools, and a shift toward fast nutrient cycling and bacterial-dominated decomposition (Chapela et al. 2001; Dickie et al. 2011). These changes in the soil can in turn facilitate invasion by plant species that are adapted to high-nutrient soils (Dickie et al. 2014). The spread and invasion of non-native fungi can also have cultural and social impacts, as it affects the perception of the native habitat by local people with a subsequent loss of sense of place (Dickie et al. 2016). Economically, the introduction of new species can positively impact timber production, and the use of novel edible fungi may emerge with commercial interest. However, many introduced species can also be toxic to humans that accidentally consume their fruiting bodies (Nuñez and Dickie 2014).

In the invaded range, non-native ectomycorrhizal fungi interact with native biota. Some non-native ectomycorrhizal fungi, for example, can form novel associations with native plants. Some cosmopolitan ectomycorrhizal fungi species that associate with *Pinus spp.* can also be found associated with native *Nothofagus spp.* (Dickie et al. 2010). Moreover, some non-native ectomycorrhizal fungi, such as *Amanita muscaria*, have been found to be spreading into native forests associated with native tree species (Orlovich and Cairney 2004; Dickie and Johnston 2008). The mechanisms underlying this process, together with the impact of these novel associations, are still uncertain. In any case, beta diversity of the belowground communities decreases, as the arrival of these species of cosmopolitan ectomycorrhizal fungi associated with pines increases similarity between different communities. There is also evidence that native ectomycorrhizal fungi generally do not associate with pine species (Dickie et al. 2010; Gundale et al. 2016). Intrinsic ectomycorrhizal fungi species traits define their invasion success and probably are involved in the interaction with native biota. Some non-native ectomycorrhizal fungi, for example, perform better in the invasion front due to high spore resistance, long-distance dispersal, and high spore production. Although these groups of ectomycorrhizal fungi allow pines to establish, they are then outcompeted by late-successional non-native species (Peay et al. 2010). If and how these sets of non-native species interact with each other and with native ectomycorrhizal fungi are still not clear. We also have limited information about whether non-native fungi replace native ectomycorrhizal fungi, but this is clearly a possibility (Nuñez and Dickie 2014), which would have a direct impact on plant species composition. Given the above information, there is sufficient evidence to suggest that the spatial complexity of ectomycorrhizal fungal communities has implications for forest succession, expansion, and invasion dynamics (Dickie and Reich 2005).

Undoubtedly, there is a clear need to shed light onto the mechanisms by which simplification of the belowground community could occur. More work is needed to fully understand how non-native ectomycorrhizal fungi could influence the rate of species spread and the resistance of the community to future invasions, in addition to altering a much wider range of ecosystem processes such as soil respiration and soil carbon stocks.

15.5 Conclusions

Due to the wide latitudinal and climatic gradient present in South America, there is a great variety of ecosystems and species, with high endemism and very limited distributions. The addition of large areas planted with non-native conifer species represents an important driver of biodiversity loss in these areas. Currently more than ten million hectares are covered by a small subset of non-native tree species, mainly of the *Pinaceae* family. Additionally, an undetermined area is being invaded by pines, generating a homogenization process at landscape, stand, and even microsite scales, causing ecosystem consequences across trophic levels.

The planted and invading pines are perceived by society, in most cases, as transforming agents of the landscape. The extensive and homogeneous areas planted with pines are seen as antagonistic to the diverse and increasingly threatened native ecosystems. Currently across South America, it is possible to attribute homogenization at the landscape scale to the expansion of commercial plantations. However, land use change is a much more complicated process dominated by constant degradation and deforestation of native forest (e.g., agricultural expansion, fires, forest substitution, firewood extraction) generally resulting in a final state of plantations; however, plantations are not always the initial drivers of this change (Echeverria et al. 2006; Nahuelhual et al. 2012). Thus, comprehensive landscape planning is required to maintain landscape heterogeneity and conserve the remaining patches of native ecosystems.

Pine invasions initiate largely from pine plantations. Therefore, recognizing that pine plantations provides a series of goods and services (such as wood, fiber, and fuel), there is now a pressing need to develop practices oriented toward avoiding negative externalities, such as invasion of native ecosystems. Non-native conifer plantations are an undeniable source of propagules that cause invasion in natural ecosystems, even in protected areas, with all the problems for biodiversity conservation that follow from these invasions. Most of the cultivated pine species have the potential to become invasive (Pauchard et al. 2015; Nuñez et al. 2017), but we should avoid planting those with the highest invasive potential (Rejmánek and Richardson 1996) or with international evidence of becoming invasive. Appropriate management strategies need to be developed inside and outside pine plantations in order to prevent or reduce impacts on local biodiversity (Pauchard et al. 2015; Braun et al. 2017; Nuñez et al. 2017). Less dense plantations with a well-developed understory are key to increasing diversity within plantations, as well as allowing species to disperse across the landscape. On the other hand, the establishment of pines outside the plantations should be avoided by controlling wildings, especially in riparian zones and open ecosystems.

Homogenization generated from pine invasion in natural areas is a gradual process, which can take years or decades, but it is inevitable if timely control measures are not taken. In the early stages of pine invasion, there is a reduction in the abundance of some native plant species associated with increased of canopy cover (Urrutia et al. 2013; Franzese et al. 2017) and a restructuring of the soil fungal communities initiated by the arrival of ectomycorrhizal fungi associated with the pines (Nuñez and Dickie 2014). In later invasion stages, changes occur in plant community composition resulting from the inevitable loss of native species (Abreu and Durigan 2011; Bravo-Monasterio et al. 2016). This homogenization, as a result of the loss of native species and the increase in pine abundance, brings changes to all levels of organization within the ecosystem, which can be permanent, giving rise to a novel habitat. Furthermore, this biodiversity loss can affect the stability of ecosystems by decreasing the ability of communities to respond to environmental change and disturbances (Gámez-Virués et al. 2005).

One of the current challenges in biodiversity conservation lies in how to deal with the synergy of factors that promote ecosystem homogenization. Wildfires and invasions could generate a positive feedback (Brooks et al. 2004; Mandle et al. 2011), speeding up the process of biodiversity homogenization in degraded ecosystems. Positive feedbacks between fire and invasion of pines adapted to fire (i.e., serotinous pines; e.g., Taylor et al. 2017) can be a major cause of unidirectional changes in natural ecosystems (Simberloff et al. 2010; Veblen et al. 2011), especially in those which have not evolved under a high fire frequency (Brooks et al. 2004). The main effect of invasive pines on fire risk in natural forests is related to higher flammability and fuel continuity, although over time it is expected that the amount of fine fuel could also increase (Cobar-Carranza et al. 2014). In steppe systems, invasive pines contribute to greatly elevated fuel loads in invaded areas (Taylor et al. 2017). In mediterranean and temperate ecosystems, the number and severity of wildfires are expected to increase in the future due to the current scenario of climate change, coupled with an increase in human activity, and the large concentration of pine plantations (Peña and Valenzuela 2008; McWethy et al. 2018). In the summer of 2017, more than 100,000 ha of native forest burned just within Chile, in sectors dominated by pine plantations (CONAF 2017) or with evidence of pine invasion (Bustamante and Simonetti 2005; Gómez et al. 2011). If a timely restoration and control of post-fire pine regeneration are not performed in these zones, it is expected that vegetation composition will be drastically modified and dominated by pines. This highlights the importance of analyzing the traits of the pine species that are proposed for introduction into natural areas and discouraging the use of those species whose abundances would likely be promoted by disturbances such as fire (Franzese and Raffaele 2017).

To mitigate future impacts caused by pines, and other tree plantations, it is important to develop a comprehensive landscape plan, understanding that pine plantations coexist and interact with other land uses, productive and non-productive, and, therefore, the management of these plantations must be implemented in accordance with the social and ecological context in which they are located. Along the same lines, the reduction of large areas planted with a single objective and management strategy is crucial to reduce landscape homogenization and other negative impacts associated with intensive and extensive management of tree monocultures.

Acknowledgments R.G. and A.P. were funded by CONICYT PIA APOYO CCTE AFB170008. R.G. was funded by FONDECYT 11170516. A.P. was funded by FONDECYT 1140485. J.F. was funded by Agencia Nacional de Promoción Científica y Tecnológica (PICT 2014-3466).

References

Aber J, Christensen N, Fernandez I, Franklin J, Hidinger L, Hunter M, MacMahon J, Mladenoff D, Pastor J, Perry D, Slangen R, van Miegroet H (2000) Applying ecological principles to management of the U.S. National Forests. Issues Ecol 6:1–22

Abreu RCR, Durigan G (2011) Changes in the plant community of a Brazilian grassland savannah after 22 years of invasion by *Pinus elliottii* Engelm. Plant Ecol Divers 4:269–278

Abreu RCR, Assis GB, Frison S, Aguirre A, Durigan G (2011) Can native vegetation recover after slash pine cultivation in the Brazilian Savanna? For Ecol Manag 262:1452–1459

Armesto JJ, Manuschevich D, Mora A, Smith-Ramirez C, Rozzi R, Abarzúa AM, Marquet P (2010) From the Holocene to the Anthropocene: a historical framework for land cover change in southwestern South America in the past 15,000 years. Land Use Policy 27(2):148–160

Braga EP, Zenni RD, Hay JD (2014) A new invasive species in South America: *Pinus oocarpa* Schiede ex Schltdl. Biol Invasion Rec 3:207–211

Braun A, Troeger D, García R, Aguayo M, Barra R, Vogt J (2017) Assessing the impact of plantation forestry on plant biodiversity. A comparison of sites in Central Chile and Chilean Patagonia. Glob Ecol Conserv 10:159–172

Bravo-Monasterio P, Pauchard A, Fajardo A (2016) Pinus contorta invasion into treeless steppe reduces species richness and alters species traits of the local community. Biol Invasions 18:1883–1894

Briones R, Jerez V (2007) Efecto de la edad de la plantación de Pinus radiata en la abundancia de Ceroglossus chilensis (Coleoptera: Carabidae) en la Región del Biobío, Chile. Bosque 28(3):207–214

Brooks ML, D'Antonio CM, Richardson DM, Grace JB, Keeley JE (2004) Effects of invasive plants on fire regimes. Bioscience 54:677–688

Buckley YM, Brockerhoff E, Langer L, Ledgard N, North H, Rees M (2005) Slowing down a pine invasion despite uncertainty in demography and dispersal. J Appl Ecol 42:1020–1030

Bustamante R, Castor C (1998) The decline of an endangered ecosystem: the Ruil *(Nothofagus alessandrii)* forest in Central Chile. Biodivers Conserv 7:1607–1626

Bustamante RO, Simonetti JA (2005) Is *Pinus radiata* invading the native vegetation in Central Chile? Demographic responses in a fragmented forest. Biol Invasions 7:243–249

Bustamante RO, Serey IA, Pickett STA (2003) Forest fragmentation, plant regeneration and invasion processes across edges in Central Chile. In: Bradshaw GA, Marquet PA, Mooney HA (eds) How landscapes change: human disturbance and ecosystem disruption in the Americas. Ecological studies, vol 162. Springer, Berlin/Heidelberg, pp 145–160

Carnus JM, Parrotta J, Brockerhoff EG, Arbez M, Jactel H, Kremer A, Lamb D, O'Hara K, Walters B (2006) Planted forests and biodiversity. J For 104(2):65–77

Chapela IH, Osher LJ, Horton TR, Henn MR (2001) Ectomycorrhizal fungi introduced with exotic pine plantations induce soil carbon depletion. Soil Biol Biochem 33(12):1733–1740

Cóbar-Carranza A, García R, Pauchard A, Peña E (2014) Effect of *Pinus contorta* invasion on forest fuel properties and its potential implications on the fire regime of *Araucaria araucana* and *Nothofagus antarctica* forests. Biol Invasions 16:2273–2291

CONAF (Corporación Nacional Forestal) (2017) Análisis de la Afectación y Severidad de los Incendios Forestales ocurridos en enero y febrero de 2017 sobre los usos de suelo y los ecosistemas naturales presentes entre las regiones de Coquimbo y Los Ríos de Chile. Informe Técnico. 56 p. Santiago, Chile

Cubbage F, Mac Donagh P, SawinskiJúnior J, Rubilar R, Donoso P, Ferreira A, Hoeflich V, Olmos VM, Ferreira G, Balmelli G, Siry J, Báez MN, Alvarez J (2007) Timber investment returns for selected plantation and native forests in South America and the southern United States. New For 33(3):237–255

de Oliveira SM, Boll PK, Baptista VD, Leal-Zanchet AM (2014) Effects of pine invasion on land planarian communities in an area covered by Araucaria moist forest. Zool Stud 53:19

de Villalobos AE, Zalba SM, Peláez DV (2011) Pinus halepensis invasion in mountain pampean grassland: effects of feral horses grazing on seedling establishment. Environ Res 111(7):953–959

Debinski DM, Ray C, Saveraid EH (2001) Species diversity and the scale of the landscape mosaic: do scales of movement and patch size affect diversity? Biol Conserv 98:179–190

Dickie IA, Johnston P (2008) Invasive fungi research priorities, with a focus on Amanita muscaria. Landcare research control report LC0809/027. Lincoln, New Zealand

Dickie IA, Reich PB (2005) Ectomycorrhizal fungal communities at forest edges. J Ecol 93(2):244–255
Dickie IA, Bolstridge N, Cooper JA, Peltzer DA (2010) Co-invasion by Pinus and its mycorrhizal fungi. New Phytol 187:475–484
Dickie IA, Yeates GW, St John MG, Stevenson BA, Scott JT, Rillig MC, Peltzer DA, Orwin KH, Kirschbaum MU, Hunt JE, Burrows LE (2011) Ecosystem service and biodiversity trade-offs in two woody successions. J Appl Ecol 48:926–934
Dickie IA, Stjohn MG, Yeates GW, Morse CW, Bonner KI, Orwin K, Peltzer DA (2014) Belowground legacies of Pinus contorta invasion and removal result in multiple mechanisms of invasional meltdown. AoB Plants 6:1–15
Dickie IA, Nuñez MA, Pringle A, Lebel T, Tourtellot SG, Johnston PR (2016) Towards management of invasive ectomycorrhizal fungi. Biol Invasions 18(12):3383–3395
Dickie IA, Bufford JL, Cobb RC, Desprez-Loustau ML, Grelet G, Hulme PE, Klironomos J, Makiola A, Nuñez MA, Pringle A, Thrall PH, Tourtellot SG, Waller L, Williams NM (2017) The emerging science of linked plant-fungal invasions. New Phytol 15(4):1314–1332. https://doi.org/10.1111/nph.14657
Echeverría C, Coomes D, Newton A, Salas J, Rey JM, Lara A (2006) Rapid fragmentation and deforestation of Chilean temperate forests. Biol Conserv 130:481–494
Echeverría C, Coomes D, Newton A, Rey-Benayas JM, Lara A (2007) Impacts of forest fragmentation on species composition and forest structure in the temperate landscape in southern Chile. Glob Ecol Biogeogr 16:426–439
Essl F, Moser D, Dullinger S, Mang T, Hulme PE (2010) Selection for commercial forestry determines global patterns of alien conifer invasions. Divers Distrib 16:911–921
Essl F, Mang T, Dullinger S, Moser D, Hulme PE (2011) Macroecological drivers of alien conifer naturalizations worldwide. Ecography 34:1076–1084
FAO (Food and Agriculture Organization of the United Nations) (2010) Global forest resources assessment 2010. FAO, Rome
Farley KA, Jobbagy EG, Jackson RB (2005) Effects of afforestation on water yield: a global synthesis with implications for policy. Glob Chang Biol 11:1565–1576
Franzese J, Raffaele E (2017) Fire as a driver of pine invasions in the southern hemisphere: a review. Biol Invasions 2017:1–10
Franzese J, Urrutia J, García RA, Taylor K, Pauchard A (2017) Pine invasion impacts on plant diversity in Patagonia: invader size and invaded habitat matter. Biol Invasions 19:1015–1027
Freedman B, Zelazny V, Beaudette D, Fleming T, Flemming S, Forbes G, Gerrow JS, Johnson G, Woodley S (1996) Biodiversity implications of changes in the quantity ofdead organic matter in managed forests. Environ Rev 4(3):238–265
Gámez-Virués S, Perović DJ, Gossner MM, Börschig C, Blüthgen N, de Jong H, Simons NK, Klein AM, Krauss J, Maier G, Scherber C, Steckel J, Rothenwöhrer C, Steffan-Dewenter I, Weiner CN, Weisser W, Werner M, Tscharntke T, Westphal C (2005) Landscape simplification filters species traits and drives biotic homogenization. Nat Commun 6:8568. https://doi.org/10.1038/ncomms9568
Gerrand A, Keenan RJ, Kanowski P, Stanton R (2003) Australian forest plantations: an overview of industry, environmental and community issues and benefits. Aust For 66:1–8
Gjerde I, Saetersdal M (1997) Effects on avian diversity of introducing spruce (Picea sp.) plantations in the native pine (Pinus sylvestris) forests of western Norway. Biol Conserv 79:241–250
Gómez P, Bustamante R, San Martin J, Hahn S (2011) Population structure of *Pinus radiate* D. Don in fragments of Maulino Forest in Central Chile. Gayana Botanica 68(1):97–101
Gundale MJ, Pauchard A, Langdon B, Peltzer DA, Maxwell BD, Nuñez MA (2014) Can model species be used to advance the field of invasion ecology? Biol Invasions 16(3):591–607
Gundale MJ, Almeida JP, Wallander H, Wardle DA, Kardol P, Nilsson MC, Fajardo A, Pauchard A, Peltzer DA, Ruotsalainen S, Mason B, Rosenstock N (2016) Differences in endophyte communities of introduced trees depend on the phylogenetic relatedness of the receiving forest. J Ecol 104:1219–1232

Hartley MJ (2002) Rationale and methods for conserving biodiversity in plantation forests. For Ecol Manag 155:81–95

Hartmann H, Gaëtan D, Bigué B, Messier C (2010) Negative or positive effects of plantation and intensive forestry on biodiversity: a matter of scale and perspective. For Chron 86:354–364

Hayward J, Horton TR, Pauchard A, Nuñez MA (2015a) A single ectomycorrhizal fungal species can enable a Pinus invasion. Ecology 96(5):1438–1444

Hayward J, Horton TR, Nuñez MA (2015b) Ectomycorrhizal fungal communities co invading with Pinaceae host plants in Argentina: Gringos bajo el bosque. New Phytol 208(2):497–506

Heilmayr R, Echeverría C, Fuentes R, Lambin E (2016) A plantation-dominated forest transition in Chile. Appl Geogr 75:71–82

Heinrichs S, Pauchard A (2015) Struggling to maintain native plant diversity in a peri-urban reserve surrounded by a highly anthropogenic matrix. Biodivers Conserv 24:2769–2788

Higgins SI, Richardson DM (1999) Predicting plant migration rates in a changing world: the role of long-distance dispersal. Am Nat 153:464–475

Kruger FJ, Breytenbach GJ, Macdonald IAW, Richardson DM (1989) The characteristics of invaded Mediterranean climate regions. Biological invasions. In: Drake JA, Mooney HA, di Castri F, Groves RH, Krüger FJ, Rejmanek M, Williamson M (eds) Biological invasions: a global perspective. Wiley, Chichester, pp 181–213

Langdon B, Pauchard A, Aguayo M (2010) Pinus contorta invasion in the Chilean Patagonia: local patterns in a global context. Biol Invasions 12:3961–3971

Larsson S, Danell K (2001) Science and the management of boreal forest biodiversity. Scand J For Res 16(Suppl 3):5–9

Ledgard NJ (2001) The spread of lodgepole pine (Pinus contorta Dougl.) in New Zealand. For Ecol Manag 141:43–57

Ledgard N (2003) What's wrong with wilding trees? NZ Tree Grow 24(1):18–19

Little C, Lara A, McPhee J, Urrutia R (2008) Revealing the impact of forest exotic plantations on water yield in large scale watersheds in South-Central Chile. J Hydrol 374:162–170

Losey JE, Vaughan M (2006) The economic value of ecological services provided by insects. Bioscience 56:311–323

Loydi A, Distel RA, Zalba SM (2010) Large herbivore grazing and non-native plant invasions in montane grasslands of Central Argentina. Nat Areas J 30(2):148–155

Lusk C (2008) Constraints on the evolution and geographical range of Pinus. New Phytol 178(1):1–3

Mandle L, Bufford JL, Schmidt IB, Daehler CC (2011) Woody exotic plant invasions and fire: reciprocal impacts and consequences for native ecosystems. Biol Invasions 13:1815–1827

McWethy DB, Pauchard A, García RA, Holz A, González ME, Veblen TT, Stahl J, Currey B (2018) Landscape drives of recent fire activity (2001-2017) in south-central Chile. PLoS ONE 13(8):e020119

Nahuelhual L, Carmona A, Lara A, Echeverría C, González M (2012) Land-cover change to forest plantations: proximate causes and implications for the landscape in south-central Chile. Landsc Urban Plan 107:12–20

Najera A, Simonetti JA (2010) Enhancing avifauna in commercial plantations. Conserv Biol 24:319–324

Nuñez MA, Dickie IA (2014) Invasive belowground mutualists of woody plants. Biol Invasions 16:645–661

Nuñez MA, Horton TR, Simberloff D (2009) Lack of belowground mutualisms hinders Pinaceae invasions. Ecology 90:2352–2359

Nuñez MA, Hayward J, Horton TR, Amico GC, Dimarco RD, Barrios MN, Simberloff D (2013) Exotic mammals disperse exotic fungi that promote invasion by exotic trees. PLoS One 8(6):e66832

Nuñez MA, Chiuffo MC, Torres A, Paul T, Dimarco R, Raal P, Policelli N, Moyano J, García RA, van Wilgen BW, Pauchard A, Richardson D (2017) Ecology and management of invasive pines around the world: progress and challenges. Biol Invasions 19:3099–3120

Orlovich DA, Cairney JG (2004) Ectomycorrhizal fungi in New Zealand: current perspectives and future directions. NZ J Bot 42(5):721–738

Paritsis J, Aizen MA (2008) Effects of exotic conifer plantations on the biodiversity of understory plants, epigeal beetles and birds in Nothofagus dombeyi forests. For Ecol Manag 255:1575–1583

Pauchard A, Langdon B, Peña E (2008) Potencial invasivo de *Pseudotsuga menziesii* (Mirb.) Franco en Bosques Nativos del Centro-Sur de Chile: patrones y recomendaciones. In: Mujica R, Grosse H, Muller-Using B (eds) Bosques Seminaturales: una opción para la rehabilitación de bosques nativos degradados. Instituto Forestal, Santiago, pp 89–114

Pauchard A, García RA, Zalba S, Sarasola M, Zenni R, Ziller S, Núñez M (2015) Pine invasions in South America: reducing their ecological impacts trough active management. In: Canning-Clode J (ed) Biological invasions in changing ecosystems. Vectors, ecological impacts, management and predictions. De Gruyter Open Ltd, Warsaw/Berlin, pp 318–342

Pauchard A, Escudero A, García RA, de la Cruz M, Langdon B, Cavieres LA, Esquivel J (2016) Pine invasions in treeless environments: dispersal overruns microsite heterogeneity. Ecol Evol 6(2):447–459

Pawson SM, McCarthy JK, Ledgard NJ, Didham RK (2010) Density-dependent impacts of exotic conifer invasion on grassland invertebrate assemblages. J Appl Ecol 47:1053–1062

Peay KG, Bidartondo MI, Arnold EA (2010) Not every fungus is everywhere: scaling to the biogeography of fungal–plant interactions across roots, shoots and ecosystems. New Phytol 185:878–882

Peña E, Valenzuela L (2008) Incremento de los incendios forestales en bosques naturales y plantaciones forestales en Chile. In: González-Cabán A (ed) Memorias del segundo simposio internacional sobre políticas, planificación y economía de los programas de protección contra incendios forestales: una visión global. General technical report PSW-GTR-208. United States Department of Agriculture, Forest Service, Pacific Southwesr Research Station, Albany, pp 595–612

Peña E, Hidalgo M, Langdon B, Pauchard A (2008) Patterns of spread of *Pinus contorta* Dougl. ex Loud. invasion in a natural reserve in southern South America. For Ecol Manag 256:1049–1054

Potton C (1994) A public perception of plantation forestry. NZ For 39(2):2–3

Rejmánek M, Richardson DM (1996) What attributes make some plant species more invasive. Ecology 77:1655–1661

Richardson DM (2001) Plant invasions. In: Levin S (ed) Encyclopedia of biodiversity. Academic, San Diego, pp 677–688

Richardson DM (2006) Pinus: a model group for unlocking the secrets of alien plant invasions? Preslia 78:375–388

Richardson DM, Brown PJ (1986) Invasion of Mesic mountain fynbos by *Pinus radiata*. S Afr J Bot 52:529–536

Richardson DM, Higgins S (1998) Pines as invaders in the southern hemisphere. In: Richardson DM (ed) Ecology and biogeography of Pinus. Cambridge University Press, Cambridge, pp 450–473

Richardson DM, Williams PA, Hobbs RJ (1994) Pine invasions in the southern hemisphere: determinants of spread and invadability. J Biogeogr 21:511–527

Richardson DM, Van Wilgen BW, Nuñez MA (2008) Alien conifer invasions in South America: short fuse burning? Biol Invasions 10:573–577

Rivera Y, Kretzer AM, Horton TR (2015) New microsatellite markers for the ectomycorrhizal fungus *Pisolithus tinctorius sensu stricto* reveal the genetic structure of US and Puerto Rican populations. Fungal Ecol 13:1–9

Rodríguez-Calcerrada J, Nanos N, del Rey M, López de Heredia U, Escribano R, Gil L (2011) Small-scale variation of vegetation in a mixed forest under storey is partly controlled by the effect of overstory composition on litter accumulation. J For Res 16:473–483

Rozzi R, Armesto JJ, Figueroa J (1994) Biodiversidad y conservación de los bosques nativos de Chile: una aproximación jerárquica. Bosque 15:55–64

Salas C, Donoso PJ, Vargas R, Arriagada C, Pedraza R, Soto D (2016) The forest sector in Chile: an overview and current challenges. J For 114(5):562–571
Sarasola M, Rusch V, Schlichter T, Ghersa C (2006) Invasión de coníferas forestales en áreas de estepa y bosques de ciprés de la cordillera en la Región Andino Patagónica. Ecol Austral 16:143–156
Simberloff D, Von Holle B (1999) Positive interactions of nonindigenous species: invasional meltdown? Biol Invasions 1:21–32
Simberloff D, Nuñez MA, Ledgard NJ, Pauchard A, Richardson DM, Sarasola M, van Wilgen BW, Zalba S, Zenni R, Bustamante R, Peña E, Ziller S (2010) Spread and impact of introduced conifers in South America: lessons from other southern hemisphere regions. Austral Ecol 35:489–504
Simonetti J, Grez A, Estades C (2013) Providing habitat for native mammals through understory enhancement in forestry plantations. Conserv Biol 27(5):1117–1121
Smith-Ramirez C (2004) The Chilean coastal range: a vanishing center of biodiversity and endemism in South American temperate rainforests. Biodivers Conserv 13(2):373–393
Spellerberg I (1996) Plantation forests protect our biodiversity? Too much of a generalization to be true. N Z For 40(4):5–7
Sutton WRJ (1995) Plantation forests protect our biodiversity. N Z For 40(3):2–5
Taylor KT, Maxwell BD, Pauchard A, Nuñez MA, Rew LJ (2016) Native versus non-native invasions: similarities and differences in the biodiversity impacts of *Pinus contorta* in introduced and native ranges. Divers Distrib 22:578–588
Taylor KT, Maxwell BD, McWethy DB, Pauchard A, Nuñez MA, Whitlock C (2017) *Pinus contorta* invasions increase wildfire fuel loads and may create a positive feedback with fire. Ecology 98:678–687
Tomasevic JA, Estades CF (2008) Effects of the structure of pine plantations on their "softness" as barriers for ground-dwelling forest birds in south-central Chile. For Ecol Manag 255(3):810–816
Urrutia J, Pauchard A, García RA (2013) Diferencias en la composición vegetal de un bosque de *Araucaria araucana* (Molina) K. Koch y *Nothofagus antarctica* (G. Forst.) Oerst. asociadas a un gradiente de invasión de Pinus contorta Douglas ex Loudon. Gayana Botanica 70(1):92–100
Valduga MO, Zenni RD, Vitule JRS (2016) Ecological studies reveal broad and heterogeneous impacts of exotic tree species plantations – examples from a megadiverse country. Ann Braz Acad Sci 88:1675–1688
Veblen TT, Holz A, Paritsis J, Raffaele E, Kitzberger T, Blackhall M (2011) Adapting to global environmental change in Patagonia: what role for disturbance ecology. Austral Ecol 36:891–903
Vihervaara P, Marjokorpi A, Kumpula T, Walls M, Kamppinen M (2012) Ecosystem services of fast-growing tree plantations: a case study on integrating social valuations with land-use changes in Uruguay. For Pol Econ 14:58–68
Zenni RD (2015) The naturalized flora of Brazil: a step towards identifying future invasive non-native species. Rodriguésia 66:1137–1144
Zenni RD, Ziller SR (2011) An overview of invasive plants in Brazil. Braz J Bot 34:431–446

Chapter 16
Biotic Homogenization of the South American Cerrado

Rafael Dudeque Zenni, Rafaela Guimarães, and Rosana Tidon

Abstract The South American Cerrado is a world biodiversity hotspot and at the same time is one of the fastest degrading biomes on Earth. About half of the original vegetation cover of the Cerrado has been converted to agriculture, pasture, urban area, and infrastructure. There are 223 non-native plant species known to be naturalized in the Cerrado, which is equivalent to almost 2% of the flora of the biome. Among those plants, at least 56 species are considered invasive. There are also 13 species of flies, pertaining to the family Drosophilidae, that have been introduced from other biogeographical regions. These insects are equivalent to approximately 10% of the total drosophilid insects in the biome. In this chapter, we review the primary literature on species invasions in the Cerrado and discuss how invasions might be contributing to biotic homogenization with a focus on grasses and flies. Existing research suggests that current high levels of habitat conversion and degradation, associated with low levels of conservation efforts, imply that the Cerrado is becoming dominated by non-native species, either in the form of cultivated crops and pasture, naturalized, or invasive species. It also suggests that more native species are increasingly threatened by extinction due to habitat losses. Furthermore, there are numerous abandoned and unrestored pastures inside national parks and other reserves. Through invasions and extinctions, biotic homogenization increases the similarity of biotas. Hence, there is an urgent need to increase both awareness and management efforts to reduce the threat posed by invasive species and habitat losses.

Keywords African grasses · Biological invasions · Brazil · Drosophilid · Neotropical savannah

R. D. Zenni (✉) · R. Guimarães
Setor de Ecologia, Departamento de Biologia, Universidade Federal de Lavras, Lavras, MG, Brazil
e-mail: rafael.zenni@dbi.ufla.br

R. Tidon
Departamento de Genética e Morfologia, Instituto de Ciências Biológicas, Universidade de Brasília, Brasília, DF, Brazil

R. Rozzi et al. (eds.), *From Biocultural Homogenization to Biocultural Conservation*, Ecology and Ethics 3, https://doi.org/10.1007/978-3-319-99513-7_16

16.1 Introduction

Originally covering an area of about two million km^2, the South American Cerrado is a world biodiversity hotspot and one of the fastest degrading biomes on Earth (Dirzo and Raven 2003; Strassburg et al. 2017). About half of the original vegetation cover of the Cerrado has been converted to agriculture, pasture, urban area, and infrastructure (Klink and Machado 2005). Currently, only one-fifth of the Cerrado vegetation cover is considered undisturbed (Strassburg et al. 2017). However, 40% of this remaining vegetation cover can still be legally deforested and converted to other uses (Soares-Filho et al. 2014; Strassburg et al. 2017). Furthermore, about 3.5% of the area protected by the Brazilian Forest Act (2.1 million ha) is disturbed and constitutes a deficit of natural vegetation cover (Sparovek et al. 2010). Such high levels of habitat conversion and degradation, associated with low levels of conservation efforts, may imply that the Cerrado is becoming dominated by non-native species, either in the form of cultivated crops and pasture, naturalized, or invasive species. It also suggests that more native species will become threatened by extinction due to habitat loss.

There are 223 non-native plant species currently known to be naturalized (*sensu* Blackburn et al. 2011) in the Cerrado, which is equivalent to approximately 2% of the flora of the biome (Zenni 2015). This number is based on non-native species present in natural and seminatural habitats owing to self-propagation. It does not include species that are only cultivated and that have never spread outside of cultivated areas (Zenni 2015). From the 223 naturalized species, at least 56 are considered invasive (Zenni and Ziller 2011; Sampaio and Schmidt 2014).

Regarding the fauna, there are at least seven invasive species of vertebrates in the Cerrado (Lessa et al. 2016; da Rosa et al. 2017). Invertebrate communities are poorly studied in the Cerrado, except for butterflies, social insects, and drosophilids (Oliveira and Marquis 2002). Specifically for drosophilids, almost two decades of systematic studies revealed 13 non-native naturalized or invasive species in the Cerrado (Ferreira and Tidon 2005; Döge et al. 2015).

A recent survey of protected areas under Brazil's federal jurisdiction (as opposed to state, municipal, and private protected areas) found 118 invasive populations for 56 plant and animal species across 19 protected areas (Sampaio and Schmidt 2014). These protected areas represent 43% of all the protected areas managed by the federal government in the Cerrado ecoregion. However, we lack data on the extent of these invasions and on the overall number and identities of non-native species that are present but not naturalized. Taken together, these data indicate that the vegetation configuration of the Cerrado is gaining new species via human-mediated introduction dispersal and removal of native species.

The Cerrado biome has more than 8500 native terrestrial species, of which at least 7000 are plants (Klink and Machado 2005; REFLORA 2013). The endemism rates are usually high, reaching 44% of the species for plants and 28% for amphibians (Klink and Machado 2005). According to the IUCN Red List of Threatened Species (IUCN Version 2016–3), 114 Cerrado species (59 plants and 55 animals) are vulnerable to extinction. Twenty-one species are considered critically endangered (IUCN Version 2016–3). However, these numbers may increase threefold by 2050 if current patterns of land use remain in place (Strassburg et al. 2017).

Invasions of natural habitats by non-native species are one side of biotic homogenization, whereas the extinctions of native and endemic species are the other side (Olden 2006). Through invasions and extinctions, biotic homogenization increases the similarity of biotas (Olden 2006). Here, we review the primary literature on species invasions in the Cerrado and discuss how invasions might be contributing to biotic homogenization.

16.2 Invasive African Grasses

African grasses are the most prominent species currently invading Cerrado habitats (Pivello et al. 1999; Zenni and Ziller 2011). The most ubiquitous species are *Andropogon gayanus*, *Hyparrhenia rufa*, *Melinis minutiflora*, *Urochloa decumbens*, and *Urochloa maxima* (Pivello et al. 1999; Klink and Machado 2005; Almeida-Neto et al. 2010; Sampaio and Schmidt 2014). These species also are widespread and invasive in many parts of the world. There are no estimates of the extent of the Cerrado invaded by African grasses or of the size of the area of abandoned pastures where African grasses were initially seeded and have since remained in the ecosystem. However, it could be safe to assume that African grasses are virtually present in all protected areas of Cerrado and along all the roads and degraded areas of the biome.

African grasses that became invasive in the Cerrado were introduced mostly during the nineteenth and twentieth centuries. *Melinis minutiflora* was first recorded in Brazil in the early 1800s when it was initially described as a native species (Zenni and Ziller 2011). Only later did researchers learn that the species was accidentally brought from Africa to Brazil most likely via dry ballast associated with the traffic of slaves. *Urochloa* species were intentionally introduced more recently, around the 1950s, as forage for livestock. The three originally introduced *Urochloa* species generated 20 cultivars currently in use in the Americas (Assis et al. 2003), including many areas of the Cerrado. *Andropogon gayanus* was intentionally introduced to Brazil in 1977 as an alternative forage plant for shaded areas with acidic soil, and for reseeding following fires, conditions under which *Urochloa* plants do not thrive (Meirelles and Mochiutti 1999). Finally, *Hyparrhenia rufa* probably shares an introduction history similar to *M. minutiflora*. There are no actual records of its introduction, but until the 1950s, *H. rufa* was considered native to the central regions of Brazil (Goiás and Mato Grosso) where it was known to grow spontaneously (Jardim et al. 1952). Around that same time, *H. rufa* and *M. minutiflora* were the grasses most used in pastures in the state of São Paulo (Jardim et al. 1952). Currently, *Urochloa* species seem to be the most commonly used species.

16.3 Known Impacts of African Grasses

A study showed that invasions by African grasses affect Asteraceae species occupancy in Cerrado communities (Almeida-Neto et al. 2010). According to this study, which encompassed 66 Asteraceae species, intermediate levels of invasive grass cover increased overall Asteraceae abundance, density, and species richness across the 20 Cerrado communities that were sampled. Overall, Asteraceae *abundance* and *density* were negatively affected only when African grasses reached high levels of cover (> 75%); however, Asteraceae *richness* was not negatively affected by invasive grass cover. Importantly, when the results were broken down by habitat specialization (endemic, generalist, and non-native species), there was a pronounced decrease in abundance and richness of endemic species when invasive grass cover increased, whereas generalist species were less affected, and non-native Asteraceae species were not negatively affected. The percentage of Cerrado endemic species is decreased by 50% from uninvaded to highly invaded communities (Almeida-Neto et al. 2010). By negatively affecting Cerrado endemic species while seldom or positively affecting generalist and non-native species, invasive African grasses tend to decrease ß-diversity and to increase taxonomic and functional similarity across communities.

In *Melinis minutiflora* invaded sites along Cerrado forest edges, a study found that the frequency of the invasive grass was negatively correlated with woody plant density (Hoffmann and Haridasan 2008). However, the invasion of *M. minutiflora* was not correlated with native species richness. On the other hand, the presence of *M. minutiflora* in Cerrado habitats was positively correlated with disturbance and the presence of nearby roads.

In a core Cerrado area where *M. minutiflora* was experimentally removed for a 5-year period (Sato et al. 2013), the removal of the invasive species did not affect community diversity and equitability metrics (Fig. 16.1a, b). However, species richness and plant density were affected by the removal of *M. minutiflora*. While richness decreased in control areas compared to removal areas, plant density doubled 5 years after the removal of *M. minutiflora* (Fig. 16.1c, d). *M. minutiflora* has been shown to outperform native species from the Cerrado, although it might not have been able to competitively exclude native species (Martins et al. 2011). This might explain why the removal of the invasive species resulted in dramatic increases in plant density, but not in species richness.

African grass invasions in the Cerrado may also produce negative edge effects in Cerrado woodland formations or Cerradão. When present on the edge of Cerradão woodland communities, African grasses have been shown to affect the composition and structure of the native plant community for up to 20 m inside woodland Cerrado formations (Dodonov et al. 2013, 2016). However, owing to the low number of case studies, the potential edge effect of invasive African grasses on Cerradão formations is still poorly understood, and further studies are particularly needed to clarify the role of invasive grasses in the Cerradão.

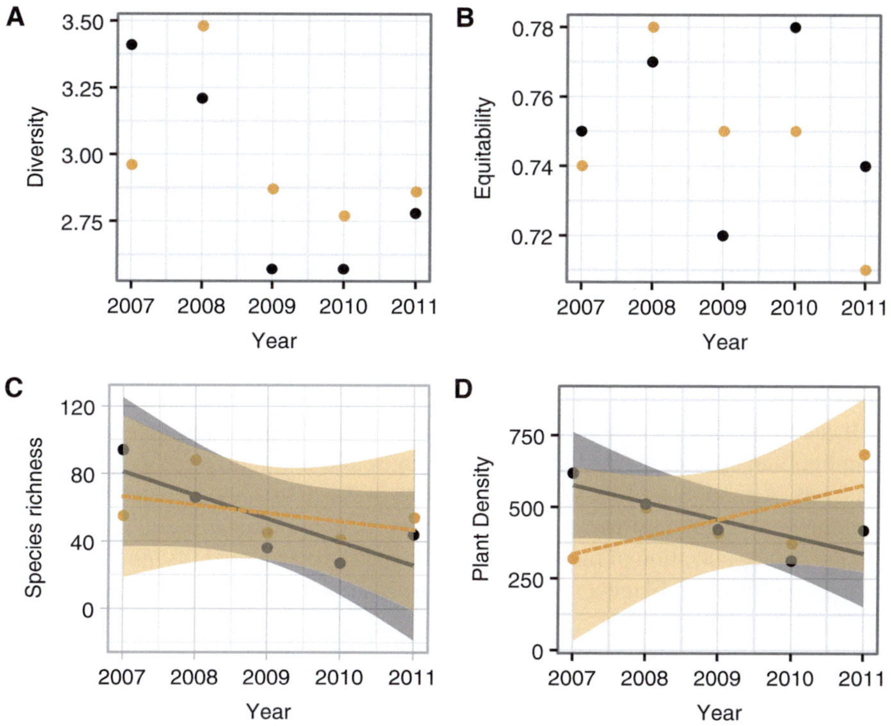

Fig. 16.1 Results of a 5-year removal experiment in the Cerrado. Orange dots and dashed lines represent areas where the invasive African grass *Melinis minutiflora* was experimentally removed; brown dots and solid lines represent control areas. Dots are mean values for each year, and shades represent standard errors. (**a**) Shannon diversity, (**b**) equitability, (**c**) species richness, and (**d**) plant density during a 5-year invasive species removal experiment. The 2007 data were collected immediately prior to *M. minutiflora* removal. (Data originally from Sato et al. 2013)

16.4 Invasive Flies

In the Cerrado, systematic studies have revealed the presence of 135 drosophilid species: 122 of Neotropical origin and 13 introduced from other biogeographical regions. Some of these exotic species (e.g., *Drosophila melanogaster* and *D. simulans*) possibly arrived in Brazil in the sixteenth century, unintentionally transported by slave ships. Other species probably arrived in Brazil later via human-mediated accidental transport in trade ships and airplanes. In the second half of the twentieth century, four new species arrivals in the Neotropics were properly recorded during the earlier stages of invasion: *Zaprionus indianus*, *Drosophila malerkotliana*, *D. nasuta*, and *D. suzuki*. Most of the exotic species now occurring in the Cerrado region are cosmopolitans (Bächli 2017) and are currently present in anthropic as well as in natural areas (Ferreira and Tidon 2005; Döge et al. 2015).

The non-natives *Drosophila simulans* and *Zaprionus indianus* currently are among the three most abundant drosophilid species in the Cerrado (the third species is the Neotropical *Drosophila willistoni*). During the rainy season, when environmental conditions are favorable and populations of most drosophilids expand, these two non-native species together can represent 90% of the total number of flies in the community. Although population growth rates depend on complex interactions between biotic and abiotic factors, these two non-native species seem to be regulated primarily by rapid and direct responses to resource availability (Döge et al. 2015). In natural vegetation, *Z. indianus* infests green fruits that are not used as resources by most of the other drosophilids (to the best of our knowledge). Such capacity could help to explain why *Z. indianus* (first recorded in Cerrado in 1999) is today the dominant drosophilid species in many drosophilid assemblages in the Cerrado.

In sum, drosophilid assemblages in the Cerrado corroborate that a few non-native cosmopolitan species are expanding their ranges and naturalizing in different environments. Although we lack specific studies to attest to this, the hyperabundance of non-native species may result in replacement of geographically endemic species and the homogenization of the invertebrate fauna. However, quantifying the changes in drosophilid assemblage's richness and structure demands long-term monitoring studies.

16.5 Other Invasive Species with Homogenizing Potential

Besides the African grasses and drosophilid species which are invasive and hyperabundant in the Cerrado, other species represent potential invasion risk given their history in other regions, for instance, the conifers *Pinus oocarpa* and *P. elliottii* and the blackfern *Pteridium arachnoideum* (Abreu and Durigan 2011; Miatto et al. 2011; Braga et al. 2014; Guerin and Durigan 2015). Invasion by *P. elliottii* causes diversity losses in grassland Cerrado (Almeida et al. 2010; Abreu and Durigan 2011). *Pinus oocarpa* is also invasive in the Cerrado (Braga et al. 2014), but its effects on biodiversity have not been studied; we may assume, however, that its effects are similar to other pine invasions (Garcia et al. 2018).

Studies have shown that *P. arachnoideum*, a fern species native to the central and southern regions of the Cerrado, has a strong invasive potential in degraded areas by forming monospecific dominance with fire-resistant rhizomes. The species decreases richness and structure of native plant communities (Guerin and Durigan 2015) and negatively affects long-term natural regeneration processes by reducing the diversity and abundance of the native seed soil bank (Miatto et al. 2011). By causing such effects, native *P. arachnoideum* invasion favors biotic homogenization owing to its positive interaction with fire and other agents of environmental degradation.

16.6 Mechanisms of Invasion Leading to Biotic Homogenization

Invasive plant species tend to be habitat generalist, have high competitive ability, and fast growth rates when compared to species that do not invade (Van Kleunen et al. 2010; Speek et al. 2011; Zenni et al. 2016). While such traits can help explain the growth and spread of invasives, it does not explain how invasions contribute to the homogenization of invaded communities. For Cerrado habitats, fire may be a mechanism fostering non-native species establishment and encroachment. For instance, it has been shown that wildfires have a lesser effect on plants and seeds of the invasive grass *Andropogon gayanus* than on native species (Marinho and Miranda 2013). In that study, all plant species were negatively affected by fire, but *A. gayanus* had lower mortality rates. This allowed the species to recolonize the area much faster after the fire, outcompeting native species and largely hindering their regeneration. A similar trend was observed for the non-native *Urochloa decumbens* under experimental burnings (Gorgone-Barbosa et al. 2015).

Fires are a common feature shaping Cerrado ecosystems. Fires quickly eliminate competing organisms, open environmental space, and release a flush of nutrients into the environment (Rossi et al. 2014; Gorgone-Barbosa et al. 2015). However, non-native species with more resistance or tolerance to wildfires can quickly recover and outcompete native species. As a consequence, after burning cycles, it could be expected that invasive species could dominate the ecosystems, whereas native species could become less abundant leading to biotic homogenization.

Another mechanism that could enhance biotic homogenization is allelopathy or the release of phytotoxins by invasive plants, which can enhance competitive ability of the invader and negatively affect sensitive native species (Bais et al. 2003). Several invasive grasses in the Cerrado, such as *U. decumbens* and *M. minutiflora*, exhibit allelopathic effects (Barbosa et al. 2008) that may be able to exclude native species. Also, some invasive species can change soil properties and microbial communities (Silva 2012), affecting the establishment and survival of native species. However, some native species have also been shown to produce phytotoxins that negatively affect invasive plants (Lopes 2016). The potential of allelopathy to reduce native biodiversity and increase biotic homogenization exists, but it seems largely community specific.

The main use of introduced African grasses is pasture for livestock. Pastures are usually formed in rural regions in close proximity to natural ecosystems, which facilitate the spread of non-native species to native habitats when disturbances occur. Moreover, there are some instances in the Cerrado ecoregion where protected areas were created on previously privately owned cattle ranches. The conversion of those ranches into protected areas resulted in the abandonment of hundreds of hectares of planted pastures that will not recover without active ecological restoration efforts (Sampaio and Schmidt 2014; Lopes 2016). Abandoned pastures may also serve as sources for invasions of surrounding Cerrado habitats. Thus, the close association of African grasses with human socioeconomic activities also promotes the spread of these non-native species, vastly increasing propagule pressure.

16.7 Final Remarks

The decrease in species abundance and density with rare events of local extinction of native species, alongside edge effects on woodland communities, seems to be the most common current scenario for the impact of invasive non-native grasses in the Cerrado. Even though there have been no records of global extinction of native plant species in the Cerrado, there is research demonstrating that endemic species are more vulnerable to competition with non-native invaders. Based on the current evidence, it would be safe to assume that the invasion of Cerrado communities by non-native species, especially African grasses, is leading to biotic homogenization by reducing dissimilarity among natural communities.

The biotic homogenization of the Cerrado by the invasion of African grasses, along with the frequent existence of abandoned and unrestored pastures inside national parks and other preserves (Horowitz et al. 2013), may be damaging to biodiversity conservation awareness efforts. As it was the case for *H. rufa* in the 1950s (Jardim et al. 1952), there is a real risk that people may see African grasses as an integral part of the Cerrado biodiversity, ultimately opposing invasion management efforts and reducing support for ecological restoration of areas occupied by non-native species. The cultural dimension of biotic homogenization, where people fail to distinguish between habitats dominated by native and non-native species, can cause problems for long-term conservation of the biodiversity of the Cerrado. There is an urgent need to increase both awareness and management efforts to reduce the threat posed by invasive species.

References

Abreu RCR, Durigan G (2011) Changes in the plant community of a Brazilian grassland savannah after 22 years of invasion by *Pinus elliottii* Engelm. Plant Ecol Divers 4(2–3):269–278

Almeida RS d, Cielo-Filho R, Souza SCPM d, Aguiar OT d, Baitello JB, Pastore JA, Kanashiro MM, Mattos IF d A, Franco GADC, Lima CR d (2010) Campo Sujo Úmido: fisionomia de Cerrado ameaçada pela contaminação de *Pinus elliottii* Engelm. na Estação Ecológica de Itapeva, Estado de São Paulo. Revista do Intituto Florestal 22(1):71–91

Almeida-Neto M, Prado PI, Kubota U, Bariani JM, Aguirre GH, Lewinsohn TM (2010) Invasive grasses and native Asteraceae in the Brazilian Cerrado. Plant Ecol 209(1):109–122

Assis GML d, Euclydes RF, Cruz CD, Valle CB d (2003) Discriminação de espécies de *Brachiaria* baseada em diferentes grupos de caracteres morfológicos. Rev Bras Zootec 32:576–584

Bächli G (2017) TaxoDros: the database on taxonomy of Drosophilidae. Available at: http://taxodros.unizh.ch/

Bais HP, Vepachedu R, Gilroy S, Callaway RM, Vivanco JM (2003) Allelopathy and exotic plant invasion: from molecules and genes to species interactions. Science 301(5638):1377–1380

Barbosa EG, Pivello VR, Meirelles ST (2008) Allelopathic evidence in Brachiaria decumbens and its potential to invade the Brazilian Cerrados. Braz Arch Biol Technol 51:625–631

Blackburn TM, Pyšek P, Bacher S, Carlton JT, Duncan RP, Jarošík V, Wilson JRU, Richardson DM (2011) A proposed unified framework for biological invasions. Trends Ecol Evol 26(7):333–339

Braga EP, Zenni RD, Hay JD (2014) A new invasive species in South America: *Pinus oocarpa* Schiede ex Schltdl. BioInvasion Rec 3(3):207–211

da Rosa CA, Curi NH d A, Puertas F, Passamani M (2017) Alien terrestrial mammals in Brazil: current status and management. Biol Invasions

Dirzo R, Raven PH (2003) Global state of biodiversity and loss. Annu Rev Environ Resour 28(1):137–167

Dodonov P, Harper K, Silva-Matos D (2013) The role of edge contrast and forest structure in edge influence: vegetation and microclimate at edges in the Brazilian Cerrado. Plant Ecol 214:1345–1359

Dodonov P, Braga A, Harper K, Silva-Matos D (2016) Edge influence on plant litter biomass in forest and savanna in the Brazilian Cerrado. Austral Ecol 42(2):187–197

Döge J, Valadão H, Tidon R (2015) Rapid response to abiotic and biotic factors controls population growth of two invasive drosophilids (Diptera) in the Brazilian savanna. Biol Invasions 8:2461–2474

Ferreira L, Tidon R (2005) Colonizing potential of Drosophilidae (Insecta, Diptera) in environments with different grades of urbanization. Biodivers Conserv 14:1809–1821

García RA, Franzese J, Policelli N, Sasal Y, Zenni RD, Núñez M, Taylor K, Pauchard A (2018) Non-native pines are homogenizing the ecosystems of South America. In: Rozzi R, May R, Chapin SF, Massardo F, Gavin M, Klaver I, Pauchard A, Núñez MA, Simberloff D (eds) From biocultural homogenization to biocultural conservation. Springer, Dordrecht, pp 254–263

Gorgone-Barbosa E, Pivello VR, Bautista S, Zupo T, Rissi MN, Fidelis A (2015) How can an invasive grass affect fire behavior in a tropical savanna? A community and individual plant level approach. Biol Invasions 17(1):423–431

Guerin N, Durigan G (2015) Invasion impact by *Pteridium arachnoideum* (Kaulf.) Maxon (Dennstaedtiaceae) on a neotropical savanna. Acta Bot Bras 29(2):213–222

Hoffmann WA, Haridasan M (2008) The invasive grass, Melinis minutiflora, inhibits tree regeneration in a Neotropical savanna. Austral Ecol 33(1):29–36

Horowitz C, Martins CR, Walter BMT (2013) Flora exótica no Parque Nacional de Brasília: levantamento e classificação das espécies. Biodiversidade Brasileira 3(2):50–73

IUCN (Version 2016–3). Red list of threatened species

Jardim WR, Moraes CL, Peixoto AM (1952) Contribuição para o estudo da composição e digestibilidade do capim jaraguá (*Hyparrhenia rufa* (Nees) Stapf.). II Reunião Interamericana da Produção Animal, Bauro, São Paulo

Klink CA, Machado RB (2005) A conservação do Cerrado brasileiro. Megadiversidade 1(1):147–155

Lessa I, Corrêa Seabra Guimarães T, de Godoy Bergallo H, Cunha A, Vieira EM (2016) Domestic dogs in protected areas: a threat to Brazilian mammals? Natureza & Conservação 14(2):46–56

Lopes PG (2016) Alelopatia em *Lepidaploa aurea* (Asteraceae) como ferramenta de restauração ecológica : potencial para o controle de gramíneas exóticas invasoras no Cerrado. MSc, Universidade de Brasília

Marinho M d S, Miranda HS (2013) Efeito do fogo anual na mortalidade e no banco de sementes de *Andropogon gayanus* kunth. No Parque Nacional de Brasília/DF. Biodiversidade Brasileira 3(2):149–158

Martins CR, Hay JDV, Walter BMT, Proença CEB, Vivaldi LJ (2011) Impacto da invasão e do manejo do capim-gordura (*Melinis minutiflora*) sobre a riqueza e biomassa da flora nativa do Cerrado sentido restrito. Rev Bras Bot 34(1):73–90

Meirelles PR d L, Mochiutti S (1999) *Andropogon gayanus*: alternativa para formação de pastagens no cerrado amapaense. Recomendações técnicas. Macapá Amapá Embrapa Amapá 8:1–3

Miatto R, Silva I, Silva-Matos D, Marrs R (2011) Woody vegetation structure of Brazilian Cerrado invaded by *Pteridium arachnoideum* (Kaulf.) Maxon (Dennstaedtiaceae). Flora 206:757–762
Olden JD (2006) Biotic homogenization: a new research agenda for conservation biogeography. J Biogeogr 33(12):2027–2039
Oliveira O, Marquis R (2002) The cerrados of Brazil: ecology and natural history of a neotropical savanna. Columbia University Press, New York
Pivello VR, Shida CN, Meirelles ST (1999) Alien grasses in Brazilian savannas: a threat to the biodiversity. Biodivers Conserv 8:1281–1294
REFLORA (2013) Lista de Espécies da Flora do Brasil
Rossi RD, Martins CR, Viana PL, Rodrigues EL, Figueira JEC (2014) Impact of invasion by molasses grass (*Melinis minutiflora* P. Beauv.) on native species and on fires in areas of campo-cerrado in Brazil. Acta Bot Bras 28:631–637
Sampaio AB, Schmidt IB (2014) Espécies exóticas invasoras em unidades de conservação federais do Brasil. Biodiversidade Brasileira 3(2):32–49
Sato MN, Miranda HS, Aires SS, Aires FS (2013) Alterações na fitossociologia do estrato rasteiro de uma área de campo sujo, invadida por *Melinis minutiflora* p. Beauv., submetida a corte anual. Biodiversidade Brasileira 3(2):137–148
Silva MRSSd (2012) Diversidade de comunidades bacterianas de solo de cerrado em resposta a diferentes alterações dos ecossistemas. PhD, Universidade de Brasília
Soares-Filho B, Rajão R, Macedo M, Carneiro A, Costa W, Coe M, Rodrigues H, Alencar A (2014) Cracking Brazil's forest code. Science 344(6182):363–364
Sparovek G, Berndes G, Klug ILF, Barretto AGOP (2010) Brazilian agriculture and environmental legislation: status and future challenges. Environ Sci Technol 44(16):6046–6053
Speek TAA, Lotz LAP, Ozinga WA, Tamis WLM, Schaminée JHJ, van der Putten WH (2011) Factors relating to regional and local success of exotic plant species in their new range. Divers Distrib 17(3):542–551
Strassburg BBN, Brooks T, Feltran-Barbieri R, Iribarrem A, Crouzeilles R, Loyola R, Latawiec AE, Oliveira Filho FJB, Scaramuzza CA d M, Scarano FR, Soares-Filho B, Balmford A (2017) Moment of truth for the Cerrado hotspot. Nat Ecol Evol 1:0099
Van Kleunen M, Weber E, Fischer M (2010) A meta-analysis of trait differences between invasive and non-invasive plant species. Ecol Lett 13(2):235–245
Zenni RD (2015) The naturalized flora of Brazil: a step towards identifying future invasive non-native species. Rodriguésia 66:1137–1144
Zenni RD, Ziller SR (2011) An overview of invasive plants in Brazil. Braz J Bot 34(3):431–446
Zenni RD, da Cunha WL, Sena G (2016) Rapid increase in growth and productivity can aid invasions by a non-native tree. AoB Plants 8:plw048

Chapter 17
Taxonomic and Phylogenetic Homogenization Across US National Parks: The Role of Non-native Species

Daijiang Li, Julie L. Lockwood, and Benjamin Baiser

Abstract US national parks are firmly engrained into the culture of the country. Since the first national park was established in 1872, these protected areas have been central to biodiversity conservation, education, and recreation in the United States. However, national parks face many of the same threats to biodiversity as non-protected areas, including non-native species invasion. As invasive species spread and replace native species across national parks, the composition, function, and evolutionary history of these ecological communities continue to change. To better understand the impact of non-native species on ecological communities in US national parks, we compared species composition for bird and plant assemblages in 244 and 241 national park units, respectively, with and without non-native species. We specifically ask if the establishment of non-native species has resulted in taxonomic and phylogenetic homogenization across national parks. We show that the establishment of non-native plants in US national parks has resulted in both taxonomic and phylogenetic homogenization. However, the establishment of non-native birds has led to phylogenetic differentiation in spite of taxonomic homogenization, indicating that taxonomic and phylogenetic homogenization are decoupled. As the uniqueness of regional floras and faunas is lost through the process of homogenization, not only is biodiversity lost but so are cultural resources and values that define the United States.

D. Li (✉) · B. Baiser
Department of Wildlife Ecology and Conservation, University of Florida, Gainesville, FL, USA
e-mail: bbaiser@ufl.edu

J. L. Lockwood
Department of Ecology, Evolution, and Natural Resources, Rutgers University, New Brunswick, NJ, USA
e-mail: julie.lockwood@rutgers.edu

R. Rozzi et al. (eds.), *From Biocultural Homogenization to Biocultural Conservation*, Ecology and Ethics 3, https://doi.org/10.1007/978-3-319-99513-7_17

Keywords Beta diversity · Birds · National parks · Phylogenetic diversity · Plants · Species invasions

17.1 Introduction

> The nation behaves well if it treats the natural resources as assets which it must turn over to the next generation increased; and not impaired in value. – Theodore Roosevelt (1910)

> Thousands of tired, nerve-shaken, over-civilized people are beginning to find out that going to the mountains is going home; that wildness is a necessity; and that mountain parks and reservations are useful not only as fountains of timber and irrigating rivers, but as fountains of life. – John Muir (1901, p. 15)

US national parks are firmly engrained into the culture of this country. Since the first national park was established in 1872, they have been central to conservation, education, and recreation in the United States (Sellars 2009). Within the US national park system lie diverse and iconic landscapes such as the geological formations at Arches National Park in Utah, the glacier-carved peaks of Glacier National Park in Montana, and the River of Grass in the Florida Everglades (Byerly 1996), as well as iconic species like the bison (*Bison bison*) of Yellowstone, the towering sequoia (*Sequoiadendron giganteum*) in Kings Canyon National Park in California, and the sleepy American alligator (*Alligator mississippiensis*) of the Everglades. National parks mean different things to different people (Mayo 1975). To some they are a safe haven for plants, animals, and wilderness. Even though they may not visit, the mere existence of such places, animals, and wilderness is righteous, invigorating, and quintessential to the culture of the United States. To others, summer's arrival means packing the family car and setting out on the open road to participate in an American rite of passage. Yet for others, these parks represent places where not only species can get away from development and civilization, but they are also places where people can immerse themselves in sights, sounds, and smells of wild nature (Runte 1997). Regardless of different perceptions of US national parks, their popularity has never been greater with a record high of 330 million people visiting in 2016 (NPS Stats, https://irma.nps.gov/Stats/Reports/National). Furthermore, US national parks are an international destination with over 13 million international visitors in 2015 (about 35.4% of total visitors, U.S. Travel Association).[1]

While the US national parks set aside land for some of the most awe-inspiring landscapes and species in North America, simple park designation does not ensure the preservation of such natural resources (Parrish et al. 2003). National parks face many of the same threats to biodiversity as non-protected areas, including non-native species invasion (Allen et al. 2009; Stohlgren et al. 2013; Abella 2014). This is exemplified by the greater than 3700 non-native plant species that have been identified in US national parks (Allen et al. 2009). While the majority of non-native

[1] https://d3a2exhi9k8v0e.cloudfront.net/sites/default/files/Media%20Root/Document/NPS_Overseas_Highlights_V1%20(1).pdf

species are not considered "major" threats to US national parks (Hiebert and Stubbendieck 1993), there are textbook examples of non-natives that have become invasive and are greatly altering ecosystems. The invasive white pine blister rust (*Cronartium ribicola*) has greatly increased tree mortality across the Western United States including within national parks (Van Mantgem et al. 2009). The exotic pest hemlock woolly adelgid (*Adelges tsugae*) is an aggressive invader in eastern North America that threatens to eliminate eastern hemlock (*Tsuga canadensis*) from its range, which includes 21% of all parks in the US national park system (Abella 2014). Eastern hemlock is a foundation species, which means that it determines community composition (i.e., which species are present) by physically structuring ecosystems and modulating ecosystem processes (Ellison et al. 2005). As a result, the loss of hemlock has cascading effects on food web structure (Baiser et al. 2013), arthropod assemblages (Sackett et al. 2011), and avian community composition (Tingley et al. 2002). The list of harmful invaders also includes species such as the invasive Burmese python (*Python molurus bivittatus*), which has caused severe mammal declines in Everglades National Park (Dorcas et al. 2012; McCleery et al. 2015).

It is clear that non-native species can have detrimental effects on the ecosystems they invade. At larger scales (e.g., continental), non-natives can also fundamentally alter patterns of biodiversity. As non-native species establish in new locations and either replace or join native species, ecological communities become more similar in their composition (Olden and Poff 2003). As a result, non-native species play a major role in the process of biotic homogenization (McKinney and Lockwood 1999; McKinney 2004; Qian and Ricklefs 2006). Biotic homogenization has implications for US national parks that are valued for their uniqueness and representation of local flora and fauna.

While the homogenization of plants and animals across US national parks undermines the biological distinctiveness of these important resources, it may have implications beyond simply changing the species that park visitors encounter. Non-native species carry a set of traits and evolutionary history to the invaded locale. If non-native species become widespread and reach multiple locales, they introduce similar traits and evolutionary histories to the invaded assemblages that can result in functional and/or phylogenetic homogenization. Furthermore, because successful non-natives are often nonrandom in regard to taxonomy (Lockwood 1999; Cadotte et al. 2006; Pysek et al. 2017) and tend to possess similar trait values (Pyšek and Richardson 2008; Blackburn et al. 2009), disparate locales that receive different species can still functionally and/or phylogenetically homogenize (Baiser and Lockwood 2011). These two aspects of homogenization are important yet understudied components of global change (Olden et al. 2004) and can give contrasting information regarding changes in biodiversity (Monnet et al. 2014). Critically, functional and phylogenetic homogenization may hold the key to predicting the ecosystems of the future because they quantify the functional and evolutionary potential of ecosystems to adapt to changing conditions.

Here we assess phylogenetic and taxonomic homogenization of bird and vascular plant assemblages in US national parks. We use species lists for over 240 park units to explore how non-native species have altered biodiversity patterns, asking if non-native species have led to an increase in taxonomic similarity (i.e., taxonomic homogenization) and phylogenetic similarity (i.e., phylogenetic homogenization) across all US national parks and across only those parks within the same ecoregion.

17.2 Methods

17.2.1 Study Sites

The National Park Service (NPS) of the United States is actively assessing and managing resources for more than 270 park units through its Inventory and Monitoring (I&M) Program (Fancy et al. 2009). The I&M Program performed an inventory of species occurrence in the 2000s for these parks with the help of local taxonomic experts. During these inventories, species' provenances (native or non-native) were assigned and then carefully reviewed before merging them into the NPSpecies database (https://irma.nps.gov/NPSpecies/). We used the occurrence and provenance of vascular plants and birds from the NPSpecies dataset because these two groups are well studied and likely have the most complete and accurate records. We only kept records listed as "present" and with a clear provenance as either "native" or "non-native." For birds, we also removed species that were tagged as "vagrant" and then cross-validated all other species in terms of their provenance with an up-to-date avian invasion atlas dataset (GAVIA; Dyer et al. 2017). We extracted a list of successful introduced species (tagged as "breeding" or "established") to the United States from GAVIA. We then verified non-native species in the NPS dataset with this list. For both plants and birds, we removed parks from Hawaii since they are far away from other parks and have not been classified into ecoregions (more details in the ecoregions subsection). In total, we obtained a list of 14703 plants from 241 park units and a list of 729 birds from 244 park units. For more details about the dataset, see Li et al. (2018).

17.2.2 Phylogenies

In order to assess changes in phylogenetic similarity across US national parks, we situated the species extracted above with previously published phylogenetic trees. Phylogenetic trees detail the evolutionary history of a given set of organisms based on their DNA. For plants, we built a phylogeny using Phylomatic v4.2 (Webb and Donoghue 2005) and the supertree "zanne2014," which is an up-to-date phylogeny for >32,000 species based on 7 genes and maximal likelihood approximation (Zanne et al. 2014).

For birds, we used the phylogeny generated by Jetz et al. (2012). They provided 10,000 phylogenies using Bayesian analyses and documented their detailed methods at http://birdtree.org. In this study, we randomly selected 100 phylogenies since these are very similar and thus gave qualitatively the same results regarding phylogenetic diversity (Baiser et al. 2018). We then used the average results from these 100 phylogenies for all avian analyses in this study.

17.2.3 Ecoregions

National parks are widely distributed across the United States (Fig. 17.1). Because homogenization is likely to be scale dependent, we chose to explore changes in taxonomic and phylogenetic similarity within and across ecoregions (i.e., continental scale). Ecoregions are areas where the type and nature of environmental resources, land use, and vegetation are generally similar (Omernik 1987; Omernik and Griffith 2014). North America has been divided into 15 broad level I, 50 level II, and 182 level III ecoregions (https://www.epa.gov/eco-research/ecoregions-north-america). For this study, we focused on the 15 broad level I ecoregions. We downloaded the ecoregion shapefiles from the US Environmental Protection Agency (EPA) website and then clipped these shapefiles to just include those that are in the United States (Fig. 17.1). Because Hawaii is not mapped in the shapefile, we excluded Hawaiian parks from our analyses, resulting in 254 parks in total (240 in the conterminous states and 14 in Alaska). When a park was located on the boundary between two ecoregions, we put that park in both ecoregions when conducting analyses at the ecoregion scale. We removed three ecoregions (water, tropical wet forests, and southern semiarid highlands) from our analyses because they have few parks (< 8), resulting in 8 ecoregions in the lower 48 states and 4 ecoregions in Alaska (Fig. 17.1). Because there are only 14 parks within Alaska, we did not conduct ecoregion level analyses for them.

17.2.4 Data Analysis

We calculated pairwise taxonomic and phylogenetic distances for all unique combinations of parks. For taxonomic pairwise distance, we chose the commonly used Jaccard index, $\beta_j = (b + c)/(a + b + c)$, where a is the number of species observed in both parks and b and c are the number of species unique to each of the two parks being compared. The Jaccard index takes on a value of 1 when no species are shared between two assemblages (parks) and 0 when both assemblages (parks) have the

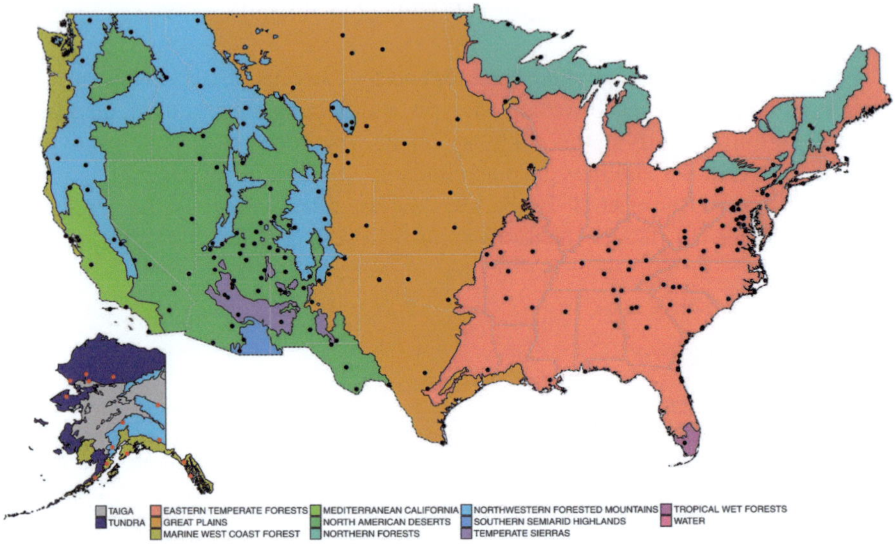

Fig. 17.1 The distribution of level I ecoregions across the United States. Dots represent national parks used in this study

same exact species composition. For phylogenetic pairwise beta diversity, we used mean pairwise distance (MPD, Webb et al. 2002), which is independent of species richness (Bello et al. 2016). For each species in one park, MPD finds the average phylogenetic distance to each species in the other park and then calculates the mean of these values. Lower values of MPD indicate that two assemblages are phylogenetically similar, while larger values indicate phylogenetic differentiation.

We calculated pairwise distances between all parks for native species only and then for native and non-natives together. To test whether pairwise phylogenetic and taxonomic similarity changed significantly after including non-native species, we conducted nonparametric paired Wilcox tests at both continental and ecoregion levels. If the average pairwise distance is lower after including non-native species, it suggests that non-native species have homogenized national parks in their taxonomic or phylogenetic composition and vice versa. Because parks in Alaska are geographically distant from the lower 48 states, we analyzed Alaskan parks separately. We further visualized phylogenetic and taxonomic similarity, for each park, by calculating its average pairwise distance with all other parks within the same ecoregion (within-ecoregions) and with all other parks that are not in the same ecoregion (across-ecoregions). Because within-ecoregions and across-ecoregions patterns were qualitatively similar, we only presented the figure for across-ecoregions below.

17.3 Results

17.3.1 Changes in Taxonomic and Phylogenetic Similarity at the Continental Scale

At the continental scale, non-native species homogenized the taxonomic and phylogenetic composition of plant assemblages across US national parks (Table 17.1, Figs. 17.2 and 17.3). This indicates that on average, parks across the United States are receiving a similar suite of non-native invaders and that these non-natives have similar evolutionary histories. The latter can be due to the establishment of the same species across parks or the establishment of different invaders that are evolutionarily closely related. Parks in the Northwestern United States had the highest levels of homogenization (Fig. 17.3). High levels of phylogenetic homogenization were found in the Pacific Northwest, Great Lakes region, and the Northeastern United States, while lower levels of phylogenetic homogenization were found in the Southeastern United States (Fig. 17.3).

Bird assemblages at the continental scale have homogenized in taxonomic composition but a lower degree that plant assemblages across parks (Table 17.1, Figs. 17.2 and 17.3). In addition, bird assemblages became less similar in their phylogenetic composition (i.e., phylogenetic differentiation) across parks. This indicates that even though similar non-native birds are establishing across parks, these birds are increasing the phylogenetic distances between park bird assemblages. The highest levels of avian taxonomic homogenization occurred in the southwestern and mid-Atlantic portions of the United States (Fig. 17.3). High levels of avian phylogenetic differentiation were found in parks across the United States; however, parks along the coast of the Southeastern United States and the Gulf of Mexico tended to have lower levels of phylogenetic differentiation (Fig. 17.3).

Table 17.1 Overall changes in taxonomic and phylogenetic composition across all parks in the conterminous United States and Alaska

Taxa	Index	Changes in US mainland	Changes in Alaska
Birds	Taxonomic	(0.6963, 0.6911) Homog.***	(0.4010, 0.4022) Diff.***
	Phylogenetic	(138.66, 139.02) Diff.***	(154.03, 154.06) Diff.***
Plants	Taxonomic	(0.9332, 0.9228) Homog.***	(0.6007, 0.6058) Diff.***
	Phylogenetic	(538.55, 535.49) Homog.***	(541.69, 540.98) Homog.***

Value pairs in parentheses are average pairwise beta diversity of native species and native plus exotic species, respectively. Note the different scales of taxonomic and phylogenetic beta diversity. Statistics and p-values (not shown) are derived from paired Wilcox tests

Abbreviations: "Diff." indicates differentiation; "Homog." indicates homogenization; "***" indicates p value < 0.001

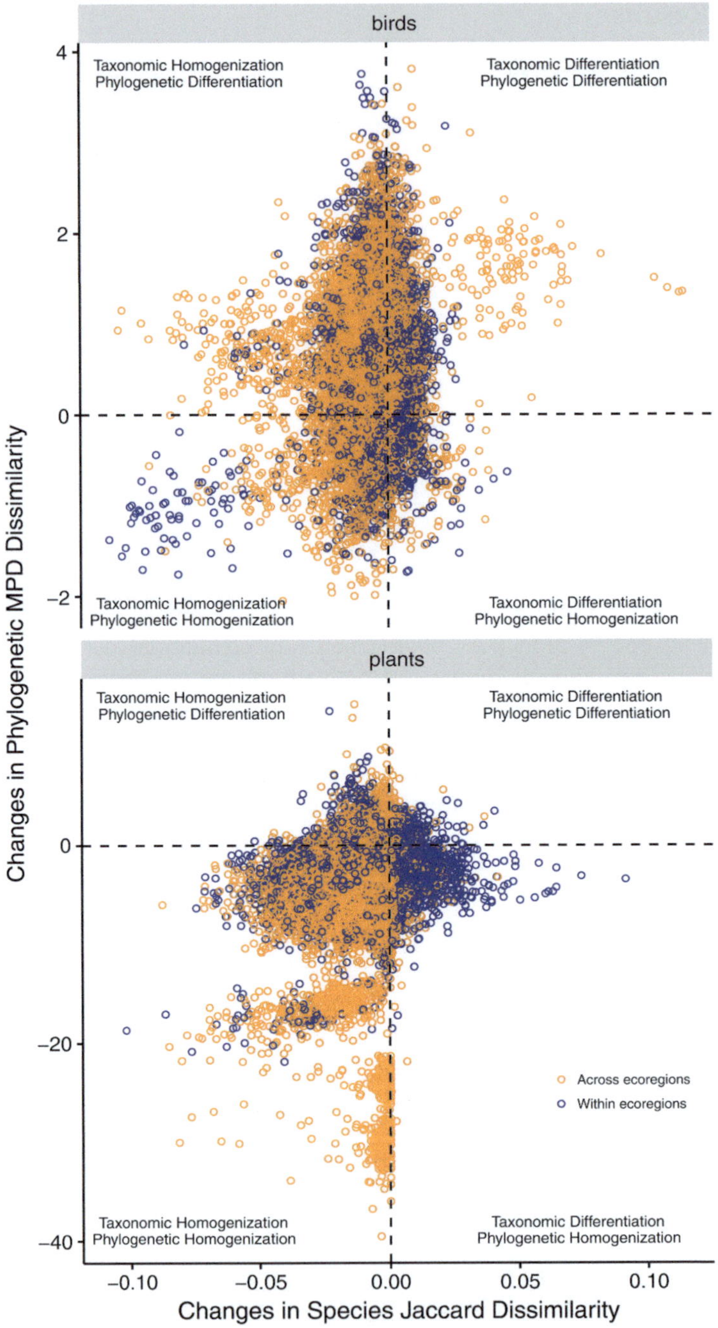

Fig. 17.2 Changes in taxonomic similarity vs phylogenetic similarity in plant and bird communities found within US national parks. Negative values indicate homogenization, whereas positive values represent differentiation

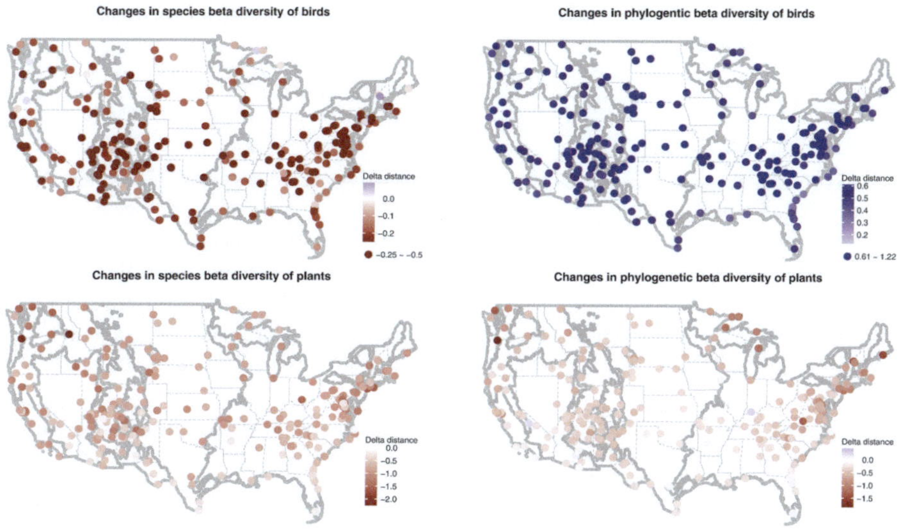

Fig. 17.3 Changes in taxonomic and phylogenetic similarity after including the presence of non-native species within national parks. Each point represents the changes in the average pairwise taxonomic or phylogenetic similarity between one park and all other parks not within the same ecoregion. For each panel, values were scaled to have mean zero and standard deviation one before calculating the changes via distance(native + exotic) – distance(native). Negative values indicate taxonomic/phylogenetic homogenization, while positive values indicate differentiation

17.3.2 Changes in Taxonomic and Phylogenetic Similarity Within Ecoregions and Alaska

Comparing parks within the same ecoregion, both plant and bird taxonomic and phylogenetic composition showed the same responses to non-native species presence as they did across the continent. For plants, all ecoregions experienced both taxonomic and phylogenetic homogenization (Table 17.2). Higher levels of taxonomic homogenization were observed in ecoregions concentrated in the Western United States (e.g., Mediterranean California, Marine West Coast Forest, Northwestern Forested Mountain ecoregions; Table 17.2; Fig. 17.1). Lower levels of taxonomic homogenization and some cases of differentiation were found in parks in the southeastern portion of the Eastern Temperate Forest ecoregion (Table 17.2; Fig. 17.1). Phylogenetic homogenization was highest in the Mediterranean California, Marine West Coast Forest, and Northern Forest ecoregions.

Bird communities in most ecoregions experienced phylogenetic differentiation and taxonomic homogenization (Table 17.2). The exceptions to this pattern were the Temperate Sierras ecoregion, which showed taxonomic and phylogenetic homogenization (with the later not showing statistical significance), and the Northern Forest ecoregion which showed significant taxonomic and phylogenetic differentiation

Table 17.2 Changes in bird and plant species and phylogenetic composition within national parks located across ecoregions

	Birds		Plants	
Ecoregion	Taxonomic	Phylogenetic	Taxonomic	Phylogenetic
Eastern Temperate Forests	Homog.***	Diff.***	Homog.***	Homog.***
Great Plains	Homog.***	Diff.***	Homog.***	Homog.***
Marine West Coast Forest	Homog.	Diff.**	Homog.***	Homog.***
Mediterranean California	Homog.***	Diff.***	Homog.***	Homog.***
North American Deserts	Homog.***	Diff.***	Homog.***	Homog.***
Northern Forest	Diff.*	Diff.***	Homog.***	Homog.***
Northwestern Forested Mountains	Homog.	Diff.***	Homog.***	Homog.***
Temperate Sierras	Homog.**	Homog.	Homog.***	Homog.***

To show the patterns better, we did not include average pairwise beta diversity of native species and native plus exotic species here. Statistics and p-values (not shown) are derived from paired Wilcox tests

Abbreviations: "Diff." indicates differentiation; "Homog." indicates homogenization; *, $p < 0.05$; **, $p < 0.01$; and ***, $p < 0.001$

(Table 17.2). Parks within the Eastern Temperate Forests and Great Plains ecoregions had the highest levels of taxonomic homogenization, while high levels of phylogenetic differentiation were spread across several ecoregions (Table 17.2). Across parks in Alaska, we observed significant biotic differentiation in bird taxonomic and phylogenetic composition (Table 17.1). Plant assemblages taxonomically differentiated and phylogenetically homogenized across Alaskan parks (Table 17.1).

17.4 Discussion

Overall, our results show that bird and plant assemblages in US national parks are becoming more similar in their taxonomic composition due to the establishment of non-native species. Non-native plants have also increased the phylogenetic similarity of plant assemblages across US national parks, while non-native birds have decreased the phylogenetic similarity of bird assemblages. These results hold at the continental and ecoregion scale. Interestingly, parks in Alaska do not conform to same taxonomic pattern and are becoming less similar in composition to one another for both birds and plants. Our results of taxonomic homogenization of birds and plant assemblages across both regional and continental scales mirror those of a meta-analysis of homogenization studies conducted worldwide (Baiser et al. 2012).

Our finding that taxonomic and phylogenetic homogenization are not necessarily coupled shows the importance of exploring multiple aspects of homogenization (Baiser and Lockwood 2011; Monnet et al. 2014). Bird assemblages experienced phylogenetic differentiation in spite of taxonomically homogenizing. This can result from at least two processes. First, it is possible that the suite of invaders that are taxonomically homogenizing a set of assemblages (e.g., birds in US national parks) have different evolutionary histories (i.e., do not have close phylogenetic relationships). As a result, a set of relatively large pairwise phylogenetic distances are added to a set of assemblages. Second, the suite of non-native species that are establishing across a set of locales may have very different evolutionary histories from the native community. Thus, pairwise distances between natives and non-native are relatively large, on average, compared to pairwise distances between natives. Both of these scenarios result in larger pairwise distances across assemblages which is the hallmark of phylogenetic differentiation.

Our results indicate that park visitors are more likely to see the same non-native species when visiting different national parks both within the same ecoregion and across the entire continental United States. For example, park visitors are highly likely to encounter the non-native plant curly dock (*Rumex crispus*) which occurs in 161 national park units or the non-native, invasive European starling (*Sturnus vulgaris*), which occurs in 204 national parks. How the homogenization of bird and plant assemblages alters the perception and recreational use of these important resources is likely dependent on the experience of individual park visitors. While some park visitors take a *laissez-faire* approach to invasive species in national parks, others recognize the threat they pose and advocate for management (Sharp et al. 2011). Visitors who come in contact with an invader may have greater concern regarding the potential threats of such species (Harvey et al. 2016). Although seeing the same non-native species at different parks due to taxonomic homogenization may not influence some park visitors, the implications of non-natives for iconic species (e.g., the loss of hemlock due to the non-native invasive woolly adelgid) may affect park visitation. For example, the decision to visit Yellowstone National Park in Wyoming and Montana was not greatly influenced by the presence of the non-native lake trout (*Salvelinus namaycush*) (Cherry and Shogren 2001). However, the prospect of a decrease in the opportunity to view birds of prey and grizzly bears (*Ursus arctos horribilis*) due to the replacement of Yellowstone cutthroat trout (*Oncorhynchus clarki bouvieri*) by lake trout did influence visitor's proclivity to visit the park (Cherry and Shogren 2001).

In our analysis, we focused on the role of non-native species in homogenizing bird and plant assemblages in US national parks using presence data. However, both species extinctions or extirpations and changing abundances can contribute to patterns of homogenization (Olden and Rooney 2006). The conceptualization of biotic homogenization as "a few winners replacing many losers..." (McKinney and Lockwood 1999) underscores the role of extinction in tandem with the spread of non-native species in the homogenization process. The Earth is experiencing extinctions at an unprecedented rate (Pimm et al. 1995). Because species that have small ranges have a greater risk of extinction (Manne and Pimm 2001; Pimm et al. 2014), it is likely that species unique to specific regions will be lost, especially when considering large spatial extents (e.g., continental or global). Thus, considering the loss of species (through extinction or extirpation) unique to each national park will likely increase the degree in taxonomic homogenization. Furthermore, the fact that extinction tends to target evolutionarily distinct species (Isaac et al. 2007) suggests that extinctions will also lead to phylogenetic homogenization.

Considering species abundance is also an important facet of documenting biotic homogenization. Anthropogenic change can lead to increases in native species that respond well to anthropogenic disturbance (Chace and Walsh 2006) and decreases (but not extirpations) in species that do not. Similarly, some non-natives may occur in high abundances across a set of locales and consequently have a greater homogenizing effect than those that are at low abundances. While quantifying homogenization based on species abundances is ideal, obtaining abundance data for a large number of locales at continental or global spatial extents is prohibitive.

National parks are essential to conservation and recreation in the United States and globally. However, these protected areas face many of the same threats to biodiversity as non-protected areas, including climate change (Hansen et al. 2014; Monahan and Fisichelli 2014; Rodhouse et al. 2016), pollution (Fakhraei et al. 2016), and poaching (Hilborn et al. 2006), and non-native species invasion. These threats are synergistic, show no sign of abating, and have the potential to completely alter ecosystems (Hansen et al. 2014). In order to "… turn over natural resources to the next generation increased; and not impaired in value" as President Theodore Roosevelt said at the beginning of the twentieth century, management of non-native species is crucial.

References

Abella SR (2014) Impacts and management of hemlock woolly adelgid in national parks of the eastern United States. Southeast Nat 13:16–45

Allen JA, Brown CS, Stohlgren TJ (2009) Non-native plant invasions of United States national parks. Biol Invasions 11:2195–2207

Baiser B, Lockwood JL (2011) The relationship between functional and taxonomic homogenization. Glob Ecol Biogeogr 20:134–144

Baiser B, Olden JD, Record S et al (2012) Pattern and process of biotic homogenization in the new pangaea. Proc R Soc Lond B Biol Sci 279:4772–4777

Baiser B, Whitaker N, Ellison AM (2013) Modeling foundation species in food webs. Ecosphere 4:1–14

Baiser B, Valle D, Zelazny Z, Burleigh JG (2018) Non-random patterns of invasion and extinction reduce phylogenetic diversity in island bird assemblages. Ecography 41: 361–374

Blackburn TM, Cassey P, Lockwood JL (2009) The role of species traits in the establishment success of exotic birds. Glob Chang Biol 15:2852–2860

Byerly A (1996) The uses of landscape: the picturesque aesthetic and the national park system. The Ecocriticism Reader: Landmarks in Literary Ecology

Cadotte MW, Murray BR, Lovett-Doust J (2006) Ecological patterns and biological invasions: using regional species inventories in macroecology. Biol Invasions 8:809–821

Chace JF, Walsh JJ (2006) Urban effects on native avifauna: a review. Landsc Urban Plan 74:46–69

Cherry TL, Shogren JF (2001) Invasive species management for the Yellowstone Lake ecosystem: what do visitors think. Yellowstone Sci 9:10–15

de Bello F, Carmona CP, Lepš J et al (2016) Functional diversity through the mean trait dissimilarity: resolving shortcomings with existing paradigms and algorithms. Oecologia 180:933–940

Dorcas ME, Willson JD, Reed RN et al (2012) Severe mammal declines coincide with proliferation of invasive burmese pythons in everglades national park. Proc Natl Acad Sci 109:2418–2422

Dyer EE, Redding DW, Blackburn TM (2017) The global avian invasions atlas, a database of alien bird distributions worldwide. Scientific Data, 4, 170041.

Ellison AM, Bank MS, Clinton BD et al (2005) Loss of foundation species: consequences for the structure and dynamics of forested ecosystems. Front Ecol Environ 3:479–486

Fakhraei H, Driscoll CT, Renfro JR et al (2016) Critical loads and exceedances for nitrogen and sulfur atmospheric deposition in great smoky mountains national park, United States. Ecosphere 7(10):e01466.

Fancy S, Gross J, Carter S (2009) Monitoring the condition of natural resources in us national parks. Environ Monit Assess 151:161–174

Hansen AJ, Piekielek N, Davis C et al (2014) Exposure of us national parks to land use and climate change 1900–2100. Ecol Appl 24:484–502

Harvey RG, Perez L, Mazzotti FJ (2016) Not seeing is not believing: volunteer beliefs about burmese pythons in florida and implications for public participation in invasive species removal. J Environ Plan Manag 59:789–807

Hiebert RD, Stubbendieck JL (1993) Handbook for ranking exotic plants for management and control. Citeseer, Denver

Hilborn R, Arcese P, Borner M et al (2006) Effective enforcement in a conservation area. Science 314:1266–1266

Isaac NJ, Turvey ST, Collen B et al (2007) Mammals on the edge: conservation priorities based on threat and phylogeny. PLoS One 2:e296

Jetz W, Thomas G, Joy J et al (2012) The global diversity of birds in space and time. Nature 491:444–448

Li D, Monahan WB, Baiser B, (2018) Species richness and phylogenetic diversity of native and non-native species respond differently to area and environmental factors. Divers Distrib 24(6):853–864

Lockwood JL (1999) Using taxonomy to predict success among introduced avifauna: relative importance of transport and establishment. Conserv Biol 13:560–567

Manne LL, Pimm SL (2001) Beyond eight forms of rarity: which species are threatened and which will be next? Anim Conserv 4:221–229

Mayo E (1975) Tourism and the national parks: a psychographic and attitudinal study. J Travel Res 14:14–21

McCleery RA, Sovie A, Reed RN et al (2015) Marsh rabbit mortalities tie pythons to the precipitous decline of mammals in the everglades. Proc Royal Society 282:20150120

McKinney ML (2004) Measuring floristic homogenization by non-native plants in North America. Glob Ecol Biogeogr 13:47–53

McKinney ML, Lockwood JL (1999) Biotic homogenization: a few winners replacing many losers in the next mass extinction. Trends Ecol Evol 14:450–453.566

Monahan WB, Fisichelli NA (2014) Climate exposure of us national parks in a new era of change. PLoS One 9:e101302

Monnet A-C, Jiguet F, Meynard CN et al (2014) Asynchrony of taxonomic, functional and phylogenetic diversity in birds. Glob Ecol Biogeogr 23:780–788

Muir J (1901) Our national parks. Houghton Mifflin, New York

Olden JD, Poff NL (2003) Toward a mechanistic understanding and prediction of biotic homogenization. Am Nat 162:442–460

Olden JD, Rooney TP (2006) On defining and quantifying biotic homogenization. Glob Ecol Biogeogr 15:113–120

Olden JD, Poff NL, Douglas MR et al (2004) Ecological and evolutionary consequences of biotic homogenization. Trends Ecol Evol 19:18–24

Omernik JM (1987) Ecoregions of the conterminous United States. Ann Assoc Am Geogr 77:118–125

Omernik JM, Griffith GE (2014) Ecoregions of the conterminous United States: evolution of a hierarchical spatial framework. Environ Manag 54:1249–1266

Parrish JD, Braun DP, Unnasch RS (2003) Are we conserving what we say we are? Measuring ecological integrity within protected areas. Bioscience 53:851–860
Pimm SL, Russell GJ, Gittleman JL, Brooks TM (1995) The future of biodiversity. Science 269:347
Pimm SL, Jenkins CN, Abell R et al (2014) The biodiversity of species and their rates of extinction, distribution, and protection. Science 344:1246752
Pyšek P, Richardson DM (2008) Traits associated with invasiveness in alien plants: where do we stand? In: Biological invasions. Springer, Berlin, pp 97–125
Pysek P, Pergl J, Essl F, Lenzner B, Dawson W, Kreft H et al (2017) Naturalized alien flora of the world: species diversity, taxonomic and phylogenetic patterns, geographic distribution and global hotspots of plant invasion. Preslia 89(3):203–274
Qian H, Ricklefs RE (2006) The role of exotic species in homogenizing the North American flora. Ecol Lett 9:1293–1298
Rodhouse TJ, Philippi TE, Monahan WB, Castle KT (2016) A macroecological perspective on strategic bat conservation in the US National Park Service
Roosevelt T (1910) Speech given at the Colorado Livestock Association in Denver on 29 Aug 1910. http://www.theodorerooseveltcenter.org/Learn-About-TR/TR-Quotes?page=3. Downloaded 17 Mar 2018
Runte A (1997) National parks: the American experience. University of Nebraska Press, Lincoln
Sackett TE, Record S, Bewick S et al (2011) Response of macroarthropod assemblages to the loss of hemlock (tsuga canadensis), a foundation species. Ecosphere 2:1–16
Sellars RW (2009) Preserving nature in the national parks: a history: with a new preface and epilogue. Yale University Press, New Haven
Sharp RL, Larson LR, Green GT (2011) Factors influencing public preferences for invasive alien species management. Biol Conserv 144:2097–2104
Stohlgren TJ, Loope LL, Makarick LJ (2013) Invasive plants in the United States national parks. In: Foxcroft LC, Pyšek P, Richardson DM, Genovesi P, editors. Plant invasions in protected areas: Patterns, problems and challenges. Springer, Dordrecht, pp 267–283
Tingley MW, Orwig DA, Field R, Motzkin G (2002) Avian response to removal of a forest dominant: Consequences of hemlock woolly adelgid infestations. J Biogeogr 29:1505–1516
Van Mantgem PJ, Stephenson NL, Byrne JC et al (2009) Widespread increase of tree mortality rates in the western United States. Science 323:521–524
Webb CO, Donoghue MJ (2005) Phylomatic: tree assembly for applied phylogenetics. Mol Ecol Notes 5:181–183
Webb CO, Ackerly DD, McPeek MA, Donoghue MJ (2002) Phylogenies and community ecology. Annu Rev Ecol Syst 33:475–505
Zanne AE, Tank DC, Cornwell WK et al (2014) Three keys to the radiation of angiosperms into freezing environments. Nature 506:89–92

Chapter 18
Homogenization of Fish Assemblages Off the Coast of Florida

Alexandrea Dagmar Safiq, Julie L. Lockwood, and Jeffrey A. Brown

Abstract Biological homogenization can alter the ecological function of systems as well as the economic value associated with those ecosystems through complex socio-ecological dynamics. The aim of this chapter is twofold: (1) to document evidence of biological homogenization across marine fish assemblages off the coast of Florida and (2) to discuss how social values may influence, and are influenced by, the biological homogenization of coastal fish assemblages. We measured biological homogenization by tracking taxonomic changes over a decade across 13 near-shore sites off the Atlantic coast of Florida. We created species-location matrices for each site, calculated recently and a decade prior, and quantitatively depicted assemblage similarity changes between sites using a hierarchical clustering algorithm. We found evidence of biological homogenization of some fish assemblages, but not all, and relatively little change in site species richness. Sites that were closer to populated coastlines, or have been subject to substantial disturbance events, are more likely to show homogenization. Protected reef sites show little evidence of homogenization. We postulate feedback mechanisms between societal values, diver practices, diver experience, and the severity of homogenization. We suggest that cultural values directly influence diver behavior, which in turn can affect assemblage homogenization. These socio-ecological feedbacks have received very little attention in the context of coastal fish conservation, but deserve attention given the perilous state of such ecosystems worldwide.

Keywords Ecotourism · Diving · Marine · Reefs · Fish · Homogenization · Florida · Caribbean

A. D. Safiq (✉) · J. L. Lockwood · J. A. Brown
Department of Ecology, Evolution, and Natural Resources, Rutgers University, New Brunswick, NJ, USA
e-mail: ads226@scarletmail.rutgers.edu; julie.lockwood@rutgers.edu

R. Rozzi et al. (eds.), *From Biocultural Homogenization to Biocultural Conservation*, Ecology and Ethics 3, https://doi.org/10.1007/978-3-319-99513-7_18

18.1 Introduction

Marine fish assemblages have been substantially altered through the effects of overfishing, habitat loss, and climate change (Levin et al. 2006; Airoldi et al. 2008; Nyitrai et al. 2012; Riegl et al. 2009). Such impacts have been measured almost exclusively in the context of loss of key species or fish biomass. Rarely has anyone measured these impacts through the lens of compositional change, which is a perspective that acknowledges that species losses and additions combine to increase or decrease spatial similarity in marine fish assemblages through time (Olden and Rooney 2006). The few studies that have adopted this perspective have highlighted the homogenizing influence of habitat simplification and phase shifts, especially among coral reefs (Thrush et al. 2006; Airoldi et al. 2008; Alvarez-Filip et al. 2015), or of heavy influxes of exotic species after the establishment of inter-ocean canals (Edelist et al. 2013). The data necessary to track longer-term changes in composition are hard to come by but especially so within marine ecosystems which are by default harder to track due to limited logistical access. Here we take advantage of a citizen science initiative to document near-shore fish assemblages to explore the extent to which we see homogenization within the Atlantic coastal waters of Florida, USA.

Florida has nearly 12,000 miles of coastline, which includes the string of small islands that lie at the southern extreme of the state called the Keys (Fig. 18.1). The near-shore marine waters of the Florida's Atlantic coast and Keys exhibit a diverse array of seabed types including significant expanses of coral and artificial reefs. These reefs, and other habitat types, are home to hundreds of fish species, some of which are important sources of commercial seafood or sold in the marine aquarium trade (Johns et al. 2014; Bruckner 2005). These fish assemblages also support a massive SCUBA and snorkel dive industry, which generates on average US$3 billion annually in the Keys alone (World Wildlife Fund Global, Accessed 12 June 2017). These reefs, and other near-shore marine habitats, thus represent significant sources of ecosystem services for the region (Lane et al. 2015).

Caribbean reefs, of which the Florida group are the northern-most members, have suffered several major disturbance events over the last decades including bleaching, the loss of top predators, and the emergence of diseases that have reduced reef structural complexity (Manzello et al. 2007; Green and Bruckner 2000). The fish that use these reefs have shown mixed responses to such changes, with some assemblages showing signs of recovery within a decade or less, with others showing strong lags whereby fish assemblages have yet to recover to their pre-disturbance composition and biomass (Alevizon and Porter 2015). There is recent evidence showing Caribbean-wide shifts in fish assemblages toward habitat generalists and away from specialists (Alvarez-Filip et al. 2015) and overall loss in fish abundance (Paddack et al. 2009). The reefs along the Florida Keys were included in these assessments; however, no one has explored whether the spatial similarity of these reefs has increased through time or if these reefs have become more similar to other non-reef habitats along Florida's Atlantic coast. We evaluate this possibility here, along with tracking how these reefs compare in their similarity to other near-shore habitats that are important for marine fishes.

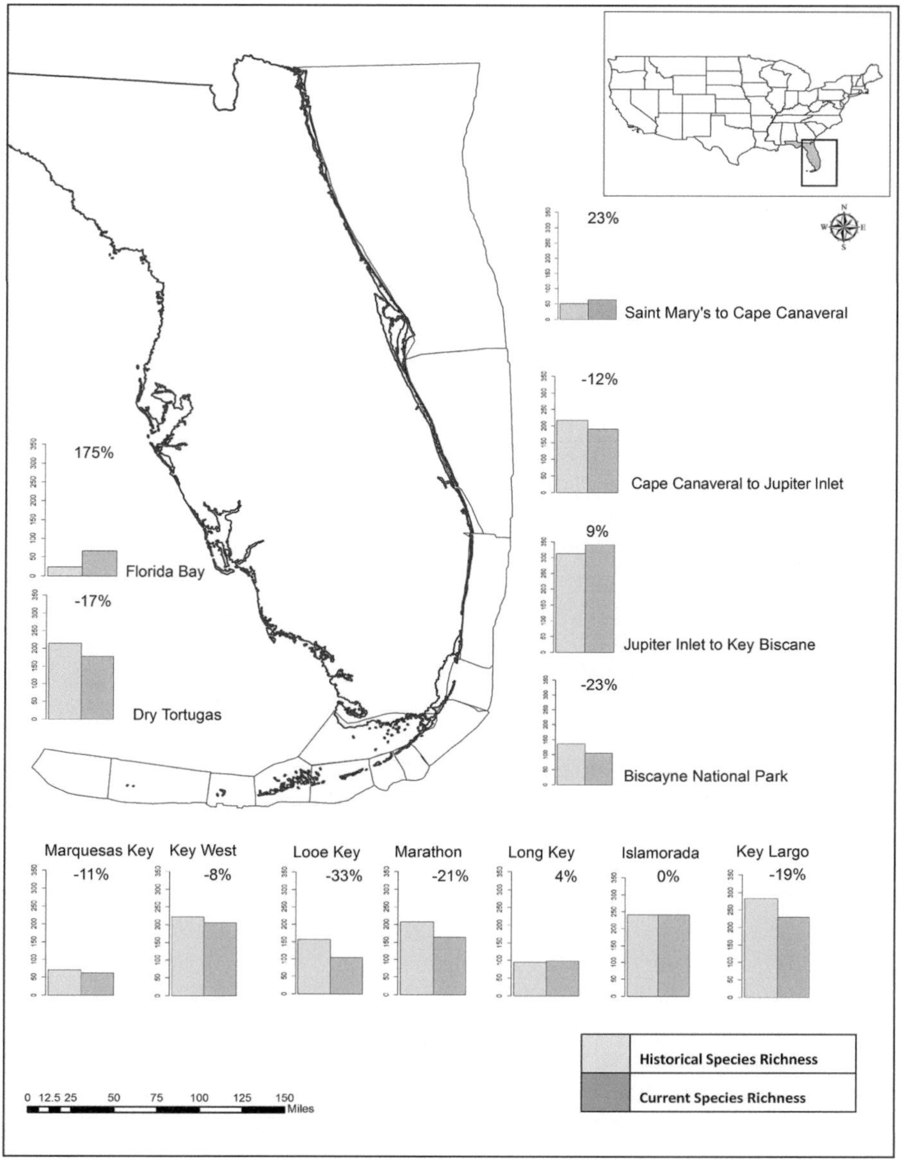

Fig. 18.1 Map of Florida showing the 13 REEF sites (sub-zones) considered in this study. Bar graphs correspond to site, with historical species richness in light gray and current species richness in dark gray. Overall changes in species richness from historical to current time frames are reported as percentages. Positive percentages indicate net gain in species richness per site, while negative percentages indicate net loss in species richness

18.2 Methods

18.2.1 Surveys

We queried the Reef Environmental Education Foundation (REEF) online database to obtain presence-absence data from across survey locations along the Atlantic coast of Florida (World Wide Web electronic publication; www.reef.org, date of download: 12 May 2017; Fig. 18.1). Our data comes from the REEF Florida East Coast and Keys (zone 3) of the tropical western Atlantic region. This zone covers survey locations from Saint Mary's River at the northern Atlantic coast boundary between Florida and Georgia, down to the Dry Tortugas, which is the western-most island group in the Keys (Fig. 18.1). Within this zone, there are 13 sub-zones within which surveys are conducted at various locations. Survey locations include natural coral reefs, artificial reefs, and other seafloor structures that attract relatively large numbers of marine fish (e.g., natural rubble, marine ridges).

In all REEF surveys, fish species presence is recorded by trained volunteer SCUBA divers using the Roving Diver Technique (RDT; Schmitt and Sullivan 1996, Hassell et al. 2013). Between 1994 and 2015, REEF volunteers conducted 37,695 RDT surveys at 1903 survey locations, summing to 34,305 dive hours with the average dive lasting 45 min. There are large differences in the total number of locations surveyed across sub-zones (from 14 to 502), with bigger sub-zones having more survey locations. The number of species detected is highly dependent on survey effort (in this case measured as dive time, and numbers of locations are surveyed); thus, we needed to standardize effort across the years within sub-zones to ensure relatively unbiased temporal assessments of compositional change. We chose years to include in our analyses so that they had dive times (effort) that were no more than 20% different in length. Based on this criterion, we selected 353 survey locations for inclusion in our analyses (20% of all sites surveyed). Initial surveys were conducted between the years of 1997 and 2004 and the later resampling surveys occurred from 2011 to 2017; thus, our data reflects on average a 20-year time span. We combined the species lists for each of these survey locations to create a scaled-up accounting of species composition across the 13 sub-zones, hereafter referred to as "sites" (Fig. 18.1). Note that there was necessarily a difference in the number of survey location information contained within each site, and thus our results cannot be fully corrected for differences in sampling intensity on species occupancy across sites.

18.2.2 Species Composition

We created two lists of species present for each of the 13 sites we analyzed. The first was composed only of species recorded as present in recent surveys ("current" time frame), and the second was only the species present in surveys at least 10 years prior ("historical" time frame). We used the cluster analysis R-package *hclust* to produce Bray-Curtis dissimilarity scores between all sites, calculated using only the historical data and then again using the current data. We depicted similarity across all 13 sites within each time frame using a dendrogram where sites joined by short branches have very similar species composition, and longer branch lengths represent increasingly divergent species composition. Clusters of sites connected by short branch lengths are referred to as supergroups as these sites have very similar species composition to each other relative to other sites in the analysis. As a coarse measure of degree of homogenization, we calculated the overall branch lengths of each of these two dendrograms (historical and current). If sites have become more similar overall in species composition (homogenized), we expected to see that the current dendrogram has a lower summed branch length than the historical dendrogram; and we would take the opposite pattern (longer summed branch lengths) as evidence of differentiation. We can also more closely track how the cluster affiliation of each site may have changed between time frames by comparing the higher-order branching patterns between these two dendrograms. We would take as evidence of homogenization a shift toward branch nodes that indicated higher similarity (shift in branch location to the right between time frames). In contrast, substantive differentiation would be characterized by an increased dissimilarity score at higher-order branching nodes (shift in branch location to the left). This more detailed comparison of the two dendrograms allows us to pinpoint sites that are homogenizing, those that are not changing, and those that are differentiating.

There are several studies that suggest that species composition of a region can change substantially through time, while species richness remains largely unchanged (Dornelas et al. 2014). There is also a presumption within many discussions about biological homogenization that sites tend to become more similar in composition due mostly to the increasing presence of ubiquitous exotic species rather than the loss of endemic native species (Olden and Rooney 2006). To explore if these patterns were present in fish assemblages within Floridian coastal waters, we calculated historical and current species richness within each site. We calculated species richness as the simple sum of all species recorded as present within each site, repeating this calculation for each time frame. We also labeled each species as either native or exotic using USGS Nonindigenous Aquatic Species database (nas.er.usgs.gov).

Changes in richness within a site over time can occur in two ways. Species can either be absent from the historical list and present in the current or vice versa. The spatial pattern by which these temporal shifts in species presence occur dictates the extent to which compositional similarities are altered (Olden and Poff 2004; Olden and Rooney 2006). Thus, for example, across-site compositional similarity can increase (homogenize) through the addition of the same species across all sites across time frames; or similarity can increase when species found only in one or a few sites disappear between time frames. Differentiation can occur if formerly common species are lost in only a few scattered sites or if species enter the presence-absence record at only one or a few sites. In order to track such changes, we recorded the number of sites in which each of the 452 species was lost or gained between time frames. We then calculated the percentage change in site occupancy between time frames by dividing the change in number of sites by the number occupied in the historical time frame. We plotted all species according to their percentage change in site occupancy, as well as the number of sites they occupied in the historical time frame. This graphic allows one to visualize how individual species contribute to homogenization or differentiation patterns observed over time.

18.3 Results

Overall, sites along Florida's Atlantic coast and Keys experienced a 67% net gain in species richness, with Florida Bay experiencing the largest increase, over the 20-year time span of our data (Fig. 18.1). Most other sites experienced a slight loss in species richness or very little change through time (Fig. 18.1). Fifty-six species were observed within at least one site in the historical time frame, but were not observed in the current time frame. Over half (29) of these "disappearing" species were originally seen only within one site, suggesting that their loss from the current time species list is due to lack of detection and not true loss. However, 11 species were found within 3 or more sites in the historical time frame but not at any site in the current time frame, which is not as likely due to lack of detection. Seventy-two species were absent in the historical time frame but were recorded as present in the current period. Over two-thirds (55) of these "appearing" species were found only within one site in the current time frame, suggesting that they were rare in the historical period and not detected and are now experiencing local population increases. Nine species appear for the first time within three or more sites in the current time frame indicating rapid increases in local populations. Only two of these species are exotics: the Pacific lionfish (*Pterois volitans*) and brassy chub (*Kyphosus vaigiensis*). These exotic species account for the largest percentage increases in number of sites occupied within Fig. 18.2. The remaining seven species are considered native to the United States according to USGS. The reasons why these, and other increasingly observed species, are now found more commonly in the near-shore marine waters of Florida is not clear but may be due to either climate change or response of a narrow suite of local fish populations that are favored by the massive habitat alterations that took place prior to the time frame of our data (Alvarez-Filip et al. 2015; Lamy et al. 2016).

Of the 324 species that were present within at least 1 site across both time frames, 90 showed no change in the number of sites they occupied, with another 96 showing <20% shifts (positive or negative) (Fig. 18.2). Most of the 82 species that expanded the number of sites they occupied through time moved into only 1 more site, but 6 increased by 3 or more sites. No species that were originally found within four or more sites expanded to occupy other sites.

In terms of species composition, total branch length for the historical dendrogram was 4.3, whereas total length for the current dendrogram was 4.5, indicating slight overall differentiation in fish assemblage composition across sites through

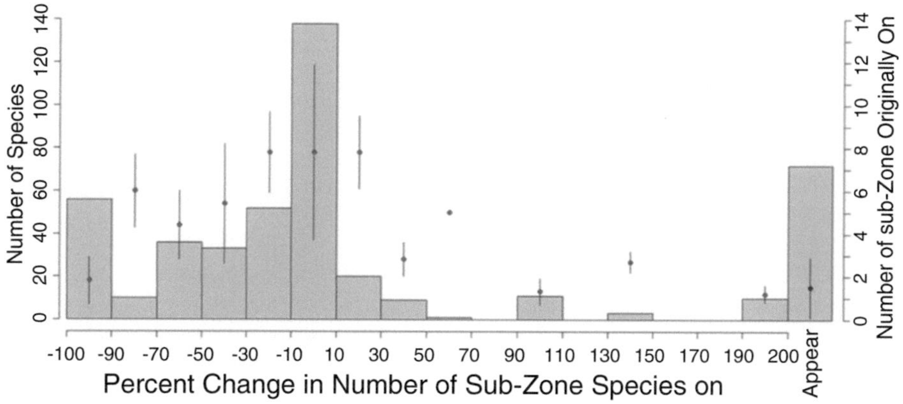

Fig. 18.2 Histogram showing the number of species that show a percentage change in number of sub-zones (sites) occupied between the historical and current time periods. Negative percentage change indicates species' range contraction, and positive change indicates species' range expansion. The y-axis indicates the number of species that fall within each percentage change bin. Also shown on the y-axis is the average number of sub-zones (sites) species within each percentile grouping occupied in the historical time period (red dots = average, bars = standard deviation). Many changes in percentage of sub-zones occupied are likely due to failure to detect species in surveys. However, large percentage decreases from species that occupied many sub-zones historically are likely true species declines, and large percentage gains from species that occupied few sub-zones historically likely represent true range expansions

time (Fig. 18.3). The Florida Bay site was consistently unique in fish assemblage composition as compared to all other sites no matter the time frame. The shift of the Saint Mary's Reef to Cape Canaveral site to become a member of the Looe Key, Biscayne National Park, Long Key, and Marquesas Key supergroup indicates that it currently has a species composition substantially more similar to the supergroup than it did historically. In contrast, the Jupiter Inlet to Key Biscayne site shifts from grouping with the other reef-dominated supergroup to becoming its own unique branch indicating that it has become much more distinct in species composition between time frames (Fig. 18.3). More subtle changes in compositional similarity occur within a supergroup consisting of the Marathon, Islamorada, Key West, Dry Tortugas, Key Largo, and Cape Canaveral to Jupiter Inlet sites (Fig. 18.3). Within this supergroup, there is a shift toward the sites becoming more similar to each other in their species composition across time frames, indicating homogenization (Fig. 18.3).

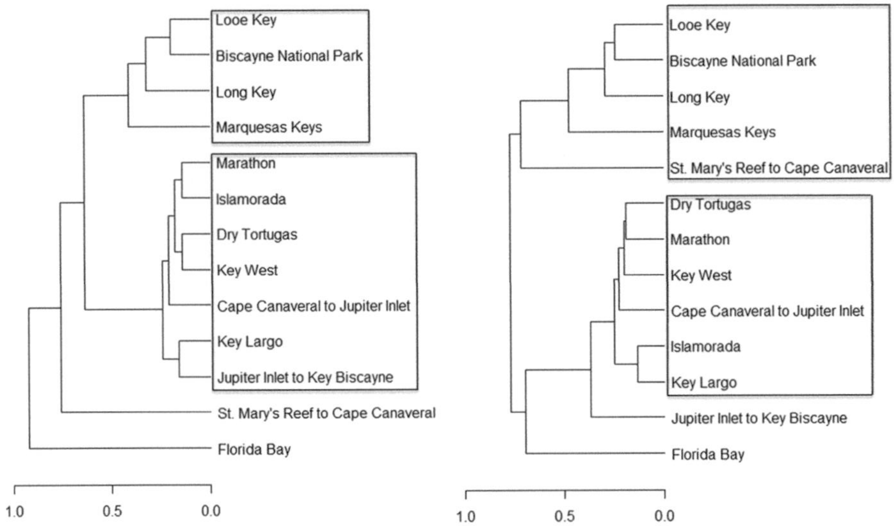

Fig. 18.3 Two dendrograms comparing fish assemblage similarity between reefs from historical (left) and current (right) time frames. Shorter branch lengths indicate higher degree of similarity, while longer branch lengths indicate lower degree of similarity, with similarity derived using Bray-Curtis scores (see main text)

18.4 Discussion

We show that fish assemblages along the Atlantic coast of Florida and the Keys have shifted in their assemblage composition through time in a spatially complex manner but without experiencing large changes in species richness. Although there has been consistent focus on the loss of species and biomass within marine ecosystems, the possibility of changes in spatial diversity has largely gone unexplored (c.f., Alvarez-Filip et al. 2015; Edelist et al. 2013). We found that the Florida Atlantic coast fish assemblage is characterized by larger-scale differentiation; however, at a smaller scale, some regions have homogenized. In particular, the Jupiter Inlet to Key Biscayne site substantially differentiated from the others through time, while a supergroup of mostly coral-dominated sites homogenized. This pattern contrasts to what we observed among another supergroup of coral-dominated sites that experienced approximately similar changes in richness through time, but did not homogenize. These non-changing reef sites are some of the more geographically isolated of the set (e.g., Marquesas Key) or are under active protection as national marine sanctuaries or national parks (e.g., Looe Key and Biscayne National Park), which may explain their stasis in fish assemblage compositions over the time frame of our investigation.

The shifts in assemblage similarity we did observe are driven mostly by species losses and not by species colonization. Although some species are expanding into new sites, this spread seems to be smaller in spatial extent than the range contractions that many other species are undergoing. The two species that spread the most across time frames were exotic species; however, we recorded a very few exotic species in our dataset. Our results suggest the need to more closely examine the fish assemblages associated with the coral reefs around the Keys, Biscayne Bay, and scattered along the most southern Florida coastline. Narrowing the habitat focus of analysis, and thus also the suite of fish species considered, should increase clarity in regard to the role of changes in coral cover, predator diversity, and ocean temperatures that may be driving homogenization across habitats.

Finally, we posit that changes in fish assemblage spatial structure will influence the attractiveness of Florida marine habitats for dive tourism. One of the main motivations for diving any area is to experience the variety of fish species that use that site, with the more species variety, the better (Bhat 2003). Homogenization of fish assemblages across sites may provide a strong feedback into the dive tourism industry by reducing the attractiveness of Florida's diving locations. From a recreational diver's perspective, homogenization may lead one to conclude that once fish at one location have been seen, there is no reason to travel to other sites that have essentially the same diversity. However, this is only one factor in a complex matrix of factors that may go into a diver's decision to move from one site to another (Biggs et al. 2015). To our knowledge, the degree to which homogenization (or differentiation) influences the attractiveness of reefs and other habitats for dive tourism has not been explored. Therefore, we see our results as providing a starting point for this analysis by documenting the degree of homogenization realized among Florida's fish assemblages, in which parts of Florida contribute most to any observed homogenization and which species are "winning" or "losing."

We envision a feedback between fish assemblage composition and the cultural connections people maintain with these assemblages as a two-way interaction (Fig. 18.4a, b). Existing research links diver activity to their efforts to conserve the species and habitats they enjoy (Dearden et al. 2007; Arin and Kramer 2002), and this provides a framework for understanding when spatial compositional changes will elicit conservation actions from divers and when it will not. For example, novice divers may fail to recognize assemblage change since they are unfamiliar with how unique any given site was previously from all others in terms of fish assemblage. They thus likely will not experience an emotional response to experiencing a homogenized (or differentiated) fish assemblage and will not go on to engage in behaviors that lead to remediation of the site (Anderson and Loomis 2012). Such divers may also fail to recognize their own behaviors are contributing to fish assemblage change or species losses, and they may have low social inclination toward habitat conservation. These feedbacks for novice divers thus either fail to stem changes that lead to assemblage change or even encourage further change (Fig. 18.4a).

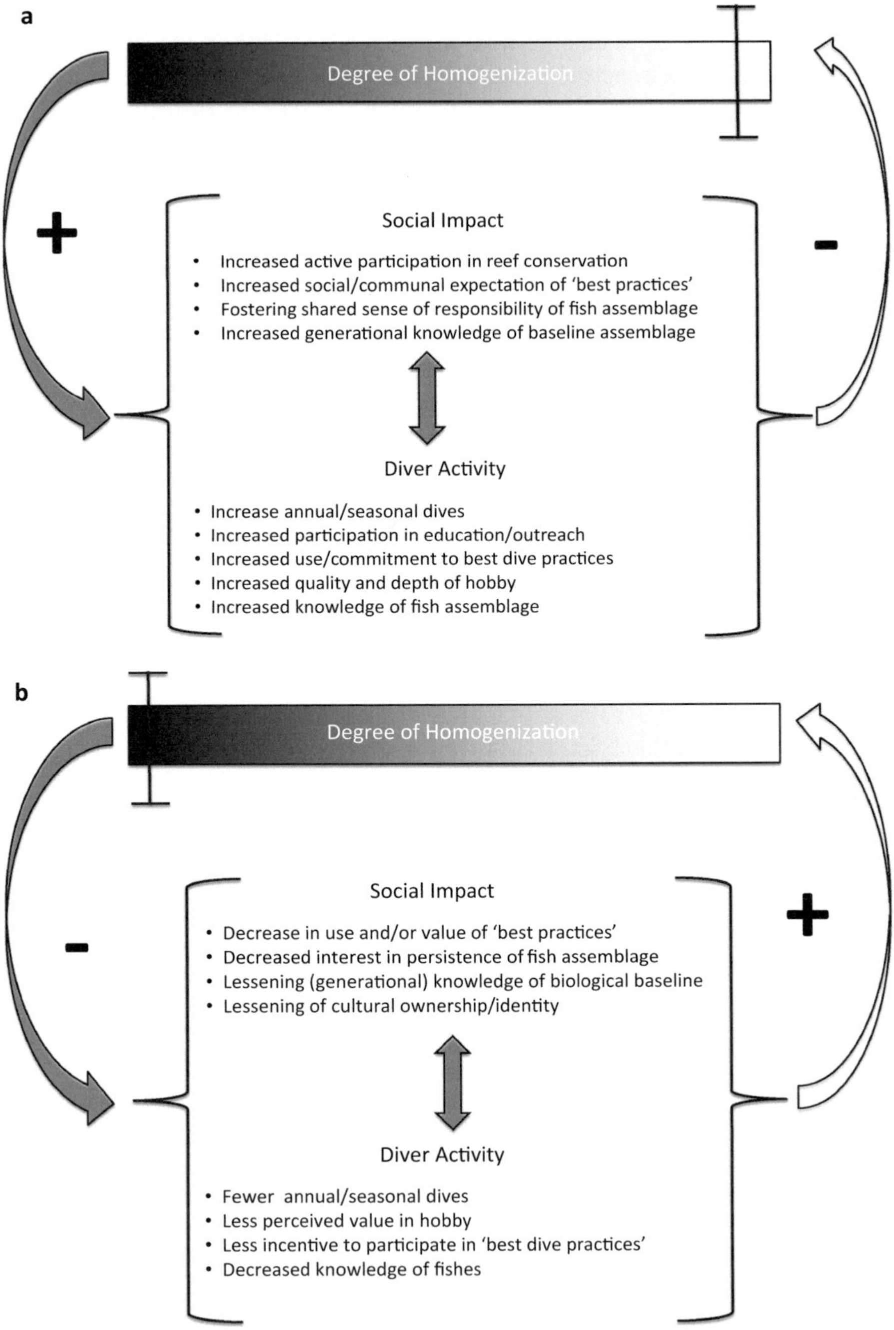

Fig. 18.4 Conceptual diagrams illustrating the interplay of SCUBA and snorkel diver cultural and social behavior and its feedback with biological homogenization. We envision that divers' Fig. 18.4

Divers that regularly visit the same sites over many years have an expectation of what a healthy baseline fish assemblage looks like (Anderson and Loomis 2012). When that baseline begins to change, more experienced divers undergo an emotional response that may result in changes of their behavior (Fig. 18.4b). Those who rely on ecotourism are some of the most politically active demographics, which is reflected in the often mandatory requirement of state natural resource managing bodies to have individuals on the boards of these stakeholders (South Atlantic Fishery Management Council. Accessed 14 June, 2017). If conservation and natural resource managers aim to find compromises that satisfy both conservationists and stakeholders, we must to take into account the cultural and economic influences that are associated with natural systems. Future research on the cultural and social effects on biological homogenization should aim to gather information from individuals who engage with fish assemblages in both a casual and dedicated manner and connect their perceptions to what species they observe to their willingness to support conservation and management actions.

References

Airoldi L, Balata D, Beck MW (2008) The gray zone: relationships between habitat loss and marine diversity and their applications in conservation. J Exp Mar Biol Ecol 366(1–2):8–15. https://doi.org/10.1016/j.jembe.2008.07.034

Alevizon WS, Porter JW (2015) Coral loss and fish guild stability on a Caribbean coral reef: 1974-2000. Environ Biol Fish 98(4):1035–1045. https://doi.org/10.1007/s10641-014-0337-5

Alvarez-Filip L, Paddack MJ, Collen B, Robertson DR, Cote IM (2015) Assemblages over decades of coral reef degradation. PLoSONE 10:e0126004

Anderson LE, Loomis DK (2012) Normative standards for coral reef conditions: a comparison of SCUBA divers by specialization level. J Leis Res 44(2):257–274

Arin T, Kramer RA (2002) Divers' willingness to pay to visit marine sanctuaries: an exploratory study. Ocean Coast Manag 45(2-3):171–183. https://doi.org/10.1016/s0964-5691(02)00049-2

Bhat MG (2003) Application of non-market valuation to the Florida Keys marine reserve management. J Environ Manag 67(4):315–325. https://doi.org/10.1016/s0301-4797(02)00207-4

Biggs D, Hicks CC, Cinner JE, Hall CM (2015) Marine tourism in the face of global change: the resilience of enterprises to crises in Thailand and Australia. Ocean Coast Manag 105:65–74

Bruckner AW (2005) The importance of the marine ornamental reef fish trade in the wider Caribbean. Rev Biol Trop 53:127–137

Fig. 18.4 (continued) experience levels will play a strong role in how they respond to homogenization, depicted as a bar above the diagram with increases from right (light) to left (dark). Diver activity levels (experienced or novice) tend to engender particular behaviors relative to adherence to "best practices," interest in fish species ecology, knowledge of biological baselines, and claiming cultural identity in the fish themselves or dive site. Given this connection, we posit that experienced divers will respond with increased conservation vigilance when realizing that fish assemblages they frequently encounter are homogenizing. The increase in conservation action will then tend to decrease homogenization levels (panel A). In contrast, novice divers may not notice that assemblages have homogenized and thus will do nothing to prevent further homogenization or even act in ways that further increase homogenization (panel B)

Dearden P, Bennett M, Rollins R (2007) Perceptions of diving impacts and implications for reef conservation. Coast Manag 35(2-3):305–317. https://doi.org/10.1080/08920750601169584

Dornelas M, Gotelli NJ, McGill B, Shimadzu H, Moyes F, Sievers C, Magurran AE (2014) Assemblage time series reveal biodiversity change but not systematic loss. Science 344(6181):296–299. https://doi.org/10.1126/science.1248484

Edelist D, Rilov G, Golani D, Carlton JT, Spanier E (2013) Restructuring the Sea: profound shifts in the world's most invaded marine ecosystem. Divers Distrib 19(1):69–77. https://doi.org/10.1111/ddi.12002

Green EP, Bruckner AW (2000) The significance of coral disease epizootiology for coral reef conservation. Biol Conserv 96(3):347–361. https://doi.org/10.1016/s0006-3207(00)00073-2

Hassell NS, Williamson DH, Evans RD, Russ GR (2013) Reliability of non-expert observer estimates of the magnitude of marine reserve effects. Coast Manag 41(4):361–380. https://doi.org/10.1080/08920753.2013.804028

Johns G, Lee DJ, Leeworthy V, Boyer J, Nuttle W (2014) Developing economic indices to assess the human dimensions of the South Florida coastal marine ecosystem services. Ecol Indic 44:69–80. https://doi.org/10.1016/j.ecolind.2014.04.014

Lamy T, Galzin R, Kulbicki M, de Loma TL, Claudet J (2016) Three decades of recurrent declines and recoveries in corals belie ongoing change in fish assemblages. Coral Reefs 35(1):293–302. https://doi.org/10.1007/s00338-015-1371-2

Lane D, Jones R, Mills D, Wobus C, Ready RC, Buddemeier RW, English E, Martinich J, Shouse K, Hosterman H (2015) Climate change impacts on freshwater fish, coral reefs, and related ecosystem services in the United States. Clim Chang 131(1):143–157. https://doi.org/10.1007/s10584-014-1107-2

Levin PS, Holmes EE, Piner KR, Harvey CJ (2006) Shifts in a Pacific ocean fish assemblage: the potential influence of exploitation. Conserv Biol 20(4):1181–1190. https://doi.org/10.1111/j.1523-1739.2006.00400.x

Manzello DP, Berkelmans R, Hendee JC (2007) Coral bleaching indices and thresholds for the Florida Reef Tract, Bahamas, and St. Croix, US Virgin Islands. Mar Pollut Bull 54(12):1923–1931. https://doi.org/10.1016/j.marpolbul.2007.08.009

Nyitrai D, Martinho F, Dolbeth M, Baptista J, Pardal MA (2012) Trends in estuarine fish assemblages facing different environmental conditions: combining diversity with functional attributes. Aquat Ecol 46(2):201–214. https://doi.org/10.1007/s10452-012-9392-1

Olden JD, Poff NL (2004) Ecological processes driving biotic homogenization: testing a mechanistic model using fish faunas. Ecology 85(7):1867–1875. https://doi.org/10.1890/03-3131

Olden JD, Rooney TP (2006) On defining and quantifying biotic homogenization. Glob Ecol Biogeogr 15(2):113–120. https://doi.org/10.1111/j.1466-822x.2006.00214.x

Paddack MJ, Reynolds JD, Consuelo A, Appeldoorn RS, Beets J et al (2009) Recent region-wide declines in Caribbean reef fish abundance. Curr Biol 19:591–595

Riegl B, Bruckner A, Coles SL, Renaud P, Dodge RE (2009) Coral reefs threats and conservation in an era of global change. In: Ostfeld RS, Schlesinger WH (eds) Year in ecology and conservation biology 2009, vol 1162. Annals of the New York Academy of Sciences, pp 136–186. doi:https://doi.org/10.1111/j.1749-6632.2009.04493.x

Schmitt EF, Sullivan KM (1996) Analysis of a volunteer method for collecting fish presence and abundance data in the Florida keys. Bull Mar Sci 59(2):404–416

Thrush SF, Gray JS, Hewitt JE, Ugland KI (2006) Predicting the effects of habitat homogenization on marine biodiversity. Ecol Appl 16(5):1636–1642. https://doi.org/10.1890/1051-0761(2006)016[1636,pteohh]2.0.co;2

Part III
Biocultural Conservation

Chapter 19
Biocultural Conservation and Biocultural Ethics

Ricardo Rozzi

Abstract Sustainable forms of co-inhabitation in this world are not only possibilities; they are actualities. However, for their expression, it is essential to undertake a twofold task: (1) to precisely and severely sanction those agents who act guided by a self-absorbed economic interest threating the sustainability of life, and (2) to decisively defend those traditions of thought and communities who favor the continuity of life in its diversity of biological and cultural expressions. For the second task to be undertaken effectively, it is critical to understand that the conservation of, and the access to, the diverse native habitats is the condition of possibility for the continuity of the diverse and sustainable life habits of communities of co-inhabitants that inhabit them. The conservation of habitats and life habits is so critical that it constitutes an ethical imperative that should be incorporated into government policies as a matter of socio-environmental justice. To implement this ethical imperative, it is essential to reorient global society to foster a culture that achieves a better integration of ontological-, ecosocial-, and ethical-biocultural foundations into education, policies, and governance. This triple integration aims to contribute to more fully understanding pressing socio-environmental problems and to more effectively implement biocultural conservation. In order to contribute to this triple integration, we offer the "3Hs conceptual lens" of the biocultural ethic to re-cognize and re-value the multiplicity of ecological worldviews, practices, and values hosted by diverse cultures in heterogeneous regions of the planet that contribute to the sustainability of life.

Keywords Education · Governance · Policy · Traditional ecological knowledge · Worldviews

---

R. Rozzi (✉)
Department of Philosophy and Religion and Department of Biological Sciences, University of North Texas, Denton, TX, USA

Sub-Antarctic Biocultural Conservation Program, University of North Texas, Denton, TX, USA

Instituto de Ecología y Biodiversidad and Universidad de Magallanes, Punta Arenas, Chile
e-mail: Ricardo.Rozzi@unt.edu

R. Rozzi et al. (eds.), *From Biocultural Homogenization to Biocultural Conservation*, Ecology and Ethics 3, https://doi.org/10.1007/978-3-319-99513-7_19

19.1 Introduction

I introduce this final part of the book with the "3Hs" formal framework of the biocultural ethic with the aim of tackling serious problems of displacements of local communities of *co-inhabitants* (humans and other-than-humans[1]) whose life *habits* and well-being are dependent on the conservation and access to their *habitats* (see Chap. 2, Rozzi 2018). Contributing to the conservation of habitats and life habits will require a double task:

(i) To deconstruct the neoliberal discourse that since the mid-20th century has been progressively and monolithically installed in formal education, political decision-making, and the media and has promoted an unsustainable culture of consumerism and individualism that threatens the life of most human and other-than-human beings.
(ii) To understand and value a multiplicity of ancient and contemporary cultures in unique and heterogeneous regions of the planet that promote harmonious forms of co-inhabitation among communities of diverse human and other-than-human beings.

Each ecological worldview is valuable in itself. At the same time, it provides a point of reference to rethink what it means to be integrally human. In the context of contemporary cosmopolitan society, this reflection urges an ethical sense of multicultural solidarity for all humanity.[2] A comprehension of the diversity of ecological worldviews helps to value the ethical multi-potentiality of the human species to coinhabit with diverse cultures and diverse species in biophysically and sociopolitically heterogeneous regions. This comprehension is the foundation for a sense of justice among different human cultures and among different biological species.

Justice has a *cultural foundation* based on a social consensus about the concepts of the good, the bad, the just, and the social virtues that demand acting in accordance with that conception and the associated practical aspects that define how relationships among people are rightly organized. Justice also has a *formal foundation*, codified in written provisions, such as the constitution of the nation states (e.g., we will discuss below the new constitutions of Ecuador and Bolivia). National constitutions define a set of rules applied by impartial judges to assess the relationships and

---

[1] I use the expression "other-than-human" to avoid the dichotomy derived from the most common expression "nonhuman." It overcomes this dichotomy for two reasons. First, it refers to the set of beings that inhabit ecosystems, biotic beings (humans, other animals, plants), and abiotic beings (rivers, rocks, glaciers). Second, the expression "other-than-human" enables us to understand that these beings inhabit not only the biophysical reality but also the images, symbols, and values of our cultures. Therefore, they are co-inhabitants in our minds as well as in the habitats shared by biocultural communities, which include biophysical and linguistic domains of reality, oneiric, and waking phases of our lives (Rozzi 2012a, b).

[2] As Uruguay's former President José Mujica has said, "we must understand that the world's indigents are not from Africa or Latin America, they are from all of humanity." Address to the 68th General Assembly of the United Nations at its New York headquarters, USA: https://gadebate.un.org/en/68/uruguay(accessed August 19, 2017)

the conflicts among individual members and institutions of society. Considering both the cultural and formal foundations of justice, it is fundamental to forge (1) ontological-, (2) ecosocial-, and (3) ethical-biocultural foundations to transform core notions of global governance that prevail today and to achieve the sense of inter-cultural and inter-species justice demanded by the biocultural ethic.

19.2 Ontological Foundations and Legal Frameworks

We can identify *ontological foundations* in ecological worldviews that broaden the horizon of ways of conceiving ecosystems, biological species, and individuals with which humans coinhabit. In the highlands of the Andes, for example, potatoes and llamas are not merely natural resources for the Aymara and Quechua cultures; potatoes and llamas are co-inhabitants who participate in rituals, agricultural, and livestock practices in the daily life of communities (Angé et al. 2018). Co-inhabitation with multiple species requires not only rational or verbal interactions but also involves corporeality, affection, and sharing everyday life (see May 2015; Mamani-Bernabé 2015). To sustain these forms of co-inhabitation, it is indispensable to build institutional and sociopolitical contexts that support land and water sovereignty, as discussed by Laura Zanotti (2018) in her chapter about the *Mẽbêngôkre*-Kayapó peoples in the Brazilian Amazon.

It is essential to examine land ownership regimes, forms of governance, and socioeconomic problems in the broad sense (if the analysis of worldviews is not to be limited to a purely theoretical exercise). A first step is to extract ontological foundations from ecological worldviews that support the concept of co-inhabitants. A second step is to link the biocultural-ontological foundations with the formulation of socio-ecological rights and laws. These rights may include each of the co-inhabitants and ecosystems and the biosphere (Nature) as a whole. In these laws, co-inhabitants should be conceived as living subjects, in contrast to the prevailing economic paradigm that conceives plants, animals, and other beings as passive objects to be studied and exploited and Nature as a whole as merely the deposit of "natural resources."

A significant case to advance in the integration of concepts that are related to the notion of coinhabitant is the Constitution of Ecuador established in 2008. This constitution is innovative because it incorporates the normative determination of nature as a subject with rights and is the first in the world to assign this legal category to nature. The legal dogmatism of positive law has been based on paradigms that serve the unlimited exploitation of the Earth. If we change the paradigm to conceive it as Mother Earth, *Pacha Mama*, or *Gaia*, then law must foster a communal sense of reciprocity, complementarity, and relationality. This worldview is expressed in Article 71 of the Constitution of Ecuador:

> "Nature, or *Pacha Mama*, where life is reproduced and occurs, has the right to integral respect for its existence and for the maintenance and regeneration of its life cycles, structure, functions and evolutionary processes."

To promote governance and actions consistent with this worldview, Article 414 states that:

> The State will adopt adequate and cross-cutting measures for the mitigation of climate change, by limiting greenhouse gas emissions, deforestation, and air pollution; it will take measures for the conservation of the forests and vegetation; and it shall protect the population at risk.

The constitution introduces these articles within a socio-environmental context, with Article 14:

> The right of the population to live in a healthy and ecologically balanced environment that guarantees sustainability and the good way of living, Sumak Kawsay, is recognized.

In the Quechua language, *Sumak Kawsay* means harmonious life among humans and between humans and the Earth. The new Ecuadorian constitution illustrates how changes in the ontological foundations can (and should) be associated with changes in the normative contents, reforms in governance, economy, ethics, and legislation. As we will see below, other South American countries, such as the Plurinational State of Bolivia, have recently introduced normative concepts and incorporated the rights of Nature into their constitutions (see Zaffaroni 2011).

19.3 Ecosocial Foundations and Governance

We can identify *ecosocial foundations* in cultures whose *social orders* are intimately interrelated with their interpretations of *ecological orders*. Traditional Hawaiian culture offers a systemic approach that implies a concept of kinship with all living beings, which is associated with a concept of good life. The Hawaiian cosmogony links a sense of evolutionary kinship among all beings (e.g., plants, animals, earth, humans) with the concept *pono* (Callicott 1997). *Pono* indicates a harmony that is reached through social and ecological practices of care and land management that provide well-being and health (Vitousek and Beamer 2013). The intimate coupling between ecological and social orders supports, and is sustained by, habits of life that are consistent with the Hawaiian concept that humans are not landowners, but rather administrators or caretakers of the land or habitat.

The concept of *pono* requires establishing relationships of reciprocity between humans and the land. Reciprocity is based on *ecological concepts* of the amount of energy that can be withdrawn from ecosystems and on *social concepts* of governance that indicate the amount of energy that must be returned to ecosystems. Reciprocity is implemented through practices of care of the land and supervision of the social structures destined to the common welfare.

To put into practice ecosocial foundations, we can interrelate ancestral biocultural systems with new proposals for an intercultural democracy. These proposals incorporate multiple democratic forms. Novel examples are represented by the new Municipalism and the School of the Commons ("Escuela de los Commons") in Catalonia (Calle-Collado 2015, 2016), the "demo-diversity" conceived by the

Portuguese sociologist Boaventura de Sousa Santos (2010), or new types of plurinational states, such as the one established by the Constitution of Bolivia in 2009.

In the Political Constitution of the Plurinational State of Bolivia approved in 2009, the concept of "living well" or *Suma Qamaña* is introduced. With a concise preamble that evokes the Aymara cosmogony, and the current political transformations, the constitution starts by affirming that:

> In ancient times mountains arose, rivers moved, and lakes were formed. Our Amazonia, our swamps, our highlands, and our plains and valleys were covered with greenery and flowers. We populated this sacred Mother Earth with different faces, and since that time we have understood the plurality that exists in all things and in our diversity as human beings and cultures. Thus, our peoples were formed, and we never knew racism until we were subjected to it during the terrible times of colonialism…
>
> We have left the colonial, republican and neo-liberal State in the past. We take on the historic challenge of collectively constructing a Unified Social State of Pluri-National Communitarian law, which includes and articulates the goal of advancing toward a democratic, productive, peace-loving and peaceful Bolivia, committed to the full development and free determination of the peoples…
>
> We found Bolivia anew, fulfilling the mandate of our people, with the strength of our Pachamama and with gratefulness to God….

The concept of *Suma Qamaña* is then included among the great ethical principles in Article 8:

I. The State adopts and promotes the following as ethical, moral principles of the plural society: *ama qhilla*, *ama llulla*, *ama suwa* (do not be lazy, do not be a liar or a thief), *suma qamaña* (living well), *ñandereko* (live harmoniously), *teko kavi* (good life), *ivi maraei* (land without evil) and *qhapaj ñan* (noble path or life).

II. The State is based on the values of unity, equality, inclusion, dignity, liberty, solidarity, reciprocity, respect, interdependence, harmony, transparency, equilibrium, equality of opportunity, social and gender equality in participation, common welfare, responsibility, social justice, distribution and redistribution of the social wealth and assets for wellbeing.

Previous constitutions of Bolivia aimed for the "integration" of the indigenous people. In contrast to a forced integration or assimilation, the current Bolivian Constitution takes the indigenous worldviews to discuss the principles, values, and goals of the State. The expression *Suma Qamaña* is presented contextualized with a set of ecological and social values. To understand the meaning of *suma qamaña*, the concept needs to be analyzed in its original Aymara language and culture.

As explained in the chapter by Xavier Albó (2018), *Qamaña* means to inhabit, to live in a specific place or environment, to dwell, and *qamasiña* means to live with someone. Albó (2018) adds that *qamaña* also is the name that is given to a sheltered place, protected against the wind and constructed in a semicircle of stones, as a resting place for shepherds while they relax or attend their flocks. *Qamaña*, then, converges with two core concepts of the biocultural ethic: habitat and co-inhabitation. Co-inhabitation considers not only human beings but also nonhuman co-inhabitants, including plants, animals, and the Earth.

Suma means beautiful, agreeable, good, friendly, and also precious, excellent, finished, and perfect; hence, it has the sense of "fullness" (Albó 2018). This concept is particularly relevant because *suma qamaña* (living well together) departs from two core prevailing capitalist concepts: individual quality of life and rights and living better. The first concept is criticized because it focuses on living as individuals rather than living in community, and living together is fundamental for growing in humanness in synchronization with Mother Earth. The second concept, "living better," is criticized first because Aymara people do not believe that it is necessary to aim for better, precisely because *suma* already includes the best possible level of life (Albó 2018). In addition, the concept of living better is embedded in a never-ending development paradigm and denies the existence of limits to growth (Chakravorty 2016). Finally, to live better than others requires exploitation, serious competition, and concentrating wealth among a few (Rozzi 2015).

The concept of *suma qamaña* offers ecosocial foundations that transform and reorient the concepts of "living better" and "quality of life" toward "living well." The translation of this concept into public policies and governance models is highly relevant at this historical moment because *suma qamaña* emphasizes the importance of harmonious relations among human communities and Nature. *Suma qamaña* provides an important intercultural and biocultural link to foster sustainability, a link that the prevailing concepts of "quality of life" and "living better" fail to make. To advance toward this goal, Alexandria Poole (2018) proposes in her chapter to broaden the current formulation of the UN Sustainable Development Goals and Agenda 2013 by incorporating biocultural concepts such as *suma qamaña* ("Buen Vivir" or living well together).

19.4 Ethical Foundations and Education

We can forge *ethical foundations* rooted in comparative philosophy, ecological sciences, anthropology, theology, history, and ethnography to build a biocultural ethic. However, to cultivate a biocultural ethic, it is necessary to fulfill two basic conditions: (1) to conserve native habitats that harbor ecological worldviews and practices and (2) to transform the prevailing education and policy systems in order to foster sustainable life habits connected to the foundations of biocultural ethics. These transformations will require a greater degree of participation of intellectuals, communities, and social movements of the global North and of the global South, the West, and the East. This greater participation will help to remove the mantle of a universal discourse, forms of governance and education that have denigrated and made invisible the vernacular worldviews, their knowledge, language, practices, and ecological values.

"Uni-versal" can be interpreted with an ironic, decolonial, hermeneutic as "one [= *uni*, in Latin] verse [= *versus*, in Latin]," as "one discourse that is applicable to all cases, everywhere, at all times" (from the Latin *universalis*) (Rozzi 2012b). Modernity has been erected on universal knowledge. This knowledge has been institutionalized in higher education through its iconic institution: the "uni-versity." The hegemonic vision of the sciences and the Eurocentric discourse taught in the universities involved a coloniality of knowledge (Mignolo 2003a). Modern globalization and its high education institution, the university, have imposed a univocal sense of reality, which has been presented as epistemologically and technologically superior. This coloniality of knowledge has led to a biocultural homogenization (Rozzi 2013).

Not only higher education, but the formal education system at all levels represents one of the main causes of loss of linguistic and cultural diversity today. Worldwide, fewer than 500 languages are used and taught in formal education, leaving out more than 90% of the world's languages (Maffi 2001). In addition, more than half of the 193 countries worldwide are officially monolingual. These educational policies are due not only to the dominance of colonial languages, such as English and Spanish, but also to internal political conflicts. For example, in Africa many states see minority languages as a threat to national unity. Home to 2092 languages, Africa is home to over 30% of the world's linguistic diversity. According to Herman Batibo (2005), unless "unmarked bilingualism" (in which two or more languages of unequal social status are treated equally) is achieved in Africa's formal education systems, minority language speakers will continue to face the dilemma of either:

(i) Abandoning their native languages (and the eco-cultural knowledge that accompanies them) to gain access to wider society, or
(ii) Conserving their languages but remaining marginalized from national affairs.

The temporal rate and geographic scale of current global cultural homogenization are unprecedented. The spread of the dominant culture is proceeding by way of linguistic assimilation as languages of stronger groups monopolize education, the media, government, and other avenues of public discourse. Still today it is possible to detect how the use of local languages and forms of knowledge is restricted. Vernacular languages are often denigrated by labeling them as primitive, even as superstitious and unfit for the present-day world (Rodney 1982; Mignolo 2003b). Analysis of the ongoing linguistic elimination of languages uncovers postcolonial patterns of biocultural homogenization (Rozzi 2012a). With the aim of overcoming these patterns of linguistic discrimination, UNESCO and numerous nongovernmental organizations signed the Universal Declaration of Linguistic Rights in Barcelona in 1996. It affirms "all language communities have equal rights." Its implementation requires halting the overriding effects of the global-uniformized educational system and fostering instead the continuance of local languages and their educational practices.

The assimilating educational system constitutes an ethical problem since it conceals the plurality of human natures. It is imperative to criticize this educational oppression. To make visible, understand, and value biological and cultural diversity, we need a *multi-versity*[3] or pluriversality not a *uni-versality.* Multi-versitality allows

[3] Both terms *multiversity* and *pluriversity* have been used in educational and other realms.

The term *multiversity* was coined by Clark Kerr, President of the University of California, in the 1960s. He defined it as a university focused on research, an inconsistent institution, not a com-an appreciation of the wisdom embedded in vernacular worldviews, and it stimulates intercultural dialogues. The revaluation of the cultures of peoples that have been made invisible in each of the continents will disclose narratives that will nourish global society with concepts for co-inhabiting sustainable and heterogeneous biocultures (*sensu* Simberloff 2018).

In their chapter, Paulo dos Reyes and Silvia Regina da Lima (2018) affirm that in Latin America the recovery of sociocultural practices of local communities, especially peasant communities and those of original peoples and African descendants, is essential for finding alternative solutions to the current global socio-environmental crisis. They underline the urgent need to overcome the negation of local cultures excerpted by a modern rationality that has been incapable of considering the wisdom, thought, and knowledge of colonized peoples as legitimate. They illustrate the value of listening and dialoguing with local and colonized communities with examples of their work with African descendants and Candomblé, an African-origin religion widely practiced in Brazil that fosters relationships of solidarity among humans as well as responsibility, care, and love for nature and the divine. Based on their extensive work in liberation theology, May and May (2018) portray in their chapter how this Latin American movement has broken the homogenizing mold of traditional church doctrine and has provided a conduit for cultural diversity and respect for local cultures. Today, liberation aesthetics are embodied in liturgies, hymnody, poetry, and visual arts that "protest against injustice, reinforce resistance, and project hope" (May and May 2018, p. 396). In several regions of Latin America, the incorporation of traditional ecological knowledge (TEK) into public environmental management and restoration policies has been achieved through participatory action research (PAR) projects. Felipe Montoya-Greenheck (2018) offers in his chapter a case of effective intercultural work in the Indigenous Territory Guaymí de Osa of the Ngöbe (a.k.a. Ngäbe, Guaymí), in southern Costa Rica. At the same time, Montoya-Greenheck ponders the challenges to sustaining this PAR project.

---

munity, but several. It did not prosper, but later, in 1982, the *Multiversity of Buenos Aires* was founded in Argentina under the leadership of Miguel Grinberg, Leonardo Sacco, and Fabricio Simonelli, who promoted the reflection group called "The Culture of the Future." In 1989, the *Franciscan Multiversity of Latin America* was founded in Montevideo, Uruguay, as an institution that combines teaching with practice, research with promotion, and reflection with affectivity, with a deep ethical commitment to a re-appreciation and re-encountering with all forms of life. Subsequently, multiversities have been founded in Malaysia (1998), Mexico (1998), India (*India International Multiversity*, 1998), Africa (*Mpambo Afrikan Multiversity*, 1999), and Spain (*Multiversity of Agroecology, Biodiversity and Cultures*, 2009) with the aim to create free, open, intercultural educational communities.

The term *pluriversity* has acquired growing relevance in two complementary realms. First, the crisis and exhaustion of the present academic model with its origins in the universalism of the Enlightenment (see Castrillón 2009; Echeverría 2012; Boidin et al. 2012). Second, the related term *pluriversality* is being progressively used by decolonial thinkers to denote the entanglement of coexisting worldviews, which today are interconnected but subject to differential power relationships (Dussel 2002; Escobar 2011; Mignolo 2011).

The final chapters of this third part of the book present four cases of biocultural conservation in Asia. National-level policy changes from the 1960s to early the 2000s have impacted social-ecological conditions in Inner Mongolia. Ruifei Tang and Michael C. Gavin (2018) introduce a novel biocultural conservation approach, based on eight principles (cf Gavin et al. 2015), which can be linked to positive social and ecological outcomes. Pandurang Hegde and George James (2018) analyze how since the 1970s the Chipko movement in the Western Himalayas of India and Appiko movement in India's Western Ghats provide examples of cases led by local people who were able to succeed in changing government's policies. Making a change from clear cutting of native forests and establishing monocultures, they were able to incorporate traditional ecological management and knowledge systems based on holistic understandings that valued and protected the life habits and habitats of local people. Mitsuyo Toyoda (2018) presents an inspiring case of revitalizing estuarine environments in Japan. Estuaries used for oyster farming had undergone serious eutrophication in the 2000s. Local fishermen and the broader community initiated grassroot experiments for environmental restoration of estuaries that were conceived as commons, proposing dynamic collaborative forms of governance.

19.5 Concluding Remark

Through its concepts and case studies, this volume invites readers to consider further the wealth of biocultural worldviews and practices existing in each of the continents. These biocultural worldviews and practices markedly contrast with the monoculture of consumption established by global hegemonic economic discourse. The current trend of disconnection of global society from biocultural diversity represents an anomalous life habit. It needs to be reoriented in light of the values and practices that are still alive in a plethora of cultures and people who have vital awareness of ways of coinhabiting with diverse co-inhabitants (including rivers, the moon, the flowers, the birds, and the diversity of life forms). Moreover, the notion of kinship between humans and other species is as present in forms of traditional ecological knowledge as it is in current scientific evolutionary knowledge (Rozzi 2004). Scientific, indigenous, religious, and other ecological worldviews are embodied in the everyday lives of communities, urban and rural, which today are organizing to resist the impact of prevailing one-dimensional economic policies.

Knowledge about the heterogeneity of habitats and life habits of their co-inhabitants contributes to overcoming current narrow economic perspectives centered on monetary indicators. This biocultural knowledge broadens the spectrum of values by reintegrating ecological, aesthetic, and ethical dimensions into the appreciation of life. Paraphrasing and mutating the motto of the World Social Forum "Another World is Possible," I propose instead "This Bioculturally Diverse World is Actual" (Rozzi 2012b).[4]

"Actual" (in contrast to "possible" or potential) is understood in an Aristotelian sense. The world that *actually* exists today encompasses a myriad of ecological worldviews and sustainable practices. These worldviews and practices are not merely a *potentiality* or "possibility" for a future world; they do exist, they are actual (not only possible), they have the right to exist, and they will foster the continuity of sustainable form of co-inhabitation.[5]

I say "this world," and not "another world," because the actuality of sustainable life habits is rooted in this planet and its multiplicity of cultures; we do not have to look for another planet or future societies. I propose that by changing the motto of the World Social Forum, we should understand that "another world" is the one governed by a one-dimensional orientation that imposes a homogenizing force, an oppressive developmental model, on biocultural heterogeneity, such world that should be understood as an anomaly, "another world" that needs to be reoriented toward a biocultural world. The biocultural ethic condemns the currently globalized anomalous world for its socio-environmental injustices. Primarily for ethical reasons, this anomalous world should be promptly transformed in order to enable the reemergence of the multiple sustainable communities that today vitally resist with their plethora of biocultural worldviews and practices. The biocultural world already exists, and biocultural conservation should help re-vitalizing it.

References

Angé O, Chipa A, Condori P, Ccoyo AC, Mamani L, Pacco R, Quispe N, Quispe W, Sutta M (2018) Interspecies respect and potato conservation in the Peruvian cradle of domestication. Conserv Soc 16:30–40

Albó X (2018) *Suma qamaña* or living well together: a contribution to biocultural conservation. In: Rozzi R, May Jr RH, Chapin FS III, Massardo F, Gavin M, Klaver I, Pauchard A, Nuñez MA, Simberloff D (eds) From biocultural; homogenization to biocultural conservation. Ecology and ethics, vol 3. Springer, Dordrecht, pp 333–342

[4]The World Social Forum (WSF) started in 2001 in Brazil, running as a "parallel forum" to the neoliberal World Economic Forum (WEF) in Davos, Switzerland. Just like market globalists, who treat the WEF as a platform to project their ideas and values to a global audience, justice globalists utilize the WSF as one of the chief production sites of their ideological and policy alternatives. The WSF brings together thousands of participants to workshops, conferences, artistic performances, and other activities on socio-environmental themes (see de Sousa-Santos 2005).

[5]With a voice of resistance, unity, and change, the Declaration of the IV Continental Summit of Indigenous Peoples and Nationalities of Abya Yala in 2005 affirms that the resistance and historical struggle of the indigenous peoples of the Americas in defense of their territories and cultural identity today extend to every corner of the continent. The Declaration culminates stating that: "Another America is Possible! Never Again an America without the Indigenous Peoples!" (http://www.cumbrecontinentalindigena.org/index_en.php; accessed March 17, 2016). (http://www.cumbrecontinentalindigena.org/index_en.php; accessed March 17, 2016).

Batibo H (2005) *Language decline and death in Africa: Causes, consequences and challenges.* Multilingual Matters Limited, Clevedon
Boidin C, Cohen J, Grosfoguel R (2012) Introduction: from university to pluriversity: a decolonial approach to the present crisis of western universities. Hum Archit: J Sociol Self-Knowl 10:2
Calle-Collado A (2015) Podemos y el auge municipalista. Sobre partidos-ciudadanía y vieja política. Empiria. Revista de metodología de ciencias sociales 32:169–190
Calle-Collado A (2016) Ciclos políticos y ciclos de movilización. Entre el 15M, Podemos y nuevos municipalismos. Historia Actual Online 40:79–94
Callicott JB (1997) Earth's insights: a multicultural survey of ecological ethics from the Mediterranean Basin to the Australian outback. University of California Press, Berkeley
Castrillón AH (2009) La Cátedra de Estudios Afrocolombianos: un espacio para reflexionar sobre la pluriversidad en los modos de vivir la afrocolombianidad. Revista Educación y Pedagogía 19(48):83–96
Chakravorty S (2016) Andean cosmovision and diplomacy for life. Strateg Anal 40:440–450
De Sousa Santos B (2005) The future of the world social forum: the work of translation. Development 48(2):15–22
de Sousa Santos B (2010) Refundación del Estado en América Latina: perspectivas desde una epistemología del Sur. Plural editores, La Paz
dos Reyes P, Silva da Lima SR (2018) Candomblé in Brazil: the contribution of African-origin religions to biocultural diversity in the Americas. In: Rozzi R, May Jr RH, Chapin FS III, Massardo F, Gavin M, Klaver I, Pauchard A, Nuñez MA, Simberloff D (eds) From biocultural; homogenization to biocultural conservation. Ecology and ethics, vol 3. Springer, Dordrecht, pp 379–391
Dussel E (2002) World-system and trans-modernity. Nepantla: Views From South 3:221–244
Echeverria J (2012) Pluralidad de la filosofía: pluriversidad versus universidad. Ontology Studies (12):373–388
Escobar A (2011) Sustainability: design for the pluriverse. Development 54:137–140
Gavin M, McCarter J, Mead A, Berkes F, Stepp JR, Peterson D, Tang R (2015) Defining biocultural approaches to conservation. Trends Ecol Evol 30(3):140–145
Hegde P, James G (2018) Challenging biocultural homogenization: experiences of the Chipko and Appiko movements in India. In: Rozzi R, May Jr RH, Chapin FS III, Massardo F, Gavin M, Klaver I, Pauchard A, Nuñez MA, Simberloff D (eds) From biocultural; homogenization to biocultural conservation. Ecology and ethics, vol 3. Springer, Dordrecht, pp 427–442
Maffi L (ed) (2001) On biocultural diversity. Linking language, knowledge, and the environment. Smithsonian Institution Press, Washington, DC
Mamani-Bernabé V (2015). Spirituality and the Pachamama in the Andean Aymara worldview. In Rozzi R, Chapin FS, Callicott JB, Pickett STA, Power ME, Armesto JJ, May Jr RH (eds) Earth stewardship: linking ecology and ethics in theory and practice. Ecology & ethics. Springer, Dordrecht. pp 65–76
May R (2015) Andean llamas and earth stewardship. In Rozzi R, Chapin FS, Callicott JB, Pickett STA, Power ME, Armesto JJ, May Jr RH (eds) Earth stewardship: linking ecology and ethics in theory and practice. Ecology and ethics, vol 2. Springer, Dordrecht. pp 77–86
May R, May J (2018) Latin American theology of liberation and biocultural conservation. In: Rozzi R, May Jr RH, Chapin FS III, Massardo F, Gavin M, Klaver I, Pauchard A, Nuñez MA, Simberloff D (eds) From biocultural; homogenization to biocultural conservation. Ecology and ethics, vol 3. Springer, Dordrecht, pp 393–403
Mignolo W (2003a) The darker side of the renaissance: literacy, territoriality and colonization. University of Michigan Press, Ann Arbor
Mignolo WD (2003b) Historias Globales/diseños globales. Ediciones Akal, Madrid
Mignolo W (2011) The darker side of western modernity: global futures, decolonial options. Duke University Press, Durham
Montoya-Greenheck F (2018) Biocultural diversity and Ngöbe people in the South Pacific of Costa Rica. In: Rozzi R, May Jr RH, Chapin FS III, Massardo F, Gavin M, Klaver I, Pauchard A,

Nuñez MA, Simberloff D (eds) From biocultural; homogenization to biocultural conservation. Ecology and ethics, vol 3. Springer, Dordrecht, pp 361–378
Poole A (2018) The UN sustainable development goals and the biocultural heritage Lacuna: where is goal number 18? In: Rozzi R, May Jr RH, Chapin FS III, Massardo F, Gavin M, Klaver I, Pauchard A, Nuñez MA, Simberloff D (eds) From biocultural; homogenization to biocultural conservation. Ecology and ethics, vol 3. Springer, Dordrecht, pp 315–331
Rodney W (1982) How Europe underdeveloped Africa. Howard University Press, Washington, DC
Rozzi R (2004) Implicaciones éticas de narrativas yaganes y mapuches sobre las aves de los bosques templados de Sudamérica austral. Ornitología Neotropical 15:435–444
Rozzi R (2012a) Biocultural ethics: the vital links between the inhabitants, their habits and regional habitats. Environ Ethics 34:27–50
Rozzi R (2012b) South American environmental philosophy: ancestral Amerindian roots and emergent academic branches. Environ Ethics 34:343–365
Rozzi R (2013) Biocultural ethics: from biocultural homogenization toward biocultural conservation. In: Rozzi R, Pickett STA, Palmer C, Armesto JJ, Callicott JB (eds) Linking ecology and ethics for a changing world: values, philosophy, and action. Ecology and ethics, vol 1. Springer, Dordrecht, pp 9–32
Rozzi R (2015) Implications of the biocultural ethic for earth stewardship. In Rozzi R, Chapin FS III, Callicott JB, Pickett STA, Power ME, Armesto JJ, May Jr. RH (eds) Earth stewardship: linking ecology and ethics in theory and practice. Ecology and ethics, vol 2. Springer, Dordrecht. pp 113–136
Rozzi R (2018) Biocultural homogenization: a wicked problem in the anthropocene. In: Rozzi R, May Jr RH, Chapin FS III, Massardo F, Gavin M, Klaver I, Pauchard A, Nuñez MA, Simberloff D (eds) From biocultural; homogenization to biocultural conservation. Ecology and ethics, vol 3. Springer, Dordrecht, pp 21–47
Simberloff D (2018) Nature, culture, and natureculture: the role of nonnative species in biocultures. In: Rozzi R, May Jr RH, Chapin FS III, Massardo F, Gavin M, Klaver I, Pauchard A, Nuñez MA, Simberloff D (eds) From biocultural; homogenization to biocultural conservation. Ecology and ethics, vol 3. Springer, Dordrecht, pp 207–218
Tang R, Gavin MC (2018) The dynamics of biocultural approaches to conservation in Inner Mongolia, China. In: Rozzi R, May Jr RH, Chapin FS III, Massardo F, Gavin M, Klaver I, Pauchard A, Nuñez MA, Simberloff D (eds) From biocultural; homogenization to biocultural conservation. Ecology and ethics, vol 3. Springer, Dordrecht, pp 405–425
Toyoda M (2018) Revitalizing local commons: a democratic approach to collective management. In: Rozzi R, May RH Jr, Chapin FS, Massardo F, Gavin M, Klaver I, Pauchard A, Nuñez MA, Simberloff D (eds) From biocultural; homogenization to biocultural conservation. Ecology & ethics. Springer, Dordrecht, pp 443–457
Vitousek P, Beamer K (2013) Traditional ecological values, knowledge, and practices in twenty-first century Hawai'i. In: Linking ecology and ethics for a changing world. Springer, Dordrecht, pp 63–70
Zaffaroni ER (2011) La Pachamama y el humano. In: Acosta A, Martínez E (eds) *La Naturaleza con Derechos. De la Filosofía a la Política*. Ediciones Abya-Yala, Quito, pp 25–135
Zanotti L (2018) Biocultural approaches to conservation: water sovereignty in the Kayapó lands. In: Rozzi R, May Jr RH, Chapin FS III, Massardo F, Gavin M, Klaver I, Pauchard A, Nuñez MA, Simberloff D (eds) From biocultural; homogenization to biocultural conservation. Ecology and ethics, vol 3. Springer, Dordrecht, pp 343–359

Chapter 20
The UN Sustainable Development Goals and the Biocultural Heritage Lacuna: Where Is Goal Number 18?

Alexandria K. Poole

Abstract On 25 September 2015, the Seventieth Session of the General Assembly in the United Nations approved the new Sustainable Development Goals (SDGs) through Resolution 70/1 "Transforming our world: the 2030 Agenda for Sustainable Development," building upon the vision of the original Millennium Development Goals (MDGs). Although intended to account for the shortfalls found in the original MDGs, the SDGs still neglect fundamental qualities of cultural sovereignty that are key for maintaining sustainable practices, values, and lifestyle habits. None of the 17 SDGs emphasize the need to protect local ecological knowledge, cultural heritage—nor its interrelation with biodiversity—as a pathway to sustainability. Further, the factors that threaten local ecological knowledge, traditional lifestyles, and alternative economic practices are absent, provoking indigenous and local peoples to argue that they, and their cultural sovereignty, remain unrecognized by this new "sustainable" development agenda. I argue that focusing on indirect drivers that undermine sustainable management practices must be explicitly addressed to address this conceptual lacuna. Indirect drivers include cultural and ethical facets of the human-nature relationship. Biocultural heritage, reflecting the diverse ways of being between human communities and their local environments, is the rich history of language, heritage, cultural memory, ecological knowledge, and values and should be explicitly be articulated as a key component to any sustainability agenda. Consequently, I propose that to accomplish the SDGs' mission, it is indispensable to include a sustainable development goal "number 18" that recognizes biocultural heritage.

Keywords Ethics · Local ecological knowledge · Biocultural heritage · Sustainable Development Goals · Human-nature relationship

---

A. K. Poole (✉)
Department of Politics, Philosophy and Legal Studies, Elizabethtown College, Elizabethtown, PA, USA
e-mail: poolea@etown.edu

R. Rozzi et al. (eds.), *From Biocultural Homogenization to Biocultural Conservation*, Ecology and Ethics 3, https://doi.org/10.1007/978-3-319-99513-7_20

20.1 Introduction: Biocultural Heritage as a Lacuna in the Sustainable Development Goals

The United Nations General Assembly passed a new Resolution on 25 September 2015 entitled "Transforming Our World, the 2030 Agenda for Sustainable Development," as "a plan of action for people, planet, and prosperity" (UNGA 2015, p. 1). This resolution was intended to develop updated goals incorporating lessons learned from the Millennium Development Goals (MDG):

> The 17 Sustainable Development Goals and 169 targets which we are announcing today demonstrate the scale and ambition of this new universal Agenda. They seek to build on the Millennium Development Goals and complete what they did not achieve. They seek to realize the human rights of all and to achieve gender equality and the empowerment of all women and girls. They are integrated and indivisible and balance the three dimensions of sustainable development: the economic, social and environmental. (UNGA 2015, p. 1)

These new goals are expansions of the original eight proposed within the larger Millennium Ecosystem Framework and are intended to encapsulate the shortfalls for a more inclusive and sustainable future. The updated Sustainable Development Goals (SDGs) reflect this intention, with stronger language such as its reframed first goal, "to eradicate poverty in all forms, everywhere" (SDG #1). However, even with such intensification in tone, the discussion fails to articulate the indirect drivers which threaten biocultural heritage, linguistic diversity, and traditional ways of life that do not fit into the dominant mode of development in the global economy (Nelson 2005; see Box 20.1 for discussion on drivers). The 17 SDGs, complete with their 169 targets, remarkably, do not explicitly articulate the importance of local ecological knowledge and cultural diversity for sustainability as a high-level priority, and these concepts are only referenced in support of goals oriented toward more specific economic development.[1,2] This absence represents a conceptual lacuna that must be addressed in order to recognize biocultural heritage within development discourse.

The loss of cultural heritage—and respectively the cultural sovereignty of distinct groups—is a major contributor to environmental degradation and incidentally the creation of poverty more generally (Escobar 2011; Hunn 2007; IPCCA 2009; Maffi 2005; Posey and Dutfield 1996). Local communities lose the knowledge and capacity to maintain sustainable life habits when their traditional habitats, values, languages, and lifestyles are displaced (Maffi 2001; Rozzi 2013; UNESCO 2010; Zent 2009). Displacements affect communities living in a diverse range of settings, as articulated by distinct groups including the small island nations, traditional and

[1] This book chapter is a revised version of the article, Poole, A. (2018) "Where is Goal Number 18: The Need for Biocultural Heritage in the UN Sustainable Development Goals," *Environmental Values*, 27(1):55–80.

[2] The closest iteration present in the SDG is 4.7, which states, "…ensure that all learners acquire the knowledge and skills needed to promote sustainable development, including, among others… [the] appreciation of cultural diversity and of culture's contribution to sustainable development" (UNGA 2015, p. 17). Compare to the specific targets proposed in Table 3.

Box 20.1 Direct and Indirect Drivers of Ecological Change: The Millennium Ecosystem Assessment

In 2000, UN Secretary-General Kofi Annan called for the scientific evaluation of the world ecosystems and the services they provide for humanity in response to the recognized need for research on human well-being in relation to ecosystems or, rather, "a major international collaborative effort to map the health of our planet" (Millennium Ecosystem Assessment 2005, p. 1–2). As an international effort involving scientists, anthropologists, ethicists, economists, policy-makers, and many more, the Millennium Ecosystem Assessment Framework (MEA) has largely popularized the "ecosystem services" model, by clarifying the services or benefits humanity receives from the natural world within policy through economic evaluation (Millennium Ecosystem Assessment 2003). This model has provided a methodology for articulating the interrelation between ecosystem functioning, ecosystem services, and the anthropogenic drivers that could alter these processes and since been utilized through forms of economic and environmental policy and sustainability education (Costanza et al. 2014; Liu et al. 2010; Schröter et al. 2014).

To address the complexities of global socio-environmental change, the Millennium Ecosystem Assessment uses the concepts of direct and indirect drivers to identify causal agents of anthropogenic change (Nelson 2005). *Direct* drivers are those agents of change that are physical or mechanical processes, whereas *indirect* drivers are those agents of change that are culturally, ethically, and socially related and formulate the conditions in which many institutional, infrastructural, and administrative direct drivers are constructed (Millennium Ecosystem Assessment 2005; Nelson 2005). Understanding the inter-dynamics among culture, values, and lifestyles within the indirect drivers is a complex task that is often demarcated to the humanities or social sciences, fragmenting causal discussions regarding management decisions and ecological knowledge across the disciplines. Carpenter et al. (2006) have argued that focusing on indirect drivers remains a pressing research need in the process of linking social to ecosystem change, stating that:

> Most research related to ecosystem services focuses on direct drivers, such as land use change or invasive species. Yet, effective management requires more attention to indirect drivers such as demographic, economic, sociopolitical, and cultural factors. (Carpenter et al. 2006, p. 258)

Despite such calls, indirect drivers have largely fallen into disuse in practice. Analysis of these indirect drivers is a particular strategy for analyzing human-nature dynamics as these indirect drivers create or "drive" the conditions that shape the world. Whether explicitly acknowledged or not, indirect drivers represent influential structures linking life habits and the habitats of the urban habitat and are therefore highly relevant for developing sustainable environmental management policies. Indirect drivers shape communication, policies,

(continued)

Box 20.1 (continued)
and administration priorities, comprising the environmental context in which an individual participates, engages, senses, and perceives the world (Millennium Ecosystem Assessment 2005; Nelson 2005; Nelson et al. 2006). Additional indirect drivers that merit analysis include any institutional entities which shape and impact ecological knowledge, cultural heritage, and everyday lifestyles such as educational practices and content, agricultural, and economic systems. These structures not only shape the cultural and ethical relationship with the environment by forming the fundamental experiences that constitute the values, priorities, and knowledge of the natural world but by overlooking the significance of indirect drivers to environmental change contributes to obscuring the human relationship and interdependence with the natural world, that is, diminishes the importance of local ecological knowledge, biocultural heritage, and land ethics for sustainable development (Rozzi 2012; Poole 2015).

Local ecological knowledge is a fundamental aspect of all cultures; however, this is harder to see explicitly in industrialized cultures that are more isolated from nature and that do not explicitly acknowledge the human-nature relationship within their fundamental values (Cocks 2010; Maffi 2007b). Michelle Cocks states that the concept of biocultural diversity should more explicitly be applied to urban and rural groups that do not fit into the categories of indigenous peoples to emphasize that the biocultural link is a very real part of the human-nature relationship and that it is present in communities that may not even recognize this coevolutionary relationship (Cocks 2006). I argue that any educational or formal institution that discounts the importance of local ecological knowledge is undermining cultural integrity and heritage and therefore should be considered an indirect drive of environmental degradation and the loss of biocultural heritage (Poole 2018; Poole 2015; Poole et al. 2013). Further, biocultural homogenization occurs when the rich interrelation of biological, cultural, and linguistic diversity is overlooked (Rozzi 2013). Consequently, there is a clear need to analyze dominant global practices to see how these drivers function, either by affirming or undermining local cultural heritage and its relation to place, as well as ways to redress deleterious trends for sustainable development. Ignoring the role of cultural heritage and ecological knowledge within the indirect drivers framework leads to further degradation of those ways of life and knowledge and could result in a collective loss and depreciation of biocultural heritage, further hindering efforts within sustainable development to articulate barriers to its existence and persistence.

Drivers have the potential to shape human activities and steer society toward certain goals (Carpenter et al. 2006; Nelson 2005; Millennium Ecosystem Assessment 2005). Therefore, developing understanding of the social-ecological dynamics and infrastructural forces underlying sustainable

(continued)

Box 20.1 (continued)
societies must be at the forefront in the development of new structures for sustainable living in the twenty-first century. Biocultural heritage, local ecological knowledge, and local management practices remain key indirect drivers, and there remain many more such forces at work within the space between ethical and cultural values and the structures that shape human society and action (IPMG 2014 Matrix for SDG's: Goals, Targets and Indicators, see pp. 7–12). As long as the loss of local ecological knowledge and biocultural heritage is not recognized as a driver of ecosystem change, policy will remain at a disadvantage to address it as a threat to sustainability. A further step must be taken to address a biocultural mode of thinking within the larger SDGs and the forces that threaten or undermine these other ways of being.

indigenous communities, as well as farming or agrarian communities in rural and urban settings (AOSIS 2016; Hunn 1999; Kahn and Kellert 2002; Orr 1992; UNDRIP 2007; Zent 2009). Each community has their own biocultural heritage, the customs, cultural memory, and values, which are tied to the places in which they live. Biocultural heritage represents not only the biogenetic diversity of landscapes but also the interrelation this diversity shares with the language, heritage, cultural memory, ecological knowledge, and values of local and indigenous communities (Argumedo and Pimbert 2008; Davidson-Hunt et al. 2012; Gavin et al. 2015; Hunn 1999; Maffi and Woodley 2012).

The biocultural approach provides a novel conceptual lens to discuss the threats to local ecological knowledge (LEK), traditional ecological knowledge (TEK), linguistic diversity, and the interwoven relationship this has to the loss of biological diversity (Maffi 2001; Rozzi et al. 2006; Rozzi 2012, 2013). Further, the biocultural approach has significant consequences for sustainable development policy and practices across the rural-urban gradient (UNESCO 2010; Zent 2009). Therefore, I argue that not only should policy discussions focus on the institutional, infrastructural, and biophysical forces (direct drivers) that impact sustainability but the cultural, sociopolitical, and ethical forces (indirect drivers) which support and reify sustainable ways of life. That is, the factors contributing to the loss of ecological knowledge and biocultural heritage serve as indirect drivers for unsustainable management practices (Poole 2015; see Box 20.1). Unless explicitly identified as an entity prioritized for its own merits, threats to cultural diversity and alternative forms of economies will remain unaddressed within development discourse (Agyeman 2005; Argumedo et al. 2011; Argumedo and Pimbert 2008; Berkes et al. 2000; Cocks 2010; Mühlhäusler 2001; Rozzi 2013).[3] Consequently, at minimum, an

[3] For a discussion on the complexity of articulating biocultural heritage or traditional ecological knowledge, see Grear 2015.

additional goal within the SDGs is necessary to make explicit this missing piece in the current articulation of sustainable development.

The coevolutionary and ecological link between biological, cultural, and linguistic diversity is currently absent within the development model, revealing the accumulative biocultural heritage of the earth's peoples to be a significant lacuna in the SDGs. Such an analysis ultimately points to the need for the additional Sustainable Development Goal #18, to consider the cultural and ethical elements that comprise a sustainable human-nature relationship: "Protect, promote, and engage biocultural heritage to reinforce and support sustainable interconnections between diverse human societies and their distinct environments."

20.2 The Sustainable Development Goals and the Inclusivity Clause

Since the 1960s, the UN and other international organizations have prioritized human well-being while responding to the environmental crisis (Brundtland 1987). However, these efforts have mostly centered on indices regarding economic prosperity and growth. Consideration of threats to biocultural heritage (local ecological knowledge, ways of life, and cultural heritage) has not been prioritized. This omission is apparent when we analyze the eight major priorities articulated by the original MDGs (Table 20.1).

Note that the interrelation between cultural and biological diversity is absent in the original MDGs. Further, none of the goals specify ways of life that would not fall into the dominant development model prevalent in the global economy today such as subsistence economies. The 2012 UN MDG Report showed that some specific targets were met, such as the "proportion of people living on less than \$1.25 per day has decreased from 47% in 1990 to 24% in 2008 (from 2 to 1.4 billion)" and "enrollment rates of primary schools increased from 58 to 76% in sub-Saharan Africa between 1999 and 2010," among others (Fehling et al. 2013). While assessments have shown some improvement in the eight concentrations proposed the MDGs, these improvements nonetheless do not address the underlying causes of these detrimental conditions. Critics of the MDGs emphasize that in many ways the

Table 20.1 2000 UN Millennium Development Goals (MDGs)

1.	Eradicate extreme poverty and hunger
2.	Achieve universal primary education
3.	Promote gender equality and empower women
4.	Reduce child mortality
5.	Improve maternal health
6.	Combat HIV/AIDS, malaria, and other diseases
7.	Ensure environmental sustainability
8.	Global partnership for development

development goals do not manage to stem the overall impacts of large-scale economies on local communities as these goals do not address the underlying causes of displacement, overexploitation, and poverty.

In order to address the shortcomings of the MDGs, at the United Nations Conference on Sustainable Development (also known as the Rio+20 which took place in Rio de Janeiro, Brazil in 2012) put together an open working group to develop a set of new Sustainable Development Goals (SDG) for a post-2015 development agenda. The new development agenda aspired to "leave no one behind." However, as pointed out by representatives from indigenous communities, the agenda continued to exclude indigenous peoples through its language and the specified targets. At the Intergovernmental Negotiations on Post-2015 Development Agenda, for instance, the Indigenous Peoples Major Group (IPMG) repeatedly called attention to the near "invisibility" of indigenous peoples in the working documents defining the SDGs (IPMG 2015). To avoid this invisibility and remediate the injustice of displacements, the IPMG Policy Brief proposed that the SDGs should provide "a unique opportunity to not only remedy shortcomings of the MDG process, but also historic injustices resulting from racism, discrimination, and inequalities long suffered by indigenous peoples around the world" (IPMG 2015, p. 1; IPMG 2014).

Nonetheless, in September 2015, a global vote approved the SDGs, with a version that did not adequately include indigenous people and the deep biocultural interrelation. While the proposed expansion of the original development goals includes more explicit considerations of social and economic justice as well as an emphasis on building sustainable practices and societies, the factors that threaten local ecological knowledge, traditional lifestyles, and alternative economic practices were not explicitly addressed in the 17 new goals. Table 20.2 lists the Sustainable Development Goals with emphasis added on "inclusive" (Table 20.2).

While the new goals were intended to be revisions of the MDGs to account for the shortfalls found in the past 15 years, no single goal emphasizes the need to protect neither local ecological knowledge, cultural heritage, nor its interrelation with biodiversity as a pathway to sustainability. Further in the SDGs there is a lack of analysis of the forces that create *un*sustainability as a culture is little developed in the policy language. To tackle this deficiency, I propose that the concept of *indirect drivers* of environmental change might be useful and that the conceptual interrelation between biological, cultural, and linguistic diversity must be recognized in these goals (Box 20.1). Indirect drivers provide a hermeneutic key to better address the original shortcomings of the MDGs. For instance, MDG #1 has been expanded from its original formulation, "eradicate extreme poverty and hunger," to SDG#1, "end poverty in all its forms everywhere." The subsequent targets do not identify tribal or rural displacement as a major driver in impoverishment of communities, nor the indirect factors that often lead to the creation of slums, urban food deserts, or other forms of environmental injustices such as those found in Flint, Michigan, United States, where the poor were exposed to lead in the water system due to a failing infrastructure.

Table 20.2 2015 UN Sustainable Development Goals (emphasis added)

1.	End poverty in all its forms everywhere
2.	End hunger, achieve food security and improved nutrition, and promote sustainable agriculture
3.	Ensure healthy lives and promote well-being for all at all ages
4.	Ensure *inclusive* and equitable quality education and promote lifelong learning opportunities for all
5.	Achieve gender equality and empower all women and girls
6.	Ensure availability and sustainable management of water and sanitation for all
7.	Ensure access to affordable, reliable, sustainable, and modern energy for all
8.	Promote sustained, *inclusive* and sustainable economic growth, full and productive employment, and decent work for all
9.	Build resilient infrastructure, promote *inclusive* and sustainable industrialization, and foster innovation
10.	Reduce inequality within and among countries
11.	Make cities and human settlements *inclusive*, safe, resilient, and sustainable
12.	Ensure sustainable consumption and production patterns
13.	Take urgent action to combat climate change and its impacts
14.	Conserve and sustainably use the oceans, seas, and marine resources for sustainable development
15.	Protect, restore, and promote sustainable use of terrestrial ecosystems, sustainably manage forests, combat desertification, and halt and reverse land degradation and halt biodiversity loss
16.	Promote peaceful and *inclusive* societies for sustainable development, provide access to justice for all, and build effective, accountable, and *inclusive* institutions at all levels
17.	Strengthen the means of implementation and revitalize the global partnership for sustainable development

The creation of poverty is caused by many factors. Disregard for these multiple causes and their relation to the dominant economy ignores its significance for cultural sovereignty and its implications for earth stewardship. Local peoples are often displaced from their land by industry, impacts of climate change, or loss of local economies, leading to mass rural migrations to urban cities. These displaced communities then go on to form shantytowns at the outskirts of large urban centers, unsupported by the larger infrastructure of the city and disconnected from their traditional livelihoods. This pattern is taking place at unprecedented rates during the last 50 years in regions of Africa, Asia, and Central and South America (Rozzi 2015a, b). Large-scale development degrades local ecosystems and displaces local peoples, leading to a break in cultural heritage and ecological place, a disruption that is both environmental and cultural. To provide an approach that could reorient these unsustainable patterns, the March 2014 IPMG's vision for the SDGs refined the primary target to eradicate poverty. Table 20.3 highlights the targets for Goal 1 adjusted in terms to be inclusive for indigenous peoples.

This nexus of the right to territory and resources is embedded in the deep relationship that indigenous people way of life, culture, and production of food depends upon the land and is identifying the conditions in which poverty is being created for

Table 20.3 Targets for Goal 1 from Indigenous People's Working Group

Goal 1 Eradicate poverty for indigenous peoples	
1.1.	Secure indigenous peoples' collective rights to land, territories, and resources
1.2.	Protect local and diverse indigenous peoples' economies and livelihoods, traditional subsistence activities, and food sovereignty
1.3.	Ensure equal and just access for indigenous peoples' economies and livelihoods, traditional subsistence activities, and food sovereignty (IPMG 2014, p. 2)

Table 20.4 Law 70 of Colombia—Chapter 2 II, Article 3

Article 3 The present law is based on the following principles:
Recognition and protection of ethnic and cultural diversity and equal rights for all cultures that compose the Colombian nationality
Respect for the integrity and dignity of the Black Communities' cultural life
Participation of the Black Communities and their organizations, without detriment to their autonomy, in decisions that affect them and in those that affect the entire nation in conformity with the law
The protection of the environment, emphasizing the relationships established by the Black Communities and nature (Congreso de Colombia 1993, p.3)

those communities. Destruction of the native habitats or access to them by local communities fractures the capacity for local peoples to provide for themselves and future generations (Argumedo and Pimbert 2008; Maffi 2007b; Poole 2015). It also overlooks the important role these communities and cultures have maintaining sustainable ways of life, discounting the role local communities play as sustainable actors when not tied to industrial practices. The above targets explicitly identify this causal relation and need.

Alleviating poverty requires avoiding the original displacement and impoverishment of these peoples, rather than simply addressing needs once communities have been displaced from their homelands and traditional ways of life. Such policy is not without precedent. A landmark decision protecting the right of ancestral peoples to maintain their lands occurred with 1993 Law of Colombia, "In Recognition of the Right of Black Colombians to Collectivity Own and Occupy their Ancestral Lands" (Escobar 2011). In this case, Afro-Colombians of the south Pacific region were facing displacement due to growing economic pressures. Table 20.4 lists the guiding principles Law 70 of Colombia put forth in Article 3 to protect their way of life and livelihoods (Table 20.4).

Indeed, it seems reasonable to expect similar language within the MDG and SDG and recommendations by the IPMG for inclusivity of indigenous and local peoples. Yet MDG #1 and its SDG adaptation do not acknowledge such underlying causes of poverty. If policy were to "eradicate extreme poverty and hunger," we should equally expect to see a goal that read "preserve the sovereignty of local peoples to their land, heritage, and property," as stated in Law 70 of Colombia. Instead, the SDG perpetuates the same theme underscoring development discourse—if not exacerbating this trend—by setting the new goal which only generalizes these principles more by reframing the goal to "end poverty in all forms everywhere" (SDG #1) without addressing its ultimate (indirect) causes. SDG does not recognize that displacing local peoples from their own lands is a major driver of poverty, nor the threats to the resilience of the land on which they live, nor threats to their ways of life, all of which undermine the capacity to continue living with the land, let alone sustainably.

Protecting heritage and sustainable lifestyles is as important as developing new strategies for adapting current development practices toward sustainability and for maintaining fundamental human rights. However, in order to prevent the continued displacement of local peoples, the causes or drivers of poverty must be acknowledged and addressed. In sharp contrast to the idea of "partnership" (see SDG Goal #17), large-scale development often displaces or undermines local ways of life that are already self-sustaining in a way that framing in terms of partnership does not address. This tendency to analyze current states of impoverishment or loss, without addressing the underlying causes, remains throughout the newly revised SDGs. As such, these terms remain a framing that continues to obfuscate the impact of dominant development practices on local communities.

A poignant example of the tendency to overlook cultures that have a close relationship with their land is present in the use of "inclusivity" throughout the language of the SDG regarding subsistence economies and education. For instance, the educational model underlying SDG Goal #3, "ensure inclusive and quality education for all and improve lifelong learning," expands on MDG's goal to "achieve universal primary education" without affirming the need to provide space for traditional knowledge systems and traditional educational practices within the formal classroom. Goal #3 does not explicitly recognize other educational methods within traditional and indigenous communities. This educational model does not account for the nuanced and complex traditional knowledge of communities, thereby discounting local ways of life and further separating knowledge relevant to local landscapes (Nabhan 2001; Zent 2009). The lack of recognition for local forms of knowledge and transmission perpetuates the displacement of local ways of life and knowing, ultimately continuing the trend to undermine subsistence ways of life.

These and other examples show that the SDGs appear to be continuing the trend of the MDGs in overlooking the important role of local ecological knowledge and biocultural heritage in managing socio-ecosystems and maintaining community engagement with environmental management. This concern had already been articulated by the Indigenous People's Major Working Group,

> The failure to recognize indigenous peoples as distinct groups under the MDGs resulted in the absence of targeted measures to address their specific situations related to poverty and severely limited the realization of favourable outcomes. Furthermore, culturally-blind implementation of the MDGs resulted in inappropriate development programmes for indigenous peoples including discriminatory actions related to education, health and basic services. If the world community truly aspires to leave no one behind, it is critical that these gaps be recognized and addressed moving forward. (IPMG 2015, p. 1)

The listed SDGs do not explicitly specify the importance of threats to biocultural diversity such as linguistic losses, disruption of the oral tradition, fragmentation of communities, or other forms of displacement and general disregard for local ways of life (Krauss 1992; Maffi 1998; Pretty et al. 2009). Targeting these indirect drivers is a significant step in addressing the forces that create cultures of unsustainability that persist within postindustrial society.

20.3 Moving Toward Sustainable Development by Addressing Biocultural Diversity

The new SDG has little language affirming alternative sustainable ways of life in relation to the culture of local communities, nor does it acknowledge the need for restorative and environmental justice surrounding industrial pollution and the transboundary politics of resource use among its specific targets. To take these recommendations seriously, these priorities should be made explicit in the SDGs. This could be accomplished with an explicit goal number 18 that emphasizes the importance of promoting sustainable values, local ecological knowledge, and biocultural heritage.

There are already valuable articulations of this need, from the Indigenous People's Working Group already mentioned (March 2014) to the Indigenous People's Climate Change Biocultural Assessment (IPCCA 2009), among others. The IPCCA provides a methodological toolkit as an adaptation of the Millennium Ecosystem Assessment model by incorporating key concepts such as "Buen Vivir" and its relation to "indigenous resilience." These principles serve as the end goal for direct and indirect drivers of environmental change.[4] The toolkit has been prepared to use by local IPCCA partners, indigenous peoples, and communities interested in indigenous and local assessments of climate change.

The biocultural framework is a focused subfield emphasizing the ethical engagement of human rights with consideration of biodiversity conservation priorities, in its biological and cultural expressions. This trans- and interdisciplinary framework has been expounded upon by indigenous communities, ecologists, philosophers, and anthropologists and articulates the deep interrelation between biological, cultural, and linguistic diversity (as is demonstrated by the works presented in this volume; see also Gavin et al. 2015; Maffi 2007a; Pretty 2011; Rozzi 2013; UNESCO 2010). The implications of biocultural diversity as a conceptual unit have ramifications for the formal structures within society, such as formal education policy and foodways and the way we manage growth, development, and concepts of well-being. Additionally, the biocultural framing also provides a mechanism to articulate ways the transmission of knowledge and meaningful engagement takes place and

[4]The guiding principles proposed by the IPCCA, which include self-determination, climate justice, food sovereignty, endogenous development, adaptation/mitigation, and Buen Vivir. See the IPCCA Toolkit, available online at http://ipcca.info/toolkit-en-ipcca-methodological-toolkit.

the ways that humans modify landscapes. Luisa Maffi commented on the purpose of framing biological, cultural, and linguistic diversity as an interdependent unit:

> … the field of biocultural diversity has not adopted the conventional academic "neutrality." From its inception, it has embraced a strong ethics and human rights component, and has promoted a vision in which the protection of human rights (both individual and collective) is intimately connected to the affirmation of human responsibilities toward and stewardship over humanity's heritage in nature and culture. In this view, the biocultural diversity of life has intrinsic value, as diversity is the expression of life's evolutionary potential, and it ought to be protected and maintained. Any damage to it ought to be remedied, and any further damage ought to be prevented. This requires a complex but necessary, and ultimately winning, balance between nature conservation and human development, and between the rights of nature and the rights of humans. (Maffi 2007a)

Maffi and other biocultural conservationists argue that traditional ecological knowledge, cultural heritages, and place-specific names capture a relationship and way of relating that serves to instruct morally, perpetuating the values particular to that language and culture. When these local practices and language are lost, the meaning and ways of relating between humans and nature are similarly disrupted—as Maffi says, "losing the link" (Maffi 1998, 2001; Persic and Martin 2008). The process of "breaking the link" between culture and place is driven by external forces which displace peoples from their native lands, disregard cultural traditions, or cause linguistic assimilation and create a disconnect in consumer and industrial cultures (Maffi 2001, 2005; Mühlhäusler 2001; Nabhan and St. Antoine 1993; Nazarea 1999; Rozzi 2015a, b; Poole 2015). These forces are often mutually reinforcing and ultimately lead to local communities losing control over their environments and elements of their culture (Maffi 2012). This disconnect or displacement process is further exacerbated as ecosystems are degraded and communities must leave their native lands or move to the market economy, away from subsistence agricultural practices. The contrast of communities that are able to live within their ecological limits and those that are not also reveals that, while it is possible for humans to be destructive toward the environment, they can also serve as successful environmental stewards who add to the biocomplexity and richness of a given space.

Throughout the MDGs and SDGs, the language does not emphasize the need to protect local ecological knowledge and local ways of life as a fundamental heritage of the local human actor even as it acknowledges this importance. Fifteen years after the proposed MDGs, an explicit articulation of the need to maintain the ethical relationship urging protection of biodiversity, the environment, and local ecological knowledge is still missing within the revision of the goals. The absence of specific and exacting language affirming cultural sovereignty in its own right or the factors that threaten to undermine TEK or LEK remain in the new SDG. Language affirming alternative sustainable ways of life in relation to the culture of local communities is largely absent, additionally overlooking policy language to address the need for restorative and environmental justice surrounding industrial pollution and the transboundary politics of resource use.

These absences should be made explicit in the SDGs and the subsequent targets listed to meet these goals. To address this, I propose to add a summary goal that encapsulates this missing piece in the sustainable puzzle: the explicit recognition of local ecological knowledge and cultural heritage to sustainability as a fundamental dimension of consideration. This summary goal could be added as:

> SDG #18: Protect, promote, and engage biocultural heritage to reinforce and support sustainable interconnections between diverse human societies and their distinct environments.

20.4 Goal #18: Biocultural Heritage: A Major Driver for Sustainable Development

In this chapter, I have argued that there is a need to consider local ecological knowledge and biocultural heritage within the SDGs—and within policy more broadly—the interlinkages between biological, cultural, and linguistic diversity. This necessitates an affirmation of the deep histories of communities that have memories of the land, its changing landscape and conditions, and the flora and fauna that coexist within these shared spaces. Additionally, sustainability discourse needs to address factors that contribute to developing ecological knowledge and environmental ethics, and policy should consider factors which prevent communities from acting in accordance with their ecological conscience or culture wherever a community lives within the rural-urban gradient. Just as the environment is facing losses through dominant development practices, communities are becoming more significantly disconnected, removed, and ignorant of the natural world and its ecological processes. In order to promote the care, management, and wisdom to live sustainably with the earth and with the many cultures that thrive on this planet, it is necessary to halt and overcome the loss of such experiences.

The disconnection of communities with the land and the confluent losses of biological and cultural diversity can be reverted. For example, in 2012, the largest dam removal in US history (to date) was initiated, demolishing the Glines Canyon Dam on the Elwha River of Washington State. Within a year, Chinook salmon returned to spawn in the rivers. Mel Elofson, who lives and works with the watershed, said of the return:

> My grandmother lived on that homestead, and she walked down there as a young woman when the dams were being built and was devastated…. She used to talk about nearly being able to walk across the river on the backs of the salmon. There were tears of sadness back then, but if all the elders were alive, there would be tears of joy. (As quoted in Slobig 2014)

When the habitats are degraded and wildlife displaced and absent, and when even the memory of ecology that once was is gone, a deterioration of the sustainable and thoughtful relationship between communities and the land occurs. This deterioration disrupts not just the experience of what once was but also of what could be. On the contrary, if the habitats are restored and the biocultural heritage of local peoples recognized, sustainable life habits and wildlife return. For these reasons, sustainable development must confront these hidden, indirect drivers of biocultural loss and overcome the drivers built into society perpetuating the loss of ecological knowledges and biocultural heritages. A step in affirming the deep connections that humans do share with place in many communities and cultures around the world is by affirming biocultural heritage in its complexity, diversity, and uniqueness. A deeper analysis of the factors that erode biocultural heritage needs to be incorporated into our policy solutions for sustainable development, and this goal needs to be at the forefront of all of our thoughts as we seek social and environmental justice for a sustainable existence within our global society.

References

Agyeman J (2005) Sustainable communities and the challenge of environmental justice. New York University Press, New York

AOSIS (2016) Alliance of Small Island States, statement by the republic of Maldives on behalf of AOSIS at the high level thematic debate on achieving the sustainable development goals, 21 April 2016. http://aosis.org/wp-content/uploads/2016/04/2016.04.21-AOSIS-Statemet-HLTD-on-Implementing-SDGs-.pdf. Accessed 9 July 2016

Argumedo A, Pimbert M (2008) Protecting farmers' rights with indigenous biocultural heritage territories: the experience of the potato park. Available via: IIED/Association ANDES, [No city given]. http://pubs.iied.org/pdfs/G03072.pdf

Argumedo, A, K Swiderska, M Pimbert, Y Song, and R Pant (2011) Implementing farmers' rights under the FAO International Treaty on PGRFA: the need for a broad approach based on biocultural heritage. Paper presented at the 4th Governing Body of the International Treaty on PGRFA, Bali, 14–18 March 2011. http://cmsdata.iucn.org/downloads/farmers_rights_paper_bali_draft5.pdf

Berkes F, Folke C, Colding J (2000) Linking social and ecological systems: management practices and social mechanisms for building resilience. Cambridge University Press, Cambridge

Brundtland GH (1987) Report of the world commission on environment and development: our common future. United Nations, New York

Carpenter SR, De Fries R, Dietz T, Mooney HA, Polasky S, Reid WV, Scholes RJ (2006) Millennium ecosystem assessment: research needs. Science 314(5797):257–258

Cocks M (2006) Biocultural diversity: moving beyond the realm of 'indigenous' and 'local' people. Hum Ecol 34(2):185–200

Cocks M (2010) What is biocultural diversity? A theoretical review. In: Bates DG, Tucker J (eds) Human ecology: contemporary research and practice. Springer, Boston, pp 67–77

Congreso de Colombia (1993) Law 70 of Colombia (1993): in recognition of the right of black Colombians to collectively own and occupy their Ancestral Lands, English Translation. (trans: Jackson NL, Jackson P). April 2007. Available via: Benedict College, South Carolina. www.benedict.edu/exec_admin/intnl_programs/other_files/bc-intnl_programs- law_70_of_colombia-english.pdf

Costanza R, de Groot R, Sutton P, van der Ploeg S, Anderson SJ, Kubiszewski I, Farber S, Turner RK (2014) Changes in the global value of ecosystem services. Glob Environ Chang 26:152–158

Davidson-Hunt IJ, Turner KL, Pareake Mead AT, Cabrera-Lopez J, Bolton R, Idrobo CJ, Miretski I, Morrison A, Robson JP (2012) Biocultural design: a new conceptual framework for sustainable development in rural indigenous and local communities. Available via: S.A.P.I.EN.S [Online], 5.2. http://sapiens.revues.org/1382

Escobar A (2011) Encountering development: the making and unmaking of the third world. Princeton University Press, Princeton

Fehling M, Nelson BD, Venkatapuram S (2013) Limitations of the Millennium Development Goals: a literature review. Glob Public Health 8(10):1109–1122

Gavin MC, McCarter J, Mead A, Berkes F, Stepp JR, Peterson D, Tang R (2015) Defining biocultural approaches to conservation. Trends Ecol Evol 30(3):140–145

Grear A (2015) The discourse of "biocultural" rights and the search for new epistemic parameters: moving beyond essentialisms and old certainties in an age of Anthropocene complexity? J Hum Rights Environ 6(1):1–6

Hunn E (1999) The value of subsistence for the future of the world. In: Nazarea V (ed) Ethnoecology: situated knowledge/located lives. University of Arizona Press, Tucson, pp 22–36

Hunn E (2007) Ethnobiology in four phases. J Ethnobiology 27(1):1–10

IPCCA (2009) The Indigenous Peoples' Biocultural Climate Change Assessment Initiative. http://ipcca.info/. Accessed 5 July 2015

IPMG (2014) The Indigenous Peoples Major Group vision and priorities for the sustainable development goals. Major group position paper. March 2014. https://sustainabledevelopment.un.org/content/documents/3447SD2015%20Position%20Paper_Indigenous%20Peoples%20MG_v1_March%202014.pdf

IPMG (2015) The Indigenous People's Major Group policy brief on sustainable development goals and post 2015 development agenda, a working draft position paper. https://sustainabledevelopment.un.org/majorgroups/indigenouspeoples

Kahn PH, Kellert SR (2002) Children and nature: psychological, sociocultural, and evolutionary investigations. MIT Press, Cambridge

Krauss M (1992) The world's language in crisis. Language 68:4–10

Liu S, Costanza R, Farber S, Troy A (2010) Valuing ecosystem services. Ann NY Acad Sci 1185(1):54–78

Maffi L (1998) Language: a resource for nature. Nat Resour 34(4):12–21

Maffi L (2001) On the interdependence of biological and cultural diversity. In: Maffi L (ed) On biocultural diversity: linking language, knowledge, and the environment. Smithsonian Institution Press, Washington, DC, pp 1–50

Maffi L (2005) Linguistic, cultural, and biological diversity. Annu Rev Anthropol 34:599–617

Maffi L (2007a) Biocultural diversity and sustainability. In: Pretty J, Ball AS, Benton T, Guiant JS, Lee DR, Orr D, Pfeiffer MJ, Ward H (eds) The SAGE handbook of environment and society. Sage Publications, London, pp 267–278

Maffi L (2007b) Bio-cultural diversity for endogenous development: lessons from research, policy and on-the-ground experiences. Paper presented at conference on endogenous development and bio-cultural diversity: the interplay of worldviews, globalization, and locality, Geneva, October 2006, pp 56–66

Maffi L (2012) Interview with Luisa Maffi 03.04.2012 "Cultures Are No Museums Specimens." The European. http://www.theeuropean-magazine.com/luisa-maffi%2D%2D2/6068-biocultural-diversity. Accessed 16 March 2015

Maffi L, Woodley E (2012) Biocultural diversity conservation: a global sourcebook. Routledge, Abingdon/New York

Millennium Ecosystem Assessment (2003) Ecosystems and human well-being: a framework for assessment. Island Press, Washington, DC

Millennium Ecosystem Assessment (2005) Ecosystems and human well-being: synthesis. Island Press, Washington, DC

Mühlhäusler P (2001) Ecolinguistic, linguistic diversity and ecological diversity. In: Maffi L (ed) On biocultural diversity: linking language, knowledge, and the environment. Smithsonian Institution Press, Washington, DC, pp 133–144

Nabhan GP (2001) Cultural perceptions of ecological interactions: an "endangered people's" contribution to the conservation of biological and linguistic diversity. In: Maffi L (ed) On biocultural diversity: linking language, knowledge, and the environment. Smithsonian Institution Press, Washington, DC, pp 145–156

Nabhan GP, St. Antoine S (1993) The loss of floral and faunal story: the extinction of experience. In: Wilson EO (ed) The biophilia hypothesis. Harvard University Press, Cambridge, pp 229–250

Nazarea V (ed) (1999) Ethnoecology: situated knowledge/located lives. University of Arizona Press, Tucson

Nelson GC (2005) Drivers of ecosystem change: summary chapter. In: Hassan R, Scholes R, Ash N (eds) Ecosystems and human well-being: current state and trends. Island Press, Washington, DC, pp 73–76

Nelson GC, Bennett E, Berhe AA, Cassman K, DeFries RS, Dietz T, Dobermann A, Dobson A, Janetos A, Levy M, Marco D, Nakicenovic N, O'Neill B, Norgaard R, Petschel-Held G, Ojima D, Pingali P, Watson R, Zurek R (2006) Anthropogenic drivers of ecosystem change: an overview. Ecol Soc 11(2):29 http://www.ecologyandsociety.org/vol11/iss2/art29/

Orr DW (1992) Ecological literacy: education and the transition to a postmodern world. SUNY Press, Syracuse

Persic A, Martin G (2008) Links between biological and cultural diversity: report of the international workshop, 26–28 September 2007. Available via: UNESCO HQ Paris. http://unescodoc.unesco.org/images/0015/001592/1592551

Poole AK (2015) Urban sustainability and the extinction of experience: acknowledging drivers of biocultural loss for socio-ecological well-being. Dissertation, University of North Texas

Poole AK (2018) Where is goal 18? The need for biocultural heritage in the sustainable development goals. Environmental Values 27(1):55–80

Poole AK, Hargrove EC, Day P, Forbes W, Berkowitz AR, Feinsinger P, Rozzi R (2013) A call for ethics literacy in environmental education. In: Rozzi R, Pickett STA, Palmer C, Armesto JJ, Callicott JB (eds) Linking ecology and ethics for a changing world: values, philosophy, and action. Ecology and ethics, vol 1. Springer, Dordrecht, pp 349–371

Posey DA, Dutfield G (1996) Beyond intellectual property rights: toward traditional resource rights for indigenous and local communities. IDRC, Ottawa

Pretty J (2011) Interdisciplinary progress in approaches to address social-ecological and ecocultural systems. Environ Conserv 38(2):127–139

Pretty J, Adams B, Berkes F, de Athayde S, Dudley N, Hunn E, Maffi L, Milton K, Rapport D, Robbins P, Sterling E, Stolton S, Tsing A, Vintinnerk E, Pilgrim S (2009) The intersections of biological diversity and cultural diversity: towards integration. Conserv Soc 7(2):100–112

Rozzi R (2012) Biocultural ethics: the vital links between the inhabitants, their habits and regional habitats. Environ Ethics 34:27–50

Rozzi R (2013) Biocultural ethics: from biocultural homogenization toward biocultural conservation. In: Rozzi R, Pickett STA, Palmer C, Armesto JJ, Callicott JB (eds) Linking ecology and ethics for a changing world: values, philosophy, and action. Ecology and ethics, vol 1. Springer, Dordrecht, pp 9–32

Rozzi R (2015a) Earth stewardship and the biocultural ethic: Latin American perspectives. In: Rozzi R, Chapin FS III, Callicott JB, Pickett STA, Power ME, Armesto JJ, May RH Jr (eds) Earth stewardship: linking ecology and ethics in theory and practice. Ecology and ethics, vol 2. Springer, Dordrecht, pp 87–112

Rozzi R (2015b) Implications of biocultural ethics for earth stewardship. In: Rozzi R, Chapin FS III, Callicott JB, Pickett STA, Power ME, Armesto JJ, May RH Jr (eds) Earth stewardship: linking ecology and ethics in theory and practice. Ecology and ethics, vol 2. Springer, Dordrecht, pp 113–136

Rozzi R, Massardo F, Anderson CB, Heidinger K, Silander JA Jr (2006) Ten principles for biocultural conservation at the southern tip of the Americas: the approach of the Omora Ethnobotanical Park. Ecol Soc 11(1):43 Available via: http://www.ecologyandsociety.org/vol11/iss1/art43/

Schröter M, Zanden EH, Oudenhoven AP, Remme RP, Serna-Chavez HM, Groot RS, Opdam P (2014) Ecosystem services as a contested concept: a synthesis of critique and counter-arguments. Conserv Lett 7(6):514–523

Slobig Z (2014) Salmon return to Washington's Elwha River for the first time in 102 years. Takepart, 17 September 2014, http://www.takepart.com/article/2014/09/17/salmon-return-elwha-river-first-time-102-years. Accessed 13 July 2015

UNDRIP (2007). United Nations General Assembly resolution A/61/L.67, United Nations Declaration on the Rights of Indigenous Peoples (UNDRIP – 13 September 2007) Available via: http://www.un.org/esa/socdev/unpfii/documents/DRIPS_en.pdf

UNESCO (2010) 2010 Declaration of bio-cultural diversity. The international conference on cultural and biological diversity for development, Montreal, Canada, 8–10 June 2010 Available via: http://www.unesco.org/mab/doc/iyb/Declaration.pdf. Accessed 7 June 2015

UNGA (2015) United Nations General Assembly resolution 70/1, transforming our world: the 2030 agenda for sustainable development, A/79/L.1 (25 September 2015). Available via: undocs.org/A/RES/70/1

Zent S (2009) Traditional ecological knowledge (TEK) and biocultural diversity: a close-up look at linkages, delearning trends, and changing patterns of transmission. In: Bates P, Chiba M, Kube S, Nakashima D (eds) Learning and knowing in indigenous societies today. UNESCO, Paris, pp 39–58

Chapter 21
Suma Qamaña or Living Well Together: A Contribution to Biocultural Conservation

Xavier Albó

Abstract Living well, *suma qamaña* in Aymara or *buen vivir* in Spanish, is the fundamental moral logic that guides Aymara and other Andean cultures and has been incorporated into the Constitutions of Bolivia and Ecuador. A linguistic analysis shows that the conceptual root of *suma qamaña* (*buen vivir* or living well) is convivial gathering or community in which basic needs are met and relationships, which even extend to the earth, are harmonious. Living well implies a strong ethical component of valuing and appreciating the distinctiveness of others, as well as spirituality or lifestyle, and so can never be just economic. Since the human factor is always in the forefront, quantifiable economic scores cannot be the exclusive measure. Measurable indicators are required for determining the quality of social relations and of relations with nature, as well as for inter-cultural, inter-gender, and inter-whatever, that show the quality of these relationships, such as equitable reciprocity and convivial living. As a starting point, the best way perhaps will be for each people and culture to develop and make explicit their dearest living values (as has been done in Bolivia with *suma qamaña*) and, from these, to specify some of the desirable relationships of convivial living that indicate their being fulfilled or violated. Finally, living well is not possible if the inequalities of the general power structure are not faced and overcome.

Keywords Aymara · Bolivia · Andean culture · Convivial life · Mother Earth

For biocultural conservation, a concept that offers an alternative to the prevailing Western development-centered approach could be *buen vivir* in Spanish or *suma qamaña* in the indigenous Aymara language. Aymara people represent a high proportion of the population of the highlands of the Andes or *Altiplano* of Bolivia and Peru, as well as parts of Chile and Argentina. *Suma qamaña* is based on the belief that true well-being ("the good life") is only possible as part of a community. It is also relevant for the concepts of coinhabitant and coinhabitation of the biocultural

X. Albó (✉)
Centro de Investigación y Promoción del Campesinado (CIPCA), La Paz, Bolivia
e-mail: xalbo@cipca.org.bo

R. Rozzi et al. (eds.), *From Biocultural Homogenization to Biocultural Conservation*, Ecology and Ethics 3, https://doi.org/10.1007/978-3-319-99513-7_21

ethic (sensu Rozzi 2012, 2015), because the community in the Aymara culture includes not only humans but also plants, animals, and the Mother Earth or *Pacha Mama* (Cfr. Mamani-Betsabe 2015; May 2015). Not only for Aymara people but also for many other native peoples around the world, nature itself must be cared for and respected as a valuable part of the community. The land should not be owned; instead, it should be honored and protected.

In Bolivia, the concept of *suma qamaña*, that is, the "good life" or "living well," has become so relevant that it is among the great "ethical-moral principles of the pluralist society" that the Constitution of the Plurinational State of Bolivia reiterates several times (Preamble and Title 1, Art. 8), including it also in articles relating to education (Art. 80) and economics (Art. 306 and 313).

What is behind this concept of *living well* for Aymara people? I will explore it gradually, expanding it as I go, first in purely linguistic terms and then in a wider framework, as the logic common to many Native American people, over against the logic of the powerful and dominate societies. Finally, I will suggest ways to express this concept of *suma qamaña* through measurable indicators comprehensible in Western culture.

21.1 The Concept of Living Well for the Aymara Culture

Sometimes translations cannot be completely accurate because they lack context. Something like that occurs with the expression "good life" or "living well" – *suma qamaña* – if it is not contextualized in the original Aymara, the language and culture in which it was coined. To understand fully the meaning of *suma qamaña* and to use it correctly, the concept needs to be analyzed and appreciated in its original Aymara language and culture.[1]

Qamaña means "to inhabit, to live [in a specific place or environment], and to dwell," and *qamasiña* means "to live with someone." *Qamaña* also is the name that is given to a sheltered place, protected against the wind, constructed in a semicircle of stones, as a resting place for shepherds while they relax or attend their flocks. We see, then, that from various angles, *qamaña* means to live, reside or dwell, rest, protect, and care for others. Although not explicit, a second meaning also insinuates convivial living with nature or the *Pacha Mama* or Mother Earth. A linguistic analysis shows that the conceptual root is *qama*, to which various suffixes can be added to give new meanings. So *qama-wi* is "residence," but *qamäwi* is, very significantly, "the meeting of people who are accustomed to getting together for conversation and having a good time." One becomes a "person" (*jaqi*) only as a (heterosexual) couple, the minimal unit for convivial living that also is the source of new life. Thus when one marries, one "becomes a person" or *jaqichasiña*, and these couples are the basis of all communal organization. The Aymara concept of *chacha-warmi*

[1] This linguistic part is based primarily on Layme 2004, as well as personal consultations with the author.

underscores that this convivial coupling also presumes both difference and complementarity between those who learn to live together well.

Qamasa, along with the gerund "living, living together," means "character and shape of being," as well as "bravery, audacity, valor, and courage." Also, it is said that such-and-such places are laden with *qamasa*, generically referring to much "energy." At these places, gestures and rituals are conducted in order to fill one's self with that energy. *Qamasa* is, therefore, "energy and vital force for living and sharing with others." This is the most explicit relation between the root *qama* as something that is very strong and alive in the *Pacha Mama* and us who inhabit her and make her our dwelling place.

The most common way in Aymara to say that someone is "rich" is *qamari*. Actually, in order to speak of rich, bourgeois, arrogant people, etc., many, Aymaras or not, tend to use this term in its plural form: *qamirinaka*. The deepest connotation would not be that of a rich person living a luxurious and arrogant lifestyle at the expense of others, but rather one who has great bounty to share and celebrate with others.

This variety of meanings of *qamaña*, mostly in reference to social gathering and ecological living, is far richer than those of *jakaña*, which also means life or to live, but only in the sense of being alive, as opposed to being dead or death. Therefore, when the Andean world, as well as other original indigenous peoples, claims that their cultures are life-affirming, they are referring not only to the physical meaning of life or living but also to this whole group of social relations within a hospitable environment. So, they also speak of "caring" and "nurturing" life, as something we do together, in families, and also in the way we care for and protect our crops and the whole natural environment. *Jakaña* means simply "to live." But "to live in peace" and "to live joyously and pleasurably" require *qamaña* (Bertonio 1984).

In the concept of *suma qamaña*, *suma* is described not only as "pretty, beautiful, agreeable, good, and friendly" but also as "precious, excellent, finished, and perfect." It also has the sense of "fullness." And *suma jaqi* means "good people, kind, and generous, who show good will to others" (De Lucca 1987; Layme 2004).

Why don't Aymaras talk much about living *better*? In their concept, it is not necessary because *suma* (or *sumaq* in Quechua) already includes "the best possible level." On the other hand, Aymaras who have thought deeply about this resist saying "better" because too often it is understood as referring to individuals or groups that are better off at the expense of others. *Suma qamasiña* is about living together well, not about some who are better off at the expense of others.

This helps us understand the meaning of *q'ara*, a term often used pejoratively in reference to wealthy white people. However, rather than skin color, *q'ara* means "being better off at the expense of others" because, in reality, *q'ara* refers to nudity, that is, lacking something fundamental. It is equivalent to "uncivilized" because it shows that the rules and objectives fundamental to social gathering (*convivencia*) are not respected. It's true that many interpret this term in the racial sense of nonindigenous people, but this meaning is only figurative and can be applied as well to Aymaras who do not conduct themselves according to the basic norms of social gathering or convivial life. At times I have asked why the rich are still called *q'ara* when in reality they don't lack anything. The answer always is "they don't have anything that's their own that they produced themselves."

Following on this, the opposite of *suma qamasiña* or "living kindly and generously" would be *q'araru tukuña* or to be naked, uncivilized, and lacking something fundamental. Obviously, if there are ways of social gathering that are good and generous, then there are others that are not. The ideal is to reinforce the former. Therefore, only to say "well" in the sense of *suma(q)* and in the context of *qamaña* connotes the good of the whole social group. It would not do simply to indicate that some are much better off than others. If "we get better," then it has to be, if possible, that "everyone gets better" because all are tied together by the bonds of social gathering.

The Ministry of Development Planning for Bolivia illuminates the concept of living well in the following manner (Seminario Internacional sobre el Vivir Bien, La Paz, 3–5 December 2009):

> Good life or living well implies access to and enjoyment of material goods in harmony with nature and people. It is affective and spiritual fulfillment in the human dimension. People do not live isolated but in families and in social and natural surroundings. It is not possible to live well if others live badly or if nature is damaged.

In the Aymara and Andean context, goods not only are "material" but also *spiritual*, such as reciprocity, close friendship, and god-parenting, as well as celebrations. These always are charged with endearing affection. You could also add that Aymara families don't live isolated but rather in intense networks with the rest of the community or *ayllu*. Likewise, and without going into detail about the integrated complex of nature, production, and the spiritual and social worlds, one would have to recognize nature as having a personal and affective character because, as the *Pacha Mama*, reciprocal relations are maintained.

The French anthropologist Dominique Temple emphasized that it is one thing "to live" and another "to live humanly," which implies social gathering, interchange, and reciprocity. To live in the sense of survival, just not being dead, is more in the line of *jakaña* which means "minimal" and so indicates what we have in common with the other forms of living beings, vegetable or animal. But beyond this, it's necessary to live as humans, with the many expressions of mutual reciprocity and affection that human life implies. This also is in line with *qamaña*, and so to do this, "living well" is *suma qamaña.*

Therefore, *suma qamaña* implies a strong ethical component of valuing and appreciating the distinctiveness of the other, as well as spirituality or lifestyle. Thus the center for developing "living together well" (*buen vivir*) can never be just economic. The human factor is always in the forefront. Economics and material welfare are always among the components of development, if for no other reason than to overcome mere survival. But now it is evident that the goal of development is not ever-increasing economic growth at the expense of social gathering among people and the Mother Earth, the "womb" of all of us.

21.2 Beyond the Rural and Aymara World

Some of these elements also can be found in the cities, within specific occupational and fraternal groups, or neighborhoods, although in this urban context, not everyone participates. Rather, only those with the most resources and power do so, so there is greater risk of emphasizing profits and the personal power and prestige of individuals, over service and the idea of everyone advancing together, according to the logic of *suma qamaña*.

There are no notable differences between communities that speak Aymara or Quechua when they share similar socioeconomic circumstances because, in reality, language doesn't change what is a basically common Andean culture. Still, the Aymara expression *suma qamaña* allows for more semantic connotations than the Quichua/Quechua *sumak/sumaq kawsay*[2] introduced into the new Ecuadorian Constitution. In effect, from a purely linguistic perspective, *kawsay* is more similar to the Aymara *jakaña*. Some Bolivian Quechua speakers have suggested to me that the translation best capturing the nuances of *suma qamaña* would be *allin kawsay* or *allin tiyakuy*.

Other indigenous original peoples also have similar ideas although they express them through other concepts. Although examples are multiple, in Bolivia as well as elsewhere, I will only sketch the case of Guaraní communities of the Chaco, the ecological and cultural opposite of the Andean world, since it is the other case mentioned in Article 8 of the new Bolivian Constitution.

The Guaraní speak above all of *ñande reko* – "our mode of proceeding" – in which a fundamental role is accorded as much to reciprocal relations and the power to share hunting, fishing, food, and drink when they are abundant, as to the relation of all this to territory, viewed now not as "Mother Earth" but as "the place and environment that provide the conditions of possibility of the Guaraní mode of being" (Medina 2001a, b, 63, interpreted by Melià 1987). The principal expression of Guaraní living well surely is *arete*, their main celebration which, now in Spanish, also is called carnaval and is celebrated only in years with good production. Above all it is a celebration of maize. With maize large earthen jars of good *chicha* are prepared, and the beer is joyously shared by all until the jars are empty while dancing in rounds and celebrating their reencounter with the *aña*, "the dead" (represented by youth adequately attired with masks) who have come out of the woods to join the fun all are having. In terms of utopia, it is necessary also to underline the myth (and historic migrations) of "the land without evil" (*ivi marãei*), that is, a beautiful land, resplendent, humid, and fertile, in which all labor together (*motirõ*) and together all share the fruits of that labor and the grand *fiesta* with abundant *chicha* (Melià 1987).

[2] In Ecuador the spelling is Quichua [*kichwa*], and the term is pronounced and written *sumak*, for lacking contrast between the palatal consonant *k* and future *q*. In Peru and Bolivia, where this contrast exists, the *q* drags the *i* toward the phonetic pronunciation [e] (without being a phoneme implying a different sense), so that in Spanish the language is called Quechua [*Qhichwa*] and the term is pronounced and written *sumaq*.

My institution, the *Centro de Investigación y Promoción del Campesinado* (CIPCA) [Center for Research and Promotion of Peasant Farmers], once had a learning experience. Our first objective, thought out from a purely technical and political logic, was to support the economic freedom of those communities that were strongly dependent on the agro-industry of Santa Cruz. The Guaraní worked several months of the year during the sugarcane harvest, forced to do so due to constant indebtedness. In addition, during their temporary absence from their lands, their underutilized territories were vulnerable to being usurped by the aggressive agro-industrial advances toward the south. We thought that the communitarian sense so strong there would focus on new agricultural and ranching activities that complemented self-subsistence, supported by their *chacos* or family plots, with production in common oriented to the market. The first "labor communities" were organized and quickly expanded to many surrounding areas. Nevertheless, when the common harvest came around, our economists had to reformulate their initial purely economic calculations because the first concern of the participants wasn't to sell their products, as had been presumed, but to share as widely as possible with all members of the community, above all maize. Only after all were satisfied were the leftovers sent to the market.

21.3 Beyond Theory

The persistence of these ideas among so many peoples leads many theoreticians and even philosophers to question the ethnocentric generalizations usually made from the First World. To develop all this would require whole articles and even books. So I will limit myself only to pointing out some approaches that can be followed.

In Bolivia, David Choquehuanca, the nation's former chancellor, has been the one most forcefully to introduce the principle of living well in the Constitution and the new Plurinational State. Likewise the concept has been utilized almost simultaneously in the Constitution of Ecuador and has been the theme of various international seminars (e.g., see ALAI 2010). At a more academic level, in Bolivia the ones to have made the greatest efforts to systematize these ideas and to locate them within a broader framework are Javier Medina and Simón Yampara and other Aymaras. The practical application of living well also has been explored, above all in the new context created by the Law for Popular Participation, passed in 1994 but, particularly in rural areas, that remained only fiction for lack of resources until its recent reconstruction and the resourcing of municipal governments.

Among Medina's numerous publications that lead to greater theorization and generalizations, I call attention especially to his trilogy that links the Aymara experience to a broader theoretical framework (2001a), reflects on the Guaraní experience (2001a), and gathers and seeks new inputs to incorporate this standpoint into the Bolivian strategy for reducing poverty (2002). Medina presents the most inclusive and holistic principles of the indigenous peoples and also those of diverse theoreticians from every latitude and then contrasts them with the purely economic orientation of the dominant models of development and planning. Although these texts at times use too abstract and elitist language, they for now are the most stimulating summary for understanding the multiple facets of *suma qamaña* generally. I will only mention the criteria suggested by Medina for building "a possible and desirable" Bolivia:

A convivial society.
A frugal but with quality of life society.
A high-synergy society.
A low-entropy society.
A balanced society.
An ecological-symbiotic society.
A dynamically networked society.
A direct, multiple-level democratic society.

It isn't by chance that a short time later, Medina also published three thick volumes of the main contributions of Dominique Temple, *Teoria de la reciprocidad* (2003). Without doubt Temple is the French theoretician that mostly has built upon the pioneering work by Marcel Mauss, *Essai sur le don* (1923–1924), and has the greatest ties to contemporary Aymara thinkers.

Nor is this thought unique. At the bottom, the UNESCO declaration of 2001 about cultural diversity as the principal source of growth in humanness, and the need to subordinate economic development to the human and cultural development of all people, also moves in this same optic. The same thing is affirmed by the various instances of the United Nations that have exiled the fallacious idea of "per capita income" and substituted it with the "index of human development." The new currents about sustainable development, eco-development, and protection of the environment go in the same direction. Those who in the past put down as being primitive the reverence and affection original indigenous people express for the Mother Earth, along with the celebrations and rituals of reciprocity they practice with her, now find support for their viewpoints in these same "primitive" people.

Occurring in all this is a double Copernican Revolution in the concept of development. First, development now isn't driven by economic growth but rather by growth in humanity. Second, Mother Earth – or rather the Cosmos, to be more inclusive – doesn't revolve around us humans, but rather we feel ourselves to be the fruit and integral part of the Mother Earth and the Cosmos, with whom we are to advance and live together harmoniously. The economic priority gives way to the human, inserted in the Cosmos, not excluding economics but giving it a much broader and inclusive meaning. Only in this way can we understand what it really means to live well and actually try to do so.

Understanding living well from the great temporal cycles of nature, many of these apparent innovative ideas do nothing other than take us back to our primal sources and origins. Isn't there also much of *suma qamaña* in, for example, the biblical and Hebrew greeting *Shalom*, meaning peace in its diverse senses, always at once individual but also collective and international, as well as well-being, balance, justice, and integral equality? Don't we also find similar ideas in many eastern religions?

21.4 Measuring the Good Life: Possible Indicators

How can we measure advances toward really living *suma qamaña*? What instruments can be used? First of all, it's necessary to emphasize that, as I have indicated, living well cannot be measured only according to certain quantifiable economic levels. Quantifiable economic scores cannot be the exclusive measure, although they can be important in situations of poverty that hardly permit survival. In this initial sense of assuring full survival (*jakaña*), international instruments that measure overcoming basic poverty, such as the unsatisfied basic need index or the vulnerability index, can serve as the foundation for measuring survival conditions and then go on toward living well and finally living well together. Amartya Sen did us a great favor when, from a viewpoint similar to living well, he succeeded in doing away with "per capita income" – famous and treasonous, easily lumping rich and poor together, and then averaging incomes – as an indicator of development. He was able to substitute it with the Human Development Index (HDI) that today is used generally.

But perhaps he is still short on human because the HDI fails to include characteristics that are truly "human" (humanist, perhaps some would say), such as those mentioned as integral for living well. His indicators only refer to individuals without judging whether or not they know how to live in relation to others. They point mostly to living and growing as individuals rather than to living well with others or convivial social gathering that is fundamental for growing in humanness in synchronization with Mother Earth.

What other indicators do we need to cover these other dimensions? We have at least two veins: one that refers to the quality of social relations and another that refers to the quality of relations with nature, the Mother Earth. In the first vein, we could include, perhaps, even some of the governability index developed by the United Nations Development Program (UNDP), particularly those that appear in the subindex for participation and political stability, in the sense of "good government" (a better term than abstract "governance"). In the second, some of the indicators associated with sustainable development could be included, and others that are developed in what the Constitution of Ecuador of 2008 calls, suggestively, "rights of nature."

In both veins attempts have been made at quantification, at least to ordinal scale. Nevertheless, if we move to the more affective aspects of convivial living, it's difficult to think of universal and transcultural indicators. Surely it will be necessary to design indicators or more quantitative approximations for each social and cultural reality. As a starting point, the best way perhaps will be for each people and culture to develop and make explicit their dearest living values (as has been done in Bolivia with *suma qamaña*) and, from these, to specify some of the desirable relationships of convivial living that indicate their being fulfilled or violated. These relationships can be applied, for example, to the internal life of the community, to the administration of justice, or even to the joint celebration of rituals of production.

At some point it will be necessary to include types of inter-cultural, inter-gender, and inter-whatever indicators that show the quality of relationships, equitable reciprocity, convivial living, etc., among those who are distinct for one or another reason, for example, for their place of origin, gender, religious or political affiliation, identity and cultural habits, social position, and economic status, among others. The difference between subjects and groups to which the diverse interlocutors belong is almost always part of relationships of convivial living. For the Aymara this begins with the complementary couple *chacha-warmi* because it is the base relationship that, for the Aymara, makes a person (*jaqi*).

In another corner of the world, the Kingdom of Bhutan, a little Buddhist country in middle of the Himalaya Mountains, so far from yet at the same time similar to Andean Bolivia, we find another inspiring example. Only a few years ago, it was decided to base their system of indicators on something comparable to our *suma qamaña*, but there they are called Gross National Happiness (GNH). According to Dasho Karma Ura (2009), these are distributed among the nine following dimensions:

Psychological well-being.
Use of time.
Community vitality.
Culture.
Health.
Education.
Environmental diversity.
Level of life.
Good government.

21.5 Final Warning

One of the fundamental obstacles to achieving living well, in a way that reaches all, including the ethical, affective, and social inclusiveness components, is, obviously, the dominate structure of power, whether in its economic or political dimension, which places one group at the top of the candle which occupies all space for itself, impairing others. In the face of this, the equally obvious consequence is that it isn't possible to achieve the objective of living well without influencing the social, political, and economic power structure in order to prevent this. Only a change in the socioeconomic pyramid can facilitate a similar structural change in our system of life together. This is the toughest bone to gnaw!

This is not to say, however, that everything can be reduced to sociopolitical or economic factors. Other dimensions, molded by the capacity to connect constructively with others who are distinct, have their own spheres and specific tasks. For example, perhaps some socialist regimes were able to reduce social and economic differences. But when they tried to reduce or resolve problems as exclusive problems of social class, thus ignoring other dimensions, a boiling broth was created for inter-ethnic conflict that exploded in many scenes shortly after these regimes fell. But also, on the other hand, it will be impossible to resolve these other problems of life together or convivial living if the inequalities of the general power structure are not faced and overcome.

References

ALAI (Agencia Latinoamericana de Información) (2010) Sumak kawsay: recuperar el sentido de la vida. América Latina en Movimiento 452

Bertonio L (1984) Vocabulario de la lengua aymara. Introduction by X Albó and F Layme. CERES, IFEA and MUSEF, Cochabamba, Bolivia

Bolivia (2009) Constitución Política del Estado. La Paz, Bolivia. Available from: http://www.oas.org/dil/esp/Constitucion_Bolivia.pdf

De Lucca M (1987) Diccionario práctico Aymara-Castellano y Castellano-Aymara. Los Amigos del Libro, La Paz and Cochabamba, Bolivia

Karma Ura D (2009) Gross national happiness. Center for Bhutan Studies and GNH Research, Thimphu, Bhutan. Available from: www.grossnationalhappiness.com

Layme F (2004) Diccionario bilingüe aymara castellano. Consejo Educativo Aymara, La Paz, Bolivia

Mamani-Bernabé V (2015) Spirituality and the Pachamama in the Andean Aymara worldview. In: Rozzi R, Chapin FS III, Callicott JB et al (eds) Earth stewardship: linking ecology and ethics in theory and practice. Springer, Dordrecht, pp 65–76

Mauss M (1923–1924) Essai sur le don. Forme et raison de l'échange dans les societés archaiques. L'Année Sociologique, Paris

May RH Jr (2015) Andean llamas and earth stewardship. In: Rozzi R, Chapin FS III, Callicott JB et al (eds) Earth stewardship: linking ecology and ethics in theory and practice. Springer, Dordrecht, pp 77–86

Medina J (2001a) La comprensión indígena de la buena vida. PADEP-GTZ, La Paz, Bolivia

Medina J (2001b) Ñande teko: La comprensión guaraní de la buena vida. PADEP-GTZ, La Paz, Bolivia

Medina J (2002) La vía municipal hacia la buena vida. PADEP-GTZ, La Paz, Bolivia

Melià B (1987) La tierra sin mal de los guaraní. Suplemento Antropológico, Asunción

Rozzi R (2012) Biocultural ethics: recovering the vital links between the inhabitants, their habits, and habitats. Environ Ethics 43:27–50

Rozzi R (2015) Earth stewardship and biocultural ethics: Latin American perspectives. In: Rozzi R, Chapin FS III, Callicott JB, Pickett STA, Power ME, Armesto JJ, May RH Jr (eds) Earth stewardship: linking ecology and ethics in theory and practice. Springer, Dordrecht, pp 87–112

Temple D (2003) Teoría de la reciprocidad, three volumens. PADEP-GTZ, La Paz, Bolivia

Chapter 22
Biocultural Approaches to Conservation: Water Sovereignty in the Kayapó Lands

Laura Zanotti

Abstract This chapter engages with biocultural approaches to conservation to affirm the growing need and application of frameworks to address power, rights, and indigenous perspectives in local to global environmental problems. Through a case study that focuses on Mẽbêngôkre-Kayapó peoples' projects for water sovereignty, I describe the sociopolitical and institutional contexts and cultural perspectives that shape Kayapó engagements with aquatic environment. While river and riparian habitats are central to Kayapó livelihoods and political goals, Kayapó understandings of aquatic landscapes are not central to national and international visions of sustainable Amazonian futures. These visions fashion the Amazon as a conservation-development mosaic instead of a bioculturally diverse homeland. I argue that bicultural approaches are particularly well-suited to highlight these different perspectives on policy and practice and to shift conversation paradigms in a way that can accommodate Kayapó Peoples' and other Indigenous Peoples' strategies to retain a good or *mejkumrej* life.

Keywords Biocultural · Kayapó · Rights · Power · Indigenous worldviews

22.1 Introduction

This chapter engages with biocultural approaches to conservation to affirm the growing need of frameworks to address local to global environmental problems. In a review of the current state of scholarship addressing human-environmental linkages, Nightingale suggests that both relational and interactional approaches have emerged over the past several decades (West et al. 2015). Interactional approaches are those approaches where "'social' and 'ecological' components are considered to

L. Zanotti (✉)
Department of Anthropology, Center for the Environment, Purdue University, West Lafayette, IN, USA
e-mail: lzanotti@purdue.edu

R. Rozzi et al. (eds.), *From Biocultural Homogenization to Biocultural Conservation*, Ecology and Ethics 3, https://doi.org/10.1007/978-3-319-99513-7_22

be distinct but interacting' the focus is therefore on 'linked' or 'nested systems and the interactions across spatial and temporal divides" (West et al. 2015, np). Approaches that fit into this category, for example, would include socio-ecological systems and sustainability science work. Relational approaches are "processed-based" and "insist that entities only come into being in relation to each other, and therefore that it is impossible to clearly distinguish social and ecological 'components' of a system" (West et al. 2015, np). Approaches that fit into this category include actor-network theory and other science and technology studies paradigms. Whereas both types of approaches are present in scholarly, activist, and policy circles, interactional approaches tend to dominate. However, there is a growing interest in making relational approaches more visible and prominent and to identify frameworks that integrate interactional and relational approaches together. I suggest that biocultural approaches can address interactional-relational divides and in doing so have potential to challenge dominant practices of conservation to change "the social field of action" in a way that renders power, rights, and placed-based worldviews and practices as central, rather than supplemental, to conservation (Wolf 1990, p. 587; Gavin et al. 2015; Charnley et al. 2017).

Biocultural approaches have grown over the past several decades, and proponents of these approaches constantly work to disrupt the dominant discourse of "people-free" and "people-centered" divide in conservation. The goal has been to draw attention to the developed frameworks in academia and policy circles that seek to wed conservation goals with poverty alleviation and human well-being (Gavin et al. 2015, p. 140). The goal also is to push for an overall paradigm shift in conservation that consists of a move away from a bifurcated debate to the adoption of a multifaceted, unifying but flexible framework that has applications in multiple disciplines, community spaces, and policy venues. This adoption has already started to take place on multiple fronts. For example, Davidson-Hunt et al. (2012, p. 34) describe how biocultural diversity and heritage have become centerpieces at sites of global environmental governance and championed by bodies like IUCN commissions, such as the Commission on Environmental, Economic, and Social Policy (CEESP).

Biocultural approaches also have developed in parallel to and sometimes also encompass many other approaches, frameworks, and paradigms that seek similar synergistic solutions to interdisciplinary team-based work with global impact, resonance at a community level, and directed policy outcomes. For example, resilience science, couple human-natural systems, socio-ecological systems, sustainability science, actor-network theory, and political ecology, among others, have quickly developed and been adopted in scholarly circles and policy worlds as integrative solutions that attend to the multi-sectoral, institutionally complex, and multi-scalar dynamic landscape of environmental governance (West et al. 2015). However, many of these new frameworks and approaches have not been uncritically adopted. One of the more prominent critiques of resilience science, sustainability science, and socio-ecological systems work is their lack of attention to justice, rights, and equity concern issues (Ostrom 2009, 2011; Cote and Nightingale 2012; Parkes et al. 2010; Brown 2014). Other critiques suggest that interactional, systems-based approaches, such as that of socio-ecological systems, miss the point as they are similarly embedded in realist, positivistic thinking and do not recognize or make central place-based ontologies and epistemologies (Anderies et al. 2004; Schlosberg and Carruthers 2010; Elmhirst 2011; West et al. 2015). Relational approaches, on the other hand, recognize multiple epistemological and ontological approaches to conservation and increasingly address the cultural politics of difference. The cultural politics of difference, as defined in the now seminal work of Cornel West (1990, p. 93), consists of a series of "distinctive features" including to favor diversity over the monolithic and homogenous; to move toward the concrete, specific, and particular and away from the abstract, general, and universal; and to draw upon history and context to highlight the contingent, shifting, and changing fields of play. Moreover, while relational approaches have their strengths, some have been criticized for the difficulty in applying these approaches in policy venues and their initial lack of attention to people in power relationships (Goldman et al. 2011).

Because biocultural approaches draw from a diverse set of relational and interactional frameworks and paradigms "biocultural diversity and heritage, social-ecological systems theory, and different models of people-centered conservation" (Gavin et al. 2015, p. 140), they are uniquely situated to address some of the more pressing concerns of interactional approaches, that is, a much needed framework or paradigm that (1) makes equity and rights integral to its implementation, (2) seeks to transform paradigms within conservation practice, and (3) simultaneously translates across different scholarly, policy, and community divides. Moreover, as biocultural approaches seek to link "human life with the diversity of beings, considered as coinhabitants with whom humans co-constitute their identities and attain well-being" (Rozzi 2013, p. 10), they also directly engage with relational paradigms that consider human and nonhumans as equally important to addressing global environmental change. Rozzi (2013, p. 11) expands that researchers should draw attention to the "3Hs," or the life habits, habitats, and coinhabitants, which "involve biophysical, symbolic-linguistic, and institutional-sociopolitical-technological domains." These, like other relational approaches, are centered on emergence ("the whole is more than the sum of its parts") and socio-material worlds ("that the world is made up of associations of human and nonhuman elements") (Müller and Schurr 2016, p. 217). However, it does so in a way that recognizes the central importance of shifting not only the conditions of conservation but the "social fields" in which it is practiced. In this way, Rozzi (2013, p. 11) further emphasizes that it is not simply an instrumental exercise to analyze the attributes of each domain but "to consider the relations of power and differential responsibilities regarding the causes and solutions of environmental problems." For example, Rozzi (2012) describes the widespread processes of biocultural homogenization which has led to abstract, universal, and static notions of peoples and places, or, in other words, a process that upholds rather than disrupts a politics of difference. As such, biocultural approaches, inclusive of biocultural heritage, ethics, and rights, not only examine multi-scalar arrangements and power relationships but do so in a way that can directly engage with goals for socio-environmental justice (Schlosberg 2004, p. 517).

In the sections that follow, I apply biocultural frameworks to analyze the conditions in which Mẽbêngôkre-Kayapó peoples, an indigenous peoples of the Brazilian Amazon, struggle for projects for a good life or *mejkumrej* life. Specifically, I draw upon more recent projects of Kayapó communities to strive for water sovereignty to emphasize how attention to power, rights, and cultural difference is critical to addressing global environmental change. I first introduce A'Ukre village of Kayapó peoples and the habitats, life habits, and communities of coinhabitants that shape Kayapó life. Next, I describe the institutional and sociopolitical contexts in which Kayapó engage with various actors to support their lifeways and struggle for water sovereignty. I then explore two domains of Kayapó worlds as they relate to water sovereignty, mythic events and changing fishing practices. In doing so, I emphasize the strength (*tyx*) of Kayapó governance strategies and the *kukradjà* or knowledge systems of Kayapó peoples that are essential to a good life. I acknowledge that by framing this discussion within the politics of water sovereignty, I reify water as a separate category; however, I do so with a hope to enliven the "microspheres of negotiation" (Anand 2011, p. 544) that trouble the persistent water/land/nature/culture divisions. I close by circling back to questions of water sovereignty as they relate to biocultural futures. I suggest that biocultural approaches can make central the role that indigenous worldviews play in solving issues of power and rights and global environmental problems.

22.2 A'Ukre

Work presented in this chapter is based on partnering with the Mẽbêngôkre-Kayapó (people of the watery place – Kayapó) community of A'Ukre over the past 10 years (Zanotti 2016). A'Ukre was founded in 1979 on a gentle curve of the Riozinho River, a tertiary tributary of the Xingu River (Fig. 22.1). The over 350 residents in A'Ukre carry out a series of practices that are generative[1] of community life. Daily and seasonal activities are structured around ceremonial rounds, subsistence practices, media engagement, and governance agendas. Community members make a living from a mixed market and subsistence-based livelihood, with a heavy emphasis on and preference for local foodways. Food-getting and life-giving activities revolve around embodied, experiential, and kin-based horticultural practices, foraging, fishing, and hunting in the forested and savannah and aquatic environment near their home (Zanotti 2016) (Figs. 22.2, 22.3 and 22.4). Cash income comes from social welfare programs developed by the Brazilian federal government (e.g., *Bolsa Família*), handicraft production (Figs. 22.5 and 22.6), sustainable development projects, and a handful of salaried positions in the community related to education, healthcare, and other NGO work in the region (see Zanotti 2009 for a detailed

[1] I use Overing's (2003, p. 302) definition of generative practices as the "fundamental relation of everyday skills and practice to the social process (as opposed to the relation of the dramatic ritual, such as 'I pronounce you man and wife', to a social structure)."

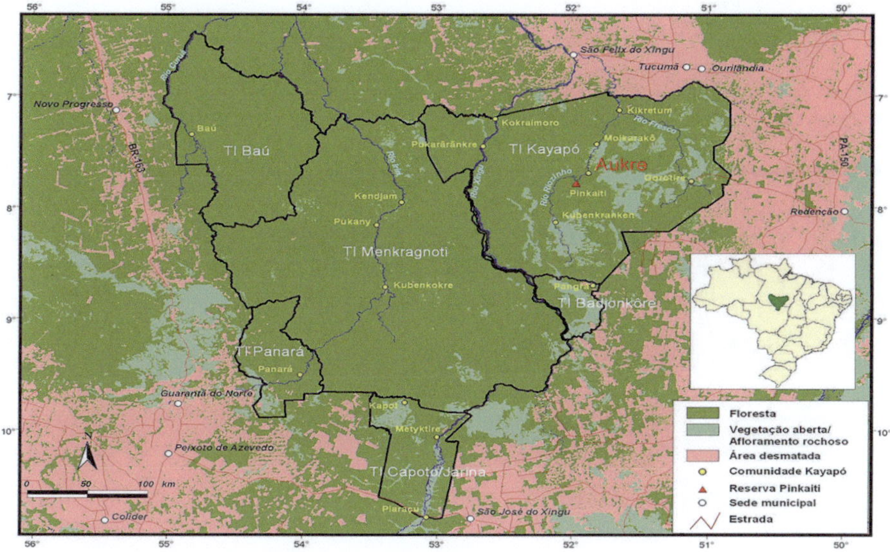

Fig. 22.1 Kayapó indigenous lands

Fig. 22.2 Forested area or bá. (Photo by Laura Zanotti)

description of these dynamics). Age grades (groupings of individuals based on their gender, biological phases through life, and other significant markers in individuals' lifespan), political societies, and other gendered and generational institutions and familial ties form the basis of community governance and stewardship, guiding political, ceremonial, and quotidian action (Fisher 2003). Kayapó peoples also have been particularly active in global, national, regional, and local politics from highly

Fig. 22.3 Savannah or kapôt (recently burned). (Photo by Laura Zanotti)

Fig. 22.4 Riozinho River or ngô. (Photo by L. Zanotti)

visible social activism to decision-making forums to diplomatic confrontations. Out of the 42 households in the village, 38 participated in this work, inviting me to consider water politics through everyday, routine, and ceremonial practice.

A'Ukre is one of over 40 villages in the Kayapó indigenous lands, a federally demarcated indigenous reserve in the southern portion of Pará state and northern portion of Mato Grosso, Brazil (Zimmerman et al. 2001). The Kayapó lands are

Fig. 22.5 Men's artisinal work

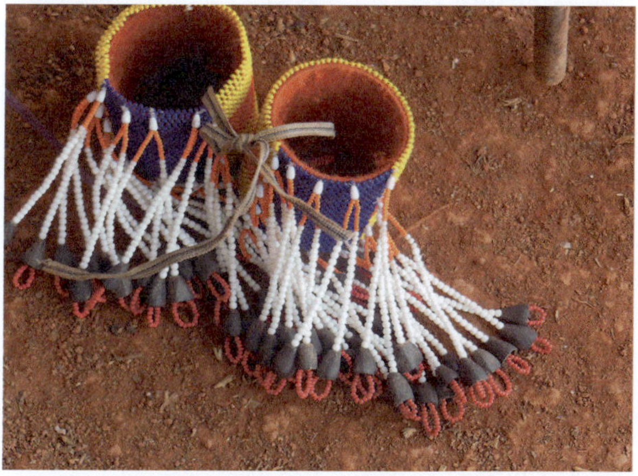

Fig. 22.6 Beaded work done by women artisans

composed of intersecting freshwater (*ngô*), savannah (*kapôt*), and forest (*bá*) environments, which Kayapó peoples further distinguish in different zones based on their proximity or distance to the community, different biophysical features, and past inhabitants or Kayapó modification of forested and savannah environments (e.g., abandoned gardens, seed dispersal practices), historic and more recent memories of places (e.g., old village sites), resource-harvesting areas, or prominence of certain species or features (e.g., place where the parakeets gather) (see Posey 1985 for an overview of Kayapó categories of different ecological zones; Zanotti 2014, 2016). Within this context, rivers and riparian habitats provide a foundation for Kayapó life habits in a community of coinhabitants that includes humans, fishes, animals, plants, waters, and other nonhuman beings.

22.3 Institutional and Sociopolitical Contexts

In this section I engage with Rozzi's (2013) proposal to consider "institutional-sociopolitical-technological domains" to describe the different contexts that shape and are shaped by Kayapó peoples' struggles to water sovereignty. I specifically draw attention to Kayapó peoples' interactions with (1) conservation initiatives and discourses and (2) national contexts. This discussion will set the stage to demonstrate the way in which power, rights, and Kayapó worldviews are addressed – or not – in these different spaces and to highlight the ongoing challenges to water sovereignty.

22.3.1 Conservation Initiatives and Discourses

Conservationist concerns became a stronghold in the Amazon region in the 1980s, offering different pathways to the Kayapó peoples for water sovereign futures. These concerns began during the Brundtland era of conservation politics which turned away from fortress models and invoked more than a decade of community conservation initiatives around the world (Vaccaro et al. 2013). Social and justice movements, especially in Latin America, grew with platforms against emerging regulatory structures and values (commodification and privatization) that continued to erode recognition of difference or emancipatory forms of cultural politics (Escobar 2015). After the 1987 Brundtland Report and the 1992 United Nations Conference on Environment and Development (UNCED) in Rio, a growing global consensus among conservation practitioners, policies, and governments supported the devolution and decentralization of environmental governance to shift toward co-management regimes, community-based programs, and the integration of democratic, participatory principles (Charnley and Poe 2007). The institutional changes invoked during this period also were enacted as neoliberal governance regimes were on the rise (Harris and Roa-Garcia 2013, p. 20). In Brazil, the devolution of governance to state and municipal levels led to the rise of nongovernmental and other civil society organizations in the Amazon. This process subsequently ushered in a new era of "multi-scale" partnerships (Ros-Tonen 2008, p. 1483). These partnerships emerged against the backdrop of an established and then emerging environmental policy regime in the country, which was focused on forests, sanitation, and protected areas (Neves 2016).

Conservation alliances began in A'Ukre in the 1990s that provided pathways to support the community's livelihoods, political activities, and knowledge exchange (Zanotti 2016). For A'Ukre, this manifested through an alliance with the international environmental organization, the Conservation International (CI), to create community-based conservation initiatives in the area centered on sustainable development programs, territorial retention, and conservation-based research initiatives (Zanotti 2009). Since its founding in 1992, the CI project both expanded to other villages and changed hands. The Wild Foundation Project, the Environmental

Defense Fund, and the International Conservation Fund of Canada serve as the main international partners after the Conservation International program ended around 2008. A'Ukre villagers now partner and engage intensely with international NGOs and the Kayapó NGO, the Associação Floresta Protegida (Protected Forest Association), with an emphasis on territorial, water, and environmental management, sustainable development, and local governance strategies (http://florestaprotegida.org.br/). Two initiatives focus on different water-based projects or political struggles: one supports Kayapó political strategies to address Belo Monte, detailed below, and the other supports a sports fishing venture. In addition to NGO work, educational and health services are offered in A'Ukre, which are funded by federal programs, including but not limited to the National Indian Foundation (FUNAI).

Despite these long-term engagements with partnerships, different discourses around the environment persist that differently recognize Kayapó worldviews or the critical importance of water versus forest systems to their life. Although recognition of water politics has recently changed with the Belo Monte dam project, conservation initiatives and actors have defined the Kayapó lands, political platforms, socioeconomic projects, and community partnerships in a variety of ways to their benefit (Zanotti 2015). To conservation scientists, seasonal tropical rivers and inland streams, such as those found within the Kayapó area, provide valuable habitats for fish and aquatic animals as well as "deliver important ecosystem services to people" (Silvano et al. 2008, p. 242). Many conservation scientists also argue that Kayapó and other indigenous peoples provide critical services to ecosystems or "actions humans have taken in the past and currently that modify ecosystems to enhance the quality or quantity of the services they provide, whilst maintaining the general health of the cognized ecosystem over time" (Comberti et al. 2015, p. 247). Yet, for the Kayapó peoples their lands cannot be reduced to conservationist visions of biodiverse, "serviced" landscapes or national visions of industrial, extractivist landscapes punctured by socio-environmental futures. For example, the river and riparian habitats within the Kayapó lands are not considered "resources" or "variables" by the Kayapó peoples but rather living landscapes in which life habits, habitats, and coinhabitants coexist. *Kukradjà* (ways of knowing, culture, or tradition) and notions of beauty, strength, and power are key tenets that structure Kayapó lifeways (Turner 2003).

22.3.2 National Contexts

The Kayapó peoples' fight for water sovereignty is currently mediated by the following policy-based, legal, and juridical spaces in Brazil and beyond: indigenous health and sanitation (well water maintenance within the community), land rights (territorial rights to lands and thus waters), mega-developmental and infrastructural projects (e.g., Belo Monte Dam), and indigenous rights and global environmental governance (e.g., right to water, Aichi Biodiversity Targets) (see Neves 2016, p. 504). In this section., I focus specifically on the Belo Monte dam project, although other national and international developments, such as agro-industrial industries, mining interests, and other infrastructural projects (roads), have also only intensified over the past several decades (Zimmerman et al. 2001; Fearnside 2012). The construction of the Belo Monte dam in particular has been a constant threat to Kayapó river and riparian habitats. The Belo Monte dam was seeded in nationalist projects that sought to modernize the Amazon in the post-World War II era and was later fortified by World Bank backed development programs in the 1970s and 1990s (Fisher 1994). During this time, the Kayapó peoples, then emerging from an intense period of contact and "pacification" and animated by a growing indigenous and environmental movement, were able to place a moratorium on the dam complex with large-scale protests and private meetings with officials (Turner 1995). Soon thereafter, subsequent political and economic shifts after the democratization of Brazil led to the first election of Workers' Party candidate Luiz Inácio Lula da Silva (Lula) followed by Dilma Rousseff. Despite promising platforms that supported a pro-poor state and evidence from pioneering conservation policies that for a time halted rapid and widespread deforestation in the Amazon region, Workers' Party policies began to backtrack on indigenous peoples, forest, and large-scale development issues (Zanotti 2015). Around the turn of the twenty-first century, the Belo Monte dam complex along the Xingu River was one of the many dam developments proposed as "clean" energy that would further fortify endogamous growth and energy security, and when Dilma took the presidency, she did not diverge from the vision of her predecessor. The third largest dam in the world, which construction began in 2011, Belo Monte was officially inaugurated in 2016 (Puentes Riaño 2016).

Projects like the Belo Monte dam complex are problematic for Kayapó desires for water sovereignty: the dam as a large-scale mega-development project that is proposed as a public good and a project that is in favor of national sovereignty and security (Zanotti 2015). Kayapó peoples along with other indigenous groups, local communities, concerned citizens, and NGOs began to wage a massive campaign against the dam starting the early 2000s, forging unprecedented multiethnic alliances, organizing large-scale national and international protests, writing open letters condemning the dam, and issuing innumerable lawsuits on unconstitutional violations that dam development has spurred (Turner and Fajans-Turner 2006). Despite an independent environmental assessment that showed a considerable impact on local and indigenous peoples and lands, the Inter-American Commission on Human Rights insistence on the unconstitutionality of the project, repeated omissions of consent, public commentary and upholding of conditions of the dam license by the dam developers, and one of the worst governmental corruption scandal that Brazil has seen in its history, dam development has not halted (Zanotti 2015).

Kayapó peoples' struggles against Belo Monte are part of a longer history of demanding their rights at the national and international level. For example, they were foundational in the indigenous rights' language that was included in the landmark 1988 Brazilian constitution, have consistently engaged with FUNAI and other agencies to demand accountability for ensuring their rights, are politically active in generating statements affirming their rights, are participating in small-scale and large-scale protests, and are engaged at sites of international governance (Turner

1995; Zanotti 2015, 2016). Nationally, Article 231(1) recognizes indigenous peoples' stewardship, and "original right to the lands traditionally occupied by them" has been one of the most powerful for indigenous peoples (da Costa 2014, pp. 14, 18). Subsequent constitutional articles, also ratified in 1988, guarantee the right to comment and participate in development initiatives and have been key spaces from which to make their rights visible.

Changing national dynamics and proposed constitutional changes also serve to undermine indigenous rights and environmental protections. For example, the proposed militarization of FUNAI and PEC 215 is just one of the several changes being pushed through as Workers' Party President Dilma Rousseff was impeached (finalized in August 2016) and the unelected, right wing president Michel Temer took power. PEC 65, which removes and reduces environmental and social protections in place for large-scale infrastructural development in the country, and PEC 241 (now 55), which would freeze public spending on welfare and public services (health, education, etc.) for 20 years, are just some of the alarming developments that are sweeping Brazil. This comes on the back of changes made to the Forestry Code in 2012 which resulted in the "relaxing of the status of protected areas and of environmental licensing rules, threats to indigenous areas and the unsatisfactory Complementary Law 140/2011, which remains in force" (Neves 2016, p. 503). Internationally, local claims have been further legitimized by the United Nations Resolution 64/292 to the right to water and sanitation and Article 25 of the United Nations Declaration on the Rights of Indigenous Peoples (http://www.un.org/esa/socdev/unpfii/documents/DRIPS_en.pdf). The Kayapó peoples in general and the residents of A'Ukre in particular have found that these spaces for rights-based claims and political activism that draws attention to their lifeways are as essential to their projects for water sovereignty.

22.4 Cultural Perspectives: Mythic Events and Fishing Practices

22.4.1 Mythic Events

In this section, I move away from broader sociopolitical and institutional contexts to examine the mythic[2] events and fishing practices as two sites where residents of A'Ukre engage in different relationships with water. I relay one commonly told

[2] In this chap. I use the term "myth" to refer to Kayapó traditional stories. There has been much debate within the literature on the use of "myth" versus "story" to describe oral narrative traditions. Many scholars are moving toward "stories" to confront criticism that "myth" designations imply "fictive" and "ahistoric" indigenous oral traditions (Dundes 1984; Schrempp and Hansen 2002). I retain the English term myth, which approximates how residents of A'Ukre distinguished between myths and stories, where myths held different spatiotemporal dimensions and contributions to oral storytelling and ceremonial traditions in ways that stories – or more current historic events – did not.

Kayapó myth to show how these oral narratives are an anchor from which experiential and embodied knowledge are imparted across space and time. Kayapó peoples draw from a rich corpus of mythic events and performative storytelling techniques to transmit knowledge: elders tell stories at night in front of hearths, senior men impart stories to younger age grades in the men's house, Kayapó school teachers relate myths to students in multicultural educational contexts, and leaders draw upon the salience of myths, stories, and songs as they interweave with political activist efforts. These are the oral narratives that are hidden in national and international agendas but critical to Kayapó biocultural worlds, myths which are essential to understanding human-environmental linkages. The myth I describe as an example here is fondly called the myth of Bira among community members.

There are many versions of the myth of Bira. As published in Wilbert and Simoneau's (1978) edited volume on Gê myths, this myth is a story of a man named Bira and his subsequent interactions with the women of the village he lives in:

> Bira is an attractive man and excelled at hunting and other subsistence tasks. Other jealous men of the village decided to turn him into a tapir and successful hunted Bira, killing him in his tapir form and taking him back to the village to eat. As was custom, the hunters shared the tapir meat with all in the villagers, and the unknowing women ate Bira, whom they had loved and known well. When the women found out what happened they were quite upset with the other men of the village. Later, the men left to go hunting and while they were away the women decided to have a ceremony. During the ceremony, the women arrived at the river singing and began to turn into fish one after another. When the men came back from hunting, they were surprised to find the women gone, and the elders and children left behind did not know where they went. Looking to the river the men were equally surprised to find many species of fish now populating the waterways and became suspicious that this is where the women went. One man found fruit near the river and used this as bait, and he could find his wife this way. Others, seeing what had happened tried the same but with no success. This is the explanation why there are so many different and distinctive species of fish in the river today and why different species have their own patterns, songs, and preferences in how they traverse and engage in fluid environments.

Although a much-abbreviated version than the same myth told to me in A'Ukre, this myth demonstrates that Kayapó worldviews and ways of knowing extend beyond the reach of the state or conservation initiatives, where human and nonhuman relationships provide important experiences from which life-giving practices emerge (speciation of the rivers). It also demonstrates that the institutions that currently bound Kayapó strategies for governance of their indigenous lands, participation in community conservation relationships, and formulation of sustainable development projects are not confined to strictly "scientized" or "environmental" narratives of ecosystem services, biodiversity, or systems approaches. Rather, fish populate the river not simply based on the biotic and abiotic components of riverine environments but because of variable relationships among hunters, wives, husbands, food, rivers, fish, and desires. This story also demonstrates how biocultural approaches, which include more relational considerations like the politics of cultural difference, can make visible these forms of knowledge and meaning-making. Recognizing Kayapó myths like these is central to understanding Kayapó projects for water sovereignty that strive to support healthy peoples and landscapes in a manner that respects indigenous worldviews.

22.4.2 Fishing Practices

Similarly, as these mythic events affirm Kayapó *kukradjà* (knowledge/culture) and customary interactions between human and nonhuman others, the habitats, life habits, and cohabitants related to fishing practices also serve as poignant reminders of biocultural interfaces. Riozinho or simply *ngô* (water), the main river adjacent to A'Ukre, contours the lives of villagers. Fishing takes place in the "port areas" near the village, which I identify as the three banks of the river that women, men, and children primarily use to dock their boats, carry out daily bathing and washing activities, play soccer, and perform other routines. Fishing activities, experiences, and knowledge are passed down through "situated action," "mutual enskillment," and "direct engagement with everyday tasks" (Pálsson 1998, p. 53). Part of being beautiful *(meij)* and strong (*tyx*), two Kayapó-valued characteristics include extensive knowledge of a proper and responsive engagement with other coinhabitants that collectively formulate the landscape. That is, by the time a certain Kayapó individual has gone through adolescence and subsequently had their first child, there are ideally a series of tasks that both men and women should have gained experience in that involve ceremonial, political, and subsistence practices. These activities are organized around different age grades, political societies, matriuxorilocal households, wage work, and other personal and collective identity markers (Fisher 2003). Fishing is one of those tasks that take place in fluid landscapes that demonstrate the importance of the "3H's" in considering biocultural worlds.

A key part of fishing practices includes the techniques and tactics that structure and shape individual and community relationships while out on the water or in the forest near ponds and streams. Hooks, lines, weights, nets, sticks (to bait fish with), empty 2-liter coke bottles, vines (to string fish on), palms (to make baskets to place fish in), and canoes are the materials that formulate interactions with fish. In addition, bows and arrows are typically used almost exclusively with timbó fishing (Fig. 22.7). Timbó fishing is a style of fishing practice where community members

Fig. 22.7 Timbó vine for fishing. (Photo by Laura Zanotti)

gather and then submerge beaten poisonous (timbó) vines in small ponds or lakes. As one Kayapó young man described to me, the poison from the vines makes the fish "slow" and "dumb," which then makes it easy to harvest the fish in the enclosed area with bow and arrow or through grabbing them (see Teixeira et al. 1984). In the dry season, net fishing and timbó fishing predominate in partially dried-up riverbeds, newly formed forest ponds, and key spots in the river, although only a handful of residents own store-bought fishing nets (Fig. 22.4). The more common and prevalent fishing technique is the one described above, where hook, lead, weight, and line are used to fish by hand and the slight flick of the wrist sends the hook several meters out, and hopefully, because of a slow tug then jerk back when something catches. Women's and men's timbó fishing events are filled with song and ritual actions that served to honor the ceremony and the life-giving activity.

Learning and mastering different skill sets to be strong (*tyx*), creatively incorporating new practices, and having healthy waters to fish in are all facets of life that the Kayapó interlocutors I worked with described as key capacities they wanted to retain. The need for store-bought fishing materials demonstrates and emphasizes the ongoing importance of sustainable development initiatives derived from community conservation partnerships and mixed market economies for fishing livelihoods. Moreover, upholding rights to their territorial homeland and transboundary threats to their lands and waters also are critical to retain the fishing practices that serve as ritually and nutritionally important to their lifeways. Similarly, recognition of Kayapó myths, which animate landscapes and structure coinhabitant relationships, is essential to the ways in which Kayapó peoples perceive of biocultural intersections.

22.5 Conclusion

Through drawing attention to the sociopolitical and institutional contexts and cultural perspectives that influence Kayapó projects for water sovereignty, I have highlighted in this chapter the role that power, rights, and indigenous worldviews play in the politics of conservation. For the Kayapó peoples, rivers are part of the habitats, life habits, and coinhabitants that are essential for a good life and the well-being or strength of the community. The river is part of a mythic past, is a space of abundance and danger, and is an orienting point in the community. It is also a site of creative transformation as fishing techniques and river routes change over time. The rivers are not separate from social life but an embedded part of bioculturally diverse worlds. These cultural perspectives and practices related to water formulate the "social and reciprocal relationships with natural phenomenon" that constitute biocultural landscapes (Hogue and Rau 2008, p. 291). Approaching the Kayapó lands from a local perspective emphasizes the critical importance of aquatic landscapes and rivers as part of conservation considerations. This perspective is critical for the Kayapó peoples given the strong ties between rivers and local livelihoods and longstanding cultural links with rivers.

Biocultural approaches are much needed in conservation. Biocultural approaches are critical for resolving the weaknesses already identified in both interactional and relational approaches to global environmental change, especially recognizing the politics of difference, attention to rights, and the linkages among the "3H's" of the biocultural ethic (*sensu* Rozzi 2012). Considering multiple dimensions of local livelihoods, the scales at which they are enacted (local perspective and practices, community conservation partnerships, national contexts, etc.) and the translational possibilities for policy worlds are powerful ways in which to recognize Kayapó projects for a good life and water sovereignty. The politics of difference plays a central role in this endeavor and allows for different possibilities of indigenous interventions in conservation to emerge. Indigenous groups have shown how their knowledge systems are divergent from the production of Western scientific knowledge but that does not make them irrelevant to the management of and governance of resources (Agrawal 1995; Berkes et al. 2000). Kayapó conceptualizations of and practices within their landscapes are central to sustainable Amazonian futures as well as alternative projects for addressing global grand challenges. Water sovereignty, especially in relation to the Kayapó peoples' ability to advocate for the politics of difference, will remain critical as threats to their aquatic environments and lifeways persist.

References

Agrawal A (1995) Dismantling the divide between indigenous and scientific knowledge. Dev Chang 26(3):413–439

Anand N (2011) Pressure: the Politechnics of water supply in Mumbai. Cult Anthropol 26(4):542–564

Anderies JM, Janssen MA, Ostrom E (2004) A framework to analyze the robustness of social-ecological systems from an institutional perspective. Ecol Soc 9(1):18 [online] URL: http://www.ecologyandsociety.org/vol9/iss1/art18/

Berkes F, Colding J, Folke C (2000) Rediscovery of traditional ecological knowledge as adaptive management. Ecol Appl 10(5):1251–1262

Brown K (2014) Global environmental change I: a social turn for resilience? Prog Hum Geogr 38(1):107–117

Charnley S, Poe M (2007) Community forestry in theory and practice: where are we now? Annu Rev Anthropol 36:301–353

Charnley S, Carothers C, Satterfield T, Levine A, Poe MR, Norman K, Donatuto J, Breslow SJ, Mascia MB, Levin PS, basurto X, Hicks C, García-Quijano C, St. Martin K (2017) Evaluating the best available social science for natural resource management decision-making. Environ Sci Policy 73:80–88

Comberti C, Thornton TF, Echeverria VW, Patterson T (2015) Ecosystem services or services to ecosystems? Valuing cultivation and reciprocal relationships between humans and ecosystems. Glob Environ Chang 34:247–262

Cote M, Nightingale AJ (2012) Resilience thinking meets social theory situating social change in socio-ecological systems (SES) research. Prog Hum Geogr 36(4):475–489

da Costa A (2014) Sustainable dam development in Brazil: the roles of environmentalism, participation, and planning. In: Scheumann W, Hensgerth O (eds) Evolution of dam policies. Springer-Berlag, Berlin, pp 12–53

Davidson-Hunt I, Turner KL, Mead ATP, Cabrera-Lopez J, Bolton R, Idrobo CJ, Miretski I, Morrison A, Robson JP (2012) Biocultural design: a new conceptual framework for sustainable development in rural indigenous and local communities. SAPIEN S Surv Perspect Integr Environ Soc 5(2):34–45

Dundes A (ed) (1984) Sacred narrative: readings in the theory of myth. University of California Press, Berkeley

Elmhirst R (2011) Introducing new feminist political ecologies. Geoforum 42(2):129–132

Escobar A (2015) Degrowth, postdevelopment, and transitions: a preliminary conversation. Sustain Sci 10(3):451–462

Fearnside PM (2012) Belo Monte dam: a spearhead for Brazil's dam building attack on Amazonia. Global Water Forum. Available at: http://www.globalwaterforum.org/2012/03/19/belo-monte-dam-a-spearhead-for-brazils-dam-building-attack-on-amazonia/

Fisher WH (1994) Megadevelopment, environmentalism, and resistance: the institutional context of Kayapó indigenous politics in Central Brazil. Hum Organ:220–232

Fisher WH (2003) Name rituals and acts of feeling among the Kayapó (Mebengokre). J R Anthropol Inst 9(1):117–136

Gavin MC, McCarter J, Mead ATP, Berkes F, Stepp JR, Peterson D, Tang R (2015) Defining biocultural approaches to conservation. Trends Ecol Evol 30(3):140–145

Goldman MJ, Nadasdy P, Turner MD (2011) Knowing nature: conversations at the intersection of political ecology and science studies. University of Chicago Press, Chicago

Harris LM, Roa-García MC (2013) Recent waves of water governance: constitutional reform and resistance to Neoliberalization in Latin America (1990–2012). Geoforum 50:20–30

Hogue E, Rau P (2008) Troubled water: ethnodevelopment, natural resource commodification, and neoliberalism in Andean Peru. Urban Anthropol 37(3–4):283–327

Müller M, Schurr C (2016) Assemblage thinking and actor-network theory: conjunctions, disjunctions, cross-fertilisations. Trans Inst Br Geogr 41(3):217–229

Neves EMSC (2016) Institutions and environmental governance in Brazil: the local governments' perspective. Revista de Economia Contemporânea 20(3):492–516

Ostrom E (2009) A general framework for analyzing sustainability of social-ecological systems. Science 325(5939):419–422

Ostrom E (2011) Background on the institutional analysis and development framework. Policy Stud J 39(1):7–27

Overing J (2003) In praise of the everyday: trust and the art of social living in an Amazonian community. Ethnos 68(3):293–316

Pálsson G (1998) Learning by fishing: practical engagement and environmental concerns. In: Berkes F, Folke C, Colding J (eds) Linking social and ecological systems: management practices and social mechanisms for building resilience. Cambridge University Press, Cambridge, pp 48–66

Parkes MW, Morrison KE, Bunch MJ, Hallström LK, Neudoerffer RC, Venema HD, Waltner-Toews D (2010) Towards integrated governance for water, health and social–ecological systems: the watershed governance prism. Glob Environ Chang 20(4):693–704

Posey DA (1985) Indigenous management of tropical forest ecosystems: the case of the Kayapo Indians of the Brazilian Amazon. Agrofor Syst 3(2):139–158

Puentes Riaño A (2016) Belo Monte and how NOT to produce energy in the 21st century. Amazon Watch Available at: http://amazonwatch.org/news/2016/0505-belo-monte-and-how-not-to-produce-energy-in-the-21st-century

Ros-Tonen MA, Van Andel T, Morsello C, Otsuki K, Rosendo S, Scholz I (2008) Forest-related partnerships in Brazilian Amazonia: there is more to sustainable forest management than reduced impact logging. For Ecol Manag 256(7):1482–1497

Rozzi R (2012) Biocultural ethics: recovering the vital links between the inhabitants, their habits, and habitats. Environ Ethics 34(1):27–50

Rozzi R (2013) Biocultural ethics: from biocultural homogenization toward biocultural conservation. In: Rozzi R, Pickett STA, Palmer C, Armesto JJ, Callicott JB (eds) Linking ecology and

ethics for a changing world: values, philosophy, and action. Ecology and ethics, vol 1. Springer, Dordrecht, pp 9–32
Schlosberg D (2004) Reconceiving environmental justice: global movements and political theories. Environ Polit 13(3):517–540
Schlosberg D, Carruthers D (2010) Indigenous struggles, environmental justice, and community capabilities. Glob Environ Polit 10(4):12–35
Schrempp G, Hansen W (eds) (2002) Myth: a new symposium. Indiana University Press, Bloombington
Silvano RAM, Silva AL, Ceroni M, Begossi A (2008) Contributions of ethnobiology to the conservation of tropical rivers and streams. Aquat Conserv Mar Freshwat Ecosyst 18:241–260
Teixeira JRM, Lapa AJ, Souccar C, Valle JR (1984) Timbós: ichthyotoxic plants used by Brazilian Indians. J Ethnopharmacol 10(3):311–318
Turner T (1995) An indigenous People's struggle for socially equitable and ecologically sustainable production: the Kayapó revolt against extractivism. J Lat Am Anthropol 1(1):98–121
Turner T (2003) The beautiful and the common: inequalities of value and revolving hierarchy among the Kayapó. Tipití: J Soc Anthropol Lowland South America 1(1):1–16
Turner T, Fajans-Turner V (2006) Political innovation and inter-ethnic alliance: Kayapo resistance to the developmentalist state. Anthropol Today 22:3–10
Vaccaro I, Beltran O, Paquet PA (2013) Political ecology and conservation policies: some theoretical genealogies. J Polit Ecol 20:255–272
West C (1990) The new cultural politics of difference. October 53:93–109
West S, Galafassi D, Haider J, Marin A, Merrie A, Ospina-Medina D, Schill C (2015) Critically reflecting on socio-ecological systems research. Resilience Science Blog. Retrieved from: http://rs.resalliance.org/2015/02/11/critically-reflecting-on-social-ecological-systems-research/
Wilbert J, Simoneau K (1978) Folk literature of the Gê Indians, vol 1. UCLA Latin American Center Publications, Los Angeles
Wolf ER (1990) Distinguished lecture: facing power—old insights, new questions. Am Anthropol 92(3):586–596
Zanotti L (2009) Economic diversification and sustainable development: the role non-timber forest products play in the monetization of Kayapó livelihoods. J Ecol Anthropol 13(1):26–41
Zanotti L (2014) Political ecology of movement: trekking and territoriality among the Kayapó. J Polit Ecol 21:108–126
Zanotti L (2015) Water and life: hydroelectric development and indigenous pathways to justice in the Brazilian Amazon. Polit Groups Identities 3(4):666–672
Zanotti L (2016) Radical territories in the Brazilian Amazon: the Kayapó's fight for just Liveihoods. University of Arizona Press, Tucson
Zimmerman B, Peres CA, Malcolm JR, Turner T (2001) Conservation and development alliances with the Kayapó of South-Eastern Amazonia, a tropical forest indigenous people. Environ Conserv 28(1):10–22

Chapter 23
Biocultural Diversity and Ngöbe People in the South Pacific of Costa Rica

Felipe Montoya-Greenheck

Abstract This article reviews a participatory action research (PAR) project in the Indigenous Territory Guaymí de Osa of the Ngöbe (a.k.a. Ngäbe, Guaymí) in southern Costa Rica. This PAR project was carried out by researchers from the University of Costa Rica, the Austrian Institute for Latin America, and the Tropical Research Station La Gamba, to study traditional ecological knowledge (TEK) of the Ngöbe people and facilitate its incorporation into public environmental management and restoration policies and initiatives in the Corcovado-Piedras Blancas Biological Corridor. The results of this multifocal research produced more questions than answers regarding the nature and the role of Ngöbe TEK for environmental conservation. It did, however, reinforce the importance of promoting biocultural diversity over the hegemonic tendencies of homogenization inherent in neoliberal capitalism as the current rendition of modernity, if we are to find sustainable solutions to the social and environmental problems and growing pressures and threats associated with this globally dominant system.

Keywords Participatory action research · Biocultural conservation · Traditional ecological knowledge · Guaymí de Osa Indigenous Territory · Ngöbe · Ngäbe

In 2010 we began a participatory action research (PAR) project for the integration of Ngöbe (also referred to as Ngäbe or Guaymí) communities in the conservation and management of ecosystems in the Golfo Dulce region of Costa Rica's South Pacific.[1] One of the main objectives of this project was to study ecological categories and environmental management systems of Ngöbe communities in Costa Rica and incorporate them into public environmental management and restoration

[1]The project Biocultural Diversity in Costa Rica's Pacific Coast (BDCR) was initiated by the Austrian Institute for Latin America and the University of Costa Rica with participation of the Tropical Research Station (TRS) La Gamba.

F. Montoya-Greenheck (✉)
Faculty of Environmental Studies, York University, Toronto, ON, Canada
e-mail: fmontoya@yorku.ca

R. Rozzi et al. (eds.), *From Biocultural Homogenization to Biocultural Conservation*, Ecology and Ethics 3, https://doi.org/10.1007/978-3-319-99513-7_23

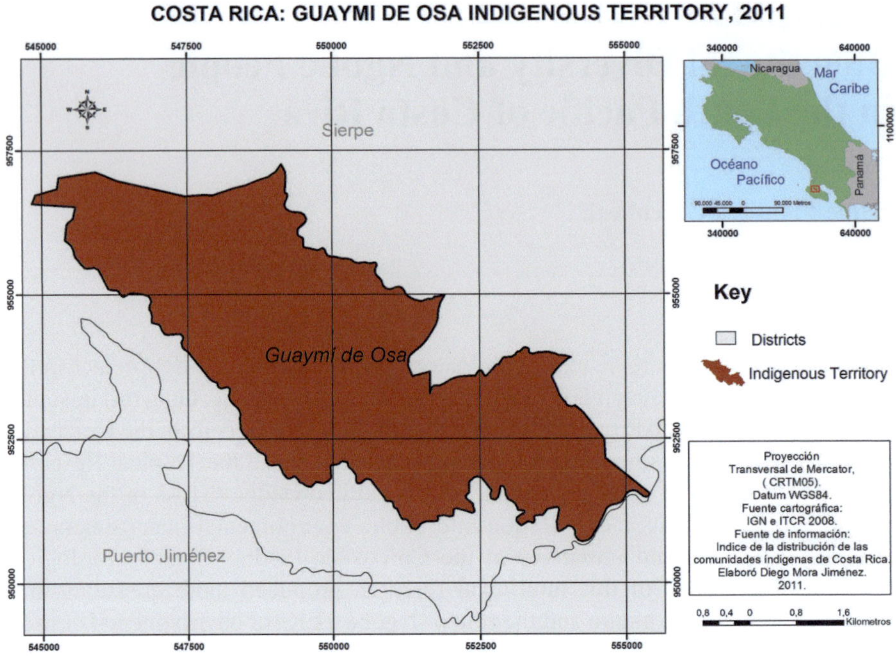

Fig. 23.1 Map of Guaymí de Osa Indigenous Territory. (Adapted from Diego Mora Jiménez in 2011)

policies and initiatives in the Corcovado and Piedras Blancas national parks in the South Pacific region of Costa Rica, a region of very high biodiversity (Fig. 23.1).

It is noticeable that where indigenous communities reside in Costa Rica is largely consistent with areas where forest cover remains relatively intact (Borge-Carvajal 2006). Before the arrival of the Europeans to the region, the indigenous populations coexisted with a mostly intact forest. In the territory that today is Costa Rica, at the time of European contact, the forest cover was more than 98% (Keog 1984). European settlers, with the introduction of their own crops, cattle, and land-use practices, began to alter the indigenous systems of stable coexistence with the natural forest, forcing the escape of indigenous communities who survived contact to refuge regions where European land-use systems were more difficult to deploy, far from the fertile valleys and in steeper mountainous terrain with denser forest cover where soils were less able to sustain the cultivation of annual crops. In these harsher environments, indigenous populations continued to survive and eke out livelihoods without destroying the natural environment that surrounded them (Camacho-Nassar 1995). In the rest of the territory, Europeans established their crops and grazed their cattle at the expense of forested lands. But it was not until the establishment of cash crops for export, such as coffee, sugarcane, and bananas in the mid- and late 1800s,

Table 23.1 Forest cover in Costa Rica 1522–1987

Year	Forest cover
1522	99.0
1569	95.0
1802	92.3
1900	86.5
1940	75.0
1950	72.0
1960	63.4
1977	41.2
1984	32.1
1987	21.0

Adapted from Fournier (1985), FONAFIFO (2012), Sader and Joyce (1988)

that significant deforestation began to occur, a process that was further accelerated from the 1950s to the 1980s, especially with the expansion of cattle grazing for meat exports in the 1970s and 1980s (Fournier 1985, Table 23.1). Current expansion of oil palm and pineapple plantations continues to occupy land that was once forested (Maglianesi-Sandoz 2013). The contrast with those areas where indigenous communities still live is increasingly evident.

In the 1970s, the Costa Rican state had a visionary policy of establishing national parks and a protected area system. Today over 25% of the national territory is under some protection regime. In 1977, Costa Rica also passed the Indigenous Law, which recognized indigenous territories as protected areas. A total of 24 indigenous territories belonging to 8 distinct indigenous peoples today represent 334,447 hectares (or 6.54%) of the national territory (Borge-Carvajal 2012). Of these, the Ngöbe people occupy five indigenous territories in the South Pacific region of Costa Rica, including Abrojo-Montezuma created in 1980, Coto Brus in 1981, Conte Burica in 1982, Guaymí de Osa in 1990, and Altos de San Antonio in 2001.

The five Ngöbe territories maintain today an important forest cover. However, they are under pressure, surrounded by nonindigenous production systems that have degraded and even destroyed the original forest ecosystems. The extensive livestock farming, large coffee plantations, banana plantations, oil palm plantations, and the rampant expansion of pineapple plantations contrast sharply with Ngöbe production systems that coexist with the forest (Fig. 23.2).

On a smaller scale, scattered both outside and within indigenous territories, nonindigenous peasant subsistence production and coffee farms represent a third land-use pattern that also contrasts with Ngöbe productive systems. The small subsistence production and coffee farms most often involve degradation of forest cover and intensive use of agrochemicals.

Fig. 23.2 (**a**) Pineapple plantations, the fastest expanding agro-industrial monocrops in the South Pacific region (Photo Felipe Montoya in 2015), contrast sharply with (**b**) Ngöbe forest habitats in the Guaymí de Osa Indigenous Territory. (Photo Christoph Campregher in 2010)

In Costa Rica there are eight different indigenous ethnic groups, most of which have occupied for several millennia the territories in which they currently live. Among these are the Bribri, the Cabécar, and the Brunka people, who along with the Ngöbe and the Térraba reside in the South Pacific region of the country. The Térraba and the Ngöbe, however, have a different history of the occupation of their territories in Costa Rica. The Térraba were forcibly relocated by the Spaniards from their original lands on the Caribbean side of Panama during the colonial period in the early eighteenth century (Solórzano-Fonseca 2011). The Ngöbe, on the other hand, represent the largest indigenous group in Panama. Since the colonial period, they have been expansive in their settlement patterns and itinerant in their quest for productive activities, commonly working as seasonal peons on agricultural plantations (Young 1971). In this way, the Ngöbe have their own pre- and post-contact history of settlement in Costa Rica.

In this project we focused on the Ngöbe communities of the Guaymí de Osa Indigenous Territory, because this territory borders Corcovado National Park and is part of the Corcovado-Piedras Blancas Biological Corridor that the Costa Rican state has tried to promote for conservation purposes. Guaymí de Osa Indigenous Territory is about 30 km^2, located in a very humid tropical forest zone with an average temperature of 26.7 °C and an average annual precipitation of 4020 mm (Borge-Carvajal 2003). Approximately 70% of the territory is under forest cover (García 2008; Guevara 2000). The Ngöbe population is about 180 persons, who live in some 40 houses dispersed in 3 settlements (Campregher 2012, Fig. 23.3).

To reach Guaymí de Osa Indigenous Territory, you exit the Inter-American Highway at the Chacarita intersection on to Highway 245, after driving for hours surrounded by extensive monocrop plantations that vary from teak to pineapple and to oil palm, in land that had previously been deforested by transnational banana plantations and whose current crops continue to be commercialized by transnational corporate giants. Once the curves begin as you enter the mountainous terrain of the Osa Peninsula, the vegetation transitions to patches of secondary and primary forest, interrupted by pastures for cattle grazing, rice plantations, and scattered peasant farms with diverse crops. At the small town of La Palma, you turn left on a dirt road that takes you to the right bank of a 100 m wide river, that depending on the season, you can wade or ride a horse across. On the left bank begins Guaymí de Osa. A red clay trail that climbs up the mountain through primary forest takes you to the Ngöbe settlement of Alto Laguna, where traditional thatched houses, interspersed with more modern concrete houses, are connected by narrow paths (Fig. 23.4).

Most Ngöbe women and girls wear traditional garments of bright colors with traditional geometric designs. The men usually don rubber boots and a machete slung on their hips to tend to their crops of maize, beans, rice, cassava, cacao, and other fruit trees. Around the homes chickens range freely, and some homes may have a small pig sty close at hand. On weekdays you can find Ngöbe children wearing the national school uniform on their way to the single-room primary school, where they learn the national curriculum, as well as take Ngöbe language and culture classes. While the Ngöbe are in contact with nonindigenous outsiders, they maintain their language Ngäbere, which retains several dialects. Some family units maintain the traditional polygamous marriage system of one husband and two or more wives. The households practice occasional hunting of wild pigs, agoutis, and armadillos, among others, and practice fishing, to supplement the protein in their

Fig. 23.3 Ngöbe habits, habitats, and coinhabitants: (**a**) Ngöbe child caring for pig. (**b**) Ngöbe girl gathering fruit. (**c**) Ngöbe woman on red clay path in Guaymí de Osa Indigenous Territory (Photos Felipe Montoya in 2010). (**d**) Ngöbe man examines forest vegetation. (Photo Christoph Campregher in 2010)

diets. Some continue to celebrate drinking chicha made with fermented maize and practice ritual drinking of cacao, related to traditional spiritual beliefs (Grüber 2012). Others, however, who have adopted the syncretic religion of Mama Chi, might shun these practices. In Guaymí de Osa, although the Ngöbe residents are relatively few, you can find adherents of different religions, as well as speakers of diverse dialects.

It was not until we started this project that we discovered the more complex historical relationship between the Ngöbe and the Guaymi de Osa Indigenous Territory. Being expansive and itinerant, we had assumed that Guaymí de Osa was at least part

a

b

Fig. 23.4 (**a**) Ngöbe family constructing a traditional thatched building. (Photo Christoph Campregher in 2010). (**b**) Traditional thatched and concrete houses side by side in the community of Alto Laguna in Guaymí de Osa Indigenous Territory. (Photo Felipe Montoya in 2010)

of the Ngöbe ancestral territory, even if it was on the periphery of its consolidated territory in Panama. But we soon realized that Guaymi de Osa had been the subject of squatter invasions during the expropriation of properties of the Osa Productos Forestales (OPF) logging company in the 1970s (Borge-Carvajal 2003). OPF had extensive idle land holdings in the region, and with state plans to create the Corcovado National Park, squatter activity had increased dramatically as a strategy to acquire land, based on Costa Rican possessory laws that allow squatters to claim legal rights after 3 years of possession (Hartshorn 1978). Among the people who colonized these lands in the 1970s were Costa Rican peasants and Ngöbe people, some of whom came directly from Panama and others from Coto Brus, Costa Rica, where they had settled a generation before during the establishment of coffee plantations by Italian immigrants. Due to the presence of a Ngöbe community in these lands, Guaymí de Osa was eventually declared an Indigenous Territory in 1988. Thus, the Guaymí Indigenous Territory of Osa has been occupied by the Ngöbe coming from diverse regions for a little more than 40 years, i.e., only one human generation (Camacho-Nassar 1995).

It is noticeable, however, that over the course of a generation, while many other places have succumbed to destructive land-use practices by both poor peasants and profit-seeking companies that have left a trail of deforested lands with highly eroded soils, polluted water, and ecosystems emptied of their native biodiversity, the Guaymí de Osa Indigenous Territory has conserved most of its original forest cover. In this way, this Indigenous Territory continues to confront us with important research questions: What elements of the Ngöbe's way of life have allowed them to coexist with the forest that surrounds them during the period from the 1970s to the present? And can these elements be extrapolated and/or incorporated into state conservation initiatives in the Corcovado-Piedras Blancas Biological Corridor?

23.1 A Multifocal Research

In search of answers to these questions, we undertook a multifocal research of the Ngöbe's ecological categories, their traditional ecological knowledge systems, and their productive practices (Durand 2005; Alarcón-Cháires 2006; Borge-Carvajal 2007). Some members of the Ngöbe community had extensive knowledge of forest plants and their medicinal uses. With them we compiled lists of plants and their properties (Campregher 2011). Other members of the Ngöbe community could name indigenous categories to describe the terrain. We witnessed a number of agricultural plots of polycultures that included rice, maize, cassava, and a collection of fruit trees (Fig. 23.5).

With the active participation of the local people, we carried out an ethnocartographic process, mapping the points and places of importance for different life habits of the Ngöbe people in their Indigenous Territory. We documented a number of traditional stories that included forest animals and supernatural entities (Moya-Chávez 2011):

Fig. 23.5 A typical collection of fruit trees near a Ngöbe home includes oranges, bananas, cacao (*Theobroma cacao*), and pejibaye (*Bactris gasipaes*). (Photo Christoph Campregher in 2010)

> The animals of days gone by were persons, both birds and animals, and they had contact with people and they spoke with the people…They would talk and advise. In fact, they were more intelligent than us. So the day came that because of our evil ways we distanced ourselves. and today we see them just as animals…If I have one hundred head of cattle and if the jaguar comes to cause me much harm, I will call for a meeting, and it is the "owner" of the jaguar who must come personally, and I will tell him, you are causing me harm, I do not want you to cause me harm, and he has to take this advice and plant it in their mind. But that was a long time ago, now we just see an animal and we want to kill it right away, but that is how is was thousands of years ago. That was the relationship we used to have, there was a kind of respect among animals and humanity, but nowadays it is no longer this way. (Moya-Chávez 2011, Felipe Montoya translation)

We explored the relationship of the Ngöbe with the bodies of water in their territories. These are indispensible for their daily chores, such as personal hygiene, washing clothes, and food preparation. They are also an important place for leisure activities, such as swimming, fishing, and hunting, as well as seed collecting for their handcrafts (Fig. 23.6). Water also figures prominently as a contextual element in Ngöbe cosmologies (Arias 2011).

Ngöbe ritual use of cacao (*Theobroma cacao*), a tree that in other indigenous territories forms an important part of agroforestry systems of production, plays an important role in the interpretation of dreams, in detecting the spiritual quality of persons, in providing mental and physical stamina, and for therapeutic purposes (Grüber 2012, Fig. 23.7).

a

b

Fig. 23.6 (**a**) A swimming hole in Alto Laguna. (Photo Christoph Campregher in 2010). (**b**) Ngöbe handicrafts featuring some seeds commonly collected in the forest and near the rivers and streams. (Photo Christoph Campregher in 2010)

We also worked together to create educational texts with illustrations of the animals and forest plants created by Guaymí de Osa Ngöbe children and youth (Estanley 2011) and collaborated with local environmental education initiatives (Fig. 23.8).

Fig. 23.7 Drinking cacao with Ngöbe leader Luis Quirós. Cacao pod (featured prominently on the table) is a fruit which some Ngöbe continue to regard as having spiritual connotations. (Photo Christoph Campregher in 2010)

a b

Fig. 23.8 (**a**) Children of the Ngöbe community in Guaymí de Osa during an environmental education workshop in 2010. (Photo Felipe Montoya in 2010). (**b**) Ngäbere alphabet illustrated by Ngöbe children of Guaymí de Osa Indigenous Territory (Estanley 2011)

23.2 More Questions than Answers

By the end of the project, we had more questions than answers. Had we really done a survey of the worldview of the Ngöbe and their practices that provide the possibilities for a more harmonious coexistence with the surrounding forest? Had we clarified what those elements were? Or had we, rather, been searching under the wrong stone? Undoubtedly, there were obvious differences between the Ngöbe-controlled lands and the neighboring lands in the hands of corporate agro-industrial production and nonindigenous rural communities (Fig. 23.9).

Peasants whose lands adjoin the Indigenous Territory clear-cut their plots, leaving them "clean" to be able to plant their maize or to allow grasses to grow for their livestock grazing (Fig. 23.9). Large rice plantations and even teak forest plantations do not incorporate native forest into their production systems. Meanwhile, Guaymí de Osa remains mostly under forest, with some small patches without forest cover for the orchards of their homes. It is significant that the Ngöbe community in Guaymí de Osa does not have cattle, because livestock is one of the main reasons why the forests of Costa Rica were originally cut down and continued to do so between 1950 and 1985 (Fournier 1991).

Furthermore, it is interesting to note that much of the land in the Chiriquí Province in western Panama from where the Costa Rican Ngöbe originally came from is cattle land, and much of it is totally deforested! In their migration from the

Fig. 23.9 Indigenous forested land (left side of the picture) and nonindigenous land use involving deforestation (central and right side of the picture) in Guaymí de Osa Indigenous Territory. (Photo Felipe Montoya in 2010)

pastures of Chiriquí in Panama to the rainforests of Osa Peninsula in Costa Rica, the productive practices of the Ngöbe people experienced a transformation. This disclosed a much more complex scenario for the research on Ngöbe traditional ecological knowledge (TEK) and productive practices that explain the conservation of the forests in Guaymí de Osa. Indeed, how did Ngöbe people manage to conserve the forests?

During an environmental education workshop organized by a Ngöbe elder for children of the Guaymí de Osa Indigenous Territory, to which we were invited, the components of Ngöbe TEK turned out to be even more diffuse and less linearly clear to us. This stimulated new questions in our research. With most of the children from the community gathered, the Ngöbe elder proceeded to pass on to them not the traditional indigenous knowledge of how to care for the land, along with the plant and animal coinhabitants of the land, but rather, he used materials prepared by the Ministry of the Environment Energy, Mines and Telecommunications (MINAET). The Ngöbe elder tried to explain the official concepts of "conservation of natural resources" and "sustainable development" used by the state, sometimes encountering language in Spanish that he did not fully understand, let alone transmit to the children themselves. If his main objective had been, as he said, "to convey to the youth our tradition of protecting the environment," why, then, had he chosen to teach concepts used by the state rather than Ngöbe concepts and categories? Could it be that the colorful posters provided by the MINAET gave his presentation an aura of authority and prestige, which perhaps homemade materials and locally developed perspectives would not do? Or might it be that the Ngöbe of Guaymí de Osa really do not have discrete categories about the "environment" as we tend to understand it, their worldview being, rather, something more integral that does not allow separating the "environment" from the rest of life and existence?

Were our attempts to discover the indigenous categories of landscape and sustainable methods of production a search for artificial artifacts of our own construction? Or perhaps, thinking along another line, was the conservation of the forest in Guaymí de Osa simply an epiphenomenon of a low population density or a "low standard of living" with few requirements for a life of subsistence without an incessant search for accumulation or luxuries? And if this were the case, could it be considered an indigenous environmental management strategy that could be shared with state conservation strategies? But how, then, could we explain the differences with the subsistence strategies of rural peasants in the Guaymí de Osa Indigenous Territory who do not exhibit the same conservation benefits evident in the Ngöbe land-use systems?

Toward the end of our research, a report by anthropologist Carlos Borge-Carvajal (2010) presented information on financial resources injected into the indigenous territories. This report included Ngöbe territories that had been supported by NGOs dedicated to environmental conservation and payments for environmental services (PES) of the forest cover in the indigenous territories, implemented by the state through the National Forest Financing Fund (FONAFIFO). Although Borge did not specifically mention the Guaymí de Osa Indigenous Territory, it could be that this support system of NGOs and PES was operating there and be a determining factor in favor of conservation in this indigenous territory. Could this explain the relatively little agriculture practiced by the Ngöbe people in Guaymí de Osa? And then, could this be considered a viable indigenous strategy or were we simply completing the circle to find that the conservation initiatives of NGOs and the state were actually responsible for the protection of the forest in this Indigenous Territory? And then, could this system be creating a paternalistic dependency relationship that might in the long run undermine a sustainable traditional indigenous way of life and ultimately have more destructive effects on the natural environment once the foreign funds dried up?

In summary, despite having achieved our specific goal of collecting Ngöbe ethnoecological information in Guaymí de Osa, we cannot say for sure that this represents an ancestral indigenous worldview or a traditional Ngöbe environmental management system that could be incorporated into state conservation efforts. In this way, too many questions remain to be examined before we can provide reliable answers.

23.3 Some Learned Central Elements

There are some central elements discovered by our research, which we can affirm are fundamental. First, the importance of protecting, maintaining, and enhancing biocultural diversity, even in complex and dynamic contexts, cannot be overstated (Maffi 2010; UNESCO 2007; Rozzi 2013). The exploration of diverse lifeworlds, especially where human communities coexist with the surrounding environment without resulting in its degradation, is not only a valid enterprise but vital one. The fact that these lifeworlds, characterized by a sustainable coexistence between humans and their environment, are perhaps more complex to explain than with a limited exposition about their worldview or with the discovery of a land-use practice does not invalidate these research initiatives. It may well be that only by exploring these points can we discover that they may be insufficient to explain a conservation situation. Second, it seems indispensible to investigate history, economics, and politics not only locally but at multiple scales, including the national level and even globally, to fully understand what we find in Guaymí de Osa. Finally, it may be only a matter of continuing to dig deeper to reveal nuances and subtleties, in order to find that a fuller understanding of complex and evolving indigenous epistemologies, ontologies, and axiologies is key to a harmonious and sustainable life in nature, with nature understood as community, not as commodity. It might very well be that the answers to these questions are rooted in the forested landscapes originally inhabited by the Ngöbe people, far away from the Guaymi de Osa territory, and it is their TEK and ancestral land-use practices that have resurfaced. Whatever the case may be (and the research methodologies and results may be different in each locality, in each indigenous territory, or for each ethnic group), the search for alternatives to the current hegemonic worldview and practices of domination focused exclusively on the accumulation of economic gains, regardless of the means of exploitation and of extraction, represents one of the most urgent quests of our day.

23.4 Final Considerations

The globalization of the neoliberal capitalist economic regime increasingly dominates the political-economic landscapes and decision-making capacities across the globe by means of its effective accumulation and concentration of wealth and power. This global regime is based in great measure on the homogenization, standardization, and industrialization of production processes and outputs so as to service economies of scale for the maximization of profits (Rozzi 2015; Montoya-Greenheck 2013), to benefit an increasingly select few (Oxfam 2017). Among the casualties of this homogenization onslaught is the decline of biocultural diversity. This includes the decline of biological diversity, or "coinhabitants" of the landscape (Rozzi 2013), through the violent appropriation of the components necessary for their reproduction, survival, and expression of their lifeworlds, as exemplified by continued deforestation, contamination, and habitat destruction, in order to service the overarching objective of profit maximization. The homogenization onslaught also includes the decline of cultural diversity through rendering redundant and obsolete human lifeways that do not conform to or service the capital accumulation project. Diverse lifeways of peoples in landscapes of "habits" and "habitats" (Rozzi 2013) are objectified and instrumentalized as warehouses of "natural resources" to exploit, as factories for the production and sale of commodities, and as dumpsites where to cast away the wastes (Montoya-Greenheck 2015).

The homogenization onslaught also includes the decline of biocultural diversity. It severs the ties that bind human societies to specific territories and landscapes that have been cogenerative of localized lifeways, which either predate neoliberal capitalism or creatively generate alternatives to it. Following the process of biocultural homogenization (sensu Rozzi 2013) discussed in this volume, these traditional life habits and habitats are alternately perceived as obstacles to the capitalist project, as resources to be exploited by it, or as dumping grounds for the waste products of the capitalist production process. Subsistence lifeways such as those of the Ngöbe people in Guaymi de Osa, by their mere existence, challenge the hegemonic paradigm. We should celebrate their persistence, as well as recognize the importance of state policies that contribute to their presence, keeping guard, nonetheless, of the incessant hegemonic tendencies to render them as obsolete, as resources to be exploited for profit, or as dumping grounds, as when, for example, their worth is reduced to mere carbon sinks for greenhouse gasses.

If the process of biocultural homogenization were simply the advance of one aesthetic favoring standardization over another aesthetic with a penchant for diversity, the issue would be merely a matter of taste or of opinion. But the issue is much more complex and broader in scope. On the one hand, it is a matter of ethics, of right and wrong. Questions like who gets to live and who must die are a matter of socio-environmental justice (Rozzi 2013) and are intimately bound within the homogenization-diversity clash. Decisions on what is right and what is wrong are based on values. Some of these values are shared by all humanity, such as the right to life, liberty, justice, and well-being. This raises questions about what is meant by well-being, and who gets to define it, well-being for whom and for how long? The sustainable development discourse tells us we should consider future generations in our search for well-being.

Therefore, to achieve this, we had better have the pertinent data, the correct analyses, and the appropriate techniques in order to guarantee that we get it right for the future generations. So, the homogenization-diversity clash, besides being a matter of ethics, is also a matter of objective science and appropriate technology. To this effect, we can argue that biocultural diversity offers us raw materials for facing uncertain futures. It also provides numerous alternatives to the current prevailing system based on the homogenization of values (the profit motive over all other values), the homogenization of lifeways (modernity and its constituent consumerism as the only viable lifeway), and the homogenization of food production (the capitalist corporate food regime as the dominant agriculture). This biocultural homogenizing system has proved itself catastrophic with respect to a vital number of Earth system boundaries and incompetent at achieving its own promises of eliminating poverty, feeding the world, and providing well-being for all.

There are still places, however, where the prevailing logic of biocultural homogenization has not yet penetrated completely and where life forms still hang on to lifeways that surrender to other values, meanings, and practices. The project entitled Biocultural Diversity in the Southern Region of Costa Rica has been an attempt to join in solidarity. Our project has used tools and instruments proper to our own craft as anthropologists, in partnership with Ngöbe people in their territories. In these spaces of biocultural collaboration, where there still predominates a diversity of life forms and lifeways not yet reduced to profit-making inputs or to profit-saving basins, it is possible to explore together alternatives of sustainability of sociocultural well-being. While we came out with more questions than what we started with, one thing became even clearer: the capacity of diverse human groups to find sustainable lifeways of harmonious coexistence with nonhuman nature, given the most basic conditions to do so. There can be no better argument to promote policies that favor biocultural diversity, if we are to overcome the multiple socio-ecological crises fueled by the homogenizing project ultimately engendered by an increasingly narrow and elitist rendering of modernity and capitalism. The Ngöbe of Guaymi de Osa, either through their traditional ecological knowledge or by their creative adaptation to new environments, despite their being subject, as we are all, to the hegemony of neoliberal plutocratic capitalism, still have much to teach us with their particular lifeways sustainably immersed in one of the most biologically diverse areas of the neotropics.

References

Alarcón-Cháires P (2006) Riqueza ecológica versus pobreza social. Contradicciones y perspectivas del desarrollo indígena en Latinoamérica. In: Cimadamore AD, Eversole R, McNeish J-A (eds) Pueblos indígenas y pobreza. Enfoques multidisciplinarios, Programa CLACSOCROP. CLACSO, Buenos Aires

Arias S (2011) Diversidad en el agua del pueblo Ngöbe. Proyecto Diversidad Biocultural en el Pacífico Sur de Costa Rica. UCR-LAI, San José, Costa Rica

Borge-Carvajal C (2003) Etnografía de los Ngöbe (Guaymí) de Costa Rica. Unpublished manuscript. RNT-ORGUT

Borge-Carvajal C (2006) Plan de acción del PSA indígena. IPP 165. FONAFIFO. Costa Rica

Borge-Carvajal C (2007) Consulta en los Territorios Indígenas del Pacifico de Costa Rica del Programa de Catastro y Registro en su componente II ABRE. Regularización del Catastro y Registro Nacional. On line http://www.uecatastro.org/componente2/InformefinaldeABRE.pdf

Borge-Carvajal C (2010) Resultados del PSA en los Territorios Indígenas apoyados por el PPD-GEF/PNUD: Casos de Conte Burica, Guaymí de Coto Brus y Bajo Chirripó. On line http://theredddesk.org/countries/costa-rica

Borge-Carvajal C (2012) Costa Rica: estado de la educación en territorios indígenas. Informe Final. Cuarto Informe del Estado de la Educación. Estado de la Nación en Desarrollo Humano Sostenible. CONARE. Costa Rica

Camacho-Nassar C (1995) En la frontera del siglo XX la exclusión de los guaymies en Costa Rica. Universidad de Costa Rica, San José

Campregher C (2011) Etnobotánica Ngäbe: conocimiento y uso de plantas en el Territorio Indígena Osa, Costa Rica, 2010–2011. Master's thesis in Natural Resource Management. UNED. San José, Costa Rica

Campregher C (2012) Uso de plantas y otros recursos no maderables del bosque en el Territorio Indígena Guaymí de Osa. Unpublished report. Proyecto Diversidad Biocultural en el Pacífico Sur de Costa Rica. UCR-LAI, San José, Costa Rica

Durand L (2005) Los mitos y la conservación ambiental. Revista Lider 13:215–225

Estanley D (2011) Abecedario y Silabario Ngöbe. Proyecto Diversidad Biocultural en el Pacífico Sur de Costa Rica. UCR-LAI, San José

FONAFIFO (2012) Estudio de cobertura forestal de Costa Rica 2009–2010. Fondo Nacional de Financiamiento Forestal. MINAE. Costa Rica

Fournier LA (1985) El sector forestal de Costa Rica: Antecedentes y perspectivas. Agron Costarric 9(2):253–260

Fournier LA (1991) Desarrollo y perspectiva del movimiento conservacionista costarricense. Editorial de la Universidad de Costa Rica, San José

García R (2008) The Osa biological corridor in the context of the Mesoamerican biological corridor. In: Weissenhofer A et al (eds) Natural and cultural history of the Golfo Dulce region, Costa Rica. Stapfia 88, Linz

Grüber B (2012) Cacao (Theobroma cacao) y su función curativa desde la perspectiva de los ngäbe de Costa Rica. Unpublished report. Proyecto Diversidad Biocultural en el Pacífico Sur de Costa Rica. UCR-LAI, San José, Costa Rica

Guevara M (2000) Perfil de los Pueblos Indígenas de Costa Rica. Report for the World Bank. Final Report. San José, Costa Rica

Hartshorn G (1978) Letter to Peter Martin, Executive Director, Institute of Current World Affairs. New Hampshire. http://www.icwa.org/wp-content/uploads/2015/09/GSH-4.pdf

Keog RM (1984) Changes in the forest cover of Costa Rica through history. Turrialba 34(3):325–331

Maffi L (2010) Biocultural diversity conservation. A Global Sourcebook. Earthscan, London

Maglianesi-Sandoz MA (2013) Desarrollo de la piñeras en Costa Rica y sus impactos sobre ecosistemas naturales y agro-urbanos. Biocenosis 27(1–2):62–70

Montoya-Greenheck F (2013) Stories for a new world. Emerging paradigms series. Part I. Biocultural diversity: an emerging paradigm in a changing world. Langscape 2(2):10–17

Montoya-Greenheck F (2015) Necro-Capitalismo vs Vida-Mundos: La Esperanza de la Diversidad Bio-Cultural. Congreso XXXIV ACELC: Panamericanismos Críticos: Solidaridades, Resistencias, Territorios. San José, Costa Rica, Junio 8–10, 2015

Moya-Chávez C (2011) Acercamiento a los conocimientos orales en los pueblos indígenas de la Casona y Alto Laguna, el caso de la tradición oral sobre animales. Proyecto Diversidad Biocultural en el Pacífico Sur de Costa Rica. UCR-LAI, San José, Costa Rica

Oxfam (2017) An economy for the 99%. Oxfam Briefing Paper. https://www.oxfam.org/sites/www.oxfam.org/files/file_attachments/bp-economy-for-99-percent-160117-en.pdf

Rozzi R (2013) Biocultural ethics: from biocultural homogenization toward biocultural conservation. In: Rozzi R, Pickett STA, Palmer C, Armesto JJ, Callicott JB (eds) Linking ecology and ethics for a changing world: values, philosophy, and action. Ecology and ethics, vol 1. Springer, Dordrecht, pp 9–32

Rozzi R (2015) Implications of the biocultural ethic for earth stewardship. In Rozzi R, Chapin FS III, Callicott JB, Pickett STA, Power ME, Armesto JJ, May Jr. RH (eds) Earth stewardship: linking ecology and ethics in theory and practice. Ecology and ethics, vol 2. Springer, Dordrecht. pp 113–136

Sader SA, Joyce AT (1988) Deforestation rates and trends in Costa Rica, 1940 to 1983. Biotropica 20(1):11–19

Solórzano-Fonseca JC (2011) La rebelión de los indígenas bajo la dirección de Pablo Presbere (Talamanca 1709–1710). Cuadernos de Antropología 21:1–30

UNESCO (2007) Links between biological and cultural diversity. On line http://unesdoc.unesco.org/images/0015/001592/159255e.pdf

Young P (1971) Ngawbe. Tradition and change among the Western Guaymí of Panama. University of Illinois Press, Urbana

Chapter 24
Candomblé in Brazil: The Contribution of African-Origin Religions to Biocultural Diversity in the Americas

Paulo José dos Reyes (Pai Paulo José de Ogun) and Silvia Regina da Lima Silva

Abstract In Latin America, the search for alternatives to environmental crisis means recovering the sociocultural practices of local communities, especially peasant communities and those of original peoples and African descendants. Modernity, organized by empirical evidence, abstract analytical thinking, and pure reason, thereby excluding the emotional and sacred, was imposed as part of the colonial enterprise. Colonial modernity effectively negated the ways and thought of local communities, even though they had demonstrated through deep time social and environmental sustainability. This negation occurred because modern rationality has been incapable of considering the wisdom, thought, and knowledge of colonized peoples as legitimate, therefore impeding their possibilities of future projection based on their own reason and historical projects. To listen and to dialogue with colonized peoples presuppose a place where an honest and profound exchange of wisdom can take place. It implies moving from "only one truth," the paradigm of scientific modernity, to an understanding that there are "other paradigms." In this chapter, embedded in the "paradigm of the other," we assume as our place of reflection, the traditional pathways of African descendants, particularly of Candomblé, an African-origin religion widely practiced in Brazil. In this religious tradition, tree leaves have souls, the spirits speak, the ancestors feast, and life is reproduced in harmonious, loving, caring, and respectful relationships. If the environmental crisis is to be confronted, the relationship to the earth needs to be understood from other paradigms. These emerge from the diverse experiences that have been made invisible during so many years of recent history. It is here that Candomblé can contribute a distinct understanding of, and relationship with, the Divine, an understanding that consequently brings forth relationships of responsibility, care, and love for nature.

P. J. d. Reyes (Pai Paulo José de Ogun)
Babalorixa or religious leader of the Candomblé community of ILÉ AşÉ ÒGÚN ÀLÁKÒRÓ, Salvador, Brazil

S. R. d. Lima Silva (✉)
Departamento Ecuménico de Investigaciones, San José, Costa Rica

R. Rozzi et al. (eds.), *From Biocultural Homogenization to Biocultural Conservation*, Ecology and Ethics 3, https://doi.org/10.1007/978-3-319-99513-7_24

Keywords Paradigm of the other · Slavery · *Quilombo* · *Orișha* · Nature

Egba, egba, enigba lati bereFeran aye: "Anytime is the time to begin to love nature."

24.1 Introduction

In Latin America, the search for alternatives to environmental crisis means recovering the sociocultural practices of local communities, especially peasant communities and those of original peoples and African descendants. Modernity has taught us to look forward for finding solutions to the many problems of today. We have been told that "the best is yet to be." Yet in Latin America, if we approach environmental questions from the socio-environmental and biocultural viewpoints, our ancestral traditions invite us to move in a different direction: to hear the voices of our ancestors and to learn their practices and knowledge, while we keep our feet deeply planted in our own time, with questions and problems we face today.

"Learning from" and "listening to" suppose an epistemological perspective that is anchored in the "paradigm of the other" (see Mignolo 2003 where this concept is developed), a perspective that connects with the validity of autonomous and autochthonous reason, thought, and future projection, a validity that the imperial and colonial expansion of modernity denied (Mignolo 2003). This negation occurred because modern rationality has been incapable of considering the wisdom, thought, and knowledge of colonized peoples as legitimate, therefore impeding their possibilities of future projection based on their own reason and historical projects. Modernity, organized by empirical evidence, abstract analytical thinking, and pure reason, thereby excluding the emotional and sacred, was imposed as part of the colonial enterprise. At the same time, "value" was defined in strictly economic terms. The imperial project was to make everyone "modern." Consequently, the traditional beliefs and practices of colonized peoples were subjugated, or even abolished, despite their having demonstrated through deep time sustainability, both environmentally and socially.

In this chapter, embedded in the "paradigm of the other," we assume as our place of reflection the traditional pathways of African descendants, particularly that of Candomblé, an African-origin religion widely practiced in Brazil. In the first part, we propose a dialogue through which we can recover the concepts and philosophies that make explicit the intimate and unique relation of Candomblé with nature. In the second part, we visit a Candomblé community that is situated on the periphery of the province of Rio de Janeiro, and that was recently legally recognized as "*Quilombo* land," that is, descendants of "Quilombos" (Cruz 2016). These were refuges for African slaves who had rebelled against Portuguese over lords or who had otherwise escaped servitude on the large agricultural estates. These settlements, such as the Quilombo de Palmares (1580–1719), were governed entirely (and clandestinely) by Africans and mixed African-Native Americans, sometimes joined by mestizos and indigenous people, and were organized collectively.[1] From this community, it is possible to learn new and different forms of living, a spirituality that assumes nature as one's lap, as a place of celebration and ritual, of caring, as if one were living wrapped in the arms of both the Divine and the Human.

Egba, egba, enigba lati bereFeran aye, "Anytime is the time to begin to love nature," affirms a Yoruba proverb (Interview with Babalorixá: Pai Paulo José de Ogún [Paulo José dos Reyes], 8–17, May 2017). Yoruba is a linguistic and ethnic group in West Africa, found especially in Nigeria, Benin, and Togo. Their cultural and religious belief in a common origin ties together the various Yoruba groups, as they all consider themselves to be descendants of the mythical creator *Odùduwá.* In the Americas, Yoruba people are found in Brazil, Colombia, Venezuela, Panama, Mexico, Puerto Rico, Cuba, Dominican Republic, Trinidad, and North Carolina (USA). In Brazil, they arrived during the final period of slavery (eighteenth and nineteenth centuries), especially in the north and northeast regions of Bahía and Pernambuco. In Rio de Janeiro, their influence is seen mainly through their African-origin religious "houses" for worshipping the *Orichas* (see Elbein dos Santos 1977, pp. 26–33). *Egba, egba, enigba lati bereFeran aye* tells us that this is our time to love, to be caring, and to share the experiences of people who live on the margins of mainstream society, yet who have conserved the secret of the sacred – a sacredness that, far from limiting human endeavor, reconnects it with the created universe and extends the possibility of continuing to exist.

24.2 The "Other Place"

To listen and to dialogue with Candomblé presuppose a place where an honest and profound exchange of wisdom can take place. It implies moving from "only one truth," the paradigm of scientific modernity, to an understanding that there are "other paradigms." For Walter Mignolo, the "other paradigms" are:

> the diversity and "diversality" [as opposed to the "universality"] of critical forms of analytical thinking and of future projects based on the histories and experiences marked by coloniality … [T]hat paradigm is diverse, it has no reference author, nor origin. The connection is that which is shared by those who have lived or learned in their bodies, the trauma, the unconscious lack of respect, being ignored … of how the body feels the nothingness of the values of progress, of the well-being that has been imposed on the majority of the inhabitants of the planet, who in that moment have to "relearn to be." (Mignolo 2003, p. 20)

In this perspective, what would "diversality" be? It would be, utopically, the hegemony of diversality as a "universal project." It is about the "other paradigm" because it implicitly carries the negation of the "abstract universality" of the modern project that constantly makes coloniality invisible. For our reflection, this perspective of the "other paradigm" is important because it makes explicit the political and

[1] For a discussion of Africa in the Americas following 1492, see Mann (2011).

geopolitical dimensions of the contribution of African-origin religions to the environmental theme.

Today, contributions for confronting the planetary environmental crisis are emerging from the Candomblé communities that were initially formed by African persons, enslaved in the context of colonialism. So now the "place" of exclusion and invisibleness is the place that is looked to for learning what modernity, in its eagerness for domination, is unable to solve. Therefore, we enter into dialogue with another culture and another religion, with wisdom that has been excluded from the established wisdom of modernity. It is an "other place" because it requires a distinct sensibility in order to be assimilated and understood. Here, in that place, tree leaves have souls, the spirits speak, the ancestors feast, and life is reproduced in harmonious, loving, caring, and respectful relationships. The proverbs, sayings, foods, and music all teach us. So, we learn in and from Candomblé!

24.3 Brazilians Sing

Music: "Salve as folhas" (Save the leaves) Maria Betania
Sem folha não tem sonho (Without leaves there are no dreams)
Sem folha não tem vida (Without leaves there is no life)
Sem folha não tem nada (Without leaves there is nothing)
Quem é você e o que faz por aquí (Who are you and what are you doing here?)
Eu guardo a luz das estrelas (I guard the light of the stars)
A alma de cada folha (The soul of every leaf)
Sou Aroni (I am Aroni)
Cosi euê (Without leaves)
Cosi orixá (There is no *orixá*)
Euê ô orixá (I am the *orixá*)
Sem folha não tem sonho (Without leaves there are no dreams)
Sem folha não tem festa (Without leaves there are no celebrations)
Sem folha não tem vida (Without leaves there is no life)
Sem folha no tem nada (Without leaves there is nothing)
Eu guardo a luz das estrelas (I guard the light of the stars)
A alma de cada folha (The soul of every leaf)
Sou Aroni (I am Aroni)

24.4 African Descendants in Latin America

Descendants of Africans represent 30% of the population of Latin America and the Caribbean; in some countries, such as Brazil, they are half or more of the entire population. Geographically, they are found especially in Brazil, Colombia, Ecuador, Peru, Venezuela, Panama, Honduras, Nicaragua, Costa Rica, Cuba, Haiti, Dominican Republic, and other Caribbean islands, with smaller populations in other countries (Bello and Rangel 2002; Anton et al. 2009).

In all these countries, African descendants suffer social exclusion and poverty because of racism and discrimination. Racism is strongly manifested in educational opportunities, jobs, and access to land. In relation to education, African descendants have difficulty both to access and to remain in the educational system. In addition, curricular texts do not include the historical and cultural perspective and experience of African descendants. Racial segregation is more than evident in the labor market where African descendants obtain only the least remunerated forms of labor.

It is in access to land that one finds the principal roots of the socio-environmental problems that confront black communities in Latin America. It was not until the end of the 1980s and the beginning of the 1990s that countries such as Brazil and Colombia effected legislative changes related to land possession that had positive repercussions for black communities, even though the laws contained many ambiguities. In the case of Brazil, the constitution of 1988 mentions the "surviving remnants" of Quilombos without clarifying the term: Article 68 (Title X "Transitory constitutional dispositions") reads: "The surviving remnants of Quilombo communities that are occupying their lands, are recognized as definitive property and the State must emit the respective property titles to them." Although the constitution was approved in 1988, this article was not debated until 1995 – the year celebrating the third centennial of Zumbi de Palmares, the country's most important historic black leader. It was on this occasion that this law was judicially regulated.[2] In Rio de Janeiro, African descendants live mostly in the urban periphery and in the rural outskirts of neighboring cities. In addition to their situation of poverty and discrimination, they experience high levels of violence and mortality especially among young men, a true genocide that demonstrates the persistence of racism in Brazilian society (Debate 2017; Smith 2017).

The persistence of racism and discrimination is part of the social background in which the Candomblé communities are immersed. The conditions imposed by slavery, together with the different forms of resistance to slavery, were the origin of this religious experience. Candomblé was, thus, a place of rebellion and transgression, by proposing new forms of relationships among blacks that arrived from distinct parts of the African continent, as well as with nature and the new land that received them.

24.5 Crossing the Waters: Slavery and Resistance

Understanding the history of black people in Brazil is fundamental for understanding the emergence of Candomblé. This religion is present throughout Brazil and assumes diverse forms depending on the geographical origin of the African slaves: Banto, Jefe, Nago, or Ijeje. Each form has its own dynamic and experiences (later in this chapter, we will discuss one of these specific communities). From the sixteenth century through the nineteenth century, millions of Africans were brought to Brazil

[2] See Arruti and Hibridação (1999).

as slaves since the national economy was based on slave labor. In order to assure the domination of white overlords, efforts were made to eliminate everything that connected Africans to their history, their past, and their African roots, thus reducing them to mere commodities.

Still, even though they came as slaves, many Africans preserved their culture and religion. They brought with them on slave ships their "tools": seeds, plants, customs, and religious experience. However historical, geographical, economic, and political circumstances shaped Candomblé in ways that distinguish it from the original African religion. It is now effectively Brazilian, but with African habits and customs – such as language, ritual offerings, and the Orisha pantheon – that have taken on new meanings in the new context. So, it is a Brazilian religion, rooted in Africa, with cultic practices originated by African descendants (Interview with Babalorixá, Pai Paulo José de Ogun [Paulo José dos Reyes], June 2017).

Originally this black religion was practiced clandestinely. It was a place that promoted resistance, capable of maintaining African descendants' links to Africa and recreating, in various forms, relations among themselves, even remaking family ties. No longer was the family part of a clan, as in Africa, but rather part of a "holy family" formed out of the religious community or *Àṣé*. Candomblé was a means for recreating families that had been completely torn apart by slavery (Cido de Osún Eyin 2002, p. 63). Therefore, the Quilombos and African religion were places where black people could resist the negation of their history and identity. These were affirmed by promoting life, dignity, community and social relationships, and the struggle for freedom. Quilombos and religion constituted concrete expressions of resistance, where history and culture were guarded and the dignity of black people maintained. Both reinforced connections to land, community, and nature.

24.6 Candomblé: "Without Leaves There Are No *Oriṣhas*"

Candomblé is the equilibrium among the elements of nature. It is oriented by the practice of unconditional love. It is religious space, free of preconcepts. It receives every one as they are. Each person is conceived in relation to ancestors that are understood, in Africa as well as Brazil, in two forms: the *Egúngún* (those who has passed through the experience of death) and the *Oriṣha* (or divinized ancestors) (Cido de Osún Eyin 2002, p. 57).

The religion is founded on worshipping the *Oriṣha*. In life, the *Oriṣha* were exceptional beings, the bearers of *Àṣé* (vital force), and that were transformed into divine ancestors. The *Oriṣhas* in Africa are related to a family, as an ancestor, and are connected to a specific geographical region. In Brazil, the gathering of blacks of diverse ethnic groups into a single *senzala*, the slave quarters on the old ranches, each with its own *Oriṣha*, promoted the creation of the ample pantheon of *Oriṣhas* celebrated in the Afro-Brazilian cults today.

The *Oriṣhas* are connected to the elements of nature: the water of rivers and oceans, the forests and rocks of the earth, fire, lightening, thunder, storms, and air.

Candomblé is deeply related to nature. The *Orişhas* are the living forces of nature, the clearest manifestation of life. Life made the *Orişhas* eternal and therefore they are manifested in the living. As the life force, the *Orişhas* not only manifest themselves in African descendants but also in every living person, who, at the same time, can identify themselves with an *Orişha* (Cido de Osún Eyin 2002, p. 57).

If we ask about the relationship between Candomblé and the environment, the answer is that all of Candomblé is related to the environment. A Yoruba proverb expresses that deep connection between religion and nature: "Without leaves, there are no *Orişhas*." Leaves are not just leaves but the whole of nature that makes possible this ancestor cult. The four elements are always present in Candomblé: earth, fire, water, and air. They are interconnected and are in constant movement, that is, Candomblé is movement. Creation myths reveal this intimate relationship with nature (Cido de Osún Eyin 2002, pp. 29–54). Thus, this religion proposes a living relationship between humans and nature that goes beyond African descendants. Because of this, we find, in this religious experience that emerges from local communities, alternatives for confronting the environmental crisis that threatens the planet.

The following are some of the *Orişhas* and their relationship to particular dimensions of life and the elements of nature that keep their presence (Interview with Babalorixá: Pai Paulo José de Ogun [Paulo José dos Reyes], June 2017; Cido de Osún Eyin 2002, pp. 77–271):

- *Eşu*: Communicator that breaks established paradigms; is a principle of transformation and communication; precedes the order of the universe; and is represented by a spiral
- *Ógun*: Related to iron and steel, to metallurgy and technology; represents the earth and its minerals; opens roads in forests, streets, and highways; *Orişha* of abundance that makes humans evolve, always respecting fellow human beings and nature
- *Oşoóşi*: Son of the woods and forests and the father of leaves; represents nature as the preserver of life, plant, and animal; lord of the woods and forests; the ancient hunter who assures human survival but who uses only a single arrow in order to kill no more than that which is necessary for survival, thus avoiding wastage
- *Osanyìn*: Knows the secrets that awaken the power and force of leaves, knowledge that is unknown to other *Orişhas*; represents the leaves and plants that the earth, fertilized by rain, produces
- *Òşùmàrè*: *Orişha* of all movements and natural cycles; lives in heaven and visits the earth by traveling on the rainbow; related to the sky, earth, and rainbow
- *Naná Buruku*: Divinity present at the beginning of creation; the oldest divinity related to the human trajectory on earth; still water, the water of life and death; represented by still waters and clay
- *Oşun*: Represents the politicization of women; the first woman to receive honorary titles; the queen of rivers, mother of hunters; fresh water, rivers, rapids, falls, and lagoons symbolize her

- *Yánsàn*: Related to fire; represents contradictory elements, born of water and fire, and queen of wind, lightening, and storms
- *Yemoja*: Mother of the world who sustains humanity; worshiped (in Brazil) as the mother of the sea; unarrested water that extends widely and unites peoples
- *Ṣangó*: The *Oriṣha* of justice; natural fire

As can been seen, the *Oriṣhas* are related to the elements of nature. This relationship is manifested in the lives of the members of the community. Numerous rituals mediate this relationship. Offerings are made in places associated with an *Oriṣha* and are a way of strengthening the sons and daughters of that *Oriṣha*, that is, the community. Herbs that cleanse and revitalize different dimensions of the body and make possible the recovery of health are used in bathing. Even the kitchen is a place where the elements of nature acquire great significance. In Candomblé, food – ritual eating – is an important vehicle of communication, of strength, and of the integration of human life (Caetano de Sousa Junior 2009). Food marks every day of the life of Candomblé communities. Food "is in everything … in the *terreiro* [the physical place where rituals and ceremonies are conducted] everyone eats" (Caetano de Sousa Junior 2009, p. 81). The kitchen is converted into sacred space. In it the elements of nature are prepared and transformed in order to be offered and served at the community table.

Clearly, this theological framework is the basis for the contributions of Candomblé to the environment. As Janet do Nascimento (Iya Aja Jemi, personal communication), member of the *ILÉ AṢÉ ÒGÚN ÀLÁKÒRÓ*, explains:

> Candomblé is directly related to the environment. At the spiritual level, the worshipped divinities are represented in nature, "Kosi Ewé Kosi Orisa"—without leaves there are no *Oriṣhas*. On the part of persons devoted to this religion, a very deep and caring respect exists for all the environment because it offers us the things we need for the rituals that are performed. There is a dialogue between cult and nature. It's possible for us who believe in the *Oriṣhas* to perceive clearly the answers and the ways nature conspires in our worship.

As we have seen, in Candomblé:

> we find important teaching about caring for nature. The lives of members of this religion are lived in deep communion with nature. Daily life is marked by baths, tea, meals, and offerings passing through this profound encounter with whole of creation. The preservation of water, herbal gardens, the recycling of ceremonial objects, are some examples of this responsible and caring action that manifests the search for a healthy life, seeking to be self-sustainable, and taking advantage again and again of what nature offers. (Interview with Babalorixá: Pai Paulo José de Ogun [Paulo José dos Reyes], June 2017)
>
> *Egba, egba, enigba lati bere Feran aye,*
> *(Anytime is the time to begin to love nature)*

Alternative principles for practicing alternative ways of living with nature are found in Candomblé. "Regarding the preservation of the environment, in the community we are educated to conserve and protect it. This is done through the community garden, planting vegetables, and organizing seminars for building awareness among members of the broader community (*Egbé*), as well as among the local people. People are urged not to contaminate rivers, the sea, nor lakes," explains Janet do

Nascimento (Iya Aja Jemi, personal communication). Cultivating the soil, using fully all that is available, even the house where one lives is contemplated as more than physical space – the environment or nature is part of everything. The recovery and preservation of herbs and the sacred space of gardens all speak of responsibility for, and preservation of, the environment. Recycling even can be seen in decorations. For the celebration of the community's 21st anniversary, trash from throughout the area was reused as decoration. This was beautifully expressed as responsibility and environmental care.

How are these principles worked out in the daily life a community?

24.7 *ILÉ AşÉ ÒGÚN ÀLÁKÒRÓ*: Quilombo de Mongaba (Kilomba) – Alternative Community, Liberating Practices

We want to share the experience of a particular community, the *ILÉ AşÉ ÒGÚN ÀLÁKÒRÓ*. Pai Paulo de Ogun is the Babalorixa of this community that is part of the Quilombo. This community, with its projects and daily practices, is "another paradigm" that points toward a distinct relation to nature but also "another paradigm" of the concept of nature, of religion, and of social responsibility and political commitment that are part of religious practices.

The *ILÉ AşÉ ÒGÚN ÀLÁKÒRÓ* community is located in Magé, Province of Rio de Janeiro. During the period of slavery, the place was a hospital situated along the old Gold Road, between Rio de Janeiro and Minas Gerais, Pai Paulo de Ogun explains. In 2016, the community was recognized as "Quilombo de Bongaba," at the provincial level by the Association of Surviving Remnants of Quilombos of the Province of Rio de Janeiro. Currently, it is in the process of being recognized at the national level by the Palmares Foundation. "Part of the land has long been occupied by descendants of slaves, but another part has been invaded by others. We have organized to recover our roots and to recover our territory," says Pai Paulo de Ogun. One project is the construction of a school and a cultural library that will contribute to recovering the history of the Quilombo. Recognition as a Quilombo has created a cultural reference for the region as the approximately 50 people who live there are keeping alive Quilombo traditions.

Throughout the year, the Quilombo carries out different activities, such as monthly conversations about different issues, and dances and games of African origin, among other cultural activities (Fig. 24.1).

The activities that are conducted in the Quilombo de Bongaba are marked by their sociopolitical commitment. This flows from spirituality, as Pai Paulo de Ogum and others comprehend their experience of Candomblé. Since this experience is rich and broad, we will limit ourselves to aspects that are most directly connected to the environmental commitment.

Fig. 24.1 Cultural activity in *ILÉ AṣÉ ÒGÚN ÀLÁKÒRÓ*. Rituals celebrating nature and aspects of daily life constantly remind the community of the life force of the *Oriṣhas*. (Photo Silvia Regina da Lima Silva)

Inspired by the African Ubuntu proverb, "A person is a person through other persons," the religious community is organized into core groups that are to serve both inside and outside the community. These core groups are inspired by the lives of the *Oriṣhas*. There is a correspondence between the kind of service offered and the *Oriṣha* to which the core group is dedicated, based on aspects of the life of the *Oriṣha*. These core groups give attention to:

NACO – Communication
NADETEC – Unemployment and technological training
NAS – Health
NACON – Consciousness raising or awareness
NAECO – Ecology
NAI – Elderly people
NAF – Family
NAM – Women
NAJU – Legal matters
NAJO – Youth
NAP – Psychology
NACRI – Children

These core groups are composed of community members who are professionals in the area of service or who are especially interested in the theme; all service is voluntary. Three of these core groups in particular are worth mentioning: (1) NAS (health), (2) NACON (consciousness raising or awareness), and (3) NAECO (ecology).

1. NAS (Health). This core group is devoted to *Omolu*, the *Orişha* that is considered to be a medical doctor because this divinity cures sickness. It is dedicated to cultivating medicinal herbs for the community. The objective is to use the herbs, roots, and cereals, grown in the sacred garden, for healing. By covering the body from head to toe with these plants, it is believed that their healing properties are transmitted to the body. Heads, for example, are believed to be inhabited by diverse forms of unhealthy patterns that can affect the physical body as illness. This vision of healing sees the body and health holistically.
2. NACON (Consciousness raising or awareness). *Oşoosi* directs this core group because the divinity is the *Orişha* of hunting and also of awareness and consumption. The group's objective is to orient community members about being wise consumers. It helps people to develop emotional intelligence about how to consume rationally, eliminate waste, and assure that goods are multiplied and well administered.
3. NAECO (Ecology). Inspired by *Osanyin*, who provides wise leadership through phytotherapy (herbalism) and sustainable food chains, this group's purpose is to develop community awareness about the preservation of nature. It organizes the annual seminar on the environment, *Egba, egba, enigba lati beréFeran ayé* (anytime is the time begin to love nature). The seminar includes talks and concrete actions such as reforestation.

One of the ways, then, that the principles of Candomblé are worked out in the daily life of a community is through these core groups. These are actively involved in putting into practice many of the principles or teachings of Candomblé in very concrete ways. At the same time, they engage community members in lifestyles that actively embody these principles (Fig. 24.2).

Fig. 24.2 Outdoor activity during the annual seminar on the environment, "Egba, egba, enigba lati beréFeran ayé." Every year the *ILÉ AşÉ ÒGÚN ÀLÁKÒRÓ* community organizes an environmental seminar which includes not only theory but also practical activities oriented toward improving the environment such as planting trees. Children and youth are especially involved in the seminar. (Photo Silvia Regina da Lima Silva)

24.8 Conclusion

The environmental question in Brazil is caught in the conflict between economics and ecology. If on one hand there is progressive environmental legislation, on the other belief in unlimited economic growth continues to be strong. A collective consciousness about the environmental impacts of economic policy is lacking. Faith in unlimited economic growth especially dazzles the economic elites. Although the last 15 years have witnessed important social welfare advances, it also is clear and lamentable that more has not been achieved, such as an effective agrarian reform. Additionally, the governments of this period, like those before them, have demonstrated serious deficiencies in environmental governance.

The relationship to the earth, the natural environment, needs to be understood from other paradigms. These emerge from other places, from the diverse experiences that have been made invisible during so many years. It is here that Candomblé can contribute a distinct understanding of, and relationship with, the Divine, an understanding that consequently brings forth relationships of responsibility, care, and love for nature. This is one of the conclusions that we come to at the end of our brief survey of the richness of Candomblé as a way of understanding. The brevity of this reflection makes us fearful that we have not treated with the respect, depth, and seriousness that this religious tradition deserves. However, it is difficult to find a way of introducing and presenting these experiences to people who possibly have never had contact with this religion. We hope we have been both deeply respectful and informative. This journey has revealed how words, spirits, and practices of commitment to the lives of people and the care for the planet, seen not in competition or mutually exclusive but as part of one movement to rescue life, are emerging from the peripheries, from "other places" and "other wisdoms."

As we mentioned at the beginning, in order to understand and dialogue with the Candomblé experience and its relationship to nature, it is necessary to look to another place and another logic. Abandoning the reductionist logic of scientism is required. One must become sensitive to others and occupy, even if only briefly, that "other place" – people who, far from their own land and people, were capable of maintaining connections to, and the pertinence of, their cultural traditions and to recreate their religion. Religion reties not only continents separated by slavery and ocean waters but reunites, remaking the deep, inseparable connection of human beings as part of nature.

> "Anytime is time to begin to love nature."
>
> *Egba, egba, enigba lati beréFeran ayé.*

Acknowledgment This chapter was written jointly, based on Lima da Silva's visits to the community in January 2017 and interviews she conducted with Pai Paulo de Ogun during May and June 2017. We consider the community of *ILÉ AşÉ ÒGÚN ÀLÁKÒRÓ* to be the proper subject of our narrative and experiences that we share in this chapter.

References

Anton J, Bella A, Del Popolo F, Paixão M, Rangel M (2009) Afrodescendientes en América Latina y el Caribe: del reconocimiento a la realización de derechos, Serie 87, Población y desarrollo. CEPAL/UNFPA, Santiago de Chile

Arruti J, Hibridação M (1999) Segmentação e mobilização política de indios e negros: notas explicatorias a partir dos campos brasileiro e colombiano. Trabajo presentado en el XXII Encuentro Anual de la Associacão Nacional de Pós-graduacão e Pesquisa en Ciencias Sociais (ANPOCS), 19–23 de octubre, 1999, Caxambu. Available via CEPAL%20No%2076%20Exclusion%20 indigenas%20y%20afros.pdf

Bello A, Rangel M (2002) La equidad y la exclusión de los pueblos indígenas y afrodescendientes en América Latina y el Caribe. Revista de la CEPAL 76(April):39

Caetano de Sousa Junior V (2009) O Banquete Sagrado. Editora ATALHO, Salvador

Cido de Osún Eyin (2002) Candomblé – a panela do segredo. Editora Arx, São Paulo

Cruz C (2016) Nos 129 anos da Lei Aurea, quilombo en Magé é símbolo de resistência e discute políticas para o negro. Periódico Extra. Extra.globo.com (25 de julio). https://extra.globo.com/noticias/rio/nos-129-anos-da-lei-aurea-quilombo-em-mage-simbolo-de-resistencia-discute-politicas-para-negro-21335559.html?topico=baixada-fluminense-rj

Debate (2017) Jóvenes negros principales víctimas de violencia en Brasil. Debate.com.mx. May 6. Available at https://www.debate.com.mx/ahora/Jovenes-negros-principales-victimas-de-violencia-en-Brasil-20170605-0423.html

Elbein dos Santos J (1977) Os Nago e a morte. Editora Vozes, Petrópolis

Mann CC (2011) 1493. Uncovering the new world Columbus created. Alfred A. Knopf, New York

Mignolo WD (2003) Historias Globales/diseños globales. Ediciones Akal, Madrid

Smith C (2017) Battling anti-black genocide in Brazil. NACLA Rep Am 49(1):41–47

Chapter 25
Latin American Theology of Liberation and Biocultural Conservation

Roy H. May Jr. and Janet W. May

Abstract The theology of liberation emerged specifically as a theology for Latin America. It broke with ages-old ways of doing and being by giving voice and visibility to the poor and indigenous people long silenced and made invisible by official doctrine and ecclesiology. Cultural identity and diversity underlie the theology of liberation, first by articulating the geographical (and therefore sociopolitical) context for doing this theology and then by planting the subject as the central axis for reflection. As a consequence, it has stimulated theological, liturgical, and artistic expressions reflecting diverse racial and ethnic and other social contexts, thereby breaking the homogenizing mold of traditional church doctrine and Christian thought. Cultural aesthetics, particularly as an expression of the spirituality of liberation, has been embodied in liturgies, hymnody, poetry, and visual arts. These art forms protest against injustice, reinforce resistance, and project hope. This theology is a force for cultural diversity and local respect in an ever-increasingly homogenized world order.

Keywords Liturgy · Hymnody and *coritos* · Literature · Visual arts · Local aesthetics · Theological diversity

During the late 1960s and throughout the 1970s, the theology of liberation emerged specifically as a theology for Latin America – as a *Latin American* expression of Christianity.[1] Christian theology was thought out in ancient Greece and Rome and

[1] Simultaneously, Black theology emerged from African-American Christianity in the United States, rooted deeply in the experience of slavery, and continued racial discrimination and the African cultural experience in America. It also is liberation theology, but initially had little contact with Latin America. The founding works are by James H. Cone (1969, 1970, 1975), late professor of systematic theology at Union Theological Seminary in New York City.

R. H. May Jr. (✉) · J. W. May
Departamento Ecuménico de Investigaciones, San José, Costa Rica

R. Rozzi et al. (eds.), *From Biocultural Homogenization to Biocultural Conservation*, Ecology and Ethics 3, https://doi.org/10.1007/978-3-319-99513-7_25

then, in its Western version, in Europe. It understood itself as universal: homogeneous and unchanging. Christianity encompassed in Roman Catholicism was brought to the Americas by the Spanish and Portuguese and imposed on native peoples as part of the Conquest. Not only did it understand itself as monolithic and hierarchical, it also saw itself as legitimating the Conquest. The church believed that God had given it the Americas to Christianize. Spanish intellectuals such as Juan Ginés de Sepúlveda argued that the native peoples of America were naturally inferior and so were "natural slaves." As such, they could be legitimately be enslaved. War, then, was "justified" in order to appropriate native peoples, their labor, and bring them into the church. The "Conquest" was the way the church could impose the Holy Gospel. Native religious expressions were suppressed or incorporated into Roman Catholicism if they could be reworked as part of orthodox doctrine. Although there were important exceptions, such as Dominican priest and bishop Bartolomé de las Casas – "Defender of the Indians" – who debated Ginés de Sepúlveda (Hanke 1994) and denounced Spanish cruelty and defended the humanity and rights of Indians (Gutiérrez 1993), the church aligned itself with the great landowners and the military, becoming one of the central pillars of traditional social structure. The poor were encouraged to "stay in their proper (subordinate) place" and receive their "reward" for social and ecclesial obedience in heaven (Gutiérrez 1989). The church understood itself as universal – that's what "catholic" means – so diversity in liturgy and doctrine were prohibited.

The historical scenario began to change, however, when Pope John XXIII called the Second Vatican Council. Meeting in Rome between 1962 and 1965, the church's bishops "opened the windows to let in fresh air," as the Pope requested. Although doctrine and structure were not actually changed, they were greatly liberalized (Alberigo 2006; Hahnenberg 2007). Among other reforms, the Council called for more diversity in doctrinal expression and established regional councils of bishops in order to respond relevantly to historical and geographical contexts. In Latin America, this also was a time of social ferment, with widespread social protest, social reform, and revolutionary movements, all of which, in some way or another, affected the church. Fissures were opening in this monolithic structure, and the theology of liberation infiltrated the cracks. It emerged as a way of doing theology specifically in this political and cultural context. In so doing, it broke with ages-old ways of doing and being by giving voice and visibility to the poor and indigenous people long silenced and made invisible by official doctrine and ecclesiology. Needless to say, this was controversial and even costly.

Active and influential especially during the 1970s and 1980s, this theology always was a minority within Latin American Christianity. As social movements dissipated in many countries due to political repression and changing economic and social realities, so has liberation theology become less present and influential. Still, it remains a vital part of Latin American theology and continues to be a force for cultural diversity and local respect in an ever-increasingly homogenized world order.

25.1 A *Latin American* Theology

One of the first and clearly seminal theologians to propose the theology of liberation is Gustavo Gutiérrez of Peru. His book, *A Theology of Liberation, History, Politics and Salvation* (1973), outlined a new way of doing theology and named Latin America as its specific location:

> This book is an attempt at reflection, based on the Gospel and the experiences of men and women committed to the process of liberation, in the oppressed and exploited land of Latin America. It is a theological reflection born of the experience of shared efforts to abolish the current unjust situation and to build a different society, freer and more humane. Many in Latin America have started along the path of a commitment to liberation, and among them is a growing number of Christians … Our purpose … is to reconsider the great themes of the Christian life within this radically changed perspective and with regard to the new questions posed by this commitment. This is the goal of the so-called *theology of liberation.* (Gutiérrez 1973, p. ix)

The book goes on to analyze aspects of Christian thought from this Latin American context, including sections on Latin American sociohistorical reality, liberation, and church. A major contribution is that it does theology by taking into account historical context. This contribution has core convergences with approaches to biocultural conservation and biocultural ethics (cfr. Rozzi 2012; Gavin et al. 2015). Gutiérrez outlined a new theological method, one rooted in history as opposed to deduction from preestablished doctrine. Analogously, biocultural conservation is inductive (Gavin et al. 2015), and biocultural ethics is *contextual* and *systemic* (Rozzi 2013).

For Gutiérrez, theology is to be critical reflection on praxis, "human action as the point of departure for reflection" (1973, p. 9). Continuing, he explains, "Theology is reflection, a critical attitude. Theology *follows*; it is the second step" (1973, p. 11). The first step is immersion in history and culture and vital participation in them. This participatory approach also is a core attribute of biocultural conservation because its sources are the struggles for environmental well-being and the ancestral traditions and ecological knowledge of indigenous people.

Liberation theological reflection inevitably leads to the "history of the Other" (Gutiérrez 1980, p. 370) and the centrality of subjects as the "interlocutors" for doing theology. For this theology, the subject or interlocutors are the "nonpersons," that is, "those who are not considered human beings by the existing social order: exploited classes, marginalized races, despised cultures" (Gutiérrez 1980, p. 353–354). So the "locus" are the poor, represented as the mass of indigenous people and the popular social classes made up of workers, miners, and peasant farmers. They are understood as active artisans of their own histories. This is "a *new way* to do theology," Gutiérrez argues:

> which does not stop with reflecting on the world, but rather tries to be part of the process through which the world is transformed. It is a theology which is open –in the protest against trampled human dignity, in the struggle against the plunder of the vast majority of people, in liberating love, and in the building of a new, just, and fraternal society—to the gift of the Kingdom of God. (1973, p. 15)

Other theologians and church workers, not only Roman Catholic but also from historic Protestant traditions such as Methodism and Presbyterianism, began working out of this new framework. They emphasized intellectual resources developed in Latin America: the liberation tradition in philosophy (Enrique Dussel, Arturo Roig, Horacio Cerruti, Augusto Salazar, Leopoldo Zea, Rodolfo Kusch); sociology and economics dependency theory (Raul Prebisch, Fernando Henrique Cardoso, Theotonio dos Santos, Aníbal Quijano); literature (José María Arguedas); and political thought (José Carlos Mariátegui).[2] Although classic and contemporary Christian thinkers continued to be present in theological thought, its Latin American identity became evident throughout the time period.

25.2 Theology of Liberation and Cultural Aesthetics

Cultural identity and diversity underlie the theology of liberation, first by articulating the geographical (and therefore sociopolitical) context for doing this theology and then by planting the subject as the central axis for reflection. As a consequence, it has stimulated theological, liturgical, and artistic expressions reflecting diverse racial, ethnic, and other social contexts, thereby breaking the homogenizing mold of traditional church doctrine and Christian thought. Cultural aesthetics, particularly as an expression of the spirituality of liberation (Bonnin 1982), especially has been embodied in liturgies, hymnody, poetry, and visual arts. These art forms protest against injustice, reinforce resistance, and project hope.

25.3 Liturgy

Liturgies such as the mass for the land without evil (*missa da terra sem mal*) and the *Quilombo* mass both by Pedro Casaldáliga and Pedro Tierra of Brazil are notable expressions of cultural aesthetics stimulated by the theology of liberation. The "land without evil" (or "the holy mountain") (1980) draws on the mythical and utopic vision of the Guaraní people of Bolivia and Paraguay that constantly leads them to change geographical location as they search for this holy mountain. This "land without evil" is a place for rest and happiness, for singing and dancing, for celebration and feasting, and for peace and tranquility (Albó et al. 1990, p. 284). Pedro Casaldáliga, for many years the Roman Catholic bishop of São Felix do Araguaia, Brazil, has long been identified with liberation theology and is a prolific writer, especially of poems. Pedro Tierra is a longtime social activist, writer, and government official. The publication of the land without evil mass included music by

[2] For an introduction to Latin American liberation philosophy, see Mendieta 2016. For the history of Latin American philosophical thought, including concise characterizations of the thought of these Latin American thinkers, see Dussel et al. 2009.

well-known Brazilian musician Martín Coplas and photographs by acclaimed photographer Cláudia Andujar, thus merging words, sounds, and vision.[3]

Sometime later, Casaldáliga and Tierra were joined by performing artist Milton Nascimento to write the *Quilombo* mass (1982). The *Quilombos* were autonomous communities established by escaped African slaves during the seventeenth through the nineteenth centuries in Brazil and elsewhere in Latin America (Mann 2011, pp. 332–369; see Dos Reyes and da Lima Silva 2018, pp. 380–387). This musical, then, incorporates African-origin culture and beliefs into a Christian mass.[4] However, the *Quilombo* mass was prohibited by the Vatican for being contrary to orthodox doctrine because it incorporated into a Christian liturgy aspects of traditional African spirituality.

In similar fashion, throughout Latin America local musicians and writers drew from their own cultural contexts to produce Creole and peasant masses in Argentina, Peru, Colombia, Nicaragua, and El Salvador, among others.[5] These renditions of the Roman Catholic mass incorporated popular beliefs, traditional music, and revolutionary commitment, molding the mass to specific cultural contexts and reflecting liberation theology.

25.4 Hymnody and *Coritos*

Protestantism has a long history of hymnody, so when it moved into Latin America through its missionary enterprises, it took its hymns along. These, of course, were written in European and US contexts, with words and music rooted in those cultural situations, and then transplanted into Latin America. But the theology of liberation's emphasis on acculturating Christianity to the *Latin American* reality for a *Latin American* church inspired *Latin American* hymnody. Notable are the hymns by Pablo Sosa and Federico Pagura of Argentina, Mortimer Arias of Bolivia/Uruguay, Simei Monteiro of Brazil, and Ulises Torres of Chile, to name some of the most prominent. Each of these writers was significantly engaged in the defense of human rights and ministries among the poor in their countries, and they molded hymn verses around these themes, crafting the underlying theology from liberation theology. Likewise the lyrics were set to secular tunes traditional to the writers' own cultures.

---

[3] See https://www.youtube.com/watch?v=UQ6Sai3BErs

[4] See https://www.youtube.com/watch?v=3D2AvQPtzkg

[5] See: https://www.youtube.com/watch?v=SsbJIVweail; https://www.youtube.com/watch?v=O-DalDIfiwY; https://www.youtube.com/watch?v=QE_5WBNF3Zg&list=RDQE_5WBNF3Zg&index=1; https://www.youtube.com/watch?v=QE_5WBNF3Zg; https://www.youtube.com/watch?v=rICObuAJlAk

https://www.youtube.com/watch?v=BHDs2D5yDi0; https://www.youtube.com/watch?v=FcrLM1DvBLo

Progressive Roman Catholicism also incorporated congregational singing into the mass, giving rise to a whole genre of popular "coritos" based on the concerns of liberation theology and set to culturally appropriate melodies. These "coritos" – literally little choruses – do not tend to have formal authors, as they have emerged from local struggles for social justice when people put together words and local or traditional folk tunes according to the situation. Thus, they are plastic and dynamic. Many, of course, now have been written down and widely shared. These have become very influential in both catholic and protestant settings and continue to be sung in liturgical situations throughout Latin America.

25.5 Literature

The theology of liberation clearly influences the poetry of Julia Esquivel (Guatemala) and of Rubem Alves (Brazil), the prayers of Luis Espinal (Bolivia), and the literature of Ernesto Cardenal (Nicaragua).

Esquivel, a Presbyterian, studied theology and became deeply involved in human rights and other struggles for social justice. Her prophetic poetry draws on the suffering and resistance – especially of Mayan women – during the long civil war in Guatemala. Her verses are testimonial and openly theological. "They have threatened us with resurrection" she writes in a powerful poem, continuing: "To dream awake, to keep watch asleep, to live while dying, and to know ourselves already resurrected" (Esquivel 1982).

Alves, also a Presbyterian and one of the first theologians of liberation, became disenchanted with formal theology and moved into psychoanalysis and education. However, his poetry is theopoetic in which he reflects deeply on the meaning of life through themes of love, hope, sorrow, joy, reunion, and separation. Late in life he wrote children's stories. He is recognized as a leading figure in Brazilian literature. Alves died in 2014.

Originally from Spain, Espinal was a Jesuit priest and social critic in Bolivia who commented during radio programs and wrote extensively in newspaper columns. Before going to Bolivia in 1968, he authored numerous poems or "prayers," as he called them. Many of these he adapted to Bolivia and often read them as he ended his radio program. Murdered by a military government in 1980, his prayers have been published in numerous editions and have become incorporated into the corpus of Latin American Christian liberation literature.

Cardenal, recognized as a leading Latin American literary figure, actively supported the Sandinista Revolution and became minister of culture in the revolutionary government. His poems, reflecting his revolutionary and Christian commitments in Central America, peruse politics and liberation, justice and commitment, the significance of the universe, the diversity of life, and connection to nature.

Each of these literary personages draws from liberation theology inspiration and content for shaping a distinctive Latin American literature that embodies the struggles of the oppressed for liberation.

25.6 Visual Arts

The visual arts also have been shaped by liberation theology. Paulette Gayer (2014) observes:

> As the ideas of liberation theology grew within the [Christian] communities, religious art demonstrated this new symbiosis of praxis and faith. The art of Latin America became a prominent voice teaching a different idea of what it meant to be liberated by Christ. Murals in public and in churches showed the church as part of the landscape of their history. The role of the church and individuals would be represented as both oppressor and revolutionary. Paintings retold Biblical stories in modern day settings of war and oppression. Finally, works of art even substituted main Biblical characters with revolutionary heroes. This artistic journey goes from traditional to the shocking.

The popular paintings of Nicaraguan common folk, and painters such as Argentine artist Adolfo Pérez-Esquivel, express the spirituality of liberation. These paintings incorporate images from everyday life and political realities, resistance, and struggle for justice. The Nicaraguan primitivist school emerged in the 1970s from the Christian base communities on Solentiname Island in Lake Cocibolca (Lake Nicaragua), where Ernesto Cardenal and others were actively teaching Christianity through liberation theology. The community was attacked and burned by the Somoza government, and many were killed or driven away. Still, the paintings suggest a dreamed-for utopia of peace, justice, and ecological well-being visualized through vibrant cultural images. The canvases of Pérez-Esquivel are visual interpretations of New Testament teachings and Christian traditions brought into the Latin American reality and are notably sensitive to the political repression and human rights violations of Argentina. Especially moving is his portrayal of the Via Crucis or the Stations of the Cross in which he paints the suffering of Jesus and his crucifixion in the context of Argentina.[6] The final station dramatically portrays the multicultural reality of Latin American as a source of strength and hope for confronting the political repression and abuse of human rights that have been so prevalent in many countries. Pérez-Esquivel was awarded the Nobel Peace Prize in 1980 for his human rights work.

25.7 Local Liberation Aesthetics

Of course these only are examples of well-known artists and literary figures who have been influenced by the theology of liberation. However, even at the local and nonprofessional levels, one finds innovative liturgical, literary, and artistic expressions that have been stimulated by liberation theology. In different countries, murals – a time-honored art form in Latin America – depicting local culture and struggles adorn the walls of churches, especially Roman Catholic. Evangelical or

[6] See http://www.alastairmcintosh.com/general/1992-stations-cross-esquivel.pdf

Protestant churches have organized hymnody contests among their congregations and sermon-writing competitions among pastors, challenging them to incorporate cultural and political themes. Numerous local artists, sculptors, wood carvers, quilters, and writers have produced works. These foster cultural aesthetics that draw on liberation and its theology, affirming the local before the universal.

An especially important example of local artisans' creative expression of liberation theology through the arts can be found in Chile. Emerging through the support of the Vicariate of Solidarity of the Roman Catholic Church and international foundations following the 1973 Pinochet military coup, groups of women began using cloth, needles, and thread to tell their stories of suffering as well as their hopes and daily struggles. They adapted a unique Chilean art form, the *arpillera*, a quilting technique, often incorporating scraps of clothing from relatives who were incarcerated, tortured, exiled, murdered, or disappeared by the military dictatorship (for images see Onion 2014). Their work was done in secret, unsigned, in order to avoid persecution. Church workers and other volunteers smuggled these cloth pieces out of Chile, selling them so that the women had some income and also using them to unmask the violence and repression of the military regime, creating international awareness of Chilean reality and the demand for change. This creative or aesthetic expression of the Chilean women inspired people in other countries also to use cloth tapestries to tell their own stories of struggle, resistance, and hope for change (Agosín 1996).

Other examples of local art would include the brightly painted crosses of El Salvador and liturgical stoles of the Guatemala. The crosses from La Palma, Chalatenango, painted by local artisans, vibrantly depict scenes and relationships from everyday life as well as specifically Christian and biblical stories. The stoles are woven by Mayan women in Panabaj, near Santiago de Atitlán, using indigenous cultural as well as traditional Christian symbols. Both cases not only keep alive, but promote, ancient aesthetics and incorporate them as local or intercultural expressions of Christianity.

25.8 Opening the Path to Theological Diversity

Most significantly, the theology of liberation created a new theological conjuncture by stimulating, as part of its constitution, theologies from specific subjects and social locations. This is affirmed by Gutiérrez (1980, p. 375):

> To the extent that exploited classes, poor people, and despised races have become aware of the oppression that they have suffered during centuries, a new historical situation has been created. … As part of these sectors of humanity, Christian communities are endeavoring to live and to think their faith. They are reading the Biblical message of the liberating love of the Father from the poor of the world. As a consequence, different theologies of liberation are walking together: black theology and that of other minorities, theologies from Africa, Latin America and Asia, feminist theology. For the first time in many centuries, theological reflection is surging from different places that are rooted in the world of poverty and oppression.

This theological reflection often has been facilitated by workshops of representatives of these differing social locations to produce theologies directly by the "anonymous of history" (Gutierrez 1980, p. 376). Oftentimes theologians accompanied social movements to draw from that experience novel approaches to doing theology, incorporating into theological discourse the local experience. Through the years, then, *campesino* theology (Barros and Caravias 1988; Caravias 1983), Indian theology (CENAME 1991; Estermann 2006), African-descendant theology (Rodriguez da Silva 1990; Aparecido da Silva 1998), Latin American feminist theology (Aquino 1993; Gebara 1994), and among other theologies of liberation have been produced, each corresponding to different cultural and social environments.

Since Latin American liberation theology broke apart traditional theological reflection, local expressions of the theology of liberation have become a worldwide phenomenon. Subaltern cultures and other discriminated social groups, especially in the Global South, have appropriated the method of the theology of liberation to produce their own theologies: Black theology in much of Africa, Dalit theology in India, Water Buffalo theology in Thailand, Third-eye theology in China and Asia, Minjung theology in South Korea, Struggle theology from the Philippines, and Palestinian theology from the Holy Land, among others. Each of these theologies is committed to drawing from local cultural environments and sociopolitical situations to produce understandings of Christianity that promote social justice and cultural identity.

In 1976, representatives of these then emerging theologies of liberation met in Dar es Salaam, Tanzania, and founded the Ecumenical Association of Third World Theologians (EATWOT) (http://www.eatwot.net/) as a forum for exchanging ideas, sharing experiences, and producing theological discourse in liberation mold. Although primarily located in the geographical "Third World," EATWOT includes theologians from subaltern groups in the United States and other developed nations; Third World refers more to social location than to geographical location. Through international meetings and publications, EATWOT has stimulated significant theological reflection from these specific social/cultural locations (Fabella 2006).

In recent years, the theology of liberation increasingly has assumed interculturality as its central organizing hermeneutic. Interculturality emphasizes diversity and historical experience and is viewed as fundamental for the decolonization – the eradication of colonially imposed social, political, and mental structures – of peoples (Estermann 2010). Theological interculturality affirms religious pluralism and argues that there are many paths to God (Torres et al. 2007). This intercultural approach to liberation theology has become increasingly prominent in Latin America (Virgil et al. 2008; Aquino 2007) and is evident in the work of EATWOT, especially in its journal *Voices* (http://eatwot.net/VOICES/). Finally, even issues such as sexual orientation and queer theology are being taken up by the theology of liberation. As Chilean theologian Diego Irarrázaval (2011) affirms, "the theology of liberation constitutes a meta-paradigm with a polyphonic development."

25.8.1 Conclusion

The historical project of the theology of liberation advocates the self-determination of peoples and cultures that are discriminated and marginalized. In so doing, it implicitly recognizes the political rights of cultures and legitimizes their ways of thinking. By establishing the subject and historical context as the fundamental axis for doing theology, the theology of liberation has promoted cultural identity and diversity. Thus it breaks the trend toward homogenization and favors biocultural conservation. Still, although widespread, the theology of liberation never has been more than a minority movement within Latin American Christianity. Now, after years of political repression, fierce opposition by the Vatican, and the dissipation of social movements coupled to the massive homogenization of culture and economics during the neoliberal epoch, times are hardly friendly to the theology of liberation. Affirmation of local culture has been overwhelmed by the massive propagation of pentecostalized Christianity and pop Christian music associated with conservative evangelicalism and the so-called prosperity gospel from the United States. Nevertheless, as long as there are subaltern groups, liberation theology will have its place demanding that these groups not only be respected but that they have voice and vote in forging their own histories, whether that be in church or broader society.

References

Agosín M (1996) Tapestries of hope, threads of love. The arpillera movement in Chile 1974–1994. University of New Mexico Press, Albuquerque

Alberigo G (2006) A brief history of Vatican II. Orbis Books, Maryknoll

Albó X, Libermann K, Godínez A, Pifarre F (1990) Para comprender las culturas rurales en Bolivia. MEC-CIPCA-UNICEF, La Paz

Aparecido da Silva A (ed) (1998) Existe um pensar teológico negro? Paulinas, São Paulo

Aquino MP (1993) Our cry for life: feminist theology from Latin America. Orbis Books, Maryknoll

Aquino MP (2007) Feminist intercultural theology. Toward a shared future of justice. In: Aquino MP, Rosando-Nunes MJ (eds) Feminist intercultural theology. Latina explorations for a just world. Orbis, Maryknoll, pp 9–28

Barros M, Caravias J (1988) Teologia da tierra. Vozes, Petrópolis

Bonnín E (ed) (1982) Espiritualidad y liberación en América Latina. DEI, San José

Caravias JL (1983) Luchar por la tierra, Inspiraciones bíblicas para las comunidades campesinas. Ediciones Paulinas-Catena, La Paz

Casaldáliga P, Tierra P (1980) Missa da terra sem males. Music by Martín Coplas with photographs by Cláudia Andujar. Tempo e presença, Rio de Janeiro

Casaldáliga P, Tierra P (1982) Missa dos Quilombos. Music by Milton Nacsimento. Ariola, Sao Paulo

CENAME (1991) Teología India. Primer encuentro taller latinoamericano México. CENAMI and Abya-Yala, México City/Quito

Cone JH (1969, 1989) Black power and black theology. Orbis Books, Maryknoll

Cone JH (1970, 2010) Black theology of liberation. Orbis Books, Maryknoll

Cone JH (1975, 1997) God of the oppressed. Orbis Books, Maryknoll

Dos Reyes PJ, da Lima Silva SR (2018) Candomblé in Brazil: the contribution of African-origin religions to biocultural diversity in the Americas. In: Rozzi R, May RH Jr, Chapin FS III, Massardo F, Gavin M, Klaver I, Pauchard A, Núñez MA, Simberloff D (eds) From biocultural homogenization to biocultural conservation. Ecology and ethics, vol 3. Springer, Dordrecht, pp 379–391

Dussel E, Mendieta E, Bohórquez C (eds) (2009) El pensamiento filosófico latinoamericano, del Caribe y "latino" (1300–2000): historia, corrientes, temas y filósofos. Siglo XXI, Mexico City

Esquivel J (1982) Threatened with resurrection. Prayers and poems from an exiled Guatemalan. Brethern Press, Elgin

Estermann J (coord) (2006) Teología andina. El tejido diverso de la fe indígena. ISEAT y Plural Editores, La Paz

Estermann J (2010) Interculturalidad, vivir la diversidad. ISEAT, La Paz

Fabella V (2006) Ecumenical Association of Third World Theologians (EATWOT). In: Muller K, Sundermeier T, Bevans SB, Bliese RH (eds) Dictionary of mission: theology, history, perspectives. Wipf & Stock, Eugene, pp 117–120

Gavin M, McCarter J, Mead A, Berkes F, Stepp JR, Peterson D, Tang R (2015) Defining biocultural approaches to conservation. Trends Ecol Evol 30(3):140–145

Gayer P (2014) Liberation art in Latin America, the visual evangelism of a different gospel. Course paper for The Bible and the Arts, University of Sheffield. Available at: https://www.academia.edu/10458155/Liberation_Art_in_Latin_America. Accessed 28 Nov 2016

Gebara I (1994) Teologia em ritmo de mulher. Edições Paulinas, São Paulo

Gutiérrez G (1973) A theology of liberation, history, politics and salvation. Trad. C Ida and J Eagleson. [Spanish edition 1971]

Gutiérrez G (1980) Teología desde el reverse de la historia. In: Gutiérrez G (ed) La fuerza histórica de los pobres, selección de trabajos. CEP, Lima, pp 305–394

Gutiérrez G (1989) Dios o el oro en Las Indias siglo XVI. Ediciones Sígueme, Salamanca

Gutiérrez G (1993) Las Casas: in search of the poor of Jesus Christ. Orbis Books, Maryknoll

Hahnenberg EP (2007) A concise guide to the documents of Vatican II. St Anthony Messenger Press, Cincinnati

Hanke L (1994) All mankind is one. A study of the disputation between Bartolomé de las Casas and Juan Ginés de Sepúlveda on the religious and intellectual capacity of the American Indian. Northern Illinois University Press, DeKalb

Irarrázaval D (2011) "Innovación teológica en América Latina." Paper presented in the fifth international meeting of theologians and social scientists. Departamento Ecuménico de Investigaciones (DEI), 31 October – 2 November, San José, Costa Rica

Mann CC (2011) 1493 Uncovering the new world Columbus created. Alfred A. Knopf, New York

Mendieta E (2016) Philosophy of liberation. The stanford encyclopedia of philosophy, Zalta EN (ed), URL = https://plato.stanford.edu/archives/win2016/entries/liberation/

Onion R (2014) The colorful quilt squares Chilean women used to tell the story of life under Pinochet. The Vault. Available via: http://www.slate.com/blogs/the_vault/2014/09/10/history_of_quilting_arpilleras_made_by_chilean_women_to_protest_pinochet.html. Accessed 29 June 2018

Rodriguez da Silva M (1990) Teología afro-latinoamericana. Primer ensayo ecuménico para una teología negra de la liberación. Centro Cultural Afro-Ecuatoriano, Quito

Rozzi R (2012) Biocultural ethics: the vital links between the inhabitants, their habits and regional habitats. Environ Ethics 34(1):27–50

Rozzi R (2013) Biocultural ethics: from biocultural homogenization toward biocultural conservation. In: Rozzi R, Pickett STA, Palmer C, Armesto JJ, Baird Callicott JB (eds) Linking ecology and ethics for a changing world. Ecology and ethics, vol 1. Springer, Dordrecht, pp 9–32

Torres-Queiruga A, Susin LC, Sobrino J (eds) (2007) Teología del pluralismo religiosos, el nuevo paradigma. Concilium 319 (February)

Virgil JM, Tomita L, Barros M (2008) Along the many paths to God, Interreligious studies series 1. LIT Verlag, Berlin

Chapter 26
The Dynamics of Biocultural Approaches to Conservation in Inner Mongolia, China

Ruifei Tang and Michael C. Gavin

Abstract Biocultural approaches to conservation aim to provide ethically sound and effective mechanisms for confronting widespread cultural and biological homogenization. Here we use eight principles of biocultural approaches to conservation as an analytical framework to examine potential drivers of biological and sociocultural change in Inner Mongolia, China. We present two case studies to examine how national-level policy changes from the 1960s to early 2000s have either increased or decreased adherence to principles of biocultural conservation, as well as the subsequent impacts on social-ecological conditions in Inner Mongolia. We find that greater adherence to principles of biocultural conservation may be linked to potential increases in positive social and ecological outcomes. Our analysis also emphasizes that even long-standing trends in resource management institutions and their social and ecological implications can be reversed in relatively short time spans given the proper enabling conditions.

Keywords Bridging organizations · Dynamic social-ecological systems · Facilitated co-management · Government policy · Resource management institutions

---

R. Tang (✉)
CEESP (Commission on Environmental, Economic and Social Policy), IUCN, Wellington, New Zealand

M. C. Gavin
Human Dimensions of Natural Resources Department, Colorado State University, Fort Collins, CO, USA

Department of Linguistic and Cultural Evolution, Max Planck Institute for the Science of Human History, Jena, Germany
e-mail: michael.gavin@colostate.edu

R. Rozzi et al. (eds.), *From Biocultural Homogenization to Biocultural Conservation*, Ecology and Ethics 3, https://doi.org/10.1007/978-3-319-99513-7_26

26.1 Introduction

Biocultural approaches to conservation aim to provide ethically sound and effective mechanisms for confronting widespread cultural and biological homogenization (Rozzi 2013). These approaches integrate key lessons from social-ecological systems, people-centered conservation (co-management, integrated conservation and development projects [ICDPs], community-based conservation), traditional ecological knowledge, the study of common pool resources, and biocultural heritage and ethics (Gavin et al. 2015). Through this interdisciplinary synthesis that draws on nearly five decades of theory and evidence derived from multiple fields, biocultural approaches to conservation provide a framework for conservation planning and actions.

Gavin et al. (2015) suggest eight core principles of biocultural approaches to conservation (see Table 26.1). Together these principles highlight the dynamic, pluralistic, and partnership-based characteristics of all biocultural approaches to conservation Gavin et al. (2018). Although the components of biocultural conservation all derive from evidence bases in multiple fields, studies must still address the degree to which the combined set of principles of biocultural conservation can be linked to improved outcomes for biological and cultural diversity. Here we take a step toward filling this gap by using the principles of biocultural conservation as an analytical framework to examine potential drivers of biological and sociocultural change in Inner Mongolia, China. We propose that greater adherence to the principles of biocultural conservation proposed by Gavin et al. (2015) would improve the chances of conserving both biological and cultural diversity, whereas diverging from the principles would increase the possibility of biocultural homogenization.

During the period of the 1960s to early 2000s, policies of the Chinese government have driven profound changes to land and resource ownership, as well as community institutions across the country. In this chapter we examine how the national-level policy changes from the 1960s to early 2000s have either increased or decreased adherence to the principles of biocultural conservation, as well as the subsequent impacts on social-ecological conditions, in the Inner Mongolia Autonomous Prefecture. This region will be referred to as Inner Mongolia in this chapter. Inner Mongolia is a province of China, which is distinct from the Mongolian People's Republic, an independent country known as Outer Mongolia to Chinese and as Mongolia to the wider world.

26.2 Site Locations and Research Methods

We present two case studies from Inner Mongolia (Fig. 26.1): one from the Alashan area in western Inner Mongolia (Fig. 26.2) and the other from the Xiwu Qi area in eastern Inner Mongolia (Fig. 26.3).

Table 26.1 Application of biocultural conservation principles in Inner Mongolia, China

Biocultural conservation principles	Effects of government policy (1960s to early 2000s)	New opportunities for biocultural conservation (2000s onward)	
		Alashan case	Xiwu Qi case
1. Acknowledge multiple stakeholders and objectives	The policies reflected the objectives of government-appointed committees and not the traditional Mongolian elder councils that previously oversaw decisions affecting communal resources	Local community, bridging NGO, and local government are the main stakeholders; the ICDPs attempt to address each stakeholder's interests and objectives	The cooperative law empowered local herders to organize and pursue their objectives
2. Use intergenerational planning and institutions for long-term adaptive management and governance	Decision-making power was taken away from Mongolian elders, and government policies eroded intergenerational knowledge transmission	Elected community committee leads the projects and reflects traditional communal decision-making	The law provided a chance for revitalizing traditional herding practices and natural resource management institutions
3. Recognize that cultural dynamics shape conservation	Traditional cultural practices were suppressed and mandated changes in community leadership, and decision-making undermined the traditional institutions	Traditional land use practices were promoted and incorporated new tools to replace reliance on saxoul for firewood	The cooperative combines elder's traditional knowledge with other aspects of institutional arrangements that have emerged in recent decades
4. Tailor interventions to social-ecological context	The policies followed a "one-fits-all" approach, which neither recognized nor respected local social-ecological contexts	Through close consultation with communities and community control over project implementation, the ICDP approach better reflected the local social-ecological context	Participating families resumed their traditional communal lifestyle, which they believe is a more natural fit to the grassland ecosystems
5. Devise novel, diverse, and nested institutional frameworks	Policies and decisions reflected government ideology and interests and suppressed locally based institutions	Established a facilitated co-management model in which the local community, facilitating NGO facilitator, and local government formed a nested institutional framework	Vertical and horizontal institutional linkages played important roles, ensuring funding, legal assistance, and knowledge sharing

(continued)

Table 26.1 (continued)

Biocultural conservation principles	Effects of government policy (1960s to early 2000s)	New opportunities for biocultural conservation (2000s onward)	
		Alashan case	Xiwu Qi case
6. Prioritize partnerships and relation building	Local participation in decision-making diminished in favor of a centralized government system	NGO empowered local community members to control natural resource management and linked community to government agencies	NGO connected the cooperative with similar organizations and encouraged participation in regional and national cooperative conferences
7. Incorporate distinct rights and responsibilities of all parties	Government policies compromised the rights of local Mongolians and replaced traditional mechanisms for knowledge transmission	A unique mechanism to recognize resource use rights of the local community and empower community-based management was created	The cooperative used the opportunity to increase its responsibilities over pastoral management
8. Respect and incorporate different worldviews and knowledge systems	The policies actively sought to replace traditional Mongolian worldviews, beliefs, and knowledge systems	The projects drew on Mongolian traditional ecological knowledge and worldviews, as well as on new tools introduced by NGO partners	Cooperatives revitalize traditional herding practices while incorporating new models of leadership, including youth participation
A. Ecological impacts	Grassland and saxoul degradation began in the 1980s and worsened in 1990s	saxoul coverage increased	Herding rotations resumed which may lead to rotating communal pasture management and improved ecological conditions
B. Social impacts	Traditional lifestyles, institutions, cultural practices, and beliefs were suppressed; policies undermined economic independence and cultural identity of Mongolian herders	Community cohesion and community leadership were strengthened; traditional knowledge and practices were revitalized; partnerships between communities, NGO, and governmental agencies grew	Traditional institutions and collective herding practices were resumed; social cohesion and traditional leadership were strengthened; partnerships among cooperatives and NGOs grew

The focal village in Alashan had a territorial area of 667 km^2 within the Gobi Desert. Saxoul (*Haloxylon ammodendron*), the keystone large shrub or small tree species in the local Gobi Desert ecosystem, covers approximately 112 km^2 (SEE 2005). The total population in the village is 185, of which 101 are indigenous Mongols and 84 are Han, the ethnic majority in China (SEE 2005). Most families in

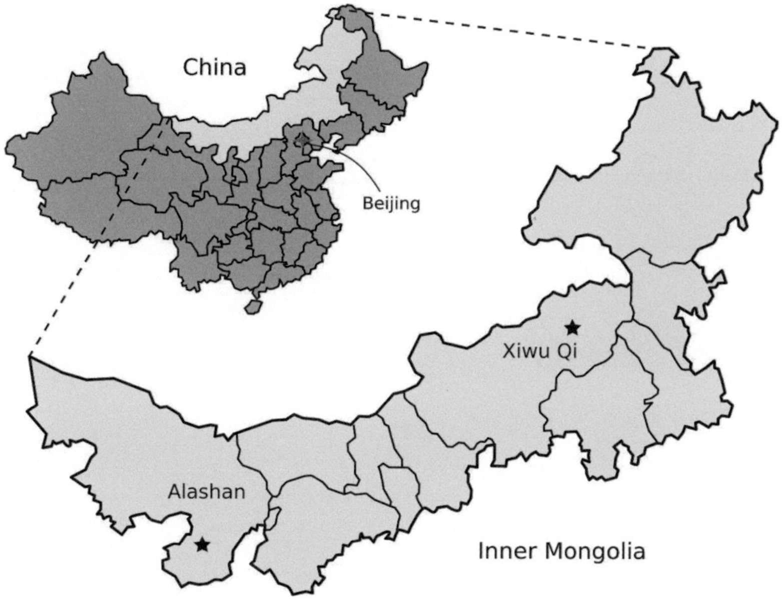

Fig. 26.1 The locations of two research sites in Inner Mongolia, China. (Map modified from Croquant 2007)

the village herd sheep, goats, and camels, and some raise pigs and chickens for supplementary income (SEE 2005). Individual households held 16.7–33.4 km^2 of land, and GDP per capita was approximately US$ 346 in 2004 (SEE 2005).

The focal village in Xiwu Qi had a total territory of 380 km^2 located in a grassland ecosystem (Fig. 26.3). The official village population was 396 within 96 households (Ao, pers. comm. 2010). However, the actual number of the accessible households during the research period was 69 (Tang 2013). Thirteen households had relocated to town due to mining activities, eight households lived in town or other cities due to employment, and another six households were herding and visiting relatives in other areas (Tang 2013). The average pasture per household was 3.3–4 km^2, and the average annual income per household was US$ 1950 (about US$ 487/person) (Tang 2013). The total number of livestock in the village was 3200, the vast majority being sheep ($n = 2800$), with the remainder consisting of goats, cows, and horses (Tang 2013).

The indigenous Mongols have lived in the Mongolian Plateau for over a thousand years (Da 2008). Between the early thirteenth and the fifteenth centuries, the Mongol Empire formed and thrived. The Yuan Dynasty, which was established by Mongols, expanded its territory to Europe, including Russia, for a brief period in the fourteenth century. In 1912, the vast Mongolian Plateau was divided into the Mongolian People's Republic and the Inner Mongolia Autonomous Prefecture of China. Under the assistance of the Soviet Union, the Mongolian People's Republic announced its independence from China, which was in transition from the Qing

Fig. 26.2 A photo from the research site in Alashan, Inner Mongolia, China, which has suffered widespread desertification. The area in the photo was grassland 50 years ago. (Photo Ruifei Tang taken in 2006)

Dynasty (also referred to as the Man Tzu Dynasty) to the Republic of China (Wang 2009). The Mongolian People's Republic remained under the Soviet Union's influence until 1992. In 1950, the newly established People's Republic of China, governed by the Chinese Communist Party, and the Mongolian People's Republic established an official diplomatic relationship (Williams 2002). Since then, the Chinese have referred to the Inner Mongolia Autonomous Prefecture, where some six million Mongolians live, as Inner Mongolia.

In both research areas, indigenous Mongolian herders have a long tradition of using communal rangelands in a sustainable manner to produce a wide array of livestock products for subsistence and trade (Bedunah and Schmidt 2004). Grasslands in Inner Mongolia present substantial challenges for herding due to limited (200 mm average), variable (>30% variation year to year), and highly seasonal precipitation, as well as unpredictable climatic disasters (Da 2003). Traditionally, Mongolian herders managed their use of grassland resources with seasonal long-distance migrations. This nomadic lifestyle allowed herders to adjust to the spatial and temporal heterogeneity of pasture resources.

Fig. 26.3 A photo from the research site in Xiwu Qi, where families herd mostly sheep in the grassland ecosystem. (Photo by Ruifei Tang taken in 2010)

The first author (Ruifei Tang) lived in both sites for data collection in 2010. As a Man Tzu Chinese,[1] who shares ancestry with Mongols, the first author was culturally accepted by the focal communities. Furthermore, due to her previous work and research experience in Inner Mongolia in 2004 and 2006, she speaks basic Mongolian and understands Mongolian cultural traditions. However, a translator assisted for in-depth communication during interviews and group discussions. The Code of Ethics of the International Society of Ethnobiology (2006) provides the ethical foundation for our research. Participants received research information sheets and signed consent forms prior to data collection. All interviewees' identities remained confidential, unless they provided expressed written consent for their identities to be shared.

The first author gathered data using qualitative research methods, including semi-structured interviews, focus group discussions, and participant observation. She spoke with local Mongolian herders, employees of environmental NGOs active in the region, experts of Mongolian ecology and culture, and government officials at each site. In each case discussions focused on obtaining information on perceived

[1]The Man Tzu originated in the twelfth century. It is the third largest ethnic minority group in China and is found in 31 provinces and regions throughout the country. More than 100,000 Man Tzu residents live in Inner Mongolia.

changes to social and ecological conditions, including the dynamics of livelihoods and natural resource management institutions, as well as the perceived drivers of any changes. We used grounded theory and open, axial, and selective coding for data analysis (Strauss and Corbin 1990). Additional details on the methods used (as well as more detailed results) for the Alashan case can be found in Tang and Gavin (2010) and Tang (2013) and for the Xiwu Qi case in Tang and Gavin (2015) and Tang (2013).

26.3 Results and Discussion

26.3.1 Government Policy Initiatives: The 1960s to Early 2000s

Since the establishment of the People's Republic of China in 1949, local Mongolians have had limited autonomy in the province of Inner Mongolia (Ao 2003; Williams 2000). Instead, the centralized government has designed, and controls, the current decision-making system. The central government has continuously sought to "uplift" indigenous Mongolians by shifting local communities from traditional nomadic pastoralism toward sedentary lifestyles (Ao 2005). From the 1960s to early 2000s, Chinese national policies had substantial impacts on traditional resource management institutions and social-ecological conditions in Inner Mongolia (Tang 2013). The *Grand Commune Policy* of the early 1960s led to the nationalization of all resources and properties (Ao 2003; Da and Si 2006). As part of the *Grand Commune Policy*, the central government claimed all land and natural resources as state property, despite previous land and property ownership. This policy ended traditional communal ownership and patterns of pasture use (see more details in Tang and Gavin 2015). In addition, beginning in the 1960s, government policies encouraged the immigration of a large number of Han Chinese, the nation's ethnic majority, into Inner Mongolia. These immigrants introduced the ideology and production methods of crop cultivation into the region (Jiang et al. 2006).

The majority of our community informants, and all expert interviewees, highlighted the fact that the commune system replaced the Mongolian traditional elder leadership with government-appointed leaders, who were usually recent Han immigrants (Tang 2013). As one elder in Xiwu Qi noted, "elders were [the] most experienced herders. Their knowledge and experience earned their reputation and leading positions in a community. All young members fully respected and were willing to follow elders' judgments and decisions … But this was ended in the 1960s."

In the 1970s and 1980s, *settlement* and *privatization* policies ended traditional nomadic practices in Inner Mongolia. The voluntary settlement policy, implemented nationwide since the 1970s, aimed to transform nomadic herders into sedentary residents. The policy was enacted in two stages: (1) de-collectivization of the agricultural sector, permitting the formation of entrepreneurs and opening the country to foreign investment, and (2) privatization and contracting out of previously state-owned industry and rural lands and relieving the central price control system (Zhao 2008). To implement the policy, the government established medical centers, veterinarian service, and primary schools to encourage herders to settle around these townships. However, the majority of our informants noted that herder families preferred the traditional nomadic lifestyle. The government eventually achieved its settlement goals using the nationwide privatization policy launched in the 1980s. The government leased land user rights to individual households for 30–90 years while claiming state ownership over all land (Ao 2003; Cao et al. 2011). In this process, the government also forced the redistribution of all previously communally owned livestock to individual households and allocated pastures to households based on livestock shares. Therefore, by the 1990s, the majority of herding families in our study sites had fenced off their lands and lived in fixed locations.

In the early 2000s, China's *Western Development Strategy* led to a range of economic development-oriented policies, which focused on 12 western provinces and prefectures (including Inner Mongolia), which the government considered to be less developed regions of China (The National Bureau of Statistics China 2010). The targeted outcomes of the strategy include the development of infrastructure (e.g., transport, hydropower plants, energy, and telecommunications), the promotion of modern education and healthcare, and the attraction of foreign investment (Hu 2005). The strategy guarantees that these 12 western provinces will receive national funding for development. In response to the strategy, regional and local governments have enforced a series of measures favoring industrial development over pastoralism. For example, according to an expert interviewee, more than a thousand mines, including coal, iron, zinc, and other metals, have opened in Xiwu Qi during this time period. The mines are state-owned, and the developers and workers are mainly Han Chinese from other provinces. Based on profits from mining (e.g., high taxation), Inner Mongolia had the fastest GDP growth in China between 2002 and 2009 and reached US$140 billion in 2009 (The National Bureau of Statistics China 2010). Urbanization has been another priority of local government in the 2000s. In 2002, local government dissolved half of the townships, most of them created in the 1970–1980s as a part of the settlement policy, and shifted social services (e.g., medical centers, veterinarian service, and primary schools) into city centers. In Xiwu Qi, for example, interviewees noted that school-aged children must attend schools in Xiwu city, and adults from each family must accompany the children, which further increases the cost of education (e.g., rent and living expenses) and contributes to financial insecurity.

26.3.2 Deviation from Biocultural Principles: The 1960s to Early 2000s

Here we outline how government policy initiatives changed the degree of adherence to principles of biocultural conservation in the two case study regions of Inner Mongolia. In general, we found evidence for a decline in the level of adherence to the principles (Table 26.1, Gavin et al. 2015).

Principle 1: Acknowledge multiple stakeholders and objectives

Biocultural approaches to conservation recognize that conservation and the management of natural resources involve partnerships among multiple stakeholder groups, each with their own set of objectives (Gavin et al. 2015). However, from the 1960s to early 2000s, the Chinese national government usurped power over major natural resource management decisions. Therefore, decisions, such as privatizing land and herds, centralizing schooling and healthcare, and industrialized development of former pasture lands, reflected the objectives of government-appointed committees and not the traditional Mongolian elder councils that previously oversaw decisions affecting communal resources.

Principle 2: Use intergenerational planning and institutions for long–term adaptive management and governance

Biocultural conservation emphasizes the need for planning across intergenerational time scales and the importance of adapting to changing social and ecological conditions. However, the consolidation of decision-making power in the younger, educated Han leaders of government-appointed committees took leadership away from Mongolian elders starting in the 1960s. These policy changes removed links to intergenerational knowledge systems that had been fundamental to land management. Prior to the 1960s, Inner Mongolia had a relatively high level of autonomy, in which intergenerational, traditional institutions played a predominant role in decision-making (Wang 2009). These institutions were characterized by elder leadership, communal lifestyles, and large-scale, four-season pasture rotations. Only from the 1960s, under the Grand Commune Policy, did herders lose their collective ownership of land and other resources, which undermined intergenerational resource management institutions (Tang 2013).

Principle 3: Recognize that cultural dynamics shape conservation

Biocultural conservation must allow for the adaptation of cultural systems that change to reflect dynamic social-ecological conditions. However, since the 1950s the central Chinese government has sought to "uplift" Mongolians from nomadic herders to sedentary residents and from "low-profit" pastoralism to a "modern" economy (Ao 2005; Da 2008) with the use of the settlement, privatization, and western development policies that have had effect of suppressing traditional cultural practices. In addition, mandated changes in community leadership and decision-making undermined traditional institutions that had been used over centuries to adapt pastoral practices to changing conditions.

Principle 4: Tailor interventions to social-ecological context

Contrary to the principle, which notes that natural resource management will be most effective when designed to match the context, the Chinese government policies discussed above followed a "one-fits-all" approach to the entire country, without consideration of the various social-ecological characteristics in different regions. By assuming that privatization, permanent settlements, consolidated decision-making, and centralized social systems could be applied universally, the government policies neither recognized nor respected local social-ecological contexts within Inner Mongolia.

Principle 5: Devise novel, diverse and nested institutional frameworks

This principle draws on the theory of institutions (as the formal and informal rules that structure interactions) (North 1991), highlighting the effectiveness of institutional arrangements that match a given context, and in which institutions at different scales act in a coherent way (Ostrom 2010). The *Grand Commune Policy* and the settlement and privatization policies implemented by the Chinese government ensured that local groups could not organize outside the confines of the ruling party and that power rested in government-selected committees (Wang 2009). As a result, the government-suppressed locally based institutions, including the traditional Mongolian nested institutions, comprised of communal ownership of grassland, elder leadership councils, and cooperation across different herding groups (Tang 2013).

Principle 6: Prioritize partnerships and relation building

Partnerships founded on the meaningful sharing of power and responsibilities for environmental stewardship (Mead 2013; Reed 2008) are at the core of biocultural approaches to conservation. However, in both case studies we examined in Inner Mongolia local participation in environmental decision-making diminished from the 1960s to early 2000s. Central government policies reduced the rights of local people to organize outside the ruling party and undermined traditional resource management institutions, thereby limiting local participation in natural resource management decisions.

Principle 7: Incorporate distinct rights and responsibilities of all parties

Biocultural conservation highlights the need to recognize the distinct rights of all parties involved, and the importance of defining the responsibilities each stakeholder group must uphold (Gavin et al. 2015). However, the *Grand Commune Policy* of the 1960s and the settlement and privatization policies of the 1980s compromised the rights of local Mongolian herders to manage land and herds communally and to maintain traditional resource management institutions, including elder councils. In addition, the government's policies replaced traditional mechanisms for knowledge transmission with mandatory government-run schools, healthcare, and other social services.

Principle 8: Respect and incorporate different worldviews and knowledge systems

This principle stresses the value of multiple forms of knowledge for increasing the adaptive capacity to cope with uncertain and dynamic social and ecological conditions (Rozzi 2007, 2013). To the contrary, the Grand Commune, settlement, and privatization policies, we discussed above actively sought to replace traditional Mongolian worldviews, beliefs, and knowledge systems, which the government considered "backward and undeveloped." For example, the settlement policy reflected the ideology of Han Chinese that has focused on agricultural production for thousands of years. The privatization policy was a mainstream ideology in the 1980s after the Chinese economy reform, which followed the logic that private ownership of properties and assets would trigger economic growth.

26.3.3 Implications for Social-Ecological Conditions: The 1960s to Early 2000s

Traditional Mongolian pastoralism and its reliance on nomadic lifestyles represented long-term adaptation of subsistence activities and social institutions to the region's grassland ecosystems (Ao 2003). A large number of our interviewees in both study locations ($n = 203$) strongly linked the imposed sedentary lifestyle mandated by Chinese government policy to grassland degradation. Traditionally, herders had rotated pastures each season, or even more frequently if required by climatic or pasture conditions, without limitations placed by property boundaries. All community interviewees across both cases ($n = 188$) emphasized that without proper rotation, pasture had no chance to recover from constant grazing, and that pasture quality had depleted since the 1970s. For example, in the Alashan case, both our interviewees and scientific studies (e.g., Guo et al. 2005a, b; Huang 2002; Peng and Xu 1996) noted the rapid loss of saxoul (*Haloxylon ammodendron*), with 82% reduction in cover within 50 years. This loss is particularly notable as saxoul is a widespread shrub species in the Gobi Desert ecosystem, where it stabilizes sand dunes and facilitates the recruitment of other plant species (Sheng et al. 2005) (Fig. 26.4).

The government's forced settlement policy also had profound negative impacts on Mongolian herders' economic independence and their traditional cultural identities. Even though the local knowledge of traditional resource use and institutions still exists, traditional resource management institutions have not been used for decades. Therefore, the local traditional biocultural system has reached a critical threshold, in which permanent loss of the knowledge will be inevitable if no immediate conservation actions are undertaken (Tang and Gavin 2010). As one local NGO leader summarized, "the Mongolian traditional belief and culture have been continuously impacted by the Han culture and so-called modern civilization. The

Fig. 26.4 A saxoul (*Haloxylon ammodendron*) shrub land in the Alashan area of Inner Mongolia, China. (Photo by Ruifei Tang taken in 2006)

agricultural methods and culture [promoted by the government] are not suitable for pastoral land. This has also changed the sustainable use and management practice of natural resources. Moreover, it has reformed local culture. The traditional respect and responsibilities to nature have almost gone away."

26.3.4 Recent Initiatives and Government Policy Changes

In the early 2000s, new developments altered the institutional landscapes in both locations we studied. In Alashan, the shift has been driven in part by actions of a prominent environmental NGO. In Xiwu Qi, changes to national policies have opened the door for new community-driven initiatives to reorganize the management of pastoral commons.

A. The Alashan Case

In the Alashan region, the NGO Alashan Ecological Society or SEE (funded by 100 Chinese entrepreneurs who seek to make positive social and environmental impacts) introduced and funded two integrated conservation and development projects (ICDPs) in 2004 (Tang and Gavin 2010). The projects have used rapid rural

Fig. 26.5 A community meeting to determine goals of the integrated conservation and development projects in the Alashan region. (Photo by Ruifei Tang)

appraisals to identify the objectives most favored by participating community members (Fig. 26.5). Based on these analyses, new projects have focused on alternative fuel sources to replace the use of saxoul as firewood. The projects have also encouraged the planting of *Cistanche deserticola*, a parasitic plant which gathers nutrients from saxoul's root system (Guo et al. 2005a). Due to its medicinal value, collecting *C. deserticola* has been a source of income for local households (Tang 2013). Since *C. deserticola* can only grow from healthy saxoul shrubs, and it is critical to keep saxoul's root system intact when digging for *C. deserticola*, the project sought to foster saxoul conservation while increasing participating families' income (Tang 2013). An elected project management committee, consisting of local Mongolian community members, now manages these projects. After providing seed funding and helping to launch the projects, the NGO now serves as a facilitator, providing technical and financial support when needed. This community-oriented model earned recognition from the local government. As the president of the Alashan local government stated (pers. comm. 2006), "SEE has brought financial investment into our region, and it has benefited our conservation and economy. Their projects have transformed the local communities … to (be) active partners (in) conservation. After all, saxoul coverage has recovered in their project areas." Based on such positive assessments, the local government began providing additional funding for the projects and promoting the model in several other villages in the Alashan region.

B. The Xiwu Qi Case

In 2007, the Chinese government enacted the cooperative law with an aim of narrowing urban-rural income disparities. The law supports cooperatives with a mechanism to officially register with the government and the new cooperatives receive tax relief. Although the law does not target traditional natural resource institutions, local communities in the Xiwu Qi region have used the new regulations to revitalize traditional practices. According to officials we interviewed at the Xiwu Qi Industry and Commerce Bureau, 29 cooperative organizations have already registered. The vast majority of these cooperatives (90%) concentrate on pastoral production with a focus on revitalizing traditional herding practices among those families that remain living in the villages.

In our study site, 12 households had formed a cooperative (Tang 2013). All members were actively involved in discussions and meetings that followed a process similar to traditional collective decision-making. The primary focus of the cooperative was to form a collective working group and revitalize herding rotation practices. The participating households removed the fences around their properties and merged the pastures and herds. The group selected the most experienced herder and his family to be the chief shepherds and provide guidance on where and when to graze. In the future, the cooperative expects all families in the village to join the organization. This would enable the revitalization of four-season herding rotations and large-scale cooperation of marketing and trade.

26.3.4.1 Toward Greater Adherence with Biocultural Principles in Two Case Studies

Principle 1: Acknowledge multiple stakeholders and objectives

In Alashan, the local community, NGO, and local government are the main participating stakeholders in the ICDP projects. In the model of facilitated co-management, in which the NGO has served as a bridging organization to link communities and government (see details in Tang 2007; Tang and Gavin 2010), the projects attempt to address each stakeholder's interests and objectives via participatory planning and management and with additional funding support from the government.

In Xiwu Qi, the new cooperative law empowered local herders to organize and pursue their unique objectives. In the case of the cooperative we studied in Xiwu Qi, the vision has been to revitalize traditional herding rotation and a communal lifestyle, conserve grassland, and strengthen social cohesion and economic development.

Principle 2: Use intergenerational planning and institutions for long term adaptive management and governance

In the Alashan case, the integrated conservation and development process began with participating community members electing their project management committee, which then pushed to reestablish the use of traditional knowledge and institutions. The committee leads the projects and reflects traditional communal decision-making, including consultation with elders.

The new cooperative in Xiwu Qi has promoted the revitalization of traditional herding practices and natural resource management institutions, including rotations of communal herds across fence-free communal pastureland, elder leadership, and collective decision-making. Community informants noted that these practices are rooted in knowledge gained over generations that ensures management choices best reflect the ecological conditions of the region's grassland. Importantly, the knowledge related to traditional herding practices and natural resource management institutions still resides with village elders. However, younger generations have not participated in these traditional practices or institutions due to the effects of the Chinese government policies outlined above (Tang and Gavin 2015). In turn, the social-ecological system is at a critical threshold, as revitalization may only be possible, while elders, who hold the knowledge about traditional practices and institutions, are still alive (Tang and Gavin 2015).

Principle 3: Recognize that cultural dynamics shape conservation

Although the ICDP projects focused on the implementation of new conservation mechanisms, including firewood alternatives and planting of *C. deserticola*, the Alashan case also included the promotion of traditional camel-herding practices (Tang and Gavin 2010).

In Xiwu Qi, the new cooperative seeks to revitalize traditional resource practices and institutional arrangements. However, many decades have passed since these practices guided pastureland management. The cooperative is now combining elders' traditional knowledge with other aspects of institutional organization that have emerged in recent decades, such as the involvement of younger and school-educated community leaders who have been appointed to leadership positions since the 1980s. In this way, the cooperative reflects both a return to traditional knowledge systems and an adaptation to new social conditions.

Principle 4: Tailor interventions to social–ecological context

In Alashan, previous national-level policies had imposed new resource management systems on Mongolian communities. This included the attempts at saxoul conservation based on restricting herding in reserves. Since 2004, the ICDP approaches promoted by SEE have better reflected the local social-ecological context through close consultation with communities during planning phases and community control over the implementation of projects. Although the successful outcome of the projects took place to a limited extent (only within the project territory), it has gained the recognition from the local government and, therefore, enables the wider use a of this social-ecological context-based conservation model in the region.

The new cooperative law has empowered local communities to define their own development trajectories without the strict oversight inherent in previous government-controlled resource management regimes. In the Xiwu Qi case, the new cooperative law has allowed participating families to resume their traditional communal lifestyle by voluntarily joining their pastures and livestock, which they argue is a more natural fit to the grassland ecosystems in the region.

Principle 5: Devise novel, diverse, and nested institutional frameworks

The ICDP projects in Alashan have established a new facilitated co-management model in which local community members, NGO facilitators, and local government have formed a nested institutional framework (Tang and Gavin 2010). The community committee implements projects with funding and technical support from the NGO and government.

In Xiwu Qi, China's cooperative law has loosened restrictions on the formation and formal recognition of local organizations. This initiative empowers local institutions, which will vary from one location to another depending on social, political, economic, and ecological conditions. Vertical and horizontal institutional linkages support the Xiwu Qi cooperative. For example, funding and legal assistance provided by the national government through regional and local governmental agencies support local cooperatives. In addition, a network of local cooperatives and NGOs share lessons learned and provide opportunities for capacity building.

Principle 6: Prioritize partnerships and relation building

In Alashan, the NGO (SEE) initiated the project and developed a close partnership with the local community. SEE empowered local community members to take control of natural resource management and linked the community committee to government agencies that now support multiple projects in the region.

The Xiwu Qi cooperative has benefited from partnerships with other recently established cooperatives. Many of these relationships have been forged by a local NGO, which connected the cooperative with similar organizations and encouraged its participation in regional and national cooperative conferences.

Principle 7: Incorporate distinct rights and responsibilities of all parties

The ICDP projects in Alashan created a unique mechanism to recognize the local community's rights for traditional access and use of saxoul shrublands. Although the NGO, government officials, including regional and county leaders and local forestry officials, and local community members all expressed a clear understanding of the rights and responsibilities each group holds for the implementation of the ICDP projects, concerns remain regarding the long-term impacts government policies have had on traditional institutions for herding and on the pathways for the transmission of traditional knowledge.

In Xiwu Qi, the new law increased the rights of local organizations to organize outside the confines of the government. The Xiwu Qi cooperative used the opportunity to increase their responsibilities over pastoral management by revitalizing traditional herding practices.

Principle 8: Respect and incorporate different worldviews and knowledge systems

The ICDP projects in Alashan drew on Mongolian traditional ecological knowledge and worldviews (e.g., the traditional use and protection of saxoul, the understanding of mutual benefits linking camel herding and saxoul health), as well as on new tools (e.g., wind and solar energy generators) introduced by NGO partners.

The formation of the Xiwu Qi cooperative and similar organizations focused on traditional knowledge and institutions brings a more diverse set of tools to the management of grassland ecosystems in Inner Mongolia, which has been dominated by government-led interventions in recent decades. The cooperative model combines the advantages of both traditional (e.g., elder leadership and communal benefit sharing) and contemporary institutions (e.g., young leadership) and is an adaptive development of the community institutions.

26.3.5 Implications for Social-Ecological Conditions

In Alashan, from the introduction of the two ICDP projects in 2004 through 2015, saxoul coverage in the region increased by 150,000 ha (IMARFA 2016; SEE 2015). In recognizing the conservation success of the ICDPs, the local government reopened some closed reserves and allowed use of saxoul by local communities. As many community interviewees ($n = 27$) and all NGO interviewees ($n = 11$) recognized, the projects also contributed to the strengthening of social cohesion within the community. As one of community elders described, "I haven't seen this kind of community meetings for more than 20 years. People are genuinely interested in the discussions and happy to contribute their ideas and knowledge. More so, everyone can have a say in the decision-making."

In Xiwu Qi, the cooperatives have reintroduced three- or four-season pasture rotation practices in the region's grasslands. This is a critical achievement for the revitalization of Mongolian traditional ecological knowledge, especially after the abandonment of seasonal rotations for 30–40 years. Certain elements of traditional community institutions and community resource governance have also been revitalized. The shift toward rotating communal pasture management additionally has the potential to improve ecological conditions (Tang and Gavin 2015; Wang 2009). However, the cooperative movement is still in an early stage, and its long-term impact on social-ecological conditions and on revitalizing traditional ecological knowledge and empowering the community requires continuous monitoring and further research.

Our results depict the dynamic nature of the social-ecological systems in Inner Mongolia over the last 60 years. From the 1960s through the 2000s, central policies of the Chinese government tended to diverge from the principles of biocultural conservation (Gavin et al. 2015). Government actions forced the privatization of land and resources, removed rights of communal organization outside of government committees, and centralized social services. The policy initiatives attempted to impose a limited set of tools for managing land, resources, and people across the diverse social and ecological landscapes of China.

Biocultural approaches to conservation emphasize the need to draw on diverse sources of knowledge and institutional arrangements and would predict that environmental management approaches that deviate from this, and other core principles, would be less likely to succeed. Here we demonstrate that diverging from the principles of biocultural conservation was associated with negative social and ecological outcomes in Inner Mongolia. The central government's institutional initiatives undermined traditional practices, knowledge systems, and the social fabric of Mongolian communities. The mismatch between government-imposed resource management institutions and the social-ecological context of the grasslands of Inner Mongolia also had negative ecological consequences hastening the loss of a keystone species and the desertification process.

However, two unconnected recent developments have opened the door to relatively rapid and substantial changes in grassland management in different parts of Inner Mongolia. The Alashan case study demonstrates the power of bridging organizations (Berkes 2009) to connect local knowledge and concerns with other forms of knowledge and funding at regional, national, and international scales. These organizations have the potential to open lines of communication and attend to historical grievances between communities and government, as well as empower local communities to pursue their self-defined goals and objectives. In Xiwu Qi, community members took advantage of recent developments in Chinese law aimed at economic development. An unintended consequence of the new law was an opportunity for Mongolian herders to revitalize traditional practices.

Although the triggers for change were different in these two case studies – bridging organization versus new law – the outcomes were similar in terms of increased adherence with the principles of biocultural conservation. The new initiatives took a more pluralistic approach to resource management by placing more power in the hands of local communities and cooperatives. These changes in grassland management also come at a critical time. Prior to the new initiatives, the social-ecological system in Inner Mongolia was arguably approaching a major threshold (Liu et al. 2007; Tang and Gavin 2010). By the late 2000s, intensive agriculture, deforestation, and mining activities dominated resource management, and a limited number of elders held the only remaining knowledge of traditional herding practices and institutions. Because prior government policies had intervened with cultural transmission pathways (e.g., through mandatory schooling in city centers), the imminent passing of this elder generation would have seen the loss of an entire system of knowledge used for managing this social-ecological system for over a thousand years. Many researchers have argued that a reduction in the diversity of knowledge systems coincides with a reduced adaptive capacity to cope with the complex and dynamic nature of social-ecological systems (Davidson-Hunt et al. 2012).

The case studies we have outlined provide evidence that greater adherence to principles of biocultural conservation may be linked to potential increases in positive social and ecological outcomes. Our analysis also emphasizes that even long-standing trends in resource management institutions and their social and ecological implications can be reversed in relatively short time spans given the proper enabling conditions. We also call for more case study-based assessment to build an evidence base for the potential of the combined set of principles of biocultural conservation to be linked to improved outcomes for biological and cultural diversity.

References

Ao RQ (2003) The changing and bring forth new ideas of the property in grassland [in Chinese]. Inn Mong Soc Sci 4:24–35

Ao RQ (2005) Institutional change and pastoral civilisation [in Chinese]. Inner Mongolia People's Press, Huhehaote

Bedunah D, Schmidt SM (2004) Pastoralism and protected area management in Mongolia's Gobi Gurvansaikhan National Park. Dev Chang 35(1):167–191

Berkes F (2009) Evolution of co-management: role of knowledge generation, bridging organizations and social learning. J Environ Manag 90(5):1692–1702

Cao JJ, Xiong YC, Sun J, Xiong WF, Du GZ (2011) Differential benefits of multi-and single-household grassland management patterns in the Qinghai-Tibetan plateau of China. Hum Ecol 39:217–227

Croquant (2007) The location of Hohhot Prefecture within Inner Mongolia Autonomous Region of China. Wikipedia Open Source. Retrieved on 6th September, 2010 via http://en.wikipedia.org/wiki/File:Location_of_Hohhot_Prefecture_within_Inner_Mongolia_(China).png

Da LT (2003) A study of the non-equilibrium ecosystem in Inner Mongolia grassland [in Chinese]. Friends Nat 3:34–39

Da LT (2008) The effects of institutional and policy changes to grassland ecosystems in Inner Mongolia [in Chinese]. In: Zheng YS (ed) A review of the environment and development in China. Social Science Academy Press, Beijing

Da LT, Si H (2006) The causes of desertification in Inner Mongolia [in Chinese]. Paper presented to the Friends of Nature Forum. Beijing

Davidson-Hunt IJ, Turner KL, Mead ATP, Cabrera-Lopez J, Bolton R, Idrobo CJ … Robson JP (2012) Biocultural design: a new conceptual framework for sustainable development in rural indigenous and local vommunities. S.A.P.I.EN.S. Surveys and Perspectives Integrating Environment and Society, (5.2). Retrieved from http://sapiens.revues.org/1382

Gavin M, McCarter J, Mead A, Berkes F, Stepp JR, Peterson D, Tang R (2015) Defining biocultural approaches to conservation. Trends Ecol Evol 30(3):140–145

Gavin MC, McCarter J, Berkes F, Mead ATP, Sterling EJ, Tang R, Turner NJ (2018) Effective eiodiversity conservation requires dynamic, pluralistic, partnership-based approaches. Sustainability 10(6):1846

Guo QS, Tan DY, Wang CL, Shi ZM, Ma C (2005a) Effects of Cistanche deserticola inoculation on undergrowth vegetation of natural Haloxylon ammodendron forest [in Chinese]. Ecology 8:15–19

Guo QS, Wang CL, Guo ZH, Tan DY, Shi ZM (2005b) Geographic distribution of existing *Haloxylon* desert vegetation and its patch character in China [in Chinese]. For Sci 5:23–27

Hu JT (2005) Public speech at the five-year anniversary of the great western development strategy by Hu Jintao [in Chinese]. Accessed on 17th September 2018 via http://www.gov.cn/ldhd/2005-02/05/content_9525.htm

Huang PX (2002) Excused irrigation vegetation and its restoration in arid area [in Chinese]. Science Press, Beijing

IMARFA (Inner Mongolia Autonomy Region Forestry Administration) (2016) The report of preventing desertification in Inner Mongolia [in Chinese]. Accessed on 17th Mar 2017 via http://www.nmglyt.gov.cn/xxgk/xwzx/lyyw/201612/t20161212_120771.html

International Society of Ethnobiology (2006) The ISE code of ethics. Retrieved on March 17th, 2010 via http://www.ethnobiology.net/what-we-do/core-programs/ise-ethics-program/code-of-ethics/

Jiang GM, Han XG, Wu JG (2006) Restoration and management of the Inner Mongolia grassland require a sustainable strategy. AMBIO J Hum Environ 35(5):269–270

Liu J, Dietz T, Carpenter SR, Alberti M, Folke C, Moran E et al (2007) Complexity of coupled human and natural systems. Science 317(5844):1513–1516

Mead ATP (2013) Sharing power: Māori and protected areas. In: Katene S, Mulholland M (eds) Governance and leadership in future challenges for Māori: He Kōrero Anamata. Huia Publishers, Wellington, pp 195–204
North DC (1991) Institutions. J Econ Perspect 5(1):97–112
Ostrom E (2010) Polycentric systems for coping with collective action and global environmental change. Glob Environ Chang Hum Pol Dimens 20(4):550–557
Peng H, Xu ZF (1996) The threatened wild plants used for medicine as Chinese medicinal herbs [in Chinese]. In: McKinnon J, Sung W (eds) Conserving China's biodiversity. Environmental Science Press, Beijing, pp 175–189
Reed MS (2008) Stakeholder participation for environmental management: a literature review. Biol Conserv 141(10):2417–2431
Rozzi R (2007) Future environmental philosophies and their biocultural conservation interfaces. Ethics Environ 12(2):142–145
Rozzi R (2013) Biocultural ethics: from biocultural homogenization toward biocultural conservation. In: Rozzi R, Pickett STA, Palmer C, Armesto JJ, Callicott JB (eds) Linking ecology and ethics for a changing world, Ecology and ethics, vol 1. Springer, Dordrecht, pp 9–32
SEE (Alashan SEE Ecological Association) (2005) The annual report of ecological protection projects, 2004–2005 [in Chinese]
SEE (Alashan SEE Ecological Association) (2015) The saxaul project [in Chinese]. Retrived on 5th March 2017 via http://www.see.org.cn/Foundation/Article/Detail/3
Sheng Y, Zheng WH, Pei KQ, Ma KP (2005) Genetic variation within and among populations of a dominant desert tree Haloxylon ammodendron in China. Ann Bot 96(2):245–252
Strauss A, Corbin J (1990) Basis of qualitative research: grounded theory procedures and techniques. Sage, Los Angeles
Tang R (2007) Linking traditional ecological knowledge and sustainable resource management of saxaul in the Alashan region, Inner Mongolia of China. Master thesis submitted to Victoria University of Wellington, New Zealand
Tang R (2013) An analysis of traditional ecological knowledge and its conservation options. PhD thesis submitted to Victoria University of Wellington, New Zealand
Tang R, Gavin M (2010) Traditional ecological knowledge informing resource management: saxaul conservation in Inner Mongolia, China. Soc Nat Resour 23(1):193–206
Tang R, Gavin M (2015) Degradation and re-emergence of the commons: the impacts of government policies on traditional resource management institutions in China. Environ Sci Pol 52:89–98
The National Bureau of Statistics China (2010) The statistical report of economy and development in Inner Mongolia, 2009 [in Chinese]. Accessed on 7th May 2010 via http://www.stats.gov.cn/tjgb/ndtjgb/dfndtjgb/t20100302_402625748.htm
Wang XY (2009) Pastoral communities under environmental pressure: case studies of six villages in Inner Mongolia. [in Chinese]. Social Sciences Academy Press, Beijing
Williams DM (2000) Representations of nature on the Mongolian steppe: an investigation of scientific knowledge construction. Am Anthropol 102(3):503–519
Williams DM (2002) Beyond great walls – environment, identity and development on the Chinese grasslands of Inner Mongolia. Stanford University Press, Stanford
Zhao RW (2008) Reflection on the economic reform of China in the past thirty years [in Chinese]. Contemp Financ Econ 9:11–17

Chapter 27
Challenging Biocultural Homogenization: Experiences of the Chipko and Appiko Movements in India

Pandurang Hegde and George James

Abstract In the 1970s, a peasant movement to save forests in the Western Himalayas of India drew international attention. The Chipko movement, as it was called, also had a decisive impact on other grassroots environmental movements. One of them, the Appiko movement, began in a small district in India's Western Ghats with the protest of local people against the state government's policy of clear-cutting natural, indigenous forests to establish monocultures of such high-revenue species such as eucalyptus and teak. Like Chipko, the Appiko movement was committed to traditional ecological management and to village sustainability. Its activism like that of Chipko was against developmental policies that homogenized the local sustainable economy within a larger centralized economy. This chapter shows that the habits and habitat of local people of the hill regions of the Western Himalayas and the Western Ghats are embedded in local indigenous knowledge systems based on holistic understandings of its ecology. For them, forests are not just a one-dimensional resource to be exploited for the benefit of commerce. They are multidimensional and ethical and take care of the needs of human and non-human actors encompassing all life forms.

Keywords Chipko · Appiko · Traditional ecological knowledge · Ecosystems · Non-violence

---

P. Hegde (✉)
Appiko-Chipko Movement, Sirsi, North Kanara, Karnataka, India

G. James
Department of Philosophy and Religion, University of North Texas, Denton, TX, USA
e-mail: George.James@unt.edu

R. Rozzi et al. (eds.), *From Biocultural Homogenization to Biocultural Conservation*, Ecology and Ethics 3, https://doi.org/10.1007/978-3-319-99513-7_27

The Chipko movement of the 1970s in the Western Himalayas is widely known as one of the most successful grassroots environmental movements of India and of the world (Weber 1988; Guha 1991, pp. 152–184). It was successful, not only in the sense that it mobilized an enormous number of villagers against government policy inimical to the interests of the villages and drew international attention to an environmental conflict in the region but because it succeeded in bringing about policy changes that protected the forests upon which the villages were dependent.

The Chipko movement also had a great influence upon movements worldwide. Its values and strategies proved relevant to other struggles to resist environmental degradation, often in regions geographically and climatically different from the Western Himalayas. Because it was a challenge to the homogenization of forest resources in a forest policy that was uninformed and unconcerned about the diversity of the forest habitat and its relation to the habits and needs of the local inhabitants, it resonates with biocultural ethics.

In the 1980s, the Chipko movement inspired another conservation initiative in the forested region of southwestern India: the Appiko movement in the state of Karnataka (Fig. 27.1).

Both Chipko in Hindi and Appiko in Kannada (the language of the state of Karnataka) mean 'to hug'. By calling itself Appiko, it identified with the Chipko concerns for biodiversity, for local access to forest resources and for the strategy of non-violent resistance. In this chapter, we examine the Chipko and Appiko movements of India as the response by forest inhabitants to the homogenization policy of a centralized system of forest use.

27.1 Challenging the Homogenization of Forest Socio-ecological Systems in the Western Himalayas

During the colonial period, the British narrowly valued forests of India as a resource (Guha 1991, pp. 35–61). Early in the nineteenth century, the forests dominated by the valuable hardwood tree teak (*Tectona grandis*) in Malabar in southwestern India were a vital resource for the burgeoning ship-building industry in the British Isles (Pande and Pande 1991, pp. 154–162). When the supply of timber for this purpose was stressed, the region of Tehri Garhwal, in the Western Himalayas, with its abundant forests of the deodar trees (*Cedrus deodara*), proved opportune. By the middle of the nineteenth century, the forests of the Himalayas were instrumental for the expansion of British industry and commerce in India and to its far-reaching railway network, which required a colossal supply of wood for fuel and railway ties.

The extractive forest policy led to indiscriminate removal of timber from the forests in Garhwal. The establishment of a government management policy of these resources, what came to be known as scientific forestry, took place only after the massive logging and the development of an effective system for river transport of timber. Its purpose was to regulate the extraction of timber to insure adequate repro-

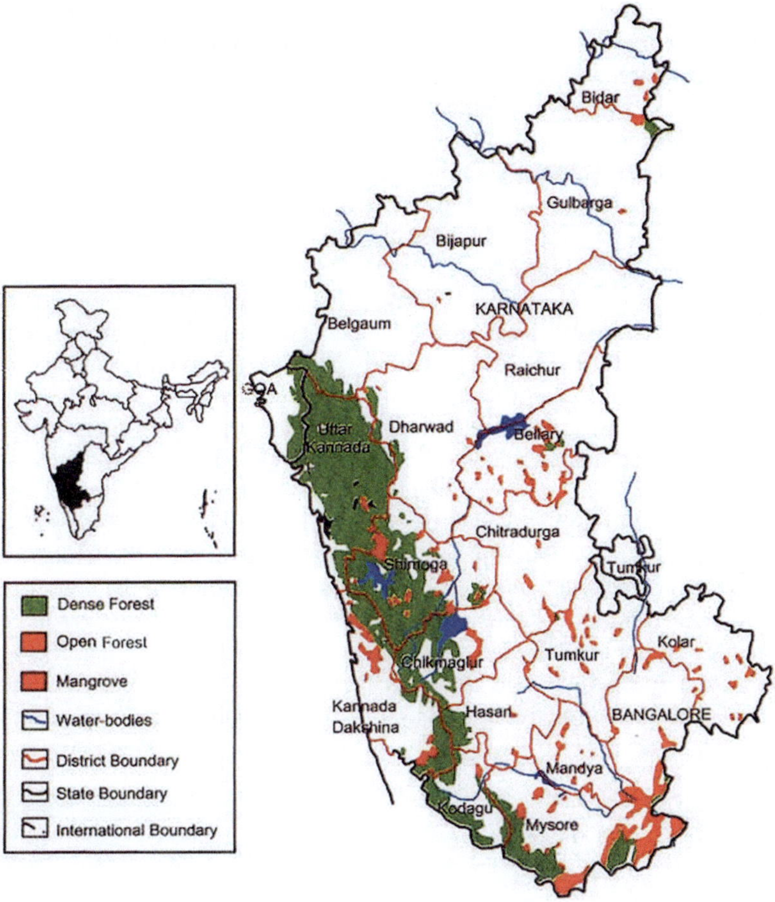

Fig. 27.1 Region of the Appiko movement with Appiko areas

duction of revenue-producing species. Increasing commercial forest management by the state reduced the traditional access of village people to the forests (Guha 1991, pp. 27–43).

Prior to colonial encroachment into the Himalayas, biological and cultural diversity of forest ecosystems in Himalayan region was the basis of an agri-pastoral system of livelihood (Guha 1991, pp. 9–34). In the hill villages, the forests provided ecosystem services for small-scale terraced cultivation of staple crops such as rice and wheat. It yields green and dry leaf fodder for the cattle and water from the streams to irrigate the crops. This economy was supported by small-scale livestock and migratory sheep rearing in the high-mountain regions.

These Himalayan forests are the main catchment of the tributaries of the Ganges and Yamuna river basins. The places where these rivers originate are the holy pilgrimage places in India. From ancient times, the devotion of the hill people to the forests and to the rivers upon which their livelihoods depended was expressed in the stream of pilgrimages to the temples at the four sources of the Ganges: Yamunotri, Gangotri, Kedarnath and Badrinath. Today the *Char Dham Yatra* (the journey to four holy places or abodes) is itself a manifestation of cultural diversity. It is a living experience of the natural diversity of the landscape and of the cultural diversity of the Hindu religious tradition. The pilgrimage route crosses the most diverse forest ecosystems in India. Today one of the main objectives of this route along the river is to experience the cultural and landscape diversity of the great Himalayan range (Haberman 2006, p. 47, Fig. 27.2).

The legacy of colonial forestry homogenized complex forest ecosystems by ignoring the worldviews of local inhabitants, which were based upon an indigenous biocultural ethics (cfr. Rozzi 2012). Nature was respected as holy, having spiritual, ecological and instrumental values. With the independence of India in 1947, the colonial extractive forest pattern continued. The new centralized forest policy was not intended to meet the demands of colonial powers but rather to generate commercial revenue for the new independent state. Thus the government initiated polices to cut trees for timber and to tap resin from the chir pine (*Pinus roxburghii*) (Guha 1991, pp. 43–47). The threat of military aggression by the Chinese in the early 1960s triggered a new wave of roadway development into the border regions of the Garhwal Himalayas. This led to further exploitation of forest resources, the development of monocultures of the chir pin and the neglect of other ecologically important tree species, such as the broadleaf Himalayan oak (*Quercus leucotrichophora*). The impact of the destruction of diversity and homogenization of the

Fig. 27.2 Natural forests of the Himalayas. (Photo Pandurang Hegde)

landscape was a reduction of the regenerative capacity of native ecosystems, especially forests, and of water and biomass needed by the village economy (Guha 1991, p. 152, Fig. 27.3).

Over the course of time, the interests of the village communities came into conflict with the commercial interests propagated by the government's forest department. Overexploitation of forests resulted in a shortage of fodder, fuel and water, thus the basis of their agricultural economy. Access to forest resources was dominated by government policy allowing forest felling by outside contractors. The forest department auctioned forests to these contractors who brought workers from outside the region leaving high unemployment rates. As resource scarcity accelerated, the people experienced hardships accessing their basic forest resources. Conflict over the forest resources forced villagers to initiate direct action to stop the felling of trees. Their effort to save the habitat on which the village economy survived was based on the philosophy of non-violence, promoted by Mahatma Gandhi. For them, the only way to stop the felling of the trees was through non-violent action.

Gaura Devi, an elderly woman from the Reni village in the Garhwal Himalaya, called for embracing the trees, to stand between the axe and the tree in order to protect them from destruction. Thus, in 1974 the Chipko movement was born. This grassroots response to halt the process of biocultural homogenization was an innovative response based on a philosophy and a worldview in which every living crea-

Fig. 27.3 A view of a teak monoculture plantation. (Photo Pandurang Hegde)

ture, including trees, is divine and has spiritual as well as instrumental value. As in the purview of biocultural ethics, they are valued as co-inhabitants (*sensu* Rozzi 2013, 2015a, b). This holistic approach is embodied in the everyday life of the village people's cultural practices and in their restraint placed on exploitation of their natural resources. Through decades of biocultural conservation by the village community, the Chipko movement was able to influence forest policy, eventually challenging the reductionist 'development' model that decimated diversity and contributed to cultural homogenization. In this process we can identify three main phases of the struggle that evolved with the grassroots experiences: economic, ecological and regenerative.

27.1.1 Chipko Economic Phase

Logging in Himalayas was an economic activity that provided a sustained supply of timber for the industries in the Gangetic Plain. Contractors, who carried out the logging, were from the cities. The Chipko activists believed that the benefits of the logging activity accrued only to the contractors and their labourers. Based on this premise, the Chipko volunteers demanded that the contractor system used in extraction of timber be stopped and be replaced with the establishment of labour cooperatives.

At first these labourer-run cooperatives were seen as the ideal system wherein the labourers, rather than the contractors, would receive the benefits of the logging activities. Local labourers would be more caring towards the forests and would follow the management systems that would be beneficial to conservation of the forest resources. It had the dual objectives of economic justice and ecological sustainability. The government agreed to this demand made by the movement, and in many areas logging contracts were awarded to cooperatives managed by labourers. They also established small-scale industrial units, such as turpentine factories. They also successfully implemented legal minimum wages and other welfare schemes for the labourers. This helped bring employment opportunities and needed income to the hill people.

The continuation of large-scale timber extraction, however, led to the deterioration of living conditions of hill women, who had to walk long distances to gather fuel wood, fodder and water. Though the women had launched the movement to conserve these natural resources, the state and the volunteers of the movement provided an economic solution that did not match the scale of the hardship faced by the women. The leaders of Chipko emphasized the economic and social goals of the movement but overlooked the problematic issue of homogenization of native habitats, life habitats and communities of co-inhabitants. They addressed the economic inequities between the contractors and the local labourers but did not address the larger issue of the homogenization of the local economy within the larger economic system. This economic phase of the Chipko movement did not address the issues of biocultural homogenization, of biocultural ethics and of biocultural conservation (cf. Rozzi 2013). Thus, both the extraction of timber and the state's control of commercial policies continued unabated.

27.1.2 Chipko: The Ecological Phase

It was in the Hemval valley in the Tehri Garwhal district that the hill women challenged the economic phase and initiated the ecological phase of the movement. In the late 1970s, the hill women were protesting the tree felling by the state forest department. As an expression of their intention to protect the trees, they tied the sacred thread, or *rakhee*, to the trees that were marked for felling. The villagers kept vigil for several days. At this time a high-ranking forest official came to the forest to meet the Chipko women. In this meeting the officer dismissed their concerns as ignorance of the purpose of the forest policy. He asked the question, 'Do you know what the forests bear?' His answer was 'timber, resin and foreign exchange'. This, for him, was the road to the development and prosperity. The village women spontaneously responded to the forest officer with a slogan they had already learned from the Chipko activists:

What do the forests bear?
Soil, Water and Pure Air
Soil, Water and Pure Air
Are the very basis of life.

This spontaneous response expresses an ethic that relies on engagement with the diversity of nature as the basis for a rural economy. The prior response of the Chipko movement had challenged the homogenization of the development but only from a commercial perspective. The activists were fighting for economic benefits, whereas the women were asking for something more: the conservation of the forest ecosystems upon which their lives depend. For village people, the basis for the development was natural diversity, the availability of the biomass for agriculture and livestock and the availability of water sources. This change in the perspective of the activists led to a new political demand: a moratorium on felling of trees in Himalayas for all commercial purposes.

This ecological phase was the toughest moment for the Chipko activists because the demand was to change the forest policy objective from *commercial* to *ecological*. After sustained actions in numerous regions in Garhwal, the movement eventually succeeded. In 1981, the government placed a moratorium on the felling of trees above an elevation of 1000 metres. This victory of the hill women inspired the numerous villagers to launch the regeneration phase to conserve the natural resources of the villages (James 2013, p. 110).

27.1.3 Chipko Regeneration Phase

With their non-violent actions, the hill women halted deforestation. They also initiated steps towards increased natural diversity through regenerating the forests in barren land. According to estimates in about 1,568 villages in the Garhwal region, women have taken control of about 20,000 hectares of barren forest land and have brought back forest greenery through the planting and regeneration of local forest species. In most of the villages, the *Mahila Mangal Dals* (women's groups) have taken the initiative to regenerate the land (Fig. 27.4). Almost all the households in the villages participate in the protection and regeneration of forest resource. Through the concept of social fencing, the village livestock is not allowed to graze inside this forest. The extraction of fuel wood and fodder is also controlled. It is based on the principle of equity and need. The villagers laid out rules for the management of the forests. Those who violate the rules are fined. They have set up their own watchwomen to enforce the rules and take care of the forests near the village.

This innovative approach to conservation of the natural resources has spread to other regions of the Himalayas and to the adjacent state of Himachal Pradesh. Women's groups have succeeded in greening the barren Himalayas with little or no support or aid from financial institutions. Their success can be attributed to grassroots conservation based on a local ecological ethic that respects the habitat pro-

Fig. 27.4 People involved in reforestation. (Photo Pandurang Hegde)

vided by Mother Earth. The decentralization of control of the natural resources has helped reduce their hardship. They are able to reap benefits from the regenerated forests as they can collect biomass like fodder and fuel wood from the regenerated forests. This phase has reinforced the concept of co-evolutionary relationships of life habits with the habitat that has reshaped a barren landscape into an asset that provides for the survival of human and non-human life systems in the fragile Himalayan ecosystem.

27.2 Challenging the Homogenization of Tropical Forests in Western Ghats, South India

The Western Ghats or *Sahyadri*, the mountain range running parallel to the West coast of India, is a UNESCO World Heritage site that has been identified as a world biodiversity hotspot (Myers et al. 2000). It hosts the catchments of the major rivers that provide irrigation for a large portion of South India and is best known for its tropical forests and high diversity of flora and fauna. Spread along 1600 km from Kerala to Maharashtra, it is one of the most important ecological regions of India. Along with ecological diversity, cultural and agricultural diversity evolved over time to enable the local people to sustain a viable livelihood.

The history of Western Ghats reveals an abundant trade in spices. In early times, the region exported pepper and other spices to the Roman Empire and beyond. The spice trade from the West coast of India to European and Arab countries was so abundant that the district of Uttar Kannada in Northwest Karnataka came to be known as 'pepper queen'. From ancient times the forests of the Western Ghats were regarded as a community asset belonging to the people of the villages. In 1801, to acquire control over its forest resources, the British government abrogated community forests rights and gave control of the forests of the Western Ghats to the British East India Company (Hegde 1988, p. 29). The forests of the Western Ghats had stands of teak hardwood that was especially valuable to the ever-increasing demand of the British Empire. Forests were felled to generate revenue for the colonial government. Later, the British proceeded to clear-cut diverse native forests and replace them with monocultures of teak to supply the increasing British desire for hardwood (Hegde 1988, p. 30).

The conversion of the 'pepper queen' to a 'timber queen' represents two contrasting worldviews. The former is based on sustainable use of diverse natural resources. The latter was based on revenue production relying upon monocultures and non-sustainable use of the resources. The costs of the latter fell largely upon the villagers. As in the case of the Himalayas, after independence from British rule, the new state government's control of the Western Ghats continued to privilege large-scale conversion of native forests into monoculture plantations. This homogenization of the forest industry had a negative impact on local agricultural economies. The decimation of biodiversity caused the drying up of the water sources and a scarcity of biomass for agricultural inputs. As agricultural yields began to dwindle, the people became concerned about the depletion of the forests.

The grassroots response to 'scientific forestry', whose purpose was to generate revenue, was to challenge the homogenization of forest resources. The villagers near the town of Sirsi, in the district of Uttar Kannada, wrote to the forest officials about the problems faced by local people and their desire to halt the process of deforestation. Dissatisfied with the government response, they heard from Chipko activists about the non-violent strategy to save the forests of the Himalayas. On Sept. 8, 1983, the villagers launched a similar movement. Initially the demand of the Appiko movement was to halt the clear felling of natural forests and the conversion of natural forests to monoculture plantations. It challenged the official government policy that entailed several processes of biocultural homogenization (Fig. 27.5).

Recognizing the widespread support for the movement, in 1985 the government of Karnataka agreed to the demand of Appiko to stop conversion of natural growth biodiverse forests into teak monoculture plantations. Yet the timber concessions given to wood-based industries continued to cause damage to the existing forest resources. In order to halt further deforestation, the movement put forth the demand for a moratorium on the felling of all green trees in the natural forests in Western Ghats. After the well-organized grassroots actions over the course of a decade, the state government agreed to change its forest policy. In 1989, the state government imposed a ban on felling of green trees in the natural forests. Largely because of the impact of the Appiko movement and the positive response of the state government, this ban continues today.

Fig. 27.5 Activities of the Appiko movement (meeting). (Photo Jayanta Bandyopathaya)

In addition to supporting this ban, the government sponsored tree-planting programmes to reforest barren land. But it did so with an exotic, fast-growing species of acacia (*Acacia auriculiformis*). Of course, this did not contribute to restoring the native biodiversity of the region. By contrast, the Appiko movement implemented reforestation programmes based on decentralized nurseries of endemic and indigenous species that lead to enhancement of biodiversity of the fragile tropical forest ecosystems. This grassroots response to monocultural homogenization is based on indigenous knowledge and the people's initiative in taking the decision to conserve the existing diversity. Decentralized decision-making led to empowerment of forest dwellers, increasing their self-esteem and fostering life habits that enhanced conservation of their regional biodiversity.

27.3 The 'Five Fs' and Non-timber Forest Products

In its reforestation programmes, the grassroots response was based on an idea that came from the Chipko movement, known as the *five Fs*. The Appiko movement plants trees that help regenerate the soil and support the domestic economy. The first is *food*. Trees provide fruit and nuts that address the nutritional needs of humans as well as wildlife. Second, they plant trees that provide *fodder* for small-scale livestock to meet the nutritional needs of domestic animals and enrich the soil through the composting of bio-waste. Third, *fibre* trees provide for manufacturing local agricultural implements and products that help the artisans earn cash for survival. Fourth, trees shed leaves that break down as forest litter and become *fertilizer* for the other plants and crops. Lastly, trees provide *fuel* for cooking food (Fig. 27.6).

The Appiko movement does not support large-scale monoculture of fast-growing species but plants trees whose branches can provide fuel wood so that there is natural regeneration and trees that survive to provide fuel in the coming years. It also engaged in rural technologies, such as biogas, to minimize the need for fuel from the forests.

Expanding beyond the five Fs concept, the Appiko movement has engaged in a great variety of strategies to address the ecological degradation that affected their village economy. A strategy that has especially enhanced the reputation and credibility of the Appiko movement was its effort to document and initiate a quantification process of the resources and the local revenue that local people generate from their use of non-timber forest products (NTFPs). Mainstream forestry remains occupied with the production of timber to generate cash income. The timber industry survives only by the clear felling or the selective felling of trees. In contrast, indigenous people all over the world have evolved ways to utilize forest diversity without cutting down the trees. They follow traditions, which evolved over the centuries, of collecting NTFPs for their own consumption and for generating income. The forest dwellers collect a variety of forest products such as mushrooms, leaves, flowers, medicinal herbs, resins, honey, wild fruits and natural fibre. They are used for food, fodder, medicine and many other needs. For indigenous forest dwellers,

a b c d e

Fig. 27.6 Examples of the *five Fs*. (**a**) and (**b**) natural fibre to weave baskets, (**c**) tubers, (**d**) wild figs and (**e**) wild food. (Photos Pandurang Hegde)

especially in remote forest regions, forests provide food and nutritional security to local communities. In addition, they provide employment opportunities to earn income by contributing to the production of needed natural products.

While the colonial rulers and the governments that emerged after independence tended to minimize the importance of non-timber forest products, recent research indicates that in India over 50 million people are dependent on NTFPs for their survival and income (Hedge et al. 1996). In fact over 50% of forest revenue and

70% of the forest export income in India are derived from NTFPs (Shiva and Mathur 1996). According to the Food and Agricultural Organization (FAO), about two billion forest dwellers are dependent on NTFPs for subsistence, income and livelihood security (Vantomme 2003).

After launching the grassroots struggle to stop deforestation in South India, the Appiko movement initiated programmes of biodiversity conservation based on indigenous knowledge. One core focus was assistance to woman-based forest groups in adding value to non-timber forest products and sustainable harvesting of the resource. A notable example is the management of a wild fruit known as kokum (*Garcinia indica*), endemic to Western Ghats. Traditionally this fruit is used in Ayurvedic (ancient Indian) medicine. It is an example of biocultural knowledge deeply rooted among the forest-dwelling communities. They also extract oil from kokum seeds which is highly valued for its medicinal properties. Kokum fruits are used to make a natural juice of medicinal value to be sold in the urban areas. Other examples include natural fibres such as bamboo and rattan (cane), which are used to make baskets and household items. With the development of appropriate skills, forest dwellers are able to learn the modern designs and earn a reasonable income from these plant products (Fig. 27.7).

Fig. 27.7 Kokum (*Garcinia indica*) tree, fruits and juice. (Photos Pandurang Hegde)

Another non-timber forest product with which Appiko has been involved has been to the revival of the tradition of indigenous beekeeping. This activity has produced multiple benefits, assuring livelihood opportunity and biodiversity conservation. Although beekeeping has been part of India's culture and economy for millennia, since India's independence, the government and other international organizations have supported beekeeping based on the exotic European bee (*Apis mellifera*). International organizations supported this exotic species because of the volume of honey and revenue it produced. However, in much of India and especially the Western Ghats where the forage areas are in the forest rather than in large monocultures of sunflower or mustard plant, *Apis mellifera* did not do well. It is susceptible to the menace of mites and other pathologies to which the native species *Apis cerana* is immune. *Apis cerana* has co-evolved with the native diversity of Western Ghats ecosystems and has been used by the people from time immemorial. While the *Apis mellifera* was selectively bred to maximize honey production, the *Apis cerana* exists among the co-inhabitants of the Western Ghats. The cultivation of the *Apis cerana* along with other life habits of the people is a part of the traditional ecological knowledge that has enabled the forest culture to survive. While the *Apis mellifera* may be the most productive, the *Apis cerana* was best adapted to the tropics and especially to the forest areas of the Western Ghats. As part of a biocultural conservation programme, the Appiko movement helped to launch 'Save Honey Bees Campaign' (SHBC) to conserve the native honey bees *Apis cerana*. Their role as pollinators of cultivated and wild diversity is still not fully known to scientists. Conservation of *Apis cerana* contributes to maintaining both biological and cultural diversity of the Western Ghats.

Work on NTFPs is based on traditional ecological knowledge of forest dwellers. It aims to enhance their confidence and avoid imposing an alien 'knowledge' from outsiders. Forest dwellers are familiar with the resource base as they grow up with these species and can identify the process of overextraction that leads to unsustainable use of the resource. The concept of biodiversity conservation is not alien but evolves through everyday actions. The financial, social and biological well-being of the forest-dwelling population is linked to the sustainable use of the resource that generates NTFPs over a long period of time.

27.4 Biocultural Conservation

The holistic integration of habits and habitat, ecosystems and cultures that can be seen in ancestral Amerindian ecological knowledge is also visible among the forest people of India. Both Chipko and Appiko movements evolved principles of forestry rooted in the understanding that the forests provide the basis for survival because of their capacity to generate the basic resources of land, water and soil. In contrast to the reductionist basis of scientific forestry that led to degradation and homogenization of landscape and culture, Chipko and Appiko focus on preserving the diversity of species and culture and its spiritual healing power to bring back the regenerative capacity of the land. The value of the diverse and intricate habits of human beings with their habitat and with the co-inhabitants of such cultures is not widely appreciated in today's globalizing society. The drive to generate ever higher levels of capital has marginalized other life forms and marginalized those societies that over time have developed remarkably sustainable strategies for the preservation and enhancement of life, as well as most other living beings. It also excludes such people and cultures from the dominant discourses of globalizing society.

Consistent with a biocultural conservation approach (*sensu* Rozzi 2013), the habits and habitat of local people, especially the hill regions of the Himalayas and Western Ghats, are embedded in local indigenous knowledge systems based on holistic understanding of its ecology. For them, forests are not just a one-dimensional resource to be exploited for the benefit of commerce as is done in the Western world. It is multidimensional and ethical. The forest takes care of the needs of human and non-human actors encompassing all life forms. It provides the spiritual basis for what Rabindranath Tagore called the culture of the forest, *aranya* culture, which advocates a harmonious relationship between humanity and nature. This biocultural initiative of conservation is linked to the livelihood resources of the people and deeply influenced by the cultural and ethical values of the local population.

The trend of biocultural homogenization in exploiting forest and natural resources was challenged by Chipko and Appiko movement through grassroots people's organizations in which the energy of the community was reoriented to protect the remaining forests, by restoring remaining native, biodiverse forest ecosystems and by initiating steps to provide sound basis for biocultural conservation programmes. These actions at the local level led to policy-level changes in forest governance while incorporating the biocultural issues at the level of government.

It is a special case in which the biocultural ethics is practiced at regional and national levels, based on the success of ecological principles that enhance the basic capital of human kind, mainly natural resources of soil, water and air. This analysis provides the basis for integration of ecological philosophy into governance of natural resources. This is an environmental philosophy in action with a sound basis provided by biocultural conservation, with the diversity of ideas from the different ecosystems in Himalaya and Western Ghats.

References

Guha R (1991) The unquiet woods: ecological change and peasant resistance in the Himalaya. Oxford University Press, Delhi

Haberman D (2006) River of love in an age of pollution: the Yamuna river of northern India. University of California Press, Berkeley

Hegde P (1988) Chipko and Appiko: how the people save the trees. Quaker Peace and Service, London

Hegde R, Suryaprakash S, Achoth L, Bawa KS (1996) Extraction of NTFPs in the forests of BR Hills. Contribution to rural income. Econ Bot 50:243–251

James GA (2013) Ecology is permanent economy: the activism and environmental philosophy of Sunderlal Bahuguna. SUNY Press, Albany

Myers N, Mittermeier RA, Mittermeier CG, da Fonseca GAB, Kent J (2000) Biodiversity hotspots for conservation priorities. Nature 403:853–858

Pande ID, Pande D (1991) Forestry in India through the ages. In: Rawat AS (ed) History of forestry in India. Indus publishing Company, New Delhi, pp 151–162

Pascal JP (1988) Wet evergreen forests of the Western Ghats of India. Institut Francais de Pondichery, Pondicherry

Rozzi R (2012) Biocultural ethics: recovering the vital links between the inhabitants, their habits, and habitats. Environ Ethics 43:27–50

Rozzi R (2013) Biocultural ethics: from biocultural homogenization toward biocultural conservation. In: Rozzi R, Pickett STA, Palmer C, Armesto JJ, Callicott JB (eds) Linking ecology and ethics for a changing world. Ecology and ethics, vol 1. Springer, Dordrecht, pp 9–32

Rozzi R (2015a) Earth stewardship and biocultural ethics: Latin American perspectives. In: Rozzi R, Chapin FS III, Callicott JB, Pickett STA, Power ME, Armesto JJ, May RH Jr (eds) Earth stewardship: linking ecology and ethics in theory and practice. Ecology and ethics, vol 2. Springer, Dordrecht, pp 87–112

Rozzi R (2015b) Implications of the biocultural ethic for earth stewardship. In: Rozzi R, Chapin FS III, Callicott JB, Pickett STA, Power ME, Armesto JJ, May RH Jr (eds) Earth stewardship. Linking ecology and ethics in theory and practice. Ecology and ethics, vol 2. Springer, Dordrecht, pp 113–136

Shiva MP, Mathur RB (1996) Management of minor forest produce for sustainability. Oxford and IBH Publishing, New Delhi

Vantomme P (2003) Compiling statistics on non-wood forest products as policy and decision-making tools at the national level. Int For Rev 5(2):156–160

Weber T (1988) Hugging the trees: the story of the Chipko movement. Penguin Books, New Delhi

Chapter 28
Revitalizing Local Commons: A Democratic Approach to Collective Management

Mitsuyo Toyoda

Abstract This chapter examines the issue of revitalizing rural environments in Japan toward ecologically sustainable communities on the basis of the field research conducted during environmental restoration activities at the estuary called Kamoko on Sado Island, Japan. This estuary, which has been used for oyster farming for more than 70 years, has undergone serious eutrophication particularly after the construction of revetment. Responding to local fishermen's desire to improve the environment, a collaborative platform was created in 2008 in order to facilitate the sharing of various knowledge and experiences and to promote grassroots experiments for environmental restoration. The key concern in the process of restoration is transforming the degrading environment to commons, viz., important shared resources. The notion of commons here contains a different implication from traditional commons that has been managed by a closed community. Alternatively, the revitalization of degraded commons requires the realization of a wider and deeper public participation through the development of an open platform. On the basis of the experiment of building an open system of commons, this chapter reports essential ideas for the growth of collaborative governance of the environment.

Keywords Public participation · Collaborative environmental governance · Commons · Sustainable development · Environmental restoration

M. Toyoda (✉)
Center for Toki and Ecological Restoration, Niigata University, Sado, Niigata, Japan
e-mail: toyoda@cc.niigata-u.ac.jp

R. Rozzi et al. (eds.), *From Biocultural Homogenization to Biocultural Conservation*, Ecology and Ethics 3, https://doi.org/10.1007/978-3-319-99513-7_28

28.1 Introduction

Although the degradation of nature is undoubtedly related to modern industry and consumptive lifestyles, it is also important to recognize that abandoning the use of local natural resources has been causing serious ecological degradation in some parts of the world. This problem is happening in many rural areas of Japan. As domestic natural resources began to be substituted by imported and artificial goods, many parts of the land came to be left without adequate maintenance. The loss of biodiversity is considerable in such areas.

One of the main focuses of Japanese environmental restoration is to reestablish sustainable resource management in rural areas by learning from traditional practices and exploring new possibilities of wise use.[1] This process inherently embraces various social issues, such as the lack of concern about local resources that results from the predominance of commodity-oriented lifestyle, the shortage of human resources necessary for proper management due to serious depopulation, and the absence of long-term vision arising from the lack of local successors. Taking these issues into consideration, we must ask how it is possible to revitalize rural environments with a view to developing ecologically sustainable communities.

In this chapter, I approach to this issue on the basis of the field research conducted during environmental restoration activities on Sado Island, Japan (Fig. 28.1). I have been engaging in the practice of field environmental philosophy on this island since 2007. One of my major projects conducted on this island is the restoration of the estuary called Kamoko. The estuary has undergone serious eutrophication. This problem is associated with, for example, the construction of concrete block revetment around the shore, sewage disposal in the watershed, and unsustainable methods for oyster farming. In addition to these issues, the weakening of people's connection with the estuary generates the problem of the lack of public concern about its condition. How to restore the estuary as *commons* is a vital topic for advancing the restoration of Kamoko.

This chapter focuses on the restoration of the Kamoko Estuary, thereby adding to the general methodological and theoretical approaches of biocultural ethics and conservation (cfr. Rozzi 2013). It illustrates how the restoration of a habitat (the estuary) enables the restoration of life habits (oyster fishing) and the return of diverse coinhabitants, including diverse forms of human cultures (fishers and other citizen in Kamoko) and biological species (oysters, wetland plants, etc.). The chapter centers on examining the process of generating a new form of commons focusing on the case study of Kamoko. The revitalization of commons is an important issue in Japan, where the restoration of secondary nature, rather than untouched nature, is

[1] I use the word *environmental (ecological) restoration* to signify human activities aimed at improving the conditions of degraded areas of our environments and bringing back the healthy functions of ecological systems. To restore generally means to bring back the original condition. The Japanese word *saisei*, which is often used as a translation of *restoration*, means more than bringing back a former condition and entails the meaning of revitalization. In this paper, I use the word *environmental restoration* to signify both the restoration and revitalization of our environments.

Fig. 28.1 Map of Japan, showing the location of Sado Island

at issue. I present the case of the restoration of Kamoko as a model example of such a challenge. Based on the brief discussion of the notion of commons, I examine both relevant and controversial aspects of various interpretations of the commons and consider the possibility of an alternative understanding of commons that facilitates collaboration among a variety of stakeholders.

28.2 Regional Characteristics of Kamoko

Kamoko is located on the east end of the central plain between the Kosado and Ōsado mountainous regions (Fig. 28.2). Its size is approximately five square kilometers, the biggest inland water on outlying islands in Japan. Kamoko was a freshwater lake until the end of nineteenth century, but as a result of connecting the lake to the ocean by constructing an aqueduct as a flood control measure in the beginning of the twentieth century, the lake became brackish. Since this change occurred, oyster farming has become the major local industry around this estuary. Small wooden oyster shacks standing along the shore create a unique landscape of Kamoko (Fig. 28.3).

This estuary used to be known for its scenic beauty even outside of the island, and its picturesque landscapes have been enjoyed through poems and paintings. However, the ecological condition of Kamoko is in a critical situation. As men-

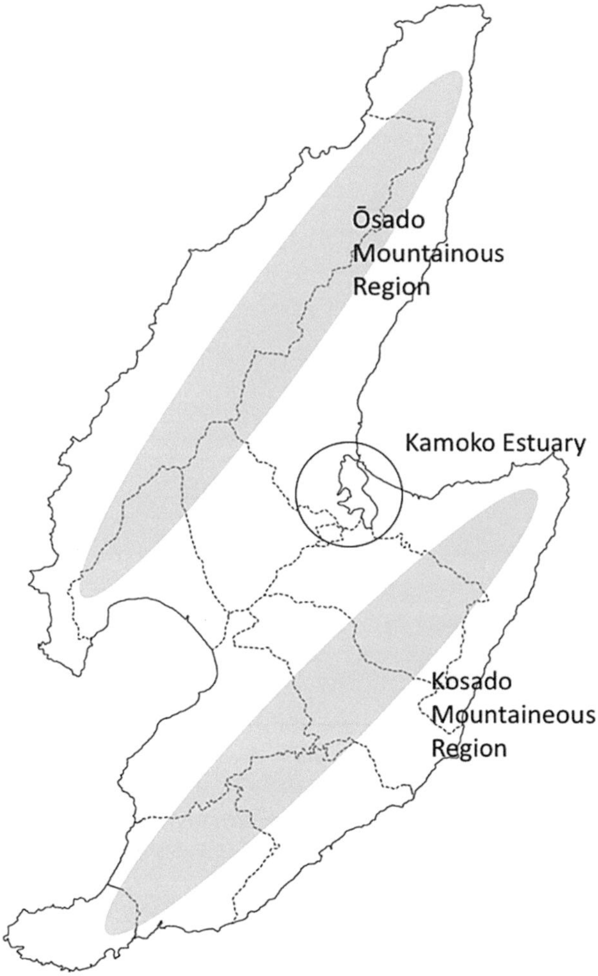

Fig. 28.2 Location of the Kamoko Estuary on Sado Island

tioned earlier, eutrophication is advancing especially after the construction of revetment about 40 years ago (Fig. 28.4). In 2009, red tide occurred in this estuary and caused a devastating damage of oyster farming. After this incident, local fishermen became frightened of the unpredictable damage caused by red tide every year. Many fishermen decided to quit oyster farming, and the local fishing industry is declining (Fig. 28.5). This tendency might result in the negative spiral of ecological health of the estuary since oysters themselves are known to have the function of filtrating water and to mitigate eutrophication.

The improvement of its environmental conditions is an urgent issue for the sustainable development of Kamoko. There are some technical solutions to be considered: the improvement of the water quality flowing into the estuary, the restoration of seaweed and reed beds on the shore, the recruitment of newly joining fishers, etc. In addition to these efforts, it is also crucial to revitalize the estuary as *commons*:

Fig. 28.3 Scenery of Kamoko Estuary. (Photo Shunsuke Shii)

Fig. 28.4 Today more than 80% of Kamoko's shore is covered by pile-sheet revetment. (Photo Mitsuyo Toyoda)

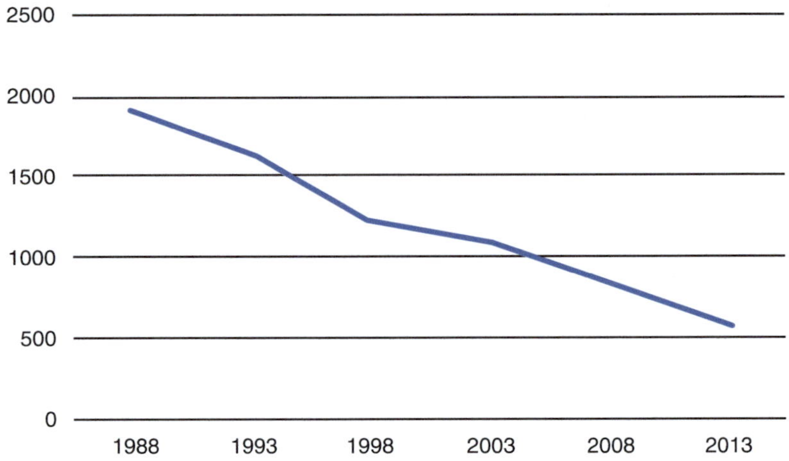

Fig. 28.5 Decline of the number of oyster rafts between 1988 and 2013

shared natural resources that are used and maintained communally and sustainably. The lack of public concern is a fundamental problem connected to the degradation of nature. In the case of Kamoko, people's relation to this estuary has changed significantly after the alteration of revetment and the spread of urbanized lifestyle. It has little connection with the everyday life of people living in the watershed. The degradation of the estuary is a serious concern for local oyster farmers but not for others. If we could alter this separation and cultivate public concern about what is happening in the estuary, collective engagement may grow. How to change people's connection with the estuary is thus an important issue to be considered.

28.3 Autonomy as the Foundation for Successful Governance of Commons

Commons signifies shared resources as well as people's right to access them. It is a key concept to learn sustainable resource management from the past and to inquire into the responsible use of natural resources and is one of the approaches for considering the possibility and difficulty of environmental governance.

The notion of commons, however, has been triggering active discussion over what approach to resource management would be sustainable. A well-known thesis concerning the difficulty of managing commons has been developed by Garett Hardin (1968) in "The Tragedy of the Commons." In this essay, he uses the example of overgrazing in order to illustrate the sustainability issues that result from overpopulation. He recognizes the danger of self-interestedness inherent in each individual being and emphasizes the danger of individual's uncontrollable exploitation of resources.

Hardin holds a skeptical view about the system of the commons and insists the need for managing resources by a coercive force outside their individual psyches, which he describes as Hobbesian "Leviathan." He thinks that the natural desire of human beings to seek one's self-satisfaction needs to be controlled by the external force. He writes, "Tragedy is the price of freedom in the commons... Survival is possible under several different politico-economic systems—but not under the system of the commons (Hardin 2001)." As Poteete et al. (2010) mention, this view is regarded as a conventional theory of collective action: the only way to solve the commons problem is to impose a "solution" from the outside. Hardin's statement has also been used as a justification for privatizing natural resources. For example, Smith (1981) argues, "the only way to avoid the tragedy of the commons in natural resources and wildlife is to end the common property system by creating a system of private property rights."

It is true that swelling population has resulted, for example, in the scarcity of water, the rapid increase of species extinction, and the aggravation of pollution. If we look at the occurrence of these problems around the world, in the case of overpopulation, the corruption of sustainable resource management seems inevitable.

However, Hardin's illustration of the tragedy of the commons has directed academic's attention to the examination of the historical and contemporary cases of the management of the commons. Counterarguments raised against Hardin's theory contain the following two points: (1) Hardin misunderstands that the commons imply free access to resources (Cox 1985; Berkes et al. 2000), and (2) sustainable resource management does not necessarily depend on external authority for controlling the users. Based on a careful examination of the users of traditional resource management, Elinor Ostrom (1990) states that local natural resources have been used by a closed community in which the boundary between users and nonusers is clearly defined. Instead of depending on external control, a group of users of a resource needs some autonomy to make their own rules for the sustainable use of this resource (Ostrom et al. 1999). While Hardin explains the tragedy of commons in terms of freedom, which he interprets as the exemption from regulation, Ostrom highlights the notion of autonomy, which combines the two contrasting meanings—freedom and governance. Freedom in this sense does not necessarily mean one's power to do whatever one wishes to do. It rather indicates the power of rulemaking necessary for the sustainable management of resources and the centrality of collective deliberation maintaining this power.

There is a reason why autonomous rather than external governance is considered adequate for sustainable resource management. Ostrom (1990) argues that the governance of common-pool resource (CPR) is closely connected with the unique setting of each case. Rules for the governance differ depending on "specific attributes of the related physical systems, cultural views of the world, and economic and political relationships that exist in the setting" (Ostrom 1990, p. 89). She further states, "Without different rules, appropriators could not take advantage of the positive features of a local CPR or avoid potential pitfalls that might be encountered in one setting but not others" (Ostrom 1990, p. 89).

Ostrom's argument is convincing because they are derived from a variety of case studies. It is however necessary to turn our eyes to the degradation of commons that is actually happening now. The problem of environmental degradation cannot be solved merely by adopting traditional system of commons. For example, although Ostrom (1990) identifies the drawing of a clear boundary between users and nonusers as one of the conditions of sustainable management of commons, such condition does not work in the situations in which the degradation is happening due to the abandonment of resources or in which nonusers are causing negative impacts on the conditions of natural resources. In some situations, we need to create alternative system of commons that differs from traditional ones.

In Japan, the malfunction of traditional commons is occurring due to several reasons. For example, the system of *iriai* for the management of lands and natural resources is weakening due to the decrease of people's dependence on local resources, the change of community and lifestyle, the modern development of the notion of property rights, and the serious depopulation and aging. Although *iriai* is an institutionalized system based on the exclusive recognition of resource users' rights and the shared strict rules of the land and resource management, such a rigid framework of *iriai* needs to be softened in some occasions. Alternative frameworks have been explored particularly with the emphasis of the notion of *sato*. This notion began to be known globally through the Satoyama Initiative presented by the Ministry of the Environment of Japan and the United Nations University Institute of Advanced Studies.[2] Sato signifies the close interrelationship of nature and culture and the importance of shared management of natural resources. It carries a more general meaning than *iriai* and has been developed as a key concept that makes us aware of the traditional relationships with natural resources and revitalizes them in present settings. In addition to *satoyama*, which means commonly used forest, such notions as *satoumi* (commonly used coastal resources) and *satogawa* (commonly used river) began to be used in order to encourage people to reinterpret the values of commonly used natural resources while taking into consideration current environmental issues such as the conservation of biodiversity and the sustainable use of energy (Yanagi 2008; Centinkaya 2009; Berque and Matsuda 2013).

In the next section, I explore the issue of developing alternative commons in light of the empirical study of the management of the Kamoko Estuary on Sado Island.

28.4 Building a Collaborative Platform for Restoring Commons

The traditional system of commons is founded on the idea that a group of users needs to have some autonomy to consider responsible ways of using resources respecting the value of sustainability. This group is closed, according to Ostrom, so

[2] See, for example, http://satoyama-initiative.org/wp-content/uploads/2013/08/The_Satoyama_Initiative_leaflet_EN.pdf (accessed on April 7, 2017).

that the right to access resources is limited to people who take care of the resource on a regular basis. Such traditional system does not necessarily function in the case of Kamoko.

This estuary is owned by Sado City, but it is in a state of non-statutory public property: Kamoko is not governed under specific laws or governmental regulations. Local users are expected to maintain the environment of the estuary through their autonomous practices. A major industry in Kamoko is oyster farming which started in the beginning of the twentieth century. Oyster farmers are regarded as the main users of the estuary. In order to revitalize their industry, fishers organized Kamoko Fisheries Cooperative (KFC) in 1956. This organization has been responsible for governing the resources of the estuary from the viewpoint of the fishing industry.

As mentioned earlier, the condition of Kamoko has been deteriorating for several reasons: the bank construction, sewage disposal, and a large amount of wastes washed on the shore. Silty sediments are accumulating at the bottom and creating an oxygen deficient environment. In view of the current severe environmental degradation of the estuary, the fishermen strongly hope to improve its environmental condition for revitalizing their industry. KFC is a principal agent in the process of revitalization. However, their effort is not sufficient for developing the sustainable management of Kamoko. Fishermen in Kamoko are of course expected to employ environmentally responsible farming methods and work on the improvement of the estuary. In addition to their effort, however, the cooperation of non-fishers is necessary in promoting the ecological restoration of the estuary, for at least two reasons. First, the degradation of the quality of water is associated with the infrastructure around the estuary as well as human activities in its watershed. Non-fishers are also responsible for the degradation of the estuary. Domestic and agricultural wastewater has been affecting the water quality. Second, the continuous effort of restoring the estuary requires the participation of people of various generations. However, aging is a serious issue affecting the fishers' community in Kamoko. The participation of younger generation carries an important meaning in acknowledging the value of sustainability and also in inheriting unique local customs and knowledge that teach us how to conserve resources and avoid risks and dangers brought by water. Unfortunately, some religious rituals and festivals that developed around the estuary have already been lost due to the decline of communal activities.

A crucial challenge is to enlarge the circle of participation for the restoration of Kamoko. To do this, it is necessary to grow this estuary as a common, that is to say, an important natural resource for local communities. This estuary used to be a widely acknowledged local common that provides a variety of benefits to local life. People obtained what they needed from the estuary (e.g., food, housing materials, etc). It also provided children with an ideal environment to learn how to swim and to have fun with their friends. Local people's close connection with the estuary has been the source of cultural inspiration: a variety of narratives and religious rituals have developed from everyday interaction. After the alteration of revetment and the spread of urban lifestyles, however, this connection has changed significantly. Most local residents just stay away from the estuary.

In order to change the situation, it was necessary to create a system to revitalize people's connection with the estuary and to restore it as a common not only for fishers but also for non-fishers. With the view to facilitating collective thinking and action toward the restoration of the estuary, a co-learning organization was established in July 2008 by a group of fishermen, governmental officials, and researchers. This organization is called KAMOKEN, the abbreviation for Kamoko Suikei Saisei Kennkysho, which can be translated as a research center for restoring Kamoko watershed. It is a platform where people who occupy various positions and belong to different generations meet, exchange ideas, and work together.

This organization was designed as an open institution in which anyone who was interested in the conditions of Kamoko and the activities of environmental conservation could participate. The mere establishment of this organization was not sufficient for the achievement of wider participation. As previously mentioned, Kamoko was not an important concern for most local residents, especially for non-fishers. The first step was, accordingly, to develop a greater recognition of the values of this estuary by examining diverse concerns and interests expressed by non-fishers and connecting the issue of restoring the estuary with a variety of topics such as education and leisure.

Another important mission of KAMOKEN was to facilitate democratic decision processes for collaboration to grow. As Arnstein (1969) argues in "A Ladder of Citizen Participation," deeper public participation is possible when people are enabled to engage in decision-making processes. A mere assembling of people, however, does not necessarily generate an adequate environment for collaboration. Through field research, I recognized the problem of *epistemic hierarchy*. In spite of a growing recognition that locally established and shared knowledge is essential to deepening our understanding of unique characteristics of each region, such knowledge is often regarded as less significant compared to academic and scientific knowledge in many environmental projects. In relation to this tendency, there is an unreasonable separation between specialists and lay people. Such unequal treatment, which I call *epistemic hierarchy*, ruins a fair circle of collaboration and hinders the active participation of people from various backgrounds. The hegemonic role of scientific knowledge limits the participation of diverse community members and forms of knowledge, not only in Kamoko but also in other oyster fishers' communities in the world, such as Baltimore in the USA (see Kingsland 2015). In the case of Baltimore, the main issue was the external intervention that neglected locally developed fishery and, as an alternative, promoted oyster culture through the privatization of aquatic resources. In the case of Kamoko, on the other hand, I recognized the problem in decision-making processes. When people find a problem to be solved, they tend to present petitions and ask governmental sectors and politicians for help. Certainly, there are cases in which governmental and political support is necessary for problem-solving, but more democratic approaches are also needed for the actual advancement of the revitalization of Kamoko. In order to create a fair multi-perspectival learning community and facilitate active public inquiry for environmental restoration, KAMOKEN coined this slogan: "all of us are teachers and all of us are learners."

Fig. 28.6 Local elementary schools began to use Kamoko as a field site for environmental education. (Photo Mitsuyo Toyoda)

Since its establishment in 2008, KAMOKEN has been growing by employing necessary measures in the pursuit of genuine democratic approach. The collaboration between fishers and non-fishers has gradually been realized through our attempts to create a fair participatory circle. The current main project of KAMOKEN is the restoration of the inlet in Akitsu region called *Kogomenoiri*. This project started in June 2010 as an attempt to restore reeds on the shore and to improve the water quality and the ecological habitat. It also contributed to creating a shore accessible to children and thus developing an area adequate for environmental studies of Kamoko (Fig. 28.6). Since there was no predetermined plan for how to restore the shore, a series of workshops have been conducted inviting fishers and non-fishers in order to build a detailed plan and review its effectiveness adaptively.

Although this restoration project has been taking place in public lands, governmental sectors are not the primary actor of the project. People from nongovernmental or nonacademic sectors are included from the process of planning and are encouraged to share ideas from different viewpoints. Children are also invited to the open forums. For example, at the workshop held on April 30, 2011, children, local fishermen, and governmental officials got together to consider possibilities to improve the condition of the estuary. Children presented their experiences of environmental studies around Kamoko and shared with other participants what they wished to do for the improvement of the estuary. Most children expressed great

Fig. 28.7 Shore of the Kogomenoiri inlet after restoration. (Photo Mitsuyo Toyoda)

interest in living things around the estuary and requested to improve the accessibility to the shore. During the workshop, one of the children suggested building a wooden boardwalk so that people would approach the estuary for a closer look. Responding to this child's suggestion, KAMOKEN carried out the project of constructing a boardwalk in July 2012 (Fig. 28.7). This area has gradually become a place where people in different ages learn from each other. I facilitate multigenerational learning, for example, by inviting elder fishermen who can teach children the traditional fishing method called *shibazuke-ryo* using brushwood. Such opportunity is important not only for passing on the knowledge from the past but also for encouraging the fishermen through the interaction with children.

Through the project of restoring the inlet, a community for collective thinking and action has been gradually growing. The actual progress in the improvement of the shore has generated a solid sense of achievement among the participants. In other words, a community has been empowered through actual collaborative processes. A variety of environmental projects have been carried out in Kamoko expanding a circle of participants such as ecological studies, participatory restoration activities, and building a boat using reeds on the shore (Figs. 28.8 and 28.9). The power of self-governance has gradually been growing in a community by engaging in a genuinely democratic approach of a bottom-up restoration project. Both fishers and non-fishers gradually came to see that they were the ones who could make a difference and recognized that they were able to do something, and they ought to do something for their environment. Kamoko is gradually revitalizing as a local common.

Fig. 28.8 At the Kogomenoiri inlet, reeds are harvested every year in cooperation with local residents. (Photo Shunsuke Shii)

Fig. 28.9 The project of making reed boats which started in 2016 was a breakthrough to expand the circle of collaboration. (Photo Tsuyoshi Oshikawa)

28.5 Conclusions

In the process of restoring natural environments in Japan, the notion of commons carries important implications. This is because most of the land in Japan has been conserved through its deep connection with human activities. Many natural resources have lost their commercial values and have receded from our everyday concern. If domestic resources are no longer to be used or sold, there is no reason for people to make considerable effort for the maintenance of surrounding natural resources. In order to restore local environments as commons, it is crucial to consider how to reconstruct values in the natural surroundings and turn them into important resources that people care for.

When we attempt the restoration of commons not only for the development of local industry but also for ecological revitalization, it is important to reformulate the system of governance so as to realize a wider and deeper public participation. If we hope to successfully promote environmental restoration, creating a new system might be effective because traditional organizations for the management of the commons are in general conservative and do not provide appropriate grounds for collaboration to grow. The example of KAMOKEN is an experiment of constructing an open system to conserve natural resources. It was introduced in order to assemble different stakeholders and to attempt together possible approaches to the sustainability of watershed.

The establishment of this institution has resulted in creating opportunities for collaboration and has been working as the first step toward cultivating new commons. A mere introduction of such organizational framework, however, is not sufficient for the development of commons. Through this experiment, it is necessary to grow KAMOKEN into an autonomous community in which people work on the conservation of Kamoko by setting adequate rules for sustainable environmental governance. To do this, I have been coordinating public forums for environmental envisioning and attempting to develop shared goals and objectives. KAMOKEN is under development in this respect. The introduction of this platform, nevertheless, has generated important impacts on the conservation of the estuary on at least two fronts.

First, public interest in Kamoko is growing due to various participatory environmental projects. After the restoration of the Kogomenoiri inlet, three local schools began to use Kamoko as a field site for environmental education. KAMOKEN is also creating opportunities to enjoy waterfront conservation activities inviting children and their parents. Many participants said that they did not have a chance to come close to the estuary before these opportunities. People learn both the current environmental problems of Kamoko and its ecological, educational, and aesthetic values and have become important stakeholders to consider the future of the estuary.

Second, KFC, in collaboration with governmental sectors and KAMOKEN, began to engage in a number of experimental projects for promoting environmental conservation and developing new methods of shellfish farming. When KFC was struggling with environmental degradation of Kamoko without support by non-fishers, petitioning was a major way of presenting to the public environmental problems that they suffer. After the establishment of KAMOKEN, their struggles began to be heard by externals, and the public interest in the estuary began to grow. Fishers who have been participating in KAMOKEN's activities have gained the confidence that they would be able to make a difference in their environment through experimental collaborative activities. As public participation has been highlighted in most environmental conservation projects, it is crucial to cultivate the power of considering possible solutions creatively, working together with various stakeholders. The establishment of KAMOKEN was therefore meaningful as it contributed to the growth of self-governance fostering fishers' motivation to experiment new ideas in collaboration with other stakeholders.

These influences indicate the gradual transformation of people's relationship with their habitat. Our challenge to collective governance has been yielding the diversity in this relationship and therefore exploiting various possibilities of co-inhabitance. By creating and weaving new connections between biological species and human cultures aruond the habitat, we continuously envision the path of sustainable ecological governance.

References

Arnstein SR (1969) A ladder of citizen participation. J Am Plan Assoc 35:216–224

Berkes F, Folke C, Colding J (2000) Linking social and ecological systems: management practices and social mechanisms for building resilience. Cambridge University Press, New York

Berque J, Matsuda O (2013) Coastal biodiversity management in Japanese satoumi. Mar Policy 39:191–200

Centinkaya G (2009) Challenges for the maintenance of traditional knowledge in the satoyama and satoumi ecosystems, Noto Peninsula, Japan. Hum Ecol Rev 16(1):27–40

Cox SJB (1985) No tragedy on the commons. Environ Ethics 7:49–61

Hardin G (1968) The tragedy of the commons. Science 162:1243–1248

Hardin G (2001) Carrying capacity as an ethical concept. Soc Contract 12(1):48–57

Kingsland SE (2015) Ecological science and practice: dialogues across cultures and disciplines. In: Rozzi R, Chapin FS III, Callicott JB, Pickett STA, Power ME, Armesto JJ, May RH Jr (eds) Earth stewardship: linking ecology and ethics in theory and practice. Ecology and ethics, vol 2. Springer, Dordrecht, pp 17–26

Ostrom E (1990) Governing the commons: the evolution of institutions for collective action. Cambridge University Press, New York

Ostrom E, Burger J, Field CB, Norgaard RB, Policansky D (1999) Revisiting the commons: local lessons, global challenges. Science 284:278–282

Poteete AP, Janssen MA, Ostrom E (2010) Working together: collective action, the commons, and multiple methods in practice. Princeton University Press, Princeton

Smith RJ (1981) Resolving the tragedy of the commons by creating private property rights in wildlife. Cato J 1:439–468

Rozzi R (2013) Biocultural ethics: from biocultural homogenization toward biocultural conservation. In: Rozzi R, Pickett STA, Palmer C, Armesto JJ, Callicott JB (eds) Linking ecology and ethics for a changing world: values, philosophy, and action. Ecology and ethics, vol 1. Springer, Dordrecht, pp 9–32

Yanagi T (2008) "Sato-umi"-a new concept for sustainable fisheries. In: Tsukamoto K et al (eds) Fisheries for global welfare and environment, 5th World Fisheries Congress. Terrapub, Tokyo, pp 351–358

Chapter 29
The Garden as a Representation of Nature: A Space to Overcome Biocultural Homogenization?

Tetsuya Kono

Abstract In this chapter, we will consider what kind of differences are there between the motives for making and maintaining gardens by glancing over the histories of gardens in Europe and East Asia and by comparing mainly two types of large gardens: French formal gardens and the Japanese *daimyo gardens*. We will see that a garden is the representation of nature as an ideal community, which could be an ever-fertile farm and orchard for European people; the universe rationally organized for absolute kings; an open, free, but prettified landscape for early English gardeners; a wild and growing habitat for Romanticists; or the islands of Xian where one can find inner Tao for Japanese daimyos. A better understanding of the aesthetic, political, and cultural values of gardens in these traditions can be helpful to better foster biocultural conservation and prevent biocultural homogenization. A garden of today should be a garden which suggests to us how to have the benefit of nature in a sustainable way, how to preserve and promote biodiversity, and how to develop a nature-loving culture in a local society.

Keywords French formal gardens · Japanese gardens · Landscape gardens · Sustainable society · Biocultural homogenization

After the summit in Rio de Janeiro in 1992, the preservation of biodiversity is considered as one of the most essential factors for sustainable development and human well-being. However, biodiversity losses have continued, and a widespread appreciation of the value of nature is still absent in the culture of global society (Rozzi et al. 2013). According to Ricardo Rozzi (2013)[1], biocultural homogenization is one

[1] It is interesting to know that Rozzi and his colleagues have led the creation of a park at the southernmost point of the Americas: the Omora Ethnobotanical Park. A central mission of this park is to conduct research, education, and conservation programs with the aim of incorporating the diversity of local and sustainable ecological practices and forms of knowledge into policy, economy, and formal education in the Cape Horn Biosphere Reserve in Chile (see Rozzi et al. 2006).

T. Kono (✉)
Department of Education, Rikkyo University, Tokyo, Japan
e-mail: tetsuyakono@rikkyo.ac.jp

R. Rozzi et al. (eds.), *From Biocultural Homogenization to Biocultural Conservation*, Ecology and Ethics 3, https://doi.org/10.1007/978-3-319-99513-7_29

of the strongest and pervasive drivers of today's rapid global environmental change and degradation. Rozzi has stated that we should put a higher value on biological and cultural diversity and their interrelationships, in short biocultural diversity, to resist the process of globalizing biocultural homogenization. He has proposed that an ethics for biocultural conservation is needed to achieve socio-environmental sustainability at the planetary level.

In this chapter, I explore how gardens can play an important role to achieve biocultural conservation, or instead gardens can also be an expression of and contribute to biocultural homogenization. In European and Asian traditions of gardens, we can find both remarkable similarities and contrasts about the design, motivations, and values associated with the gardens. A better understanding of the aesthetic, political, and cultural values of gardens in these traditions can be helpful to better foster biocultural conservation and prevent biocultural homogenization.

29.1 Introduction: Questions About Gardens

Many people make a small private garden situated at the front of their house, and almost every country and culture has public gardens. But what is a garden? Why do human beings make gardens?

Many different societies have made different types of gardens, although some societies do not have a tradition of making gardens. Many large gardens are reclaimed landmasses that were previously used as farms or hunting grounds, and some still are used. However, gardens are conceptually distinguished from the land reserved for practical uses such as agricultural lands, stock farms, and hunting grounds. Zoos are officially called zoological gardens, and numerous places that are called "gardens" in our society are actually botanical gardens. Thus, a garden is a designed place set aside for the display of nature, i.e., plants and wildlife.

If so, why do we display nature, cultivate it, and enjoy it in a garden? One obvious reason is that people's living space is detached from nature. A garden is often constructed in the city or its suburbs. Urban living provokes people to yearn for nature and wilderness. But if people in the city want to touch nature, it would be better to travel to the countryside or wild lands. Why then do they plan and construct a garden? For what purpose, not for practical reasons such as food producing and material obtaining, do human beings want to plan, design, change, and reconstruct nature? A garden imitates nature; it is a miniature of nature. Why and how does a garden imitate and display nature?

I would like to discuss human desires for nature that are expressed in garden's construction and maintenance and subsequently assess their potential for biocultural conservation. There are many different types of gardens in the history of the world, so I don't intend to develop a comprehensive theory about gardens in this short chapter. My goal is quite modest: to consider what kind of differences there are between the motives for making and maintaining gardens and the motives for "preserving nature," which in turn directs the idea of national parks or preservation areas.

I shall examine this theme by glancing over the histories of gardens in Europe and East Asia and by comparing mainly two types of large gardens: French formal gardens such as those planned by André Le Nôtre and the Japanese *daimyo teien* (大名庭園). A *daimyo* (大名) is a Japanese feudal lord who ruled vast, hereditary lands from the tenth century to the middle nineteenth century, before the decline of the feudal system of *Tokugawa Shogunate*. A *teien* (庭園) means a large garden. French gardens and Japanese gardens represent different conceptions of nature: the rationally designed universe and a Taoist paradise, respectively. Finally, we shall see that gardens show two opposing human desires and attitudes in relation to nature: one is to control nature and another is to be integrated with nature.

29.2 The Garden and Paradise

Some researchers (e.g., Iwakiri 2008) claim that gardens started as part of agricultural lands or hunting grounds. Some large English gardens and Japanese *daimyo* teien have been used as hunting grounds. Most gardens of monasteries and convents were still farms and orchards in the middle age of Europe (Hadfield 1955). On the other hand, other researchers, for example, Benoit-Mechin (1975), claim that ancient gardens in many places such as Africa, Polynesia, South America, and Asia were the sacred places for magical rituals or religious prayers. These were places in which shamans cured diseases and magicians expelled the evil spirits and summoned supernatural powers. Thus, like other forms of arts, gardens in their ancient form were not separated from practical and/or spiritual human activities. But, if gardens are a kind of art, they have a different meaning or function from agriculture and hunting-gathering as well as from magic and religion.

Ancient people in both the West and the East have made gardens to represent their images of paradise. In ancient Greek literature, many examples can be found (Giamatti 1966). Homer described Alcinous's palace as a paradise in his *Odyssey*. It was just an orchard in which various trees of pomegranates, apples, figs, pears, olives, and grapes were planted. They bear fruit throughout the year, and spring west breeze is always gently blowing. It is an image of an ideal farm. We can also find some different images of landscapes as paradise in Homer's mythology, a flower field with roses, violets, saffron, irises, hyacinth, and narcissus, and according to the story of Proserpina, we learn of a pasture where tender grass always sprout and streams and the ponds are filled with water. Thus, the prototype of the garden in the Western societies represents the image of the ideal farm, orchard, or pasture.

As Francis and Hester (1991, p. 2) pointed out, "[t]he garden has been viewed philosophically as the balancing point between human control on one hand and wild nature on the other. The garden has represented safety from the threat of wild nature or escape from barbarian outsiders. The garden has been nature-under-control, an idealization of what society believed that nature should be and should look like." For Western societies, well-maintained, ever-fertile farm, orchard, or pasture is what nature should be.

The ancient gardens in East Asia also represent the image of the paradise for Asian people. In China of the tenth century B.C., people in power had already constructed palace gardens in order to demonstrate their authority and to pursue the amenity of their lives (Clunas 1996; Kidu 1994; Nakamura 1999). Those gardens had plants, animals, as well as altars to enshrine gods and the immortals. According to *Shǐjì* (The Scribe's Records), it is Emperor Wu of Han (156 BCE–87 BC) who first constructed a garden with the style of the islands of Xian (仙山) under the influence of Taoism. The city of Xian is the oldest of the four great ancient capitals of China. A Chinese word, *Xian*, translatable as an enlightened person, spiritually immortal, transcendent, or celestial being, is a hermit of eternal youth and immortality who embodies the way of Dao. It is well known that Emperor Qin Shi Huangdi (259 BCE–210 BCE), the Qin dynasty and the first emperor of a unified China, searched the way to be eternally young and immortal. Wu of Han constructed three mountains in a pond to reproduce Chinese Taoist mythology. The mountain is often described as the base of the Eight Immortals or where they travel to have a banquet. This garden style of the islands of Xian was inherited by the end of Qing dynasty in the nineteenth century.

The islands of Xian are also a kind of paradise but very different from that of Western societies mentioned above. The islands of Xian were not an idealization of the places where we lead an ordinary life such as farms, orchards, and pastures but sacred mountains isolated from quotidian living. Here we can see that the Asian concept of gardening involves the desire to live a quiet life in seclusion, leaving the real world and without standing out. This Taoist idea of the paradise has come to be connected and fused with the Buddhist concept of the Pure Land and was inherited in the gardens of literary men in the era of Six Dynasties (in the third to sixth century).

29.3 The Artificial and the Modern Picture of Nature

Comparative studies tell us that from ancient times, India and the societies in the West built formal gardens, while the Asian societies such as China, Korea, and Japan made nonformal gardens or landscape gardens. Formal gardens in the Middle East and the Occident are regular, symmetrical, artificial, straight-lined, and geometrically constructed; nonformal gardens in East Asia are irregular, asymmetrical, seemingly natural, curved, and not geometrical.

European gardens have two origins: Ancient Egyptian formal gardens and ancient Persian gardens. Ancient Egyptian formal gardens are those with rectangle pool, raisin trellis, and symmetrical arrangement of trees, flowers, and bowers. Ancient Persian gardens have two parts: a formal garden and a hunting ground that enclosed natural or planted forest with a fence. In the ancient Greek *polis*, large gardens did not develop because there was no concentration of power and wealth and the cities were crowded with many houses and buildings. But during the Hellenistic period and the ancient Roman period, kings, emperors, and the aristocracy made many luxurious gardens. Later during the Italian renaissance, the House of Medici constructed on the hills in the suburbs of Florence formal gardens with fountains, statues, topiaries, orchards, trees, flowers, bowers, grottos, and so on, which were modeled after ancient roman villa.

Charles VIII of France (1470–1498) invited gardeners from Italy and constructed the garden of the royal Château at Amboise on the riverside of the Loire. As mentioned above, formal gardens had already been developed in Europe before their introduction to France. However, formal gardens are often identified with French gardens. It is because the gardens designed by André Le Nôtre (1613–1700), French landscape architect and gardener of King Louis XIV, monarch of the House of Bourbon from 1643 until 1715, were so outstanding and famous that the formal, geometrical style of garden is seen as almost synonymous with French gardens. He designed the park of the Vaux-le-Vicomte, the Palace of Versailles, Chantilly, Fontainebleau, Saint-Cloud and Saint-Germain, and so on (Fig. 29.1).

The French formal garden has a style based on symmetry, geometry, and regular patterns. In formal gardens, roads, paths, rows of trees, ponds, fountains, flowerbeds, and so on are organized in well-ordered, symmetrical, and regular ways; even trees and flowers are artificially pruned in geometrical forms such as triangular pyramid or cone. The gardens designed by Le Nôtre have been regarded as representing the height of the French formal garden style. Looking at Le Nôtre's model, can we say that the formal gardens of Versailles are extremely artificial, isolated from nature, and created by the desire to impose human order on nature?

Le Nôtre designed the gardens of Versailles together with King Louis XIV (cf. Arizzoli-Clémentel 2009; Baraton 2006; Iwakiri 2008; Kenna and Haskell 1997; Sugio 2009), who was trying to show that these gardens were those of the Sun King. The most remarkable feature of Versailles consists in the main axis that goes through the Hall of Mirrors to Le Bassin de Latone, mother of Apollon, and on to Le Bassin d'Apollon and the horizon. In the Hall of Mirrors of the palace, allegory disks of the Sun King and the sun with face and waved hair are displayed in a line, and the vanishing point at the horizon means infinity. The Hall of Mirrors is located in the east and the vanishing point is in the west. Therefore, every day the sun moves on the line of the principal axis of the garden of Versailles (Fig. 29.2). The sun is Apollo, Louis XIV, and goes around the world illuminating every part of it. The allegory of the sun rules every part of the Versailles palace and gardens. There is a privileged viewpoint in the king's place to look out at the gardens, and the point is the allegory of the power of infinite rule.

There are a great number of statues in the gardens. Some represent the four rivers of France and the four continents of Asia, Africa, America, and Europe. Other statues represent four seasons and three phases (morning, afternoon, and night) and the four elements, fire, water, earth, and wind. There are also statues of mythical gods, ancient kings, generals, and philosophers. The statues represent the world, space, time, and history of humans. We see the political meaning of the absolutism of Versailles.

Versailles, however, does not only show the power of an absolute king, but it also represents the way nature was conceived of in this period. Le Nôtre and Louis XIV

a

b

Fig. 29.1 (**a**) Orangerie du Chateau, Versailles. (**b**) Main garden and fountains of Fontainebleau Palace. (Photos Tetsuya Kono)

Fig. 29.2 Le Bassin de Latone of Versailles. (Photo Tetsuya Kono)

were living in the period of the rising of modern science, the so-called Scientific Revolution. Descartes emphasized in his works the use of reason, especially deductive and mathematical method, to develop the natural sciences while discarding perception as unreliable. It is by mathematical reasoning that human beings can seize the true picture of nature. Galileo Galilei (1564–1642) wrote in *The Assayer* (1623):

> Philosophy is written in that great book which ever lies before our eyes — I mean the universe — but we cannot understand it if we do not first learn the language and grasp the symbols, in which it is written. This book is written in the mathematical language, and the symbols are triangles, circles and other geometrical figures, without whose help it is impossible to comprehend a single word of it; without which one wanders in vain through a dark labyrinth. (Galilei 1623 as quoted Burtt 2003, p. 71)

Galileo maintained that mathematics provided a kind of necessary certainty that could be compared to God's knowledge. What Le Nôtre and Louis XIV realized in the land of Versailles is not an artificial paradise but a picture of nature in the modern age in Europe, and they put the king at the center of it. It represents the humanism that places humanity at the center of nature in the period of absolutism.

29.4 The Natural: Landscape Gardens and Romanticism

During the Tudor dynasty in England, Henry VII constructed formal gardens at Richmond Palace, Hampton Court Palace, and Nonsuch Palace. These and other gardens at Melbourne Hall, Blenheim Palace, Chatsworth House, and Badminton

House were constructed under the influence of Versailles. It was the rise of the "English" landscape garden, however, that marked the beginning of the decline of the formal garden style in the eighteenth century. The garden designers of the period ruined and abolished the formal gardens.

However, it is definitely wrong to suppose that Le Nôtre's garden is always formal, geometrical, and symmetrical. The gardens of Versailles are not only the places to be seen from a privileged viewpoint of the palace but also to walk in and discover unexpected scenes and landscapes. Moreover, at the request of Louis XIV, Le Nôtre designed the Great Trianon on the outskirts of Versailles. This palace has a small garden with an asymmetrical, irregular, and natural style, where natural trees remain and narrow paths curve under the trees. Le Nôtre himself said in a letter to a relative of King Louis XIV that this garden is "the most comfortable place" that provides with the pleasure to walk (Sugio 2009, p. 91). In this sense, Le Nôtre can be termed as a pioneer of landscape garden.

The innovation of the landscape garden resided in removing the fence surrounding the garden, thus making the garden and exterior nature continuous. The landscape gardeners discarded symmetry and regularity following the principle of William Kent that nature abhors a straight line (Clifford 1967, p. 154).

Geographical, social, and historical reasons to explain why formal gardens have not been developed in England have been offered. During the eighteenth century, the people of the Whig party, the liberals, had the tendency to give priority to the economy and liked going abroad. They, as the rising wealthy class, liked the grand, comfortable, as well as useful gardens. They preferred to see their own land on the whole as a garden rather than to enclose a part of the land and decorate and arrange it in a sophisticated way. The gardens of the British eighteenth century show how the idea of the monarch being the absolute ruler had given way to the idea of democracy. Gardens become more diverse, more democratic than the formal gardens commissioned by the old monarchs of England.

Additionally, the development of natural science and the discovery of the intricacies of nature provoked the idea that nature is great, profound, and true creation and that the beauty of the view of mountains, valley, forest, river, beach, and ocean was far beyond that of the man-made formal gardens. Philosophy and literature had also an influence in changing the garden's concept. In *Paradise Lost* (1667), English poet John Milton describes the Garden of Eden as a landscape garden. Anthony Ashley Cooper, third Earl of Shaftesbury and an Enlightenment thinker influenced by John Locke, also claimed that the natural view is more attractive than the inflexible gardens of kings. The concept of sublime has begun to be used to describe nature as greatness, which goes beyond all possibility of calculation, measurement, or imitation. This new vision of nature prepared the emergence of the romanticism movement of the group of poets that included William Wordsworth, Samuel Taylor Coleridge, John Keats, Lord Byron, Percy Bysshe Shelley, and others. Romanticism prefers more a "wild and mysterious nature" than the "tender, bright, and picturesque nature" that the early landscape gardeners wanted to imitate. Nature represented in early English landscape gardens was an idealized and "prettified" nature, not wilderness.

29.5 Japanese Gardens and Taoism

The Chinese style of gardens was introduced to England in the seventeenth and eighteenth century and was an important reason for the development of landscape gardens. Statesman and essayist Sir William Temple (*Upon the Gardens of Epicurus*, 1685) and architect Sir William Chambers (*Dissertation on Oriental Gardening*, 1772) introduced an interest on oriental gardens. As discussed above, China, Korea, and Japan have made solely nonformal gardens. We are able to say that many Japanese gardens are landscape gardens, but there are some fundamental differences from English gardens.

The difference is that Japanese gardens have developed in Buddhist temples so that gardens often have had religious meanings and feelings. Most Japanese gardens were not for any practical use at least in their early days. Surely you can find, in the countryside of Japan even today, many front yards where farmer families plant some fruit trees and keep hens for their own foods. Nevertheless, such front yards are not regarded as "gardens." As ancient Chinese represented the Taoist paradise and the Buddhist Pure Land, Japanese Buddhist gardens of the eleventh century such as Houjou-ji Temple (法成寺) and Byodo-in Temple (平等院) are constructed to realize the Pure Land of Amida which is described in the Buddhist scriptures. The stones, streams, ponds, trees, and flowers in the gardens symbolized the landscape of sacred mountains, rivers, ponds, and divine objects.

These temples were constructed by powerful politicians of the period. Religious belief and feelings were expressed through natural landscapes. In contrast to the ancient gardens in the West (which were the idealization of the living place such as farms, orchards, or pastures), Japanese gardens, at least in the early period, represent the imaginary sacred places to be separated from the ordinary life.

Many Japanese gardens at Buddhist temples are not so large since they are constructed in front of the temples, so they can be viewed from the inside of a room of the temples. However, during the Edo period (1600–1854), *daimyos*, Japanese feudal lords, started to construct *Kaiyu-shiki-teien* (回遊式庭園) style Japanese gardens, when they built the houses of the wealthy or powerful often in the *sukiya-zukuri* style of architecture (Nishi 2005; Shirahata 1997; Tanaka 1967; Tobita 1999, 2009). *Kaiyu-shiki-teien* style gardens referred to Japanese promenade gardens, go-round style gardens, and Japanese strolling gardens. A network of pebbled walking trails is installed around a pond in a vast site and leads visitors to prescribed viewpoints or teahouses from which they are meant to view the landscape with trees, flowers, rocks, man-made streams, hills, and ponds. The landscape varies by the change of seasons.

These gardens were an attempt to reproduce famous landscapes in miniature in order to replicate both Japanese and Chinese scenery. For example, Koishikawa-Koraku-en Garden (小石川後楽園) is a legendry go-round style gardens, one of two surviving Edo period clan gardens in modern Tokyo. Koraku-en Garden features a central pond and hills, making it perfect for a stroll. It is Yorifusa Tokugawa, *daimyo* of the Mito branch of the ruling Tokugawa family, who started to construct *Koraku-en* Garden at a separate Edo residence in Koishikawa in 1629 and that his successor, Mitsukuni Tokugawa, completed during his reign (Fig. 29.3).

Fig. 29.3 (**a**, **b**, **c**) Go-round style garden in Koraku-en. (Photos Tetsuya Kono)

Mitsukuni has been regarded as an enlightened ruler who governed the people by following the philosophy and ethics of the Chinese classics. The Japanese word "en" means garden or park; the name of the garden, "Koraku," meaning "enjoy-later-on," came from a Chinese text in Hanchuen's (范仲淹) *Gakuyoro-ki* (岳陽楼記, The history of Yueyang Tower) admired by Mitsukuni which said that those who are in power need to worry before people and enjoy after people. Mitsukuni incorporated in the designing of the garden some concepts of the Chinese Confucian scholar Shushunsui (朱舜水) of the Ming dynasty, including a garden reproduction of Seiko Lake (China), a "Full Moon Bridge," and other features with cultural origins in China.

Rikugi-en (六義園), located in Bunkyō-ku, Tokyo, is also an example of strolling garden and features a large central pond surrounded by man-made hills and forested area (Fig. 29.4). The construction of the park took place between 1695 and 1702 and was headed by Yanagisawa Yoshiyasu by permission of the fifth shōgun Tokugawa Tsunayoshi. The name "Rikugi" means "the six principles of poetry," referring to a system for dividing Chinese poetry into six categories. This system also influenced the division of Japanese Waka poetry as well.

Another example of *daimyo teien* is Kenroku-en (兼六園) Garden, thought of as one of the most beautiful Japanese gardens and located on the heights of the central part of Kanazawa and next to Kanazawa Castle (Fig. 29.5). The Maeda, *daimyo* family, maintained the garden from generation to generation over a period of nearly two centuries. The name Kenroku-en means "garden of the six sublimities," derived from the *Chronicles of the famous Luoyang Gardens* (洛陽名園記), a book by the Chinese poet Li Gefei (李格非). Six Sublimities refers to:

Spaciousness (宏大)
Seclusion (幽邃)
Artifice (人力)
Antiquity (蒼古)
Waterway (水泉)
Panoramas (眺望)

According to Chinese landscape theory, these are the six essential factors that make up a perfect garden. In the Kenroku-en Garden, water is diverted from a distant river to feed various streams and ponds including two main artificial ponds "Kasumigaike" and "Hisagoike." In the garden water features, hills, bridges, teahouses, trees, flowers, and stones are dotted to discover viewpoints and hidden nooks. Kasumigaike pond was compared to an open sea, with an island on which an ageless hermit with miraculous power was believed to live.

Daimyo gardens are more secular and practical than the gardens of Buddhist temples of the earlier periods. In Koraku-en, there is a small rice paddy, where *daimyos* enjoyed duck and goose hunting. Some daimyos also used the gardens for their research of natural history or botany as European landowners and priests did. *Daimyos* held tea ceremonies, fests, and banquets with other daimyos, guests, and subordinates in their gardens. Other *daimyos* opened their gardens to all citizens. The gardens are places for politics and society. Thus, *daimyo teiens* have practical uses. In this aspect they are not so different from European gardens, even though *daimyo teiens* still hold the atmosphere of seclusion and escape from the real world.

a

b

Fig. 29.4 (**a**, **b**) Go-round style garden in Rikugi-en. (Photos Tetsuya Kono)

Fig. 29.5 (**a**, **b**, **c**) Go-round style garden in Kenroku-en (Photos Tetsuya Kono)

Differences in the religious and philosophical concept of nature behind landscape designing still require to be examined. The *daimyo* strolling gardens make the man-made pond at the center of the site and put an island in the pond. As mentioned above, this is the representation of the islands of Xian of Taoism (Kaneko 2002; Nakamura 1999). Taoism (or Daoism) is a philosophical, ethical, and religious tradition of Chinese origin, teachings of Laozi (老子), that emphasizes living in harmony with the Tao or Dao (道). Tao can be interpreted as way, road, channel, path, doctrine, or line. However, Tao denotes something that is both the source and the force behind everything that exists.

Tao is the One, which is natural, spontaneous, eternal, nameless, and indescribable but is also something that individuals can find immanent in themselves (Laozi 2008; Kato 2002). It has been used to denote the flow of the universe or of nature. Tao is conceived rather as fluid body like water and air than as a rigid body. It is rather a changing process than a substance with identity. The word Tao is used to express never-ending pursuit or development in arts and martial arts such as *Jyu-do* (柔道), *Ken-do* (剣道), *Sa-do* (茶道, Japanese tea ceremony), *Ka-do* (華道, *Ikebana*, Japanese art of flower arrangement), and so on.

Strolling in a garden is considered as a process that is a state of Xian. Accordingly, there is no privileged viewpoint to overlook the entire landscape of the garden. This Taoism concept of nature has penetrated into the ordinary meaning of the Japanese word "nature." The meaning of "自然 (Shizen, nature)" means first of all natural, non-artificial, things and landscapes including mountains, forest, ocean, rivers, lakes, and animals. But, it also means a "spontaneous process" or "emergence from inherent, immanent process." Accordingly, the Japanese word "shizen" involves the idea of nature as a process. On the contrary, it never means "identical features" or "unchanged essence."

American transcendentalists such as Ralph Waldo Emerson (1803–1882), Henry David Thoreau (1817–1862), and John Muir (1838–1914) were influenced by this Asian philosophy of nature. Emerson introduced and translated Indian (Vedas) and Chinese philosophies of pantheistic ontology in the Transcendentalists' journal *The Dial* Emerson wrote:

> We live in succession, in division, in parts, in particles. Meantime within man is the soul of the whole; the wise silence; the universal beauty, to which every part and particle is equally related, the eternal ONE. And this deep power in which we exist and whose beatitude is all accessible to us, is not only self-sufficing and perfect in every hour, but the act of seeing and the thing seen, the seer and the spectacle, the subject and the object, are one. We see the world piece by piece, as the sun, the moon, the animal, the tree; but the whole, of which these are shining parts, is the soul[2].

For Emerson, Thoreau, and John Muir, nature was a *dojo*[3] to develop their spirituality, artistic sensitivity, and wisdom.

---

[2] "The Over-Soul" from *Essays: First Series* (1841), online text, http://www.emersoncentral.com/oversoul.htm

[3] In Japanese, *dojo* literally means "place of the Way." It is a space for immersive learning or meditation.

29.6 Preservation and Biocultural Diversity

As discussed above, formal and nonformal gardens mirror two opposing human desires and attitudes in relation to nature: one fosters biocultural homogenization and the other fosters biocultural conservation; one is to control nature and another is to be integrated with nature or to be a part of it. Nature is a community, and we want to rule it at the same time we want to be a member of it. The synthesis of these desires is to make an ideal community and belong to it. A garden is the representation of nature as an ideal community, which could be:

- An ever-fertile farm and orchard for European people
- The universe rationally organized for absolute kings
- An open, free, but prettified landscape for early English gardeners
- A wild and growing habitat for Romanticists
- The islands of Xian where one can find inner Tao for Japanese *daimyos*

A garden is also a political place. Some people tried to show in a garden their power to control nature and people. Others tried to find in a garden a place for seclusion from the real world. The pursuit of power and the wish to live in seclusion are inconsistent but coexist in our mind; even arrogant Louis XIV had both desires so that he expressed his contradiction in his gardens of Versailles and Great Trianon. A garden is also both a practical place of farming, hunting, and researching and a place for aesthetic appreciation or religious contemplation that has no practical functions.

French historian Benoit-Mechin (1975, p. 18) maintained that English landscape gardens are the opposite of the gardens as great arts, since the essence of English gardens consists in the desire to fit into a landscape by refusing any style to pull gardens up to the level of art. As we saw, English landscape gardens are in reality not untouched wilderness but a kind of prettified pasture. However, this tendency in English landscape gardens should have prepared the idea of the preservation of nature in national parks. Preservation has to be distinct from conservation. The latter seeks the proper use of nature, while the former seeks protection of nature from use. As Benoit-Mechin realized, preservation of parks, where people tried to preserve nature as untouched as possible or preserve wilderness as it is, is no longer called "gardens."

Gardens are nature in which people want to live, while preservation parks are nature in which people are prohibited to live but allowed to visit temporarily. English landscape gardens might involve a contradiction, that is, to live in land where one must not live. If the whole earth is a preservation park, human beings must be excluded from the earth. Then, if we cannot live in a preservation park, but only in a garden, what kind of garden should we want nature or planet earth to be? If a garden is the expression of human desire to live in nature as an ideal community, what kind of ideal community should it be and what kind of members should human beings be?

This is an issue of how to make a sustainable society. Some environmentalists have a tendency to highlight the importance of preservation, excluding human needs and cultural interactions with nature. Perhaps some areas must be preserved from any use by human beings, but a more important problem is how to utilize nature in an adequate, sustainable way. Rozzi (2013) proposes a solution to this problem. He introduces a biocultural ethics that value the links between the habitats (where we live), habits (how we live), and inhabitants (who we are). These "3Hs" constitute an ecosystem unit (Rozzi et al. 2008, p. 325).

The biosphere is composed of these habitat-habit-inhabitant units, where human beings have coevolved with their local ways of living and practices. However, these diverse ecosystem units are being violently destroyed by the imposition of a single globalizing colonial model leading to widespread processes of biocultural homogenization. Rozzi proposes that to counter these processes, a solution might be found in unique local life habits and habitats, since many provincial communities exhibit sustainable and respectful forms of co-inhabitation between humans and other-than-human beings. In conclusion, we will quote Rozzi's suggestion that might give a clue to find the answer to our questions of what kind of nature our ideal community is and what we should do to be a good member of the community. "Biocultural ethics proposes a decolonizing turn by problematizing the relationships between human habits, the habitats, and the communities of co-inhabitants. For this endeavor, interdisciplinary teamwork among ecologists and philosophers enables a recovery of the archaic meaning of ethos adding novel insights from scientific perspectives" (2013, p. 22).

If a garden is the representation of nature as an ideal community, what kind of garden should we have today? The ideal community of our day should be a sustainable society that promotes biocultural diversity. Thus, a good garden of today must neither be a garden which represents a universe rationally organized from a privileged viewpoint such as Versailles nor a garden which represents an ever-fertile farm and orchard useful only for human beings nor a garden which represents Taoist paradise such as *daimyo teien* but a garden which represents sustainability and coexistence of nature and human society. It should be a garden which suggests us how to have the benefit of nature in a sustainable way, how to preserve and promote biodiversity, and how to develop a nature-loving culture in a local society. Such a garden must be beautiful and intellectually interesting. It should be a garden which provides us with an aesthetic experience and an intellectual wonder of nature. Gardens show us the way to relate with nature. They can play a valuable role to implement, communicate, and project interdisciplinary work that, in the context of urban homogenization, contributes to compare different concepts of nature, reexamine and revise our own relationship with nature, and conceive a new idea for a sustainable society.

References

Arizzoli-Clémentel P (2009) Les jardins de Louis XIV à Versailles. Présentation et commentaires de Pierre Arizzoli-Clémentel. Edition Gourcuff Gradenigo, Montreuil
Baraton A (2006) Le Jardinier de Versailles. Grasset, Paris
Benoit-Mechin J (1975) L'Homme et ses Jardins. Edition Albin Michel, Paris
Burtt EA (2003) The Metaphysical Foundations of Modern Science. Dover Publications, Mineola
Clifford D (1967) A history of garden design. Frederick A. Praeger, New York
Clunas C (1996) Fruitful sites: garden culture in Ming dynasty China. Reaktion Books, London
Francis M, Hester RT Jr (1991) The meaning of gardens: idea, place, and action. The MIT Press, Cambridge
Galilei G (1623) The assayer. translated by Thomas Salisbury, as quoted in Edwin Arthur Burtt (2003) The Metaphysical Foundations of Modern Science. Dover Publications, Mineola
Giamatti AB (1966) The earthly paradise and the renaissance epic. Princeton, Princeton
Hadfield M (1955) Pioneer's in gardening. Bloomsbury, Edinburgh
Iwakiri M (2008) Yoroppa no Tei-en: Bi no Rakuen wo Meguru Tabi (European gardens: a journey to travel around the paradises of beauty). Chikoshinsho Publisher, Tokyo
Kaneko H (ed) (2002) Kodai Teien no Shiso: Shinsen Sekai heno Dokei (The thought in the ancient gardens: yearning for the world of Xians). Kadokawa Publisher, Tokyo
Kato C (2002) Furo-Fushi no Shintai: Dokyo to 'Tai' no Shiso (The eternally young and immortal body: Taoism and the thought of 'Tai'). Taishukan Publisher, Tokyo
Kenna M, Haskell ET (1997) Le Nôtre's gardens. Ram Publication, Treville
Kidu M (1994) Chugoku no Niwa: Fusui no Renkinjyutsu (Chinese gardens: alchemy of Feng Shui). Tokyodo Publisher, Tokyo
Laozi (2008) *Roshi* (*Laozi*). Translated by Kunio Hachiya. Iwanami Publisher, Tokyo
Nakamura S (1999) Konan no Niwa: Chugoku Bunjin no Kokoro wo Tazunete (The gardens of Jiangnan: a visit of the minds of Chinese literary men). Shinhyosha Publisher, Tokyo
Nishi K (2005) Nihon no Tei-en Bunka: Rekishi to Isho wo Tazunete. (Japanese garden culture: a visit of its history and design). Gakugei Publisher, Kyoto
Rozzi R (2013) Biocultural ethics: from biocultural homogenization toward biocultural conservation. In: Rozzi R, Pickett STA, Palmer C, Armesto JJ, Callicott JB (eds) Linking Ecology and Ethics for a Changing World: Values, Philosophy, and Action. Ecology and ethics, vol 1. Springer, Dordrecht, pp 9–32
Rozzi R, Massardo F, Anderson C, Heidinger K, Silander J Jr (2006) Ten principles for biocultural conservation at the southern tip of the Americas: the approach of the Omora Ethnobotanical Park. Ecol Soc 11(1):43. [online] URL: http://www.ecologyandsociety.org/vol11/iss1/art43/
Rozzi R, Arango X, Massardo F, Anderson C, Heidinger K, Moses KP (2008) Field environmental philosophy and biocultural conservation: the Omora Ethnobotanical Park educational program. Environ Ethics 30(3):325–336
Rozzi R, Pickett STA, Palmer C, Armesto JJ, Callicott JB (eds) (2013) Linking Ecology and Ethics for a Changing World: Values, Philosophy, and Action. Ecology and ethics, vol 1. Springer, Dordrecht
Shirahata Y (1997) Daimyo Tei-en: Edo no Kyoen (Daimyo gardens: feasts in Edo). Kodansha Publisher, Tokyo
Sugio S (2009) Furansu-shiki Tei-en no Miryoku: Le Nôtre no Sekai (The fascination of the French gardens: the World of Le Nôtre). Biocity Publisher, Tokyo
Tanaka M (1967) Nihon no Teien (The Japanese gardens). Kashima Publisher, Tokyo
Tobita N (1999) Nihonteien to Fukei (Japanese gardens and landscape). Gakugei Publisher, Kyoto
Tobita N (2009) Edo no Teien: Shogun kara Shomin made (The gardens in Edo: from Shoguns to the Common People). Kyoto University Press, Kyoto

Index

R. Rozzi et al. (eds.), *From Biocultural Homogenization to Biocultural Conservation*, Ecology and Ethics 3, https://doi.org/10.1007/978-3-319-99513-7

MIX
Papier aus verantwortungsvollen Quellen
Paper from responsible sources
FSC® C105338

If you have any concerns about our products,
you can contact us on
ProductSafety@springernature.com

In case Publisher is established outside the EU,
the EU authorized representative is:
Springer Nature Customer Service Center GmbH
Europaplatz 3, 69115 Heidelberg, Germany

Printed by Libri Plureos GmbH
in Hamburg, Germany